Govind Bajaj, Arup Bhattacharya, Laboni Bhowmik,
Avijit Dhar, Anup Ghosh, Gairik Acharya, Asidhara Lahiri

Application Development with SAP® Business Technology Platform

Rheinwerk Publishing

Editor Meagan White
Acquisitions Editor Hareem Shafi
Copyeditor Meagan White
Cover Design Abigail Damke
Photo Credit iStock: 1495185644/© NYC Stock Photos
Layout Design Vera Brauner
Production Hannah Lane
Typesetting III-satz, Germany
Printed and bound in the United States of America, on paper from sustainable sources

ISBN 978-1-4932-2855-3
2nd edition 2026

Rheinwerk Publishing, Inc.
2 Heritage Drive, Suite 305
Quincy, MA 02171
USA
info@rheinwerk-publishing.com
+1.781.228.5070

Represented in the E. U. by:
Rheinwerk Verlag GmbH
Rheinwerkallee 4
53227 Bonn
Germany
service@rheinwerk-verlag.de
+49 (0) 228 42150-0

Library of Congress Cataloging-in-Publication Control Number: 2026027748

Contents at a Glance

Contents

PART III Operating Applications

SAP PRESS

Application Development with SAP® Business Technology Platform

Foreword by Dr. Tom Janoshalmi

The use of technology, digitization, and new forms of working is accelerating the impact of the "Fourth Industrial Revolution." With this significant innovation-driven movement, executives have seen up to a 25 times faster shift toward the use of advanced technologies, secure data, and the creation of supply-chain redundancies. The ability to transform an organization's digital core will differentiate disruptors from those who are disrupted.

Successful digital strategies will transform current digital core operations into technology platforms with applications that enable companies to innovate faster with new business models. Forrester calls these digital operations platforms (DOPs) the foundation for business success.

Organizations desire more modern, event-driven data architectures and expect data to flow through and across heterogeneous landscapes. Today's businesses want a data strategy that enables them to remodel business processes by integrating structured and unstructured data from multiple layers of the technology stack.

The journey to the autonomous enterprise might not always be done right the first time. Common challenges include failing to define how to bridge the gap between ambitious strategic goals and operational reality, lacking consistent top-down sponsorship, and being unprepared to adopt standard processes for non-differentiating business areas.

Standard software often remains clean only until it is sold; modifications and customization then become the convenient—but very expensive—way of avoiding change. This creates frustration between IT and the business, especially if consultants (or their digital twins) end up defining what the business expects and what IT can deliver.

The opportunity to renovate or modernize the digital core of an enterprise using AI couldn't be more apparent. Combining modern data strategies with a flexible IT architecture and a modular consumption model is the best way to adapt your business to this new way of operating.

Dr. Tom Janoshalmi
IBM Global Managing Partner
June 2026

Preface

The time since the publication of the first edition of this book has been awe-inspiring. The technology world saw the democratization of AI technologies, opening up new frontiers. The rate of change has never been this fast. Enterprises worldwide are eager to adopt these capabilities to stay competitive. SAP needed to transform its capabilities to better serve these enterprises. Hence the need to come up with a newer edition of the book incorporating the latest changes brought in by SAP.

While it is certain that many specifics will change over the next few years, if not months, some fundamental concepts will remain constant. This book aims to focus on the fundamentals so you find it relevant over a longer period of time. In this era of rapid change, the evolution from labs into the productive world of enterprises requires careful consideration of factors normally ignored during initial proof of concepts. Drawing on our decades of SAP experience and recent work with the latest products at enterprises worldwide, we have shared guidance to help you with real-life best practices. We hope you enjoy reading this book and use the skills you learn to make a significant impact on enterprises worldwide.

How to Read This Book

This book is organized into the following 21 chapters, divided into four parts:

- **Part I: Getting Started**
  - **Chapter 1**
    In 2025, at SAP Sapphire, SAP presented how their platform generates a flywheel effect that accelerates business transformation. This chapter evaluates how SAP Business Technology Platform (SAP BTP) provides the foundation for a synergistic IT platform that can generate the flywheel effect. We then introduced several capabilities of SAP BTP as a platform that will be further elaborated on in subsequent chapters. We also touched upon the SAP Business AI Platform.
  - **Chapter 2**
    The latest advancements in technologies, both by SAP and its partners, have opened up a number of options for developing custom extensions. This chapter covers the different development options available for extending SAP solutions including SAP Joule for Developers.
- **Part II: Backend and Frontend Cloud Application Development**
  - **Chapter 3**
    This chapter explores cloud-native principles, multitenancy, microservices, and other concepts key to SAP cloud application design. It explains in detail the microservice architecture and how this architecture underpins SAP BTP's ability to enable cloud-native application development using SAP and other technologies. It also covers

SAP's clean core strategy and how to integrate cloud-native applications using SAP Integration Suite.

- **Chapter 4**
  SAP Business Application Studio is the main integrated development environment (IDE) for developing cloud-native applications in SAP BTP. You must know and understand the capabilities of SAP Business Application Studio to fully unlock the potential of developing innovative applications. This chapter shows you how to work with SAP Business Application Studio and develop your first cloud application with it.
- **Chapter 5**
  This chapter covers the features of SAP HANA Cloud, including its medallion data architecture, the SAP HANA Cloud central interface, the SAP HANA Cloud knowledge graph engine, and the SAP HANA Cloud vector engine.
- **Chapter 6**
  To expedite the development of cloud-based applications, SAP introduced a framework called SAP Cloud Application Programming Model, along with its own tools, guidance, and libraries so developers can focus on the business domain and functionality of the problem solution instead of on minute technical details. This chapter will help you understand the fundamentals of SAP Cloud Application Programming Model.
- **Chapter 7**
  Node.js has become popular for developing real-time, event driven, fast, scalable applications. This chapter illustrates how Node.js applications can be developed in the SAP context.
- **Chapter 8**
  Java continues to be one of the most widely adopted programming languages and is the most efficient for certain types of (high-load) applications. This chapter shows you how applications can be developed in Java to work with SAP.
- **Chapter 9**
  For the SAP community, ABAP continues to be the most popular language. This chapter shows you how ABAP can be used to develop cloud applications.
- **Chapter 10**
  Programming languages like Python are becoming increasingly critical, as they enable developers to build, integrate, and scale AI-driven solutions more efficiently on platforms like SAP BTP and other hyperscalers. This chapter shows you how to create an application using Python and Jupyter Notebook.
- **Chapter 11**
  Serverless technology has become increasingly popular for event-driven, short-running applications. This chapter illustrates the rapid development of applications using SAP's serverless environment—SAP BTP, Kyma runtime.

- **Chapter 12**
  Providing good customer experiences is a priority for all businesses. This chapter covers how you can develop engaging client-facing applications using SAP Fiori.

■ **Part III: Operating Applications**

- **Chapter 13**
  All businesses require an efficient, fast, and secure development and delivery of applications to keep up with the speed of change required. This chapter dives deeply into the different capabilities available to realize that goal. Having an efficient development and release process is insufficient; you should have efficient end-to-end DevOps practices in place.
- **Chapter 14**
  As the world has gone online, become distributed, and become hybrid, security exposure has increased as well. It's of paramount importance that an enterprise's IT is always secure. Any custom application that is developed must be secured by design. This chapter will focus on how to develop secured-by-design applications.
- **Chapter 15**
  Developing custom functionalities is no longer sufficient. You should continuously monitor if functionalities are working as they should and are generating the desired impact so that remedial actions can be taken if required, as we'll explain in this chapter.

■ **Part IV: Supporting Tools and Services**

- **Chapter 16**
  Easy low-code orchestration and visibility of work across applications and humans to achieve an end-to-end intelligent business process is an absolute essential capability of a resilient, flexible, and intelligent enterprise. This chapter illustrates how SAP Workflow Management and other workflow tools provide that capability through real-life examples.
- **Chapter 17**
  This chapter provides an introduction to SAP Build Work Zone, an explanation of its key concepts, and an overview of the SAP Build Work Zone Editor. Together, these sections provide both the background and the practical knowledge needed to understand and use the solution.
- **Chapter 18**
  As organizations move toward becoming autonomous enterprises, any manual activities that can be automated should be automated. This chapter covers the basic capabilities of SAP Build Process Automation. We will see how, through deep integration with SAP Integration Suite and enterprise applications, SAP Build Process Automation orchestrates human and system activities while enabling intelligent bots that leverage AI to understand context, process documents, make decisions, and execute actions across SAP and non-SAP landscapes.

- **Chapter 19**
  The artificial intelligence (AI) space is evolving at an unprecedented rate, with new AI models and capabilities being introduced constantly, making AI adoption across enterprise solutions imperative for survival. Driven by this paradigm shift, enterprise applications now focus more on context-driven or context-aware intelligence that is comparable to human decision-making. This chapter introduces SAP's embedded AI solution, Joule, its key capabilities, its building blocks, and how it works.
- **Chapter 20**
  SAP Business Data Cloud is a software as a service (SaaS) data platform which provides access to harmonized and governed data sets from SAP and third-party systems. SAP Business Data Cloud accelerates the AI journey by providing AI-ready data and faster time to insights. This chapter covers the architecture of SAP Business Data Cloud, its capabilities, the importance of data products, intelligent applications, and AI/ML scenarios using SAP Databricks.
- **Chapter 21**
  Having presented the different components available to develop custom processes, this chapter brings all of these options together and highlights they work together to deliver results. It discusses the Autonomous Enterprise, which includes a unified AI platform for building, contextualizing and governing agents. This chapter will also touch upon the roadmap to glimpse at how the custom developer space will evolve in the future.

We hope that the topics covered and the level of detail provided creates a perspective to help you understand, not only the *how* of custom development, but also the *why*, so you can effectively develop custom extensions.

The highlighted boxes in this book contain information that is useful but outside the context of the current topic. To help you immediately identify the type of information contained in the boxes, we have assigned symbols to each box.

[»]

Note
Boxes marked with this symbol contain information about additional topics or important content that you should note.

[+]

Tip
This symbol refers to specifics that you should consider. It also warns about frequent errors or problems that can occur.

[!]

Warning
This symbol warns you of important points that are potentially confusing or likely to be overlooked.

Acknowledgments

Numerous contributors have directly and indirectly enriched the authors with knowledge, expertise, and experience—all of which have made their way to this book in one form or another. Calling out all of our supporters is almost impossible, so we will refrain from naming specific names.

We would like to thank both SAP and IBM for allowing us to share our experience helping thousands of SAP customers through fifty-plus years of partnership, especially providing us with early access to products, infrastructure, technical know-how, and joint investments.

Special thanks to the SAP PRESS team, whose continued guidance, edits, and reviews have shaped this book.

We would like to thank our families without whose sacrifice, support, and encouragement this book would not have been possible.

PART I

Getting Started

Chapter 1
Introduction to SAP Business Technology Platform

In this chapter, we'll explore SAP's transformation flywheel and how it helps businesses stay ahead of the uncertainties they face in today's dynamic environment. We'll also examine SAP Business Technology Platform (SAP BTP) as an enabler of this transformation and discuss the paradigm shift in custom development.

Recent years have been unprecedented: The constantly changing geopolitical situation, rapid development of technology, and environmental conditions are causing significant economic and social volatility, challenging enterprises across the globe at an unprecedented scale and pace. Uncertainty on a large scale is becoming the new normal. While uncertainties can lead to the emergence of new opportunities, business environments are also becoming increasingly competitive.

The only way enterprises can handle assaults from the competition—emerging not only from the same industry but also from adjacent industries, from digital upstarts, from rapidly diversifying giants, and from nature itself—is through continuous reinvention. Enterprises should innovate not only when major technological or business inventions arise or when disaster strikes, but also during apparently calm periods when the status quo seems fine.

Resilient, intelligent, and sustainable is what all businesses should aspire to be for success. SAP defines intelligent enterprises as enterprises who run agile, integrated business processes and use advanced technologies such as artificial intelligence (AI), machine learning, and the Internet of Things (IoT). These businesses use industry best practices to build flexible value chains. They evaluate and act on customer, partner, and employee sentiment and understand how to manage their environmental impact.

Enterprises need appropriate IT platforms to support continuous transformation and reinvention. In 2025, at SAP Sapphire, SAP presented how their platform generates a flywheel effect that accelerates business transformation. This book will evaluate how SAP BTP provides the foundation for a synergistic IT platform that can generate the flywheel effect, and how enterprises can fulfill their specific needs through custom development without incurring technical debt. But first, let's look more closely at the synergistic IT platform building blocks and the flywheel effect.

1.1 Transformation Flywheel

In order to understand the flywheel effect, let's examine the intelligent enterprise in more detail and explore how it differs from the digitization activities of past decades.

Initial digitization efforts were mainly focused on building *systems of record* to capture and preserve transactions, leading to automation and increased efficiency in business operations. Subsequently, the focus expanded to developing *systems of engagement* aiming to improve the customer experience over digital channels like web, mobile, social media, etc. By recording transactions and customer interactions, a huge amount of data is generated, which led to the development of *systems of insight and intelligence* focused on mining all this data to understand what's happening and why, with the potential to predict what's next. The role of technology shifted from being an enabler of cost reduction and efficiency improvement to helping generate revenue, thus setting the stage for the transformation to intelligent enterprises.

Intelligent enterprises leverage the latest technologies like AI, blockchain, automation, IoT, 5G, edge computing, and more, applying them at scale, which results in a change to the core of the organization. With insights derived from the vast amount of data residing inside and outside the enterprise, technologies allow you to understand what's happening, why it is happening, and can predict what may happen next while recommending what you should do to stay ahead of the competition. Not only can intelligent enterprises personalize the customer experience, they can also provide an outstanding employee experience. All these goals can be achieved by collaborating with partners and engaging in the ecosystem for broader coverage and impact.

Intelligent enterprises go beyond digitization and the leveraging of technology. For an intelligent enterprise to be effective, a culture of innovation, agility, and speed must be nurtured and promoted. The entire organization and ecosystem must be carried along this journey to ensure full participation and cooperation. The building blocks of SAP's transformation flywheel are shown in Figure 1.1.

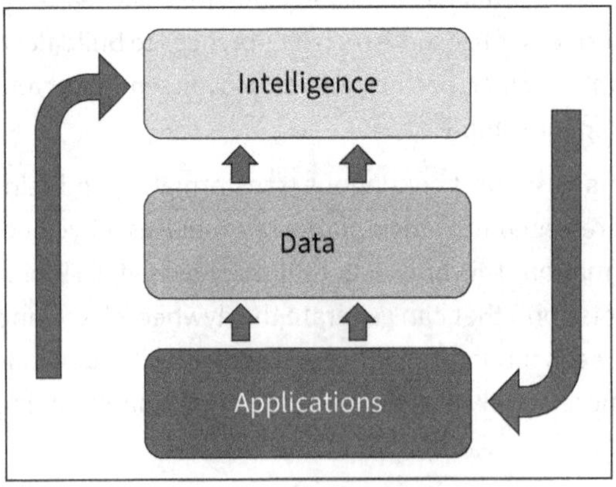

Figure 1.1: SAP's Transformation Flywheel

The flywheel consists of the following:

- **System of record or business applications**
  These are the applications where the business transactions are recorded. Examples include SAP ERP, SAP S/4HANA, SAP SuccessFactors, SAP Ariba, and others.
- **Data layer**
  This layer handles all data needs for the enterprise, whether the data is structured or unstructured, at rest, or in motion.
- **Intelligence layer**
  Powered by data, this layer provides the insights and intelligence required to improve an enterprise.

Each layer is loosely coupled with the others, each feeding off and powering the others. For example, the data layer gets its data from the application layer, both from in-house and connected system; the intelligence layer gets its input from the data layer to gain insight and drive intelligence. Insights become available when transactions occur in the application.

When these layers are aware of the needs of the other layers and communicate with each other using the same language, they can collaborate effectively, reducing waste and increasing efficiency.

With this in mind, SAP has focused on developing capabilities in each layer that follow common protocols, contain unified data models, and provide seamless native integration, in order to amplify the flywheel effect. Throughout this book, examples of these protocols in action will be provided.

Architecturally, SAP has continued to focus on cloud native capabilities supported by a cloud platform that is contextually aware of the businesses it serves. This platform is powered by AI, automation, agentic AI, and integration features, enabling organizations to benefit from the speed, flexibility, high quality, and economies of scale that cloud technologies provide.

1.2 Cloud Platforms

As discussed earlier, the cloud has brought with it a paradigm shift in the way enterprises can redesign their IT to support their digital transformation journeys, from new ways to meet infrastructure needs to the ability to develop new functionalities quickly. The cloud in its many forms (IaaS, PaaS, SaaS, etc.) can play a key role in these processes. However, for a cloud platform to effectively help an enterprise, it must have business context. Having just technical capabilities is not enough. SAP Business Technology Platform (SAP BTP) fills that gap by providing a technology platform that is aware of the business context and can therefore help deliver value to organizations quickly.

As PaaS, cloud platforms provide the ability to develop business differentiation, integrate with partners, bridge the physical-digital divide, and interact with customers in completely new ways. Thus, choosing the right platform is an important step. While most enterprises don't normally get locked into a platform now as they did in the past, having the right governance in place is essential so you can continuously evaluate the ever-evolving capabilities of various cloud platforms or new providers against the characteristics needed for the enterprise.

An enterprise's expectations about a cloud-based PaaS may be numerous, and thus, evaluation criteria will be multidimensional as well.

The first criteria is the functionality that the cloud platform offers, which can range from business functionalities to technical functionalities. Once you determine the functionalities needed by your enterprise, you can evaluate whether the nonfunctional capabilities of that platform are also appropriate. Speed, stability, scalability, high availability, disaster recovery, maturity of offerings, privacy, security, data center footprint, and SLAs are some aspects that will require closer evaluation. While PaaS are designed to hide the complexities of the underlying infrastructure and enable interaction at a higher level of abstraction, thus freeing you up from operating and maintaining the platform itself, you should understand the architecture of the platform well enough to ensure that your clients are aware of its impact on the rest of their landscape. We recommend adopting a common unifying architecture that covers the entire landscape, across on-premise, private, and public cloud footprints along with their impacts on the SaaS portfolio. Understanding your operational model and ensuring that it meets your needs are also important.

A cloud platform will generally have the following layers:

- **Security and compliance**
  All enterprises must assess the security provided by the cloud provider at each layer, including data center security; encryption and bring your own key capabilities; vulnerability assessments at all levels (e.g., network to storage to container); geofencing; continuous monitoring; and reporting of controls that support regulatory compliance needs like GDPR, HIPAA, etc.
- **Deployment**
  Be sure to check the different deployment options offered by the platform: While most PaaS offer a public cloud-only deployment model, some offer private cloud deployment models as well for clients who have special security or privacy needs.
- **Infrastructure**
  While you won't directly deal with the infrastructure of a PaaS, if the PaaS offers multiple infrastructure options, you can support special needs. For example, to run SAP HANA-based applications, a PaaS would need specialized infrastructure. Similarly, workloads from legacy applications may need to run on bare-metal servers or other appropriate hardware.

- **Runtime**
  You'll need to support the appropriate runtimes required for your applications. The platform should support as many varieties of runtimes as possible to account for future needs. Thus, the ability to handle new applications along with traditional applications is important. The ability to develop and run cloud-native applications using a microservice-based architecture managed by container-aware orchestration tools integrated to DevOps tools allows for faster, error-free releases, which is a key capability needed in a modern-day PaaS. Check for the support of open-source projects as well.
- **Development platform**
  You must ensure that the PaaS has the right development environment, supporting the right programming languages, with the right tooling for both development and DevOps, powered by the right automation and intelligence. The major usage patterns are development, middleware deployment, and integration. Low-code/no-code development has become important, with the acute shortage of skills for traditional development, and can reduce business analysts' dependence on developers.
- **Intelligence layer**
  Modern-day PaaS are not only expected to facilitate development and integration but also expected to provide capabilities to derive maximum value out of your data through analytics. Intelligence can be easily introduced by developing capabilities involving machine learning, AI, and more, as available.

1.3 Pillars of SAP Business Technology Platform

SAP BTP is positioned as the central cloud-based platform required for the key IT capability needs of any organization. This multicloud platform not only provides the technical foundation for SAP Business Suite but also helps integrate different applications, extend applications using clean core principles, and build AI-powered applications to automate and transform end-to-end business processes in a secured and scalable open way, allowing interoperability and integration with other platforms and the broader ecosystem.

Unlike other cloud platforms available in the market, SAP BTP is aware of business context and is tailored to improve business process efficiency through prebuilt business content. As shown in Figure 1.2 the key pillars of SAP BTP are as follows:

- Agent and application development
- Integration
- Artificial intelligence and data foundation

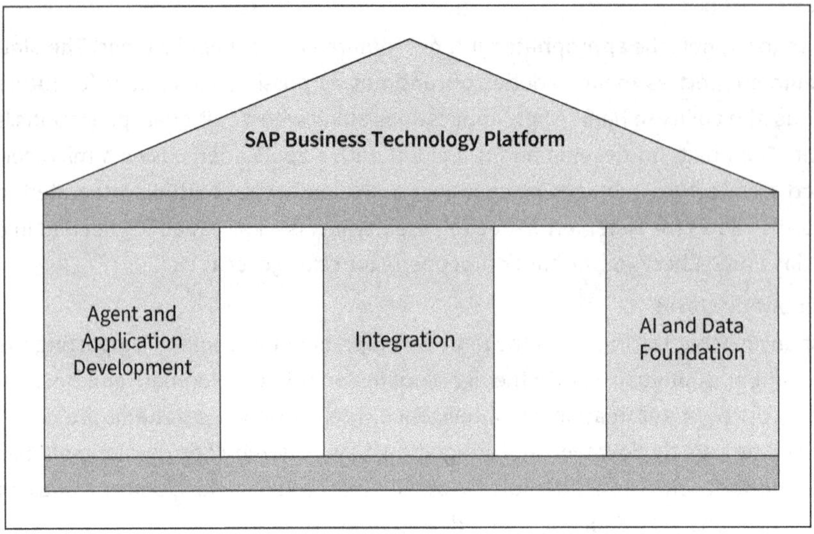

Figure 1.2: Pillars of SAP BTP

1.3.1 Agent and Application Development

SAP recognizes that many enterprises need to develop custom solutions to meet unique requirements in order to differentiate themselves from their competition.

Instead of developing code within the core ERP system (e.g., SAP S/4HANA), which causes issues during future upgrades, it is recommended to keep the core clean and develop extensions outside the core, on SAP BTP. As a result, these customer extensions are loosely coupled to the core off-the-shelf application and do not affect future upgrades. Chapter 3 discusses how to develop clean core applications or extensions. This decoupled development with SAP BTP enables the creation of purely cloud-native applications in multiple languages, thereby broadening the development community. SAP now allows development using non-SAP native tools like VS Code, Cline, Claude, Codex, Cursor, Windsurf, and others, thus opening custom application development to a much broader community. SAP BTP also offers several low-code/no-code AI-powered development options, enabling business users or "citizen developers" to develop applications. This approach means that SAP BTP can help address the huge skill shortage the industry faces today. The AI-powered development capability can also help pro-code developers.

With the advent of agentic AI capabilities, an enterprise can differentiate itself not only by developing custom applications or extending existing ones but also by changing its way of working and augmenting its human workforce with digital labor. SAP BTP also provides capabilities to develop agentic AI solutions. In short, this pillar of SAP BTP supports the following:

- Development of cloud-ready applications
- Extensions of existing applications

- Migration of classic ABAP custom code to ABAP Cloud
- Development of digital workspaces
- Automation of processes and tasks
- Development of AI agents
- Development of custom AI skills

It can also help improve developer efficiency with AI-powered development tools like SAP Joule for Developers, a natural language conversational assistant for code development that can be used in environments like Joule Studio and VS Code.

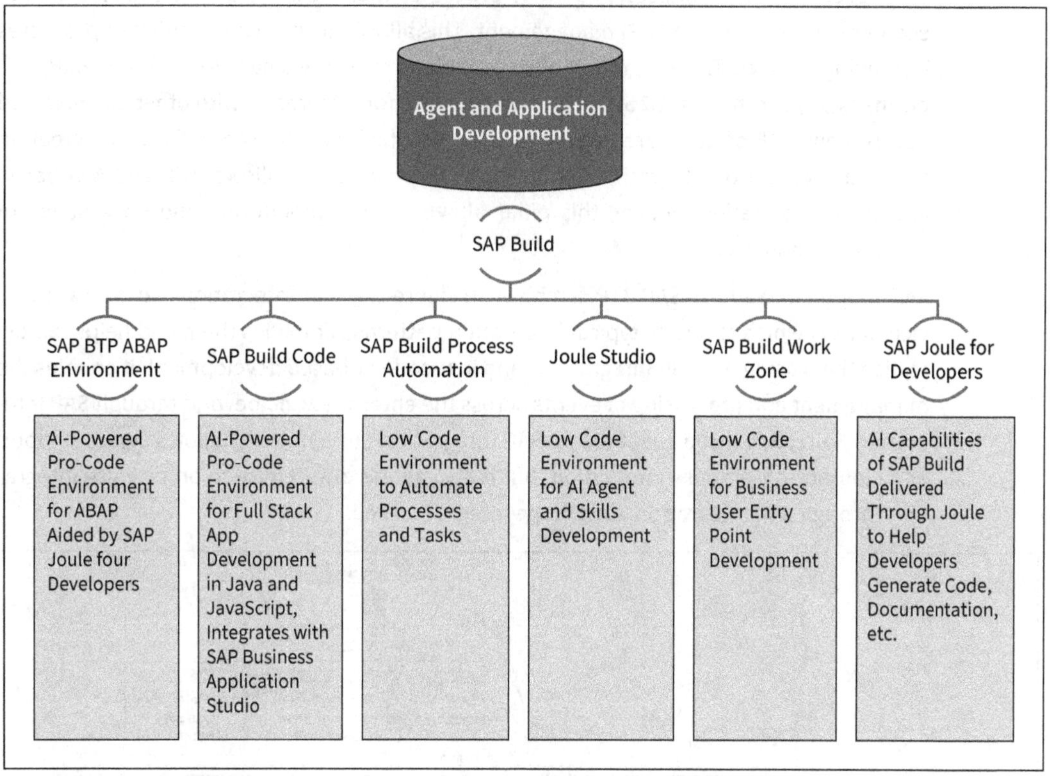

Figure 1.3: Components of the Agent and Application Development Pillar

This pillar, as shown in Figure 1.3, brings these capabilities together using *SAP Build*, a unified development and management solution. SAP Build has the following components:

- SAP BTP ABAP environment: An ABAP environment for pro-code development on SAP BTP
- SAP Build Code: An AI powered pro-code environment for full stack development in Java and JavaScript
- SAP Build Process Automation: Used for automation of processes and tasks

- Joule Studio: Used for development and deployment of custom AI agents and Joule skills
- SAP Build Work Zone: Used for low-code development of central entry points to SAP and non-SAP applications
- SAP Joule for Developers: An AI assistant for developers that improves productivity

1.3.2 Integration

The integration pillar allows applications and processes to be connected seamlessly across the enterprise and beyond using both pre-built and custom connectors. This capability supports event-driven architecture to make processes more responsive and support the API economy through secured API management. This allows flexible composition of processes leveraging services. The Integration pillar provides a robust business-to-business (B2B) and business-to-government (B2G) integration platform for integration with other partners and governments. All of these are underpinned by AI which facilitates the discovery, creation and management of integration across the enterprise. It also allows MCP and A2A-based agentic AI integration. Finally, this pillar allows for the monitoring and governance of secured integrations.

SAP Integration Suite on SAP BTP, as shown in Figure 1.4, is an integration platform as a service (iPaaS) which supports typical integration patterns, simplifies them and helps rapidly realize them using prebuilt integrations and low-code AI-based development. It enables the management and processing of events across the enterprise and beyond through SAP Integration Suite, advanced event mesh. SAP Integration Suite not only allows cloud-to-cloud and cloud-to-on-premise integration, but it also allows on-premise to on-premise integration through SAP Integration Suite, edge integration cell.

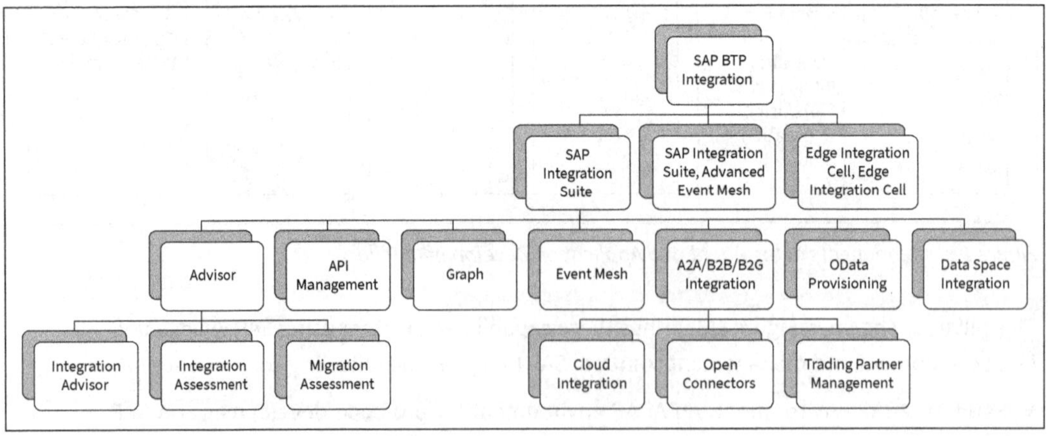

Figure 1.4: Components of SAP BTP Integration

In short, the capabilities of the integration pillar of SAP BTP are as follows:

- Cloud Integration: Used to integrate applications and processes across cloud, on-premise and hybrid landscapes using messages and APIs.
- API Management: Used to publish, promote and govern APIs in a secure enterprise ready way. It also has a developer hub that can be used to discover and utilize APIs.
- Graph: Used to expose business data via APIs as data graphs which are semantically connected.
- Event Mesh: Used to publish and consume events across the enterprise.
- Integration Advisor: Used to accelerate the creation of integration content using AI.
- Trading Partner Management: Used to manage and govern B2B interactions with other trading partners.
- OData Provisioning: Used to expose business data and processes from the backend as OData service.
- Data Space Integration: Used to exchange data within data spaces like Manufacturing-X and Chem-X, among others.
- Integration Assessment: Used to assess integration strategy and provide guidance as per SAP Integration Solution Advisory Methodology.
- Migration Assessment: Used to estimate the effort of migrating existing SAP Process Orchestration (SAP PO)-based integrations to SAP Integration Suite.
- Edge Integration Cell: Used to deploy and run integration content in secured on-premise environments,

1.3.3 Artificial Intelligence and Data Foundation

This pillar of SAP BTP allows the development, integration, and running of AI and agentic AI solutions. This component helps provide access to the right data required for applications and artificial intelligence through specialized engines such as vector databases, knowledge graphs, and SAP HANA Cloud, a cloud-native database-as-a-service (DBaaS) featuring in-memory storage and agentic memory for persistent conversation context. SAP HANA Cloud supports analytical, transactional, and multimodel applications with built-in, out-of-the-box agentic-based data analysis capabilities. It also supports being used as a data lake.

As shown in Figure 1.5, the AI and data foundation pillar of SAP BTP provides the following capabilities:

- Foundational data services like multimodel data and advanced analytics
- AI technologies and capabilities like SDKs for AI development, access to AI models, AI lifecycle management, knowledge graphs, AI orchestration, ABAP-trained models, relational pretrained models, and AI document management

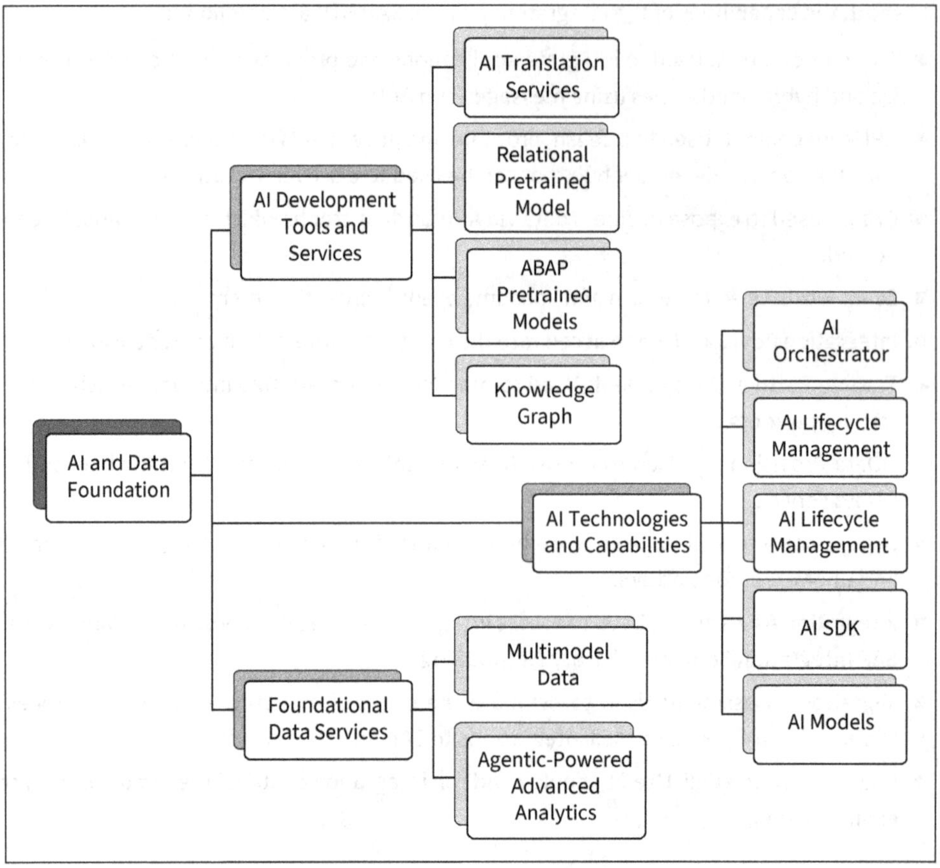

Figure 1.5: AI and Data Foundation Pillar of SAP BTP

SAP Business Data Cloud is a fully managed SaaS solution powered by SAP BTP that allows the governance and management of all data required to make well-informed business decisions. The key features of SAP Business Data Cloud include the following:

- Intelligence apps: Preconfigured and custom business content to visualize business data
- Data packages: Data products from connected SAP systems
- Central catalog: Provides visibility and enrichment of catalog assets
- System integration:
  - ERP systems like SAP S/4HANA Cloud Private Edition and SAP S/4HANA Cloud Public Edition
  - Data management systems like SAP Datasphere
  - Planning and analytics solutions like SAP Analytics Cloud
  - Partner solutions like SAP Databricks, SAP Snowflake, etc.
  - Other applications like SAP SuccessFactors
- Data governance: Allows publication of high-quality data and analytics assets

1.3.4 Deployment Models

SAP BTP supports the following deployment models, as depicted in Figure 1.6:

- Public cloud, a globally distributed shared multitenant offering across multiple global locations and global hyperscalers (Microsoft Azure, Google Cloud Platform, Amazon Web Services, etc.)
- Extended deployment for some country-specific and industry-specific requirements such as EU data residency needs
- Sovereign cloud for regulated and classified workloads
- Sovereign cloud on site (SCOS), which delivers sovereign cloud capability in a client's data center

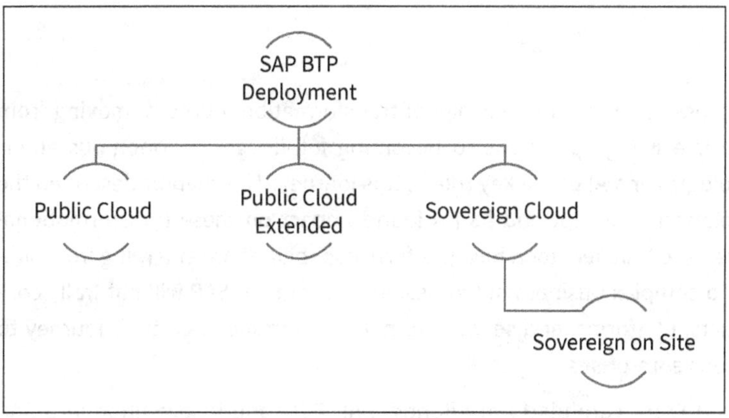

Figure 1.6: SAP BTP Deployment Models

1.4 SAP Business AI Platform

SAP elevated its cloud platform offerings to the next level by announcing SAP Business AI Platform, which brings together the capabilities of SAP BTP, SAP Business Data Cloud, and Business Transformation Management to facilitate the vision of the "Autonomous Enterprise," where business processes can run autonomously via a network of agents and assistants. This unified platform addresses major challenges preventing enterprises from successfully adopting AI, namely the lack of enterprise-specific business context for AI models, the absence of unified data across many disjointed applications, and the challenges of auditing and controlling AI. There are three pillars of the SAP Business AI Platform, as follows:

- Build: Consists of intent-driven development to create agents, apps, and workflows quickly using a wide choice of tools leveraging the new Joule Studio (v2.0). Includes connecting apps, processes, and agents across SAP and non-SAP landscapes using Model Context Protocol (MCP), Agent-to-Agent (A2A), and API Management through SAP Integration Suite. Joule Studio 2.0 allows automatic analysis of the provided business intent and the generation of requirements, specifications, code, tests, and deployments. Joule Work is a new engagement layer designed to improve productivity. A managed n8n environment will also facilitate AI orchestration and integration with Vercel to create a modern web experience.

- Contextualize and reason: This consists of using SAP-managed data products to unify SAP and non-SAP data the across landscape to give agents the data's proper business context. SAP Business Data Cloud and SAP Knowledge Graph provide the right data foundation for AI.
- Govern: This capability is used to discover, monitor, and manage agents within the context of enterprise business process using tools like SAP LeanIX and SAP Signavio.

1.5 Summary

Enterprises are progressing further in their digital transformation journeys, moving from merely digitizing their existing operations to becoming full-fledged autonomous enterprises. Cloud technologies and AI play a key role in this journey. The chapter described the emergence of the platform concept and its profound impact on these transformational journeys. The chapter also touched upon how platform capabilities are emerging from simpler functionalities to complex business outcomes. In this context, SAP will naturally continue to offer products, platforms, and services to help organizations on their journey to becoming autonomous enterprises.

This chapter reviewed SAP's core platform offering, SAP BTP, and its evolution into SAP Business AI Platform. We explored the different pillars of these platforms and how SAP is delivering the power of AI and agentic AI, contextualized with business knowledge, to transform enterprises.

The subsequent chapters in this book will go into detail about the individual capabilities of SAP BTP/SAP Business AI Platform and their roles in the development of custom solutions that will help your enterprise differentiate itself from the competition.

Chapter 2
Development Environment

In this chapter, we'll introduce you to the development environments you'll work with throughout this book. You'll learn about the architecture of SAP Business Technology Platform (SAP BTP), the different development environments provided by this platform, their key characteristics and usage, and the various services SAP delivers for cloud-native application development. Additionally, we'll introduce SAP's latest Joule-based capabilities that can significantly accelerate the design and development of enterprise-ready solutions.

In Chapter 1, we described the rationale behind SAP's foray into cloud-based development tools to meet the needs of the developer community. When SAP initially started the internal project for developing the platform as a service (PaaS), they called it Neo, and the name stuck post-release as well. Since then, SAP BTP as a platform, has evolved as a comprehensive development environment catering the need of wide spectrum of persona covering citizen developer, full-stack developers, pro-code developers, data scientists, and functional users, effectively addressing almost every diverse development and enterprise needs.

A long history of branding name changes is behind the SAP BTP brand, which launched in 2021. What initially came out as the SAP HANA Cloud Platform in 2014, SAP's PaaS for developers, had been rebranded as SAP Cloud Platform and this solution evolved to SAP BTP. The platform initially had SAP HANA and SAP MaxDB as databases. The platform had several services that developers could use to create several types of applications. SAP adopted Cloud Foundry for SAP Cloud Platform in May 2017, which is now open source and supports several other databases as well. You'll see in this chapter the imperative behind adopting Cloud Foundry, and we'll provide insights into the cloud-based paradigm while subsequent chapters will provide a more hands-on approach to leveraging this architecture.

During the course of discussion, we've kept some references to the Neo environment since some organizations who were early adopters are still using Neo. However, the Neo environment will be sunset on December 31, 2028, and SAP is no longer creating any feature enhancements for the Neo platform.

2.1 Introduction

Over the years SAP BTP has evolved significantly and provides a unified, cloud-native platform that brings together application development, process automation, data management, analytics, and artificial intelligence capabilities, which are key building blocks to lay the foundation for an autonomous enterprise. As depicted in Figure 2.1, the SAP BTP ecosystem has evolved to provide the following major functions (the current terminology was covered in Chapter 1):

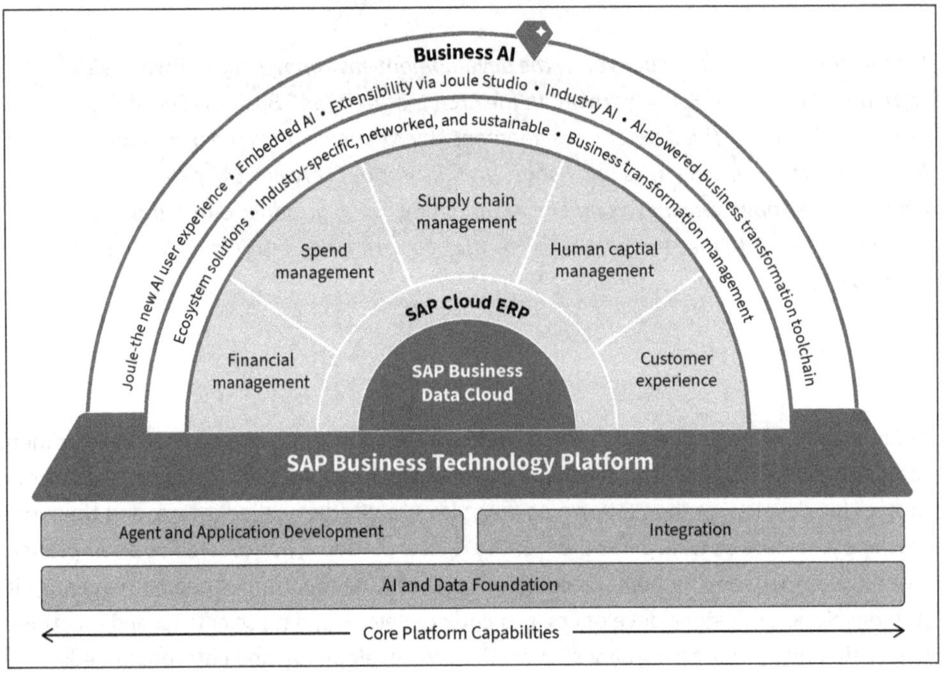

Figure 2.1: SAP BTP Overview

- **Agent and application development**
  SAP BTP now offers a comprehensive platform to create a wide spectrum of enterprise applications including cloud-native solutions, extensions for on-premise systems, and SAP's various SaaS solutions. Additionally, you can design, orchestrate, and deploy intelligent, scalable automation solutions tailored to your needs. This is enabled by SAP's latest AI assistance offering within SAP Build Code and SAP Joule for Developers.
- **Integration**
  This function includes solutions for integration scenarios and is supported by products like SAP Integration Suite, Event Mesh, cloud connectors, API Management, and SAP Business Data Cloud. Alongside AI-powered content, these solutions not only speed up integration development, but also lower ongoing support costs.
- **Artificial intelligence (AI) and data foundation**
  Over the years, AI offerings on SAP BTP have evolved significantly and the platform now offers powerful features such as generative AI, AI agents, code generation, document AI, and a wide spectrum of AI models. This enables the design and development of intelligent enterprise solutions in an accelerated manner. All of this is empowered by a strong data foundation that includes not only the core data storage but also the data processing and management services that enable enterprise applications to derive powerful

insights. Offerings like the SAP HANA Cloud vector engine, SAP HANA Cloud knowledge graph engine, Object Store on SAP BTP are the core components of this data foundation.

At the center sits SAP Business Data Cloud, serving as the foundational data hub that powers all other components in the architecture, with data as the core enabler for business processes and AI capabilities. The cloud ERP semicircle encompasses five critical business domains, including financial management, spend management, supply chain management, human capital management, and customer experience.

Ecosystem solutions are centered on industry-specific, networked, and sustainable approaches, along with business transformation management capabilities that bridge core ERP functionality with specialized industry requirements and transformation initiatives. Finally, the business AI outline layer at the top encapsulates a key set of components featuring SAP's AI strategy, which includes the Joule intelligent assistant, embedded AI, extensibility via Joule Studio, industry AI solutions, and an AI-powered business transformation toolchain.

In the following sections, we'll dive into the SAP BTP architecture and its development environments.

2.1.1 SAP Business Technology Platform Architecture

SAP BTP offers a comprehensive platform as a service (PaaS) that serves as a unified technology foundation to build, integrate, extend, and operate enterprise applications in a cloud environment. Built on a multicloud foundation with strategic support for multiple hyperscalers, including Microsoft Azure, AWS, and Google Cloud Platform, SAP BTP integrates data and analytics, artificial intelligence, application development, automation, and integration capabilities into a single unified platform featuring enterprise-grade security and governance. With a vast array of offerings, what exactly does SAP BTP's architecture look like? Let's explore the new business architecture view, as envisioned by SAP, and try to position SAP BTP within that view (see Figure 2.2).

As depicted in Figure 2.2, SAP BTP is becoming the enabler of enterprise modernization for all SAP customers who want to adopt the cloud in some form. The primary focus of this underlying architecture is the following paradigm shifts:

- SAP BTP provides the necessary foundation to support SAP's extended cloud portfolio covering core enterprise resource planning (ERP) and solution areas such as financial management, spend management, supply chain management, human capital management, and customer experience.
- Multiple major hyperscalers, including Microsoft Azure, AWS, Google Cloud Platform, and Alibaba, are supported, perfectly aligning with SAP's multicloud strategy that allows customers to mix and match different hyperscaler services alongside unified SAP BTP services.
- Enterprise application development is now accelerated more than ever with the introduction of SAP Build Code inside SAP Business Application Studio, leveraging the power

of an AI assistant to generate code, data models, scripts, and other associated artifacts. SAP Cloud Application Programming Model remains the key technology for building pro-code, enterprise-grade applications using Node.js, Java, and TypeScript, now further accelerated by AI-assisted code generation enabled through SAP Build Code.

- A key element of the ecosystem is Joule. With Joule Studio included in the SAP Build suite, developers and business experts can use SAP Build to create custom Joule agents and extend business processes with Joule skills.
- SAP HANA Cloud remains the primary cloud-native database for hosting enterprise applications and has been revamped with the latest features, including SAP HANA Cloud vector engine, SAP HANA Cloud knowledge graph engine, which are key components for building any enterprise AI solution. In addition to SAP HANA Cloud, SAP BTP also supports lightweight data storage options, including PostgreSQL, Redis, and others, through hyperscaler-based services. Object stores such as AWS S3 buckets, Azure Blob Storage, Google Cloud Storage buckets, and SAP Cloud Infrastructure containers enable storing and managing unstructured data like files, documents, images, and large binary objects.

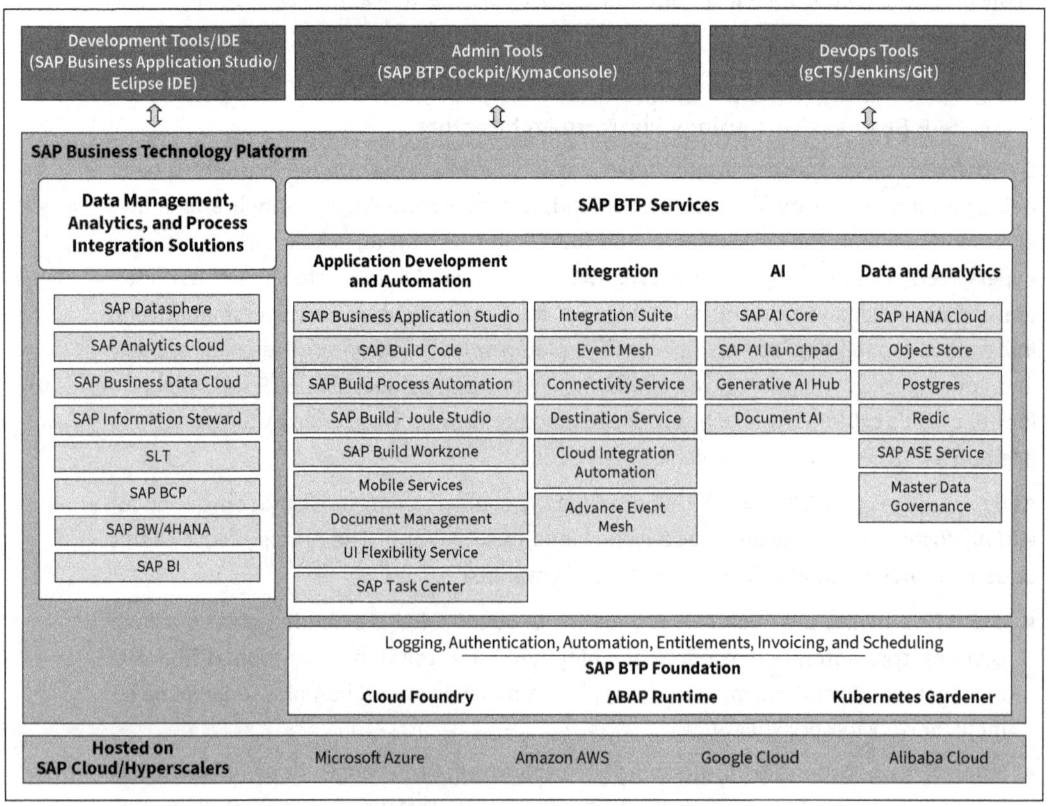

Figure 2.2: SAP BTP Architecture Overview

- With the newly revamped SAP AI Core and generative AI hub, SAO BTP now supports a wide spectrum of features to create enterprise AI solutions powered by generative AI from scratch. Secure access to over 40 foundation models from OpenAI, Google, Azure, AWS, IBM, and SAP, along with model lifecycle management through SAP AI Launchpad, makes it an ideal platform for enterprise AI adoption.

- SAP Integration Suite provides essential capabilities and services to connect and integrate enterprise business processes and applications. Event-driven architecture support and application programming interfaces (APIs) supported by REST, OData, SOAP, and other enterprise-ready protocols enable seamless integration and scalable connectivity across heterogeneous enterprise landscapes. On the other hand, secure integration with on-premise systems is ensured through principal propagation, enabling seamless authentication between systems.
- The ABAP runtime enables ABAP developers to use their existing skills and knowledge to build and run cloud applications and side-by-side extensions with the ABAP RESTful application programming model, without needing new programming language skills.
- SAP Cloud Identity Services serves as a primary identity provider, decoupling authentication from application capabilities and authorization management, including support for OAuth 2.0 and SAML 2.0 federation. It also supported federation capabilities from external identity providers such as Microsoft Entra ID, Okta, and other SAML 2.0- or OIDC-compliant systems, ensuring a hybrid identity architecture.
- With much greater openness, containerization is now supported in multiple environments including SAP BTP, Cloud Foundry environment as well as SAP BTP, Kyma runtime.
- Cloud-based enterprise application design and development is now fully accelerated with the support of SAP Joule for Consultants and SAP Joule for Developers. SAP Joule for Consultants contributes to accelerating the design phase by providing real-time expert guidance and SAP Joule for Developers helps to generate complete data models, business logic, UI components, sample data and more.

Now that we have become accustomed to SAP BTP's architecture, let's explore the key development environments available for hosting enterprise applications.

2.1.2 Development Environments

SAP BTP currently offers five different development environments including:

- SAP BTP, Cloud Foundry environment
- SAP BTP ABAP environment
- SAP BTP, Kyma runtime
- SAP BTP, serverless runtime
- SAP BTP, Neo environment (scheduled for sunset)

The serverless option is primarily used for doing short proof of concept kinds of development, which can then be scaled down to zero when not in use or scaled up, as required. This option is supported by SAP BTP, Kyma runtime.

With respect to this book, cloud-native developments signify those developments that SAP customers have created to extend and augment existing SAP functionalities, provided by SAP products whether on-premise or SaaS.

The term *low-code/no-code* has gained a lot of popularity lately because the option to create something without any code, and then also visually managing and orchestrating the application, is quite appealing to "citizen developers." Low-code/no-code tools and platforms provide powerful visual tools to help you put together an application through drag-and-drop features with minimal configuration like setting up authentication or service bindings. Additionally, Joule Studio can be used to create custom AI agents and extend Joule skills.

We'll focus primarily on the SAP BTP, Cloud Foundry environment; SAP BTP ABAP environment; and SAP BTP, Kyma runtime in this section.

2.2 SAP BTP, Cloud Foundry Environment

Cloud Foundry has become the industry-standard, open-source, cloud-based platform that can be deployed on any infrastructure, owned in-house or hosted on the cloud, called *infrastructure as a service (IaaS)*. Commercial Cloud Foundry providers can provide PaaSs like SAP BTP or IBM Cloud Foundry, among others. Since 2014, SAP has partnered with other infrastructure providers that have joined the Cloud Foundry Foundation (*https://www.cloudfoundry.org/foundation/*) so that these providers can deploy Cloud Foundry for any compliant infrastructure.

In simple terms, Cloud Foundry can be described as an operating system for a cloud platform that provides a standardization layer. The Cloud Foundry platform enables you to deploy any app or service within minutes and makes these apps or services scalable, depending on their usage. Its layer of standardization has in fact become the de facto standard for cloud platforms. The Cloud Foundry architecture offers container-based isolation between the various parts of your application, including, for instance, SAP HANA deployment infrastructure (HDI) containers (database layer plus deployment infrastructure, including SAP HANA schemas, security, etc.) or an app container in the application layer. Multiple app containers can separate microservices, for instance, in different languages.

Deploying multiple containers also improves overall performance. You can avoid vendor lock-in from an infrastructure perspective as well, as mentioned in Chapter 1, Section 1.2.

In the following sections, we'll cover the characteristics of Cloud Foundry, a concept that exists beyond SAP but has been adopted in SAP BTP for a few years now. We'll also demonstrate how this adoption has been performed and describe the Cloud Foundry services that SAP BTP provides.

2.2.1 Characteristics of Cloud Foundry

Some major characteristics that distinguish Cloud Foundry as a platform include the following features:

- Cloud Foundry is based on major open-source technologies: the Cloud Foundry application runtime and the Cloud Foundry container runtime.
- Cloud Foundry provides for container-based architecture. Thus, you can code in any language, and the container will already have an application server in that language. Thus, you can deploy a web application to any cloud infrastructure, without changing any code. This seamless portability is possible because of Cloud Foundry's container-based architecture.
- Cloud Foundry's architecture supports the use of microservices. This paradigm helps you scale services individually, as needed.

Major features of Cloud Foundry supported by SAP BTP include the following:

- **Diego runtime**
  Diego is a self-healing container management system that keeps the correct number of instances running, in Diego Cells, to avoid network failures and crashes. Diego schedules and runs tasks and long-running processes (LRPs).
- **Secure Socket Shell (SSH)**
  Cloud Foundry supports native SSH access to applications and load balancing of SSH sessions. The load balancer in your Cloud Foundry deployment provides a simple and scalable way to access container apps and other LRPs.
- **Docker**
  Docker is an open-source platform for creating containers, which are standardized packages for development, shipment, and deployment. With Docker support enabled, Cloud Foundry can also deploy and manage apps running in Docker containers.
- **Running tasks**
  Tasks running in their own containers require minimal resources. A *task* is an application or script whose code is included as part of a deployed application and runs for a finite time (unless an LRP). After the task is complete, Cloud Foundry destroys the container that was running the task, thereby freeing up resources.
- **Zipkin tracing**
  Zipkin is a distributed tracing system that application developers can use to troubleshoot failures or latency issues in microservice architectures. This ability enables tracing requests and responses across distributed systems.
- **WebSockets**
  The WebSocket protocol provides full-duplex communication over a single TCP connection and is useful for performing real-time data exchanges between a client and a server more efficiently than HTTP. Cloud Foundry uses WebSockets for the following metrics and logging purposes:

  - To stream all application event data and component metrics from the doppler server instances to the traffic controller
  - To stream application logs from the traffic controller to developers using the Cloud Foundry command-line interface (CLI)
  - To stream all application event data and component metrics from the traffic controller over the firehose endpoint to external applications or services

- **Space scoped service brokers**
  Cloud Foundry normally supports two types of service brokers: standard brokers and space-scoped service brokers. Standard brokers publish service plans either to specific or all organizations in the deployment. This capability is not supported by SAP BTP, Cloud Foundry environment yet. Space-scoped service brokers publish service plans to users with access to the space where the service broker is created.
- **Route services**
  Sometimes, application developers need to apply transformations or processing to requests before these requests reach an application (e.g., authentication, rate limiting, and caching services). In this scenario, route services are used. Developers can use these marketplace services to apply various transformations to application requests by binding an application's route to a service instance.
- **Custom domain**
  The SAP Custom Domain service allows organizations to configure their own custom domains to expose applications publicly instead of relying only on default SAP-provided subdomains (like hana.ondemand.com)
- **Joule runtime**
  Cloud Foundry serves as the primary foundational infrastructure that hosts Joule's conversational interface and integrates it with various SAP applications across the solution portfolio.
- **Python runtime**
  Python is one of the most popular languages to create AI applications and SAP BTP Cloud Foundry runtime provides comprehensive support for developing and deploying pro-code Python-based agents through its polyglot application platform. The runtime offers enterprise-grade infrastructure with built-in security, scalability, and service integration capabilities, making it ideal for hosting enterprise agents and other AI driven applications.
- **Agent development framework**
  The Cloud Foundry environment provides the ideal runtime and ecosystem to build, deploy and host any kind of pro-code/Python-based AI agents tailored to enterprise needs. The availability of the widest range of open-source libraries, including LangGraph, AG2 (AutoGen), CrewAI, Smolagents, Google ADK, Pydantic AI, and others, gives it an edge over other hyperscalers. Additionally support for native SAP Cloud SDK libraries

along with Flask/FastAPI based backend framework, makes it an ideal environment for custom agent implementations.

Some Cloud Foundry features are not yet supported by SAP BTP, including the following:

- **Container-to-container networking**
  With this feature, application instances communicate with each other.
- **Isolation segments**
  These dedicated resource pools isolate deployment workloads.
- **TCP routing**
  This feature runs applications that serve requests on non-HTTP TCP protocols. TCP routing is useful for complying with regulatory requirements that demand Transport Layer Security (TLS) to terminate as close to your apps as possible so that packets are not decrypted before reaching the application level.

2.2.2 How SAP Business Technology Platform Has Adopted Cloud Foundry

SAP BTP has been making most of the services available on Cloud Foundry. Now, it has also adopted other open-source runtimes, but Cloud Foundry was the first open-source platform that SAP adopted. The roadmap is to make all services available on Cloud Foundry and other development environments, while Neo will be phased out gradually. (Neo is not enhanced much as it's no longer on SAP's strategic roadmap.)

As a result, you can create an application in Java, Node.js, HTML5, or Python in SAP BTP or even bring your own language (BYOL) into the Cloud Foundry framework. Interestingly, Cloud Foundry enables you to use multiple languages for different pieces of your application by deploying polyglot runtime containers.

The main change in terms of designing applications built on Cloud Foundry is the adoption of the SAP Cloud Application Programming Model, which, according to SAP, is a framework of languages, libraries, and tools for building enterprise-grade services and applications. This framework also leverages widely adopted open-source technologies like Java, Node.js, Spring, and more. On top of these technologies, SAP Cloud Application Programming Model includes two more components:

- Core data services (CDS), which is the universal modeling language for both domain model and service definitions.
- Service software development kits (SDKs) and runtimes for Node.js and Java, with libraries to implement and consume services. It also includes automatic implementations.

2.2.3 Cloud Foundry Services Offered by SAP Business Technology Platform

SAP BTP now supports Cloud Foundry-based architecture. The services supported by Cloud Foundry can be obtained from SAP Discovery Center, as shown in Figure 2.3.

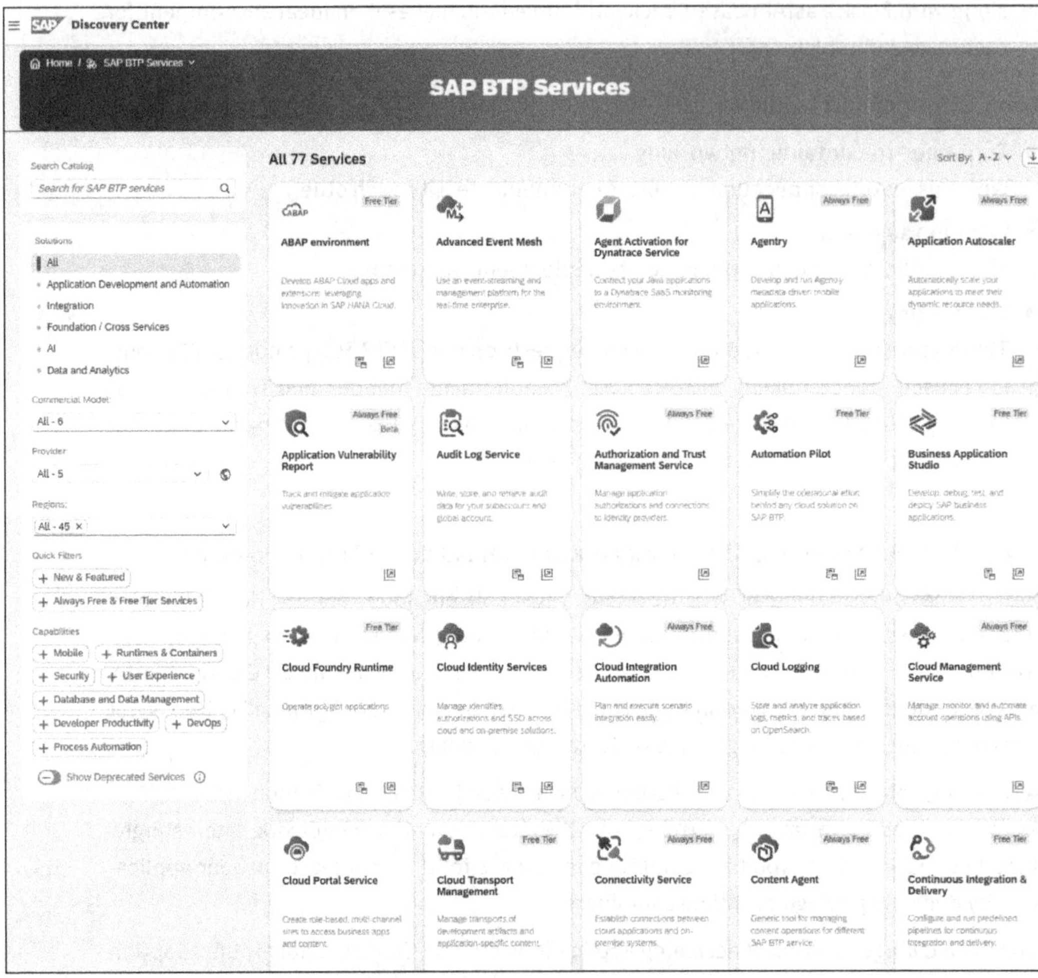

Figure 2.3: SAP Discovery Center: Cloud Foundry Services

The categorization of the services is also shown in Figure 2.3. By capability, the services are grouped in the following ways:

- **Data and Analytics**
  - Data quality service
  - SAP Master Data Governance, cloud edition
  - SAP Master Data Integration
  - Object Store on SAP BTP
  - PostgreSQL on SAP BTP, hyperscaler option
  - Redis on SAP BTP, hyperscaler option
  - SAP ASE service
  - SAP HANA Cloud

  - SAP HANA service (SAP regions)
  - SAP HANA spatial services
- **AI**
  - SAP AI Core
  - SAP AI Launchpad
  - SAP Document AI
- **DevOps**
  - Dynatrace Agent Activation for SAP BTP
  - SAP Automation Pilot
  - SAP Cloud Management service for SAP BTP
  - SAP Cloud Transport Management
  - SAP Content Agent service
  - SAP Continuous Integration and Delivery
  - SAP Job Scheduling service
  - SAP Monitoring service for SAP BTP
  - SAP Alert Notification service for SAP BTP
  - SAP Solution Lifecycle Management service for SAP BTP
  - SAP Service Manager
  - SAP Usage Data Management service for SAP BTP
- **Security**
  - SAP Authorization and Trust Management service
  - SAP Cloud Identity Services
  - SAP Credential Store
  - SAP Custom Domain service
  - SAP Data Privacy Integration
  - Identity Provisioning service
  - SAP Keystore service
  - SAP Malware Scanning service
  - OAuth 2.0 on SAP BTP
  - SAP Personal Data Manager
  - SAP Private Link service
- **Integration**
  - Cloud Integration Automation
  - SAP Connectivity service
  - SAP Destination service

- Event Mesh
- SAP Integration Suite
- SAP Master Data Integration

- **UX**
  - SAP Cloud Portal service
  - SAP Document Management service
  - SAP Forms service by Adobe
  - SAP Build Work Zone
  - UI flexibility service
- **Developer Productivity**
  - SAP Business Application Studio
  - SAP Build Code
- **Process Automation**
  - SAP Task Center
  - SAP Responsibility Management service
  - SAP Build Process Automation

2.2.4 SAP Joule for Developers

In addition to a wide spectrum of services, enterprise application development using SAP BTP, Cloud Foundry environment has become significantly faster and more efficient with the introduction of Joule powered application and code generation through SAP Build Code. Be it simple or complex, Node.js, Fiori, Java, and SAP Cloud Application Programming Model-based applications can now be generated through simple natural language prompts, avoiding the need to start from scratch. Beyond code generation, Joule can even assist you in generating the underlying data model, testing scripts, sample data, and documentation, all of which are essential at various stages of the application development lifecycle. Figure 2.4 shows a glimpse of code generated through Joule embedded in SAP Business Application Studio; for more details, see Chapter 7.

The AI assisted code generator is now embedded within SAP Business Application Studio and enabled through SAP Build Code booster execution. Also, one of the most encouraging aspects is that all these features can be explored using the SAP BTP trial environment, which remains active for 90 days, allowing developers to explore and learn without any licensing cost.

Additionally, SAP Joule for Consultants offers extra assistance during the design stage by providing information about the latest services, capabilities, and modern components you can leverage to design the solution. Using simple natural language prompts, you can either query services or offerings on SAP BTP or craft a statement representing your business scenario. Based on the given context, SAP Joule for Consultants can provide associated details, best practices, and even links to reference architecture from the architecture center.

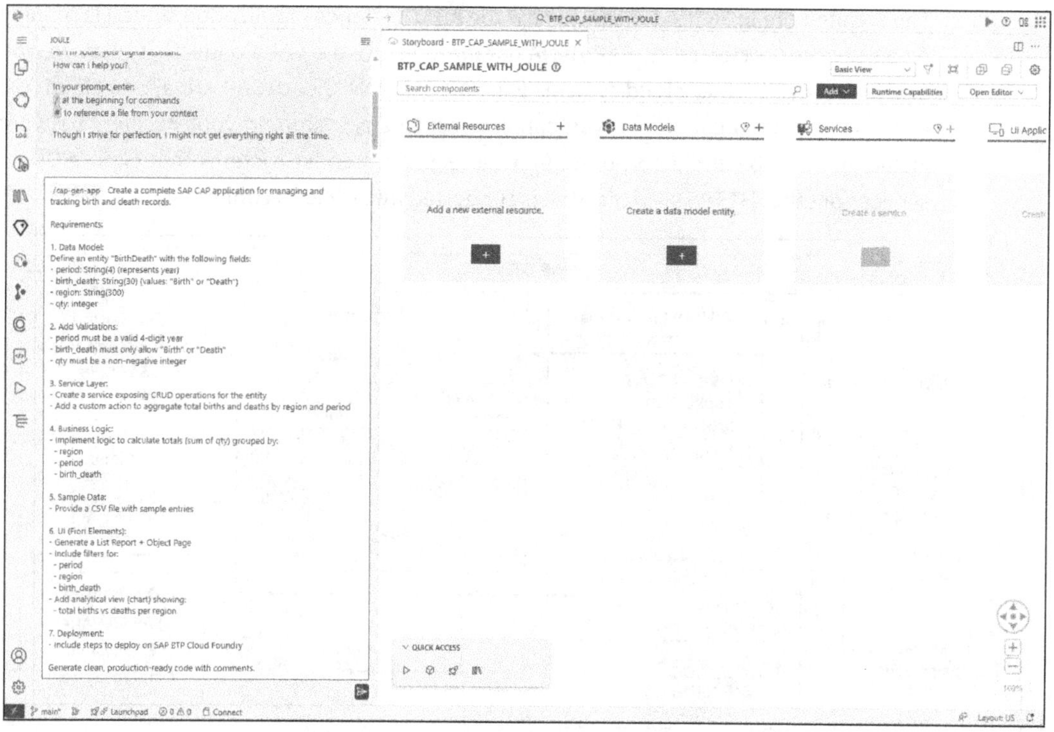

Figure 2.4: Prompt-Based Instruction

This kind of guidance at various stages of the application development lifecycle reduces research effort, improves alignment with best practices, and enables faster prototyping and implementation.

2.3 SAP BTP ABAP Environment

SAP BTP ABAP environment supports the latest version of ABAP. While the development must be performed using the ABAP perspective in the Eclipse integrated development environment (IDE) currently, the deployment of the application occurs in the ABAP server on SAP BTP. A set of development toolsets are part of the Eclipse IDE. This ABAP code as a service is operated by SAP.

Thus, SAP BTP ABAP environment has two components:

- The ABAP PaaS offering from SAP
- The ABAP platform feature set includes development tools, language version, released objects, and the ABAP RESTful application programming model

For some scenarios that are tightly coupled to an existing SAP S/4HANA model, the extension becomes difficult since it is not loosely coupled. For this scenario, the SAP BTP ABAP environment and developer extensibility have been introduced. This option is currently available in SAP S/4HANA Cloud and is planned for SAP S/4HANA.

The architecture in Figure 2.5 shows how the ABAP instance (application server) is running on a space connected to an SAP HANA database. Using the SAP Connectivity service, you can connect with other cloud systems like SAP S/4HANA Cloud, SAP Ariba, or a cloud API. Leveraging a cloud connector, you can connect to any system in your business network including those behind a corporate firewall (e.g., on-premise systems that expose OData services or other HTTPS services) that is supported by the cloud connector.

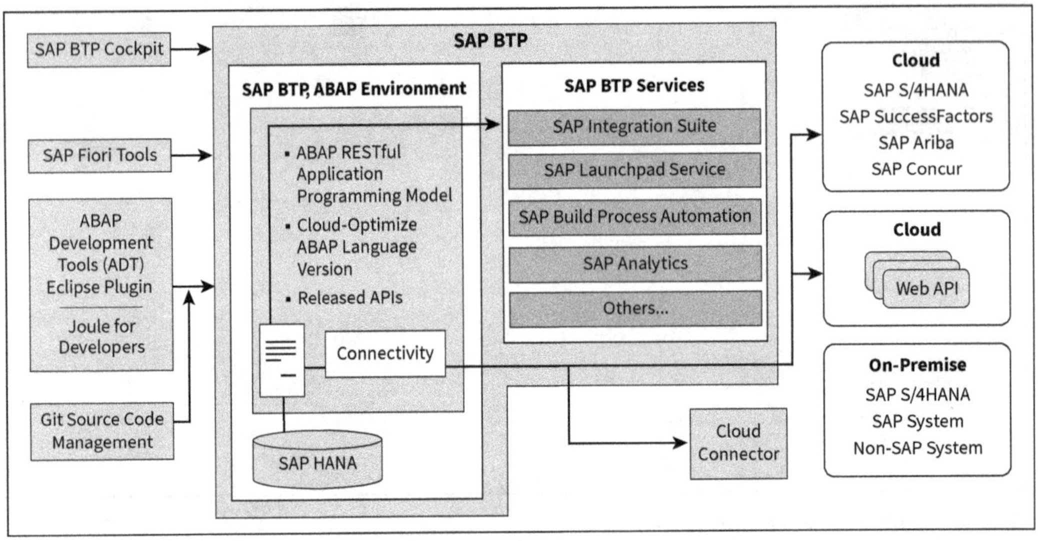

Figure 2.5: Architecture of ABAP Application Deployment

Additionally, an ABAP application can use any SAP BTP service. Within the SAP BTP cockpit, you can configure the launchpad for system administration activities using SAP Fiori apps. SAP Business Application Studio can be used to build SAP Fiori apps as well. The development is performed through ABAP Development Tools (ADT) available on the Eclipse IDE, and developments can be stored in Git repositories. Different use cases would leverage different parts of this architecture, as shown in Figure 2.6.

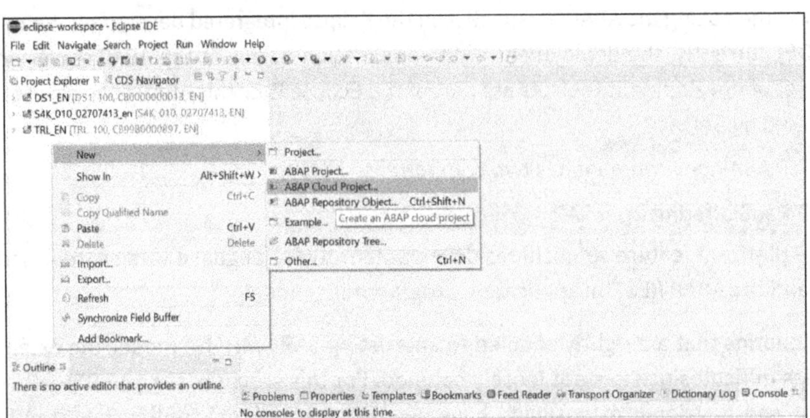

Figure 2.6: Eclipse IDE: ABAP Perspective

In this section, we'll dig more deeply into the cloud-specific features of SAP BTP ABAP environment.

2.3.1 Multitenancy in SAP BTP ABAP Environment

SAP BTP ABAP environment provides client-based support for multitenancy. This feature of multitenancy makes it possible for you to create SaaS applications that can then be used by multiple customers or by disparate business units in your organization.

When a partner wants to publish an application as SaaS using SAP BTP ABAP environment, they'll need a partner development contract. Thus, a multitenant application that is created as an add-on can be deployed to multiple consumers or tenants simultaneously. The application must be created and deployed as an add-on product version and consists of one or more software components or repositories. The ABAP environment pipeline can be used to build and publish the add-on product.

[«]

Note

For step-by-step details about the creation of add-ons, refer to **Develop and operate ABAP SaaS applications using Add-ons** at *http://s-prs.co/v550428.*

2.3.2 Key Cloud-Specific Features

SAP BTP ABAP environment, supports the cloud version of ABAP, which follows the ABAP RESTful application programming model paradigm, which leverages the following features:

- SAP Fiori is used as the frontend.
- SAP HANA is used as the database.
- CDS views are for accessing data from the backend SAP S/4HANA and for creating data models on SAP HANA databases.
- SAP BTP ABAP environment can use only released APIs and the other released objects like data dictionary objects.
- The Eclipse IDE provides tools like ADT for development. ADT shows released objects (released for usage in SAP BTP) that can be consumed by your custom developments.

Note that SAP GUI is not supported in this model.

The ABAP RESTful application programming model helps you create SAP Fiori apps and web APIs by providing standardized development flows, using best practices and guidelines to focus on the right business objects.

Common patterns using the ABAP RESTful application programming model include the following:

- Data modeling and behaviors centered around data modeling in CDS views (possible through UI annotations), behavior definitions (of your data services), and the implementation of business objects
- Business service provisioning through service definition (expose CDS views and entities) and service binding (make services available for a specific protocol)
- Service consumption through UIs like the SAP Fiori UI and web APIs

In most cases, CDS views are built on top of database tables. The SAP Gateway service builder has been replaced by this model. Some ABAP technologies are not supported, like Web Dynpro/SAP GUI.

2.3.3 SAP Joule for Developers

With SAP Joule for Developers now embedded into ABAP Development Tools (ADT), it will accelerate the daily work of ABAP developers by providing just-in-time help with code explanation, unit test generation, and application artifact exploration. This will also significantly reduce context switching and accelerate learning by closing the gap between consultant intent and technical implementation. Following traditional methods, developers spend substantial time interpreting unfamiliar objects, tracing dependencies, and seeking guidance from peers and seniors. In such instances, Joule minimizes these interruptions by offering a reliable, always-available companion capable of answering ABAP and core data services (CDS) questions (as shown in Figure 2.7), clarifying ABAP RESTful application programming model concepts, and even proposing real code based on high-level instructions.

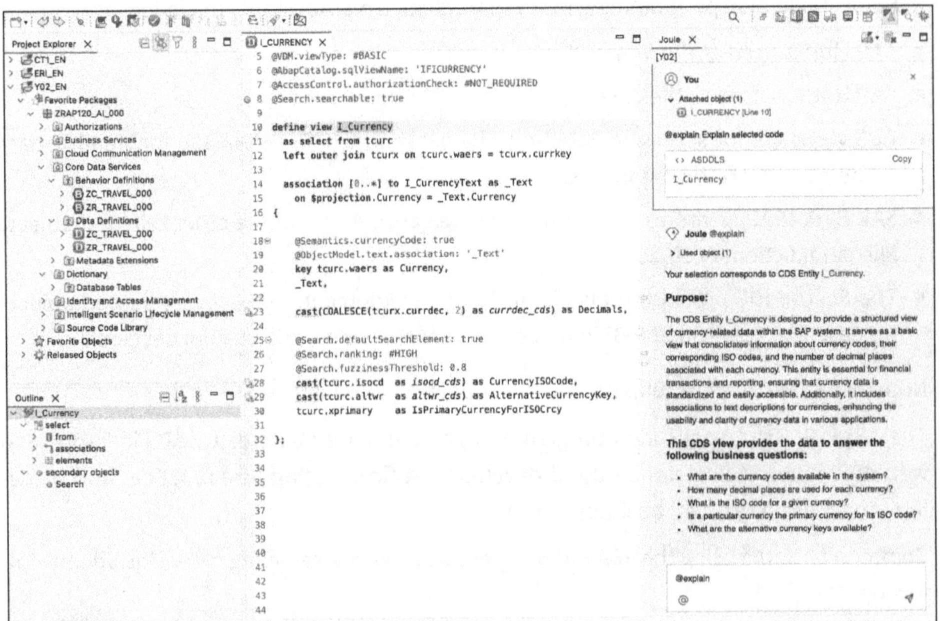

Figure 2.7: Joule Explaining a CDS View in ADT

SAP Joule for Developers is broadly accessible in the SAP BTP ABAP environment and in both SAP S/4HANA Cloud Public Edition and SAP S/4HANA Cloud Private Edition (starting with the 2025 release), with Eclipse ADT serving as the primary IDE integration point rather than the sole access method. In the SAP BTP ABAP environment specifically, once the license is activated, all development tenants in that global account are enabled automatically. Additionally, side-by-side deployment is still supported through the ADT client connecting to SAP BTP via a URL.

2.4 SAP BTP, Kyma Runtime

The open source project called Kyma uses Kubernetes clusters for container deployment, along with a set of other related activities for maintaining and scaling applications. These additional activities are now transparent to users using Kyma-managed services in the SAP BTP platform. With SAP BTP, Kyma runtime you can create your own applications or extend SAP's cloud applications.

While some differences exist between SAP BTP, Kyma runtime and the Kyma open-source project, in this book, we'll concentrate on the features and functions available in SAP BTP, Kyma runtime. Both Kyma and Kubernetes are operated by SAP, and thus, all patches and software upgrades are handled by SAP and are supported according to SAP BTP service level agreements (SLAs).

In this section, we'll describe SAP BTP, Kyma runtime in detail, including its services and functions, and review sample architecture.

2.4.1 Overview

SAP BTP, Kyma runtime, includes the following features and services:

- Serverless engine: This built-in capability enables Kyma to build serverless functions that are lightweight.
- Eventing: This feature means that events from business applications can trigger different actions.
- Service mesh: This built-in feature enables service to service communication and proxy.
- API gateway: This feature exposes APIs for consumption.
- Service catalog: This feature instantiates applications and consumes services during runtime, which can even be services from different hyperscalers, not just SAP BTP or SAP applications.
- CI/CD integration: Built-in support for SAP Continuous Integration and Delivery service with predefined pipelines
- Real-time logging: In the Kyma console, you can access logs from different systems or for a particular lambda function, described in more detail in Section 2.4.3.

- Namespaces: The Kyma console can manage different namespaces which, in the Kubernetes world, are logical groupings of workloads.

Figure 2.8 shows a snapshot of the SAP BTP, Kyma runtime console.

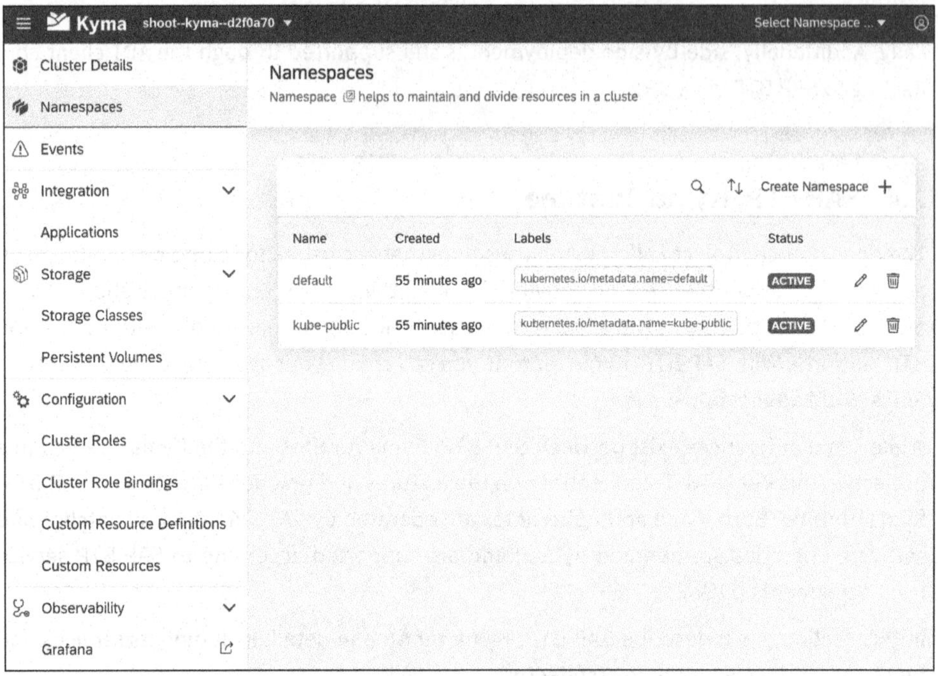

Figure 2.8: SAP BTP, Kyma Runtime: Console

The service catalog can help bind application contexts. External services or service brokers from hyperscalers can also be exposed to the service catalog to enable third-party services. The security of these connections does not need to be handled during development but instead can only be configured during runtime.

All applications that you need to connect to will be displayed in the central console of SAP BTP, Kyma runtime. To access workloads from these applications (in the form of APIs and events coming from these systems), you would go through the namespace binding to the services for the applications.

2.4.2 Creating Services

Using SAP BTP, Kyma runtime, you can create microservices in any language and run them as containerized applications. These microservices can be triggered by events from applications that are configured in the service mesh in Kyma or in Event Mesh. These microservices must also be exposed as services that can be consumed by any calling applications, as defined in configuration in the Kyma console. More details about microservice structures and their characteristics are provided in Chapter 3.

2.4.3 Creating Functions

Simply speaking, a function (also called a *lambda function*) represents a simple code snippet representing business logic that you can run without provisioning or managing servers. A function is based on the function's custom resource, which can be written in either Node.js or Python.

A function must be triggered by a business event, or its code can be consumed by other applications through an API call. A function can even consume other services from your hyperscaler. But configuration is required to bind the function to an instance of a service of an application.

The Kyma Dashboard is used for creating and configuring functions. Alternatively, you can use the command-line interface (CLI) from Kubernetes (called kubectl) for these purposes.

Figure 2.9 shows how the different components of SAP BTP, Kyma runtime that we've described in this section all work together, including how SAP BTP, Kyma runtime connects with different SAP and non-SAP systems and even with services provided by hyperscalers. SAP BTP, Kyma runtime is based on Kubernetes, which can be deployed on any major hyperscaler or on SAP's IaaS. All access to the services and functions for users are provided through the microgateway, which ensures secure connections.

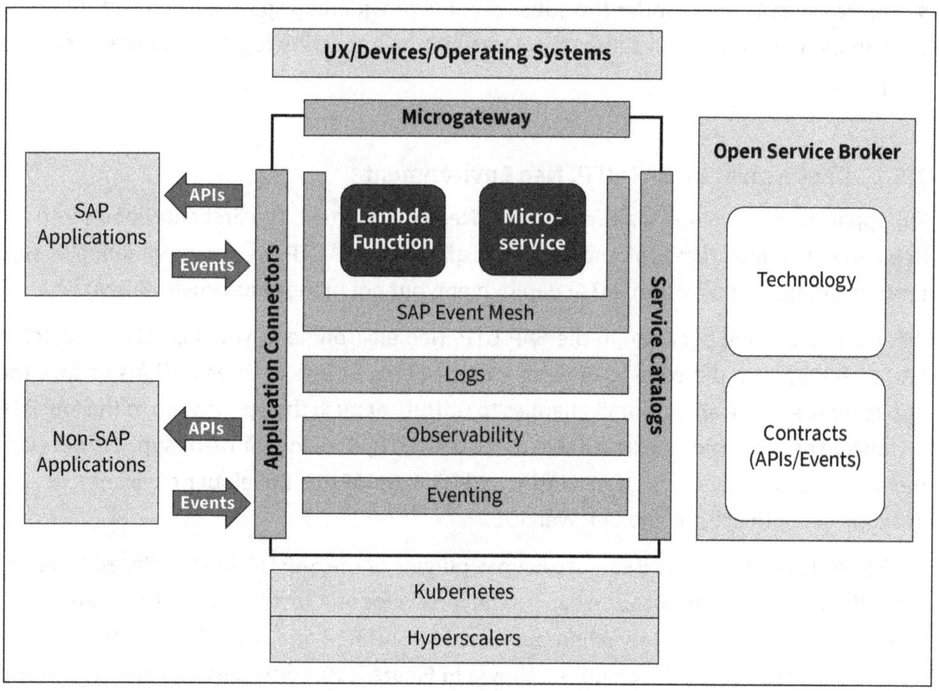

Figure 2.9: Architecture Showing Kyma Components along with SAP and Non-SAP Landscape Components

2.5 Deployment Based on Development Environment

In this section, we'll look at the deployment options for the different development environments we've discussed so far including the SAP BTP, Neo environment; SAP BTP, Cloud Foundry environment; SAP BTP ABAP environment; and SAP BTP, Kyma runtime. A few points to note as we consider our deployment options are as follows:

- The structure of SAP BTP provisioning—in terms of global accounts, subaccounts, and spaces—means that any configuration, such as connecting to a specific on-premise system or security setup, is specific to a subaccount instance of the platform. For example, if you're deploying an application specific to finance, it's better to have a separate subaccount for finance only. Using the same subaccount for all areas increases the risk of configurations being overwritten, causing conflicts and confusion.
- For each environment—development, test, and production—you need separate subaccounts. Each subaccount would be linked to the core instance of the underlying transactional system, such as an on-premise SAP S/4HANA system. Ideally, there are one-to-one links between the environments of this SAP S/4HANA instance and their corresponding subaccounts. This approach ensures testing of backend systems and security in sync, with no deployment risks.
- Further classification under the subaccount is provided by spaces, a concept inherited from Cloud Foundry. But this concept can also be applied to logical groupings of developments.

2.5.1 Deployment on SAP BTP, Neo Environment

For applications developed in the SAP BTP, Neo environment, the first entry point into the deployment architecture is often SAP Web Dispatcher. SAP BTP, Neo environment also supports multitarget application (MTA) deployment, but not in the cloud environment.

When creating an application in the SAP BTP, Neo environment, you'll create a project in SAP Web IDE, push the code to GitHub, and then choose one of three options to save the application: save locally, commit changes to GitHub, or push the application to the SAP BTP instance. The main runtimes available in the SAP BTP, Neo environment support development using Java, JavaScript, SAP HANA, and HTML5/SAPUI5. Developer environments are available on both Eclipse and SAP Web IDE.

For XS-JS development, you can use Eclipse plugins or the SAP HANA Web-Based Development Workbench. An application can be run locally for unit testing but, in this state, is not available to other users. Only when published to SAP BTP and activated will the application be ready to run on the cloud. As shown in Figure 2.10, some services are available in the SAP BTP cockpit, while the deployment and versioning functionalities are available via SAP Web IDE. Unlike other environments, SAP Business Application Studio has no roadmap to support Neo environments. With SAP BTP, Neo environment scheduled for sunset on December 31, 2028, organizations running on Neo need to migrate to the Cloud Foundry environment for greater openness, support for multiple infrastructure providers, and access to all new SAP enhancements. The recommended migration strategy includes the following four phases:

- The discover phase focuses on exploring integration artifacts, security materials, certificates, credentials, PGP keys, data stores, and third-party dependencies, and preparing an inventory.
- The plan and prepare phase involves ensuring all Neo artifacts are saved as versioned (not in draft mode), communicating endpoint changes to third parties, provisioning the Cloud Foundry tenant, and configuring OAuth authentication in both environments.
- The migrate phase executes the actual migration using SAP's provided migration pack via Postman collections or, alternatively, through manual export/import to migrate iFlows, known_hosts files, PGP keys, and security artifacts, followed by deploying and configuring iFlows in Cloud Foundry.
- The validate phase handles testing the migrated content to ensure all partner integrations remain uninterrupted before switching production traffic.

In a multitier landscape, the best practice is to pause all development before migration to avoid any double maintenance scenarios, migrate first to the non-production tenant, and then use SAP Cloud Transport Management to promote to other environments.

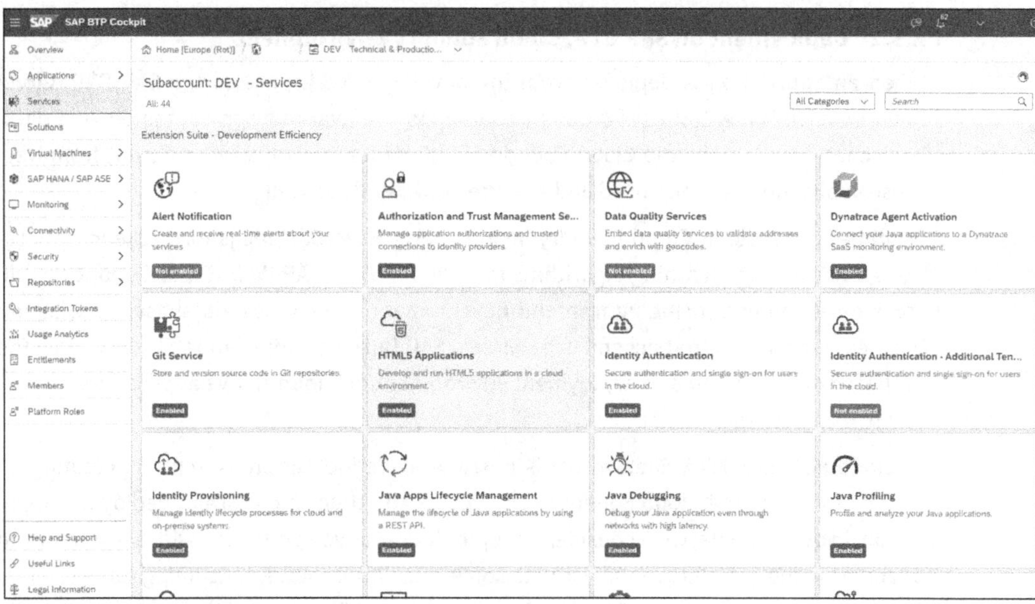

Figure 2.10: SAP BTP, Neo Environment: Deployment

Figure 2.11 shows versioning capabilities within SAP BTP, Neo environment.

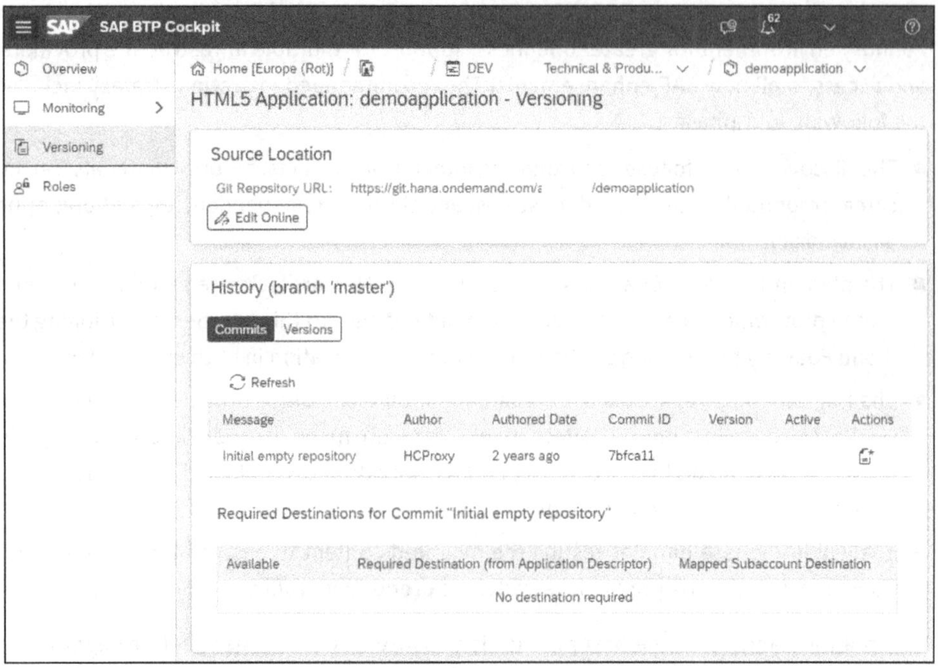

Figure 2.11: SAP BTP, Neo Environment: Versioning

2.5.2 Deployment on SAP BTP, Cloud Foundry Environment

When an application is deployed from the developer workspace to the SAP BTP, Cloud Foundry environment, it uses different containers (e.g., various app containers and database containers). Since the Cloud Foundry architecture is completely different from Neo, these applications are deployed and executed differently as well.

Runtimes available in Cloud Foundry include Java, HTML5, Node.js for backend applications, and machine learning algorithms in Python. While SAP Web IDE still supports full-stack development, including frontend development capabilities, database models, and business logic built into backend capabilities, SAP Business Application Studio is now the preferred development and deployment environment for cloud-native applications (unless you're using ABAP).

In Cloud Foundry, MTA deployment is possible. Let's look at an example of creating an application using the Cloud Foundry architecture and then using MTA to deploy the application. In our example, we'll consider a simple cloud-native application with a frontend and a schema in the SAP HANA database with some calculation views, consuming some services from other systems, perhaps SAP S/4HANA through a CDS view. How the deployment architecture should look is shown in Figure 2.12.

The AppRouter is the first point of entry when a user calls your application through a web browser. The AppRouter also handles security considerations and authenticates the user. The call goes from this application to a microservice, over an HTTP service or a REST service.

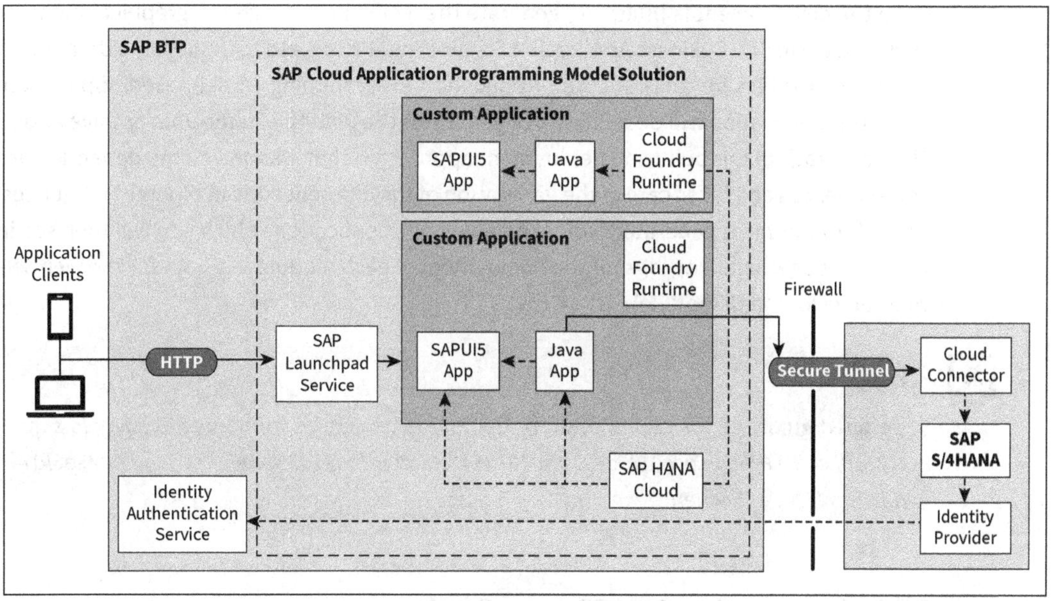

Figure 2.12: SAP BTP, Cloud Foundry Environment: Deployment Architecture

The application can be deployed into different app containers, depending on the microservices used. Cloud Foundry has container services and backing services. With each container, the core parts of the application server are deployed too. For instance, a microservice may use different languages. This multilingual operation is supported by backing services, which contain every attached resource for your application, for example, persistence objects or job schedulers.

While typical artifacts are created through the SAP Business Application Studio (or SAP Web IDE), if you use a language not yet supported by the SAP Business Application Studio/SAP Web IDE, you can connect an external development tool for any language through Git. All the code for all your applications can be stored in a Git repository. The deployed services available in Cloud Foundry can act as the central com ponent for managing the lifecycle of the application. These services enable MTAs, which will package up all the different parts of the application in a single *.yaml* file, to be transported across different landscapes. The MTA will split up the application into modules: For instance, one module may be for the SAP HANA database for database content (CDS views, calculation views), while another module may contain Java/Node.js for backend code. A third module may be an HTML5 module for the HTML frontend content. The deployment functionality also contains configuration information for security and other dependencies like services the application needs. It will take care of getting the code from the developer workspace into the relevant parts of Cloud Foundry (e.g., HDI containers into the table/views/databases, frontend/backend into app containers, etc.).

Once the development is built, you generate the *.yaml* file through the graphical menu or using the CLI in SAP Business Application Studio. You'll also need to input some details into the descriptor files for an SAP Cloud Application Programming Model-based application. Then, this application will be built, which generates the *.mtar* file in the *.mta* archive folder. The advantage of this build and deployment approach is that it knows all the dependencies and sequences and takes care of the generation of the different code in different languages. Once the *.mtar* file is generated, you can deploy this application, which you will then see in your dev space, including logs for the deployment of each module (e.g., *.srv* for the web services or *.ui* for the UI module).

[»]

Note

Detailed guidance for Cloud Foundry deployments can be found at *https://help.sap.com/docs/BTP/65de2977205c403bbc107264b8eccf4b/2a21055cc94b4a528a820f73e6fa7d69.html?q=deployment.*

2.5.3 Deployment on SAP BTP, Kyma Runtime

After the application, whether a microservice or a function is created, it is containerized using Docker containers and exposed as standard in the Kubernetes framework, and it is possible to make these scalable. For complex requirements, this is the option to be chosen. Then, the Docker image can be deployed to SAP BTP, Kyma runtime using the Kyma console. After deployment to SAP BTP, Kyma runtime, the status can be checked in the Kyma dashboard through the log output. Logs are available for each namespace and also for particular lambda functions.

Microservices can be deployed as YAML Ain't Markup Language (YAML) or JavaScript Object Notation (JSON) files. Continuous integration/continuous deployment (CI/CD) pipelines use kubectl.

A detailed view of deploying an application in SAP BTP, Kyma runtime is provided in Chapter 11.

[»]

Note

An excellent step-by-step description of the deployment of an application can be found in some public tutorials, for instance, the tutorial at *https://developers.sap.com/tutorials/deploy-to-kyma.html.*

2.5.4 Deployment on SAP BTP ABAP Environment

ABAP developments are created as software components as packages in ABAP. When you release the code, the packages are stored in the Git repository, which handles the deployment as a whole. The entire code is deployed.

Transport organizers in ADT in Eclipse are used for deploying ABAP programs to application servers configured in SAP Business Application Studio. OData services are created and then deployed to the application server in the cloud. The transports are handled using SAP Cloud ALM or Git-based CTS (gCTS).

The ABAP code can also be placed in the abapGit repository, from which the deployment can be handled. However, abapGit is currently available only as a plugin to ADT.

In later chapters, especially Chapter 6 through Chapter 12, we'll provide step-by-step instructions for developing and deploying different applications. A detailed view of deploying an ABAP application is given in Chapter 9.

2.6 Use Cases Based on Development Environment

With the cloud development paradigm gaining strength with every passing year, in the enterprise world, the usage of PaaS is increasing. SAP BTP is, of course, more than just a PaaS, with business content, industry content, and connections becoming readily available for instant consumption by customers. With the push for many customers to follow SAP's recommendation of a clean core (meaning little development on core applications like SAP S/4HANA), moving custom developments to the cloud has become imperative. So, SAP BTP is used for side-by-side extensions, for functionality enhancements of SaaS solutions, for innovations using SAP and non-SAP services, for integration, and for intelligent solutions using AI and machine learning. So, the use cases for using SAP BTP are increasing by leaps and bounds. However, from a development environment perspective, developers should have some useful guidelines about which development environment to use for which type of scenario.

One important consideration that you must assess, if possible, when codifying development guidelines at the start of any project is skill level, especially when it comes to side-by-side extensions using Cloud Foundry vis-à-vis SAP BTP ABAP environment.

2.6.1 Application Development Use Cases Supported by SAP BTP, Neo Environment

SAP's recommendation is currently to avoid SAP BTP, Neo environment (since SAP no longer focuses on and has not made many enhancements to this environment) except for the following cases:

- You've already adopted SAP BTP, Neo environment and need to extend existing applications.
- You're using services that are available only in SAP BTP, Neo environment, in the region (e.g., Europe – Frankfurt, Europe – Netherlands, Australia – Sydney, etc.) for your chosen hyperscaler. In that case, for that geography, you cannot use any other SAP BTP environment.

2.6.2 Application Development Use Cases Supported by SAP BTP, Cloud Foundry Environment

Cloud Foundry is the future environment for most applications, which is also true for SAP BTP, Cloud Foundry environment. Now that the Cloud Foundry has been in the market for quite some years, it has matured as a platform, with a long list of supported services. Some current use cases that can already be developed in Cloud Foundry include the following:

- Innovative applications using Internet of Things (IoT), machine learning, blockchains, and big data analytics.
- Building automation in processes through workflows, through RPA using SAP Intelligent RPA, or through chatbots using Joule.
- Building complex applications with a modular design, instead of a single monolithic application—in other words, an application that is a sum of smaller chunks of independent functionalities that can be created as microservices.
- For building and hosting any kind of low-code or pro-code-based agentic, custom, or class AI solutions, Cloud Foundry is the platform choice due to its enterprise-grade infrastructure with built-in security, scalability, and service integration capabilities.
- If an application has parts with different scalability requirements, then using Cloud Foundry makes sense. Also, if an application leverages services from other systems, using Cloud Foundry reduces the risk of the entire application failing because the microservice-based architecture results in a fair degree of isolation among components.
- Applications that must be built in other languages like Node.js or Python, beyond Java or HTML5 or applications needing multiple languages, must use Cloud Foundry.

2.6.3 Application Development Use Cases Supported by SAP BTP ABAP Environment

Some main drivers behind using SAP BTP ABAP environment include the following:

- Availability of ABAP skills is quite high in the SAP ecosystem, and this skill can be augmented by knowledge of cloud-specific features, including the ABAP RESTful application programming model architecture.
- Side-by-side extensibility of SAP S/4HANA on-premise or SAP S/4HANA Cloud, creating extensions of existing SAP S/4HANA functionality.
- Having a clean core, which will in turn help during upgrades and the migration of core SAP S/4HANA systems. In the clean core approach, custom developments should be done on the cloud, and SAP BTP ABAP environment is a major option for such cases. When moving existing functionality written in ABAP, this option is the preferred option, although the likelihood you'll need to modify existing code to be cloud ready is rather high.

- Newer functionality that is not available in standard SAP products but are closely bound or dependent on the processes running on SAP may consume different logic from SAP systems, which can be built using ABAP.
- Partners can develop ABAP-based applications and register them as SaaS solutions in a multitenant environment (as add-on applications) and provide these add-ons to multiple clients. Several partners have created such solution extensions, like Vistex.

2.6.4 Application Development Use Cases Supported by SAP BTP, Kyma Runtime

SAP BTP, Kyma runtime is a powerful environment with a lot of flexibility for resource optimization underneath. This is a comparatively new area for many practitioners in the SAP ecosystem, and as a result, you may encounter a steep learning curve. The increase in flexibility and power also gives more responsibility to developers, who must understand not just the code development aspects but the deployment aspects of the application, its underlying resource consumption, and so on. Some typical use cases that are a good fit for developing in this environment include the following:

- If new functionalities and a quick proof of concept must be created, then SAP BTP, Kyma runtime provides the functionality for serverless deployments. Leveraging this feature enables great flexibility in terms of scalability.
- For applications that must be containerized and deployed as microservices, SAP BTP, Kyma runtime is a great option as it provides the ability to conduct proper container-based deployments, with a lot of flexibility regarding resource consumption.
- SAP BTP, Kyma runtime is easier to use when the applications require the consumption of the services provided by the hyperscaler's platform for carrying out scaling or other non-functional requirements.

Note

For use cases that organizations are using right now built using SAP BTP, visit *www.sap.com* and navigate to **Products • Technology Platform • Business Use Cases.**

2.7 Estimating the Cost of SAP BTP Projects

All enterprise application development incurs costs, and SAP projects are no exception. To build an application on SAP BTP, you need to identify not only the building blocks but also the associated costs, which are among the most critical factors in determining return on investment (ROI). The cost of running SAP BTP services or hosting applications can vary significantly based on multiple factors, including the service plan, consumption model, usage volume, deployment landscape, scalability requirements, integration needs, and regional pricing considerations. SAP Discovery Center is a key avenue for architects, developers, and business stakeholders to explore and make informed decisions, plan budgets effectively, and assess the potential ROI of their SAP BTP-related projects.

SAP Discovery Center enables customers, partners, and developers to explore, evaluate, and accelerate the adoption of SAP BTP services. As shown in Figure 2.13, it acts as a central hub for discovering SAP BTP services, SAP Business AI-related offerings, step-by-step implementation guides in the form of SAP BTP missions, adoption frameworks, and SAP Business Data Cloud services.

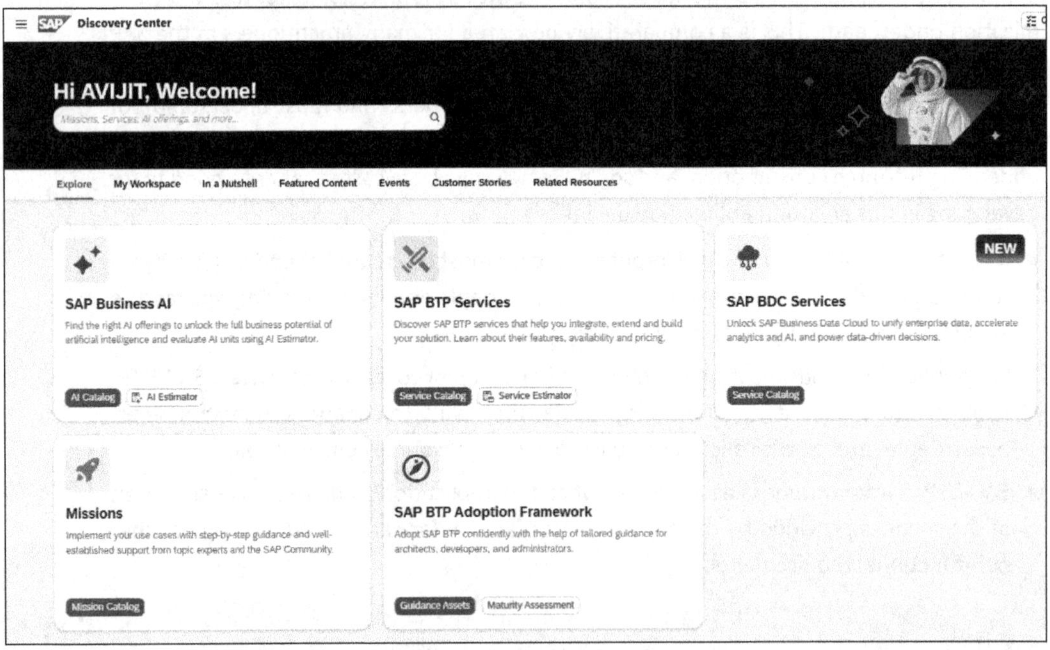

Figure 2.13: SAP Discovery Center

The **SAP BTP Services** section, as shown in Figure 2.14, lets you navigate through the latest SaaS capabilities that SAP BTP offers. As shown in Figure 2.15, you can add any service to your service estimator followed by selection of the service plan, t-shirt size, number of users, capacity units, and more, all depending on the type of service that you picked.

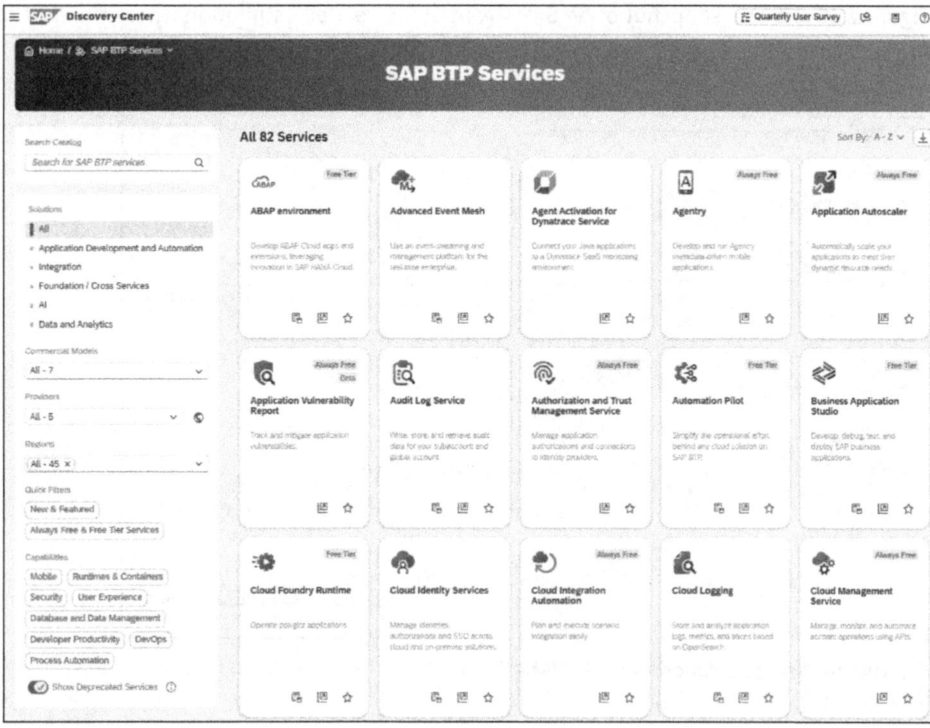

Figure 2.14: SAP Discovery Center: SAP BTP Services

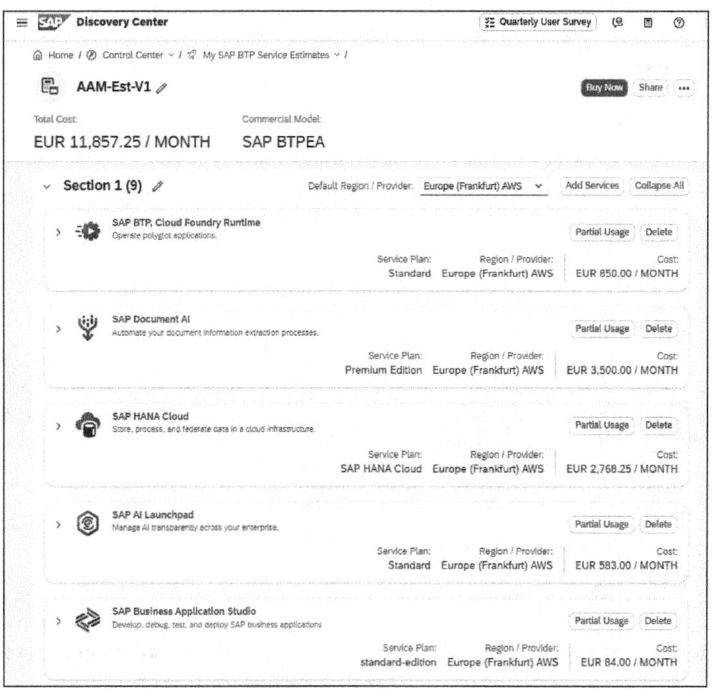

Figure 2.15: SAP BTP Service Estimator

Figure 2.16 shows a snapshot of an SAP HANA Cloud-related estimation where the cost of the service is determined based on the capacity unit.

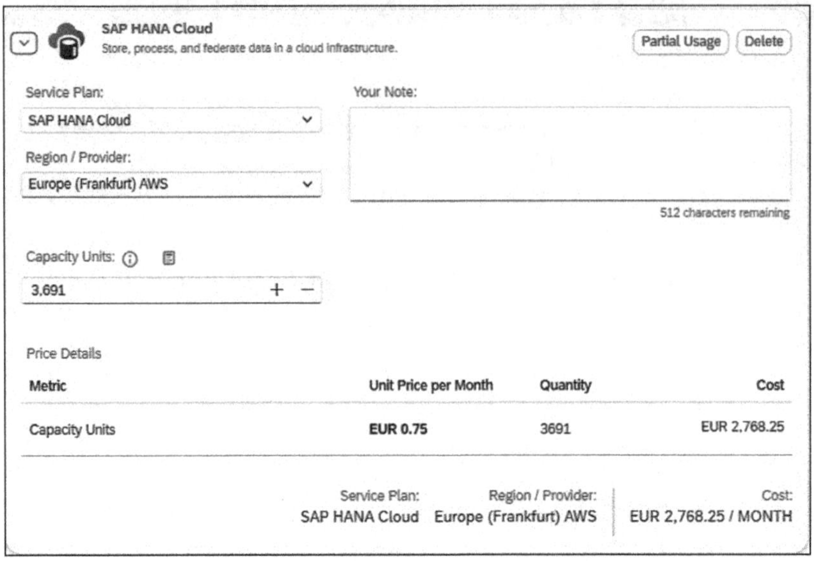

Figure 2.16: Service Estimator for SAP HANA Cloud

Figure 2.17 shows a quick snapshot of the **Capacity Unit Estimator** for SAP HANA Cloud where you can calculate the cost based on data size categories of **Large**, **Medium**, and **Small**.

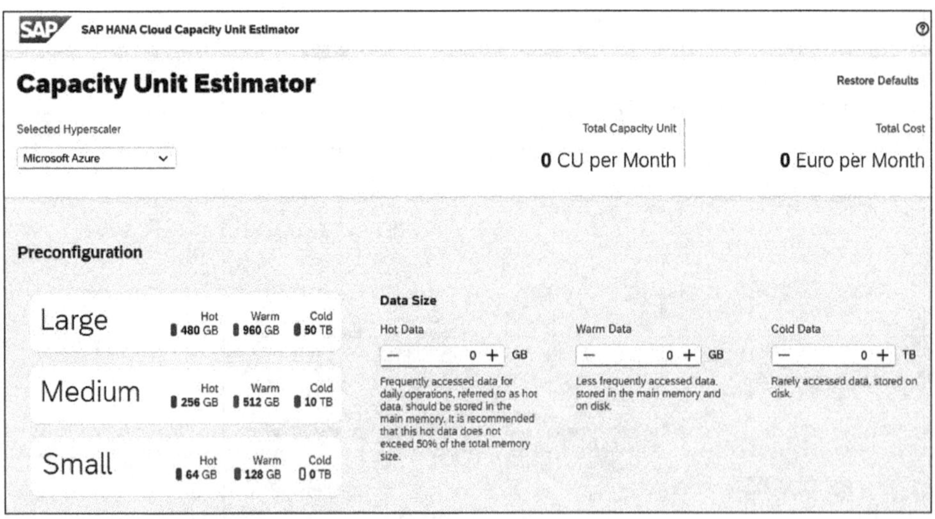

Figure 2.17: Capacity Unit Estimator

The prepared estimation can be saved as a PDF or CSV file and then shared with peer members through email.

2.8 Summary

In this chapter, we provided information about the basic concepts of SAP BTP, along with several development environments and their architectures. You learned how the platform has adopted open-source standards as a PaaS solution and enabled multiple languages—including support for any new language—through the BYOL model. We also covered using microservices to handle scalability and employing separate containers for different languages or distinct parts of an application, such as the UI or backend data model. You also learned about the applicability of different environments, and we provided recommendations for the appropriate use cases for each. Overall, SAP BTP as a platform provides all the essential components to transform an intelligent enterprise into an autonomous enterprise by combining data management, analytics, AI, integration, and extension capabilities into a single platform, including hundreds of prebuilt industry templates and solutions for SAP and third-party applications.

In the next chapter, you'll learn more about designing cloud applications in general, while in subsequent chapters we'll describe, step-by-step, how to develop your own cloud-native applications using various modeling paradigms on different development environments.

Summary

PART II

Backend and Frontend Cloud Application Development

Chapter 3

Designing Cloud Applications for SAP

The cloud phenomenon is not just about moving infrastructure but also about introducing flexibility, agility, and scale in applications by utilizing the unique power of cloud technologies. Microservices and containerization are the cornerstones of cloud-native application development and deployment. They are widely used in both Cloud Foundry and the Kyma runtime of SAP Business Technology Platform (SAP BTP) to enable agile innovation through easy composition and orchestration of cloud-native applications. This aligns with the goal of a clean core. In this chapter, you'll learn the key characteristics of cloud-native applications, principles of a clean core, basics of microservices, capabilities of SAP Integration Suite, and explore the architectural blueprint for their usage.

As discussed earlier, to support digital transformations, speed, flexibility, high quality, and low cost are absolute musts. The speed of releasing new features/functions/use cases, the speed to market for new capabilities, and the speed of adoption are essential for success. Traditional models of long-term planning and execution are no longer relevant in this fast-moving world. If an enterprise needs to change its business model or reach out to customers, it needs to do so quickly. To stop fraud, a company must analyze data quickly, even before a transaction happens. Thus, its IT infrastructure must be scalable and operate on demand. Instead of developing IT solutions themselves, companies can leverage offerings from software as a service (SaaS) providers, instead of waiting for systems/software to be procured, installed, and configured/extended to implement functionalities.

Enterprises can start configuring/extending immediately by leveraging platform as a service (PaaS) or SaaS offerings from appropriate providers. By not owning the software, companies have the flexibility to change if required. By also not owning the hardware and instead using SaaS or PaaS services on demand from cloud providers, the total cost of ownership (TCO) is lowered, and flexibility increased. This cost savings and increased flexibility provide the necessary support for maximizing the value of a digital transformation.

The cloud can also help improve quality by providing standardization. In short, the cloud provides the following benefits:

- **Speed**
  Faster provisioning of hardware software, faster development lifecycles, and faster releases.
- **Flexibility**
  Ease of change.

- **Quality**
  Better, more predictable quality through standardization.

The cloud comes in different varieties based on the following:

- **Infrastructure as a service (IaaS)**
  In the IaaS variant, hardware, networks, etc. are consumed from a wide range of possible hosting arrangements, from a private dedicated server in your own data center to a completely shared and public cloud.
- **PaaS**
  In the PaaS variant, software tools, along with the necessary hardware, are combined and provided as a service to allow you to develop and run applications quickly without the need to build your own infrastructure, software stack, etc.
- **SaaS**
  With SaaS, clients can consume applications without worrying about infrastructure.

Increasingly, enterprises consume variations of each type and have multiple clouds in their portfolios. Thus, most enterprises have a hybrid multicloud infrastructure—in other words, a mixture of on-premise systems and private and public clouds for IaaS, while consuming PaaS and SaaS offerings from multiple service providers.

In the following sections, we'll cover cloud-native applications and SAP's clean core paradigm. This chapter will also cover microservices as building blocks of composable and flexible development capability along with another key capability required for cloud applications: integration capabilities.

3.1 Cloud-Native Applications

It is evident that enterprises must look at their application strategies in a different light to maximize the value of the cloud. The cloud phenomenon is not just about moving infrastructure from dedicated, on-premise hardware to a shared infrastructure on a cloud provider's data center; it's about leveraging the capabilities made available by the infrastructure into applications. Many applications have properties that are particularly suited for the cloud. Cloud-native applications are defined as applications that are "born" in the cloud; that is, these applications leverage the functionalities of the cloud and have characteristics that support the flexibility, agility, and scale typical of cloud-based solutions.

Thus, migrating and modernizing current systems of record and systems of engagement are opportunities to accelerate connections with customers and opportunities for improved speed, scale, and cost management.

Three sets of applications are particularly relevant as an enterprise moves to the cloud:

- **Lift and shift applications**
  These applications are core stable systems that should move to the cloud as is. Typically,

an enterprise might consolidate, standardize, and/or automate their existing applications as much as possible before moving to the cloud.

- **Containerizing applications**
  These applications are reused often, may need to scale up or scale down often, and may be deployed quickly, among other things. These applications can be converted into APIs and microservices or bundled in containers. These applications are described as "cloud enabled."
- **Refactor applications**
  The advent of the cloud provides the opportunity to rewrite existing applications in new ways to make them run best on the cloud and leverage the best of capabilities offered by the cloud. These applications are described as "cloud native."

Thus, cloud-native principles and containerization techniques will continue to be important in the future even for SAP landscapes. Let's explore cloud-native applications further and then dive into microservice and container architectures.

3.1.1 What Are Cloud-Native Applications?

Applications that are designed for cloud, created on cloud-based platforms, and run on cloud infrastructures are colloquially known as *cloud-native applications*.

Cloud-native technology provides the means for organizations to build and run scalable applications in public, private, and hybrid clouds. This development is performed via containers, microservices, service meshes, immutable infrastructure, and declarative APIs to name a few relevant technologies.

As we'll discuss throughout this book, this new paradigm forces us to think differently about how we design our solutions. More ways to decompose applications are emerging beyond the three-tier model (frontend, application layer, and database layer). SAP is now front and center of this movement, thus ensuring that SAP applications benefit from this new paradigm.

3.1.2 Microservice Architectures

Microservices are an architectural style in which large applications are made up of smaller components where each component focuses on a specific functionality. Each microservice normally performs one task, and they communicate with each other using language-neutral APIs. Microservices don't need to know anything about the underlying implementation or architecture of other microservices.

The microservices architectural style is an evolution of the service-oriented architecture (SOA). However, the focus of SOA was technical integration while microservices focus on integrating business capabilities. The key design principles of microservices include the following:

- **One task**
  Microservices should focus on doing one job well. This focus allows for easier management and improved quality.
- **Standardized communications**
  Each microservice should communicate with other microservices through standard protocols like representational state transfer (REST) APIs or through explicit communication patterns.
- **Independence**
  Each microservice should be able to independently and continuously integrate and continuously develop so that the overall application can add new features/updates without impacting the whole application.
- **Individual resiliency, high availability**
  Disaster recovery clustering can be performed for each microservice instead of using one common cluster. For the entire application, this difference saves resources.

The benefits of microservices are many and include the following:

- **Scalability**
  Scaling individual microservices as needed is easier than scaling an entire application.
- **Easier maintenance**
  Given its granular nature and its focus on a single task, microservices are easier to maintain than a big monolithic application. Upgrading individual microservices can occur without having to upgrade the entire application.
  Microservices are typically aligned to one business function, and thus, its change management is easier than for a single monolithic solution.
- **Flexibility**
  Unlike monolithic applications, microservices can be developed using different technologies, that is, deployed using different frameworks like Docker, Cloud Foundry, etc.

Microservices are the key building blocks of modern cloud-native applications. However, the right granularity level must be determined. Too much granularity will incur high integration and maintenance costs. Too big in size would reduce flexibility and agility.

Special attention should be given to data management for microservices. Microservices should ensure that the data lifecycle is managed in a safe and secure way, while complying with regulations (like GDPR, HIPAA, etc.) for different types of data (e.g., data at rest versus data in motion).

Microservices should be combined with a robust automated delivery system to provide speed at scale, which is required to make the most impact. DevOps practices like continuous integration (CI)/continuous delivery (CD) should be practiced.

Operating microservices also need special consideration. While not the only option, automating the deployment of microservices for cloud-native applications is usually performed by leveraging containers, which we'll explore next.

3.1.3 Container Architectures

In a traditional model, when applications run on native hardware, a single app typically doesn't need the full resources of a single machine. Thus, most organizations run multiple apps on a single machine to avoid wasting resources. Thus, *virtual machines (VMs)* came into existence.

A VM is a software-based environment that simulates a hardware-based environment. Applications need an operating system and a set of processor cores to run, and they can run within a VM without any re-architecting. VMs provide the isolation required between apps that share the same physical resource. In this scenario, a software component called a *hypervisor* provides the layer of abstraction where the VM can work independently of the hardware. However, VMs have full operating system stacks, which make them relatively large and inefficient.

Containers are newer mechanisms that also provide isolation without needing to duplicate operating systems and system libraries. A container consists of an entire runtime environment bundled into one package. As shown in Figure 3.1, this package consists of the following components:

- An application
- All its dependencies, libraries, and other binary files
- The configuration files that are needed to run the application

Containers utilize the host's operating system, which saves space and reduces vulnerabilities. Thus, containers are more flexible and more cost effective than VMs. To sum up, while VMs virtualize hardware, containers virtualize operating systems.

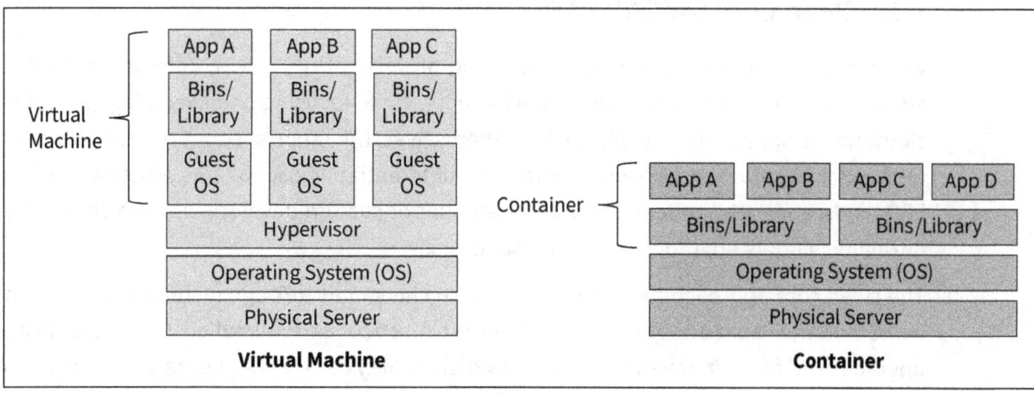

Figure 3.1: VM and Container Package

A physical server running three VMs would have a hypervisor and three separate operating systems running on top of the hypervisor. A server running three containerized applications can run a single operating system, and each container can share the operating system kernel with other containers. Since containers need less resources than VMs, more containers can run on a physical machine than VMs. Containers are portable as well. Any platform with a container engine can run containers.

Containers are easy to manage, create, and delete, and each container instance is easy and fast to start and stop. Container images are also easy to share, download, and delete. While container technology is not new, its adoption had been slow because of difficulty of use. This situation changed with the advent of standardization through Docker, Rocket, Cloud Foundry, the Open Container Initiative, and other organizations.

To further enhance the impact of containers, the orchestration and automatic management of multiple containers were required. Thus emerged container management software like Kubernetes, Docker Swarm, and others. These products help provision, manage, and scale apps. With this software, you can manage the lifecycle of containerized apps in a cluster of nodes (which are collections of machines, whether VMs or physical machines).

Apps might need other resources to run, such as storage volumes, networks, etc., to store data and to connect with other apps, among other things. Container management software allows you to manage these requirements as well.

Kubernetes is an open-source container management product started by Google and subsequently adopted by the Cloud Native Computing Foundation (CNCF) and other major industry players like IBM and Amazon. SAP uses Kubernetes Container Orchestrator as key components of the SAP BTP, Cloud Foundry environment, SAP BTP, Kyma runtime, SAP Data Intelligence, SAP Integration Suite, advanced event mesh, and SAP Integration Suite, edge integration cell.

3.2 Clean Core for SAP BTP

While multiple technologies help achieve the promise of the cloud, enterprises need to adopt a cloud mindset to realize the full benefits. While focusing on developing innovative cloud-native applications to differentiate themselves, they should adopt the concept of fit-to-standard as much as possible for their non-differentiating capabilities. Most enterprises' ERP landscapes are burdened by a huge amount of custom development, which restricts flexibility, hinders upgrades, and increases costs.

The clean core approach helps overcome these challenges through increased use of standards and the rapid adoption of industry best practices at minimal cost. It allows faster upgrades and faster innovation powered by high-quality data and AI. Loosely coupled to the standard core should be the cloud-native differentiating layer, both of which can evolve as required without impacting each other. Hence, the core can be upgraded faster, and the differentiating layer can be innovated and scaled up more quickly. SAP's clean core principles ensure that all extensions made to the standard SAP solution are both upgrade proof and cloud ready.

The following sections cover the core concept of SAP's clean core strategy, the mechanism for implementing the strategy, and the benefits of adopting it.

3.2.1 Concept

SAP's clean core focus is underpinned by the following five guiding principles:

- **Business process**
  Standardize processes to leverage industry best practices quickly and cost effectively
- **Extensibility**
  Achieve differentiation by extending standard solutions in a loosely coupled way
- **Data**
  Govern data and ensure compliance with the latest data standards
- **Integration**
  Ensure flexibility by using reliable, modern integration techniques
- **Operations**
  Improve operational efficiency through automation

While this book focuses on extensibility, all the principles are interlinked. Extensions to standard processes are made by enterprises to differentiate themselves from their competitors. These differentiations are expected to help enterprises achieve measurable goals. In order to reach these goals, extensions should work with high-quality data, be powered by automation, and supported by flexible integration capabilities. The extensibility principle is based on three fundamental rules, as follows:

1. Customer extensions should only use *released* SAP APIs, whether local or remote. SAP commits to keeping these APIs stable; thus the extension will not get adversely impacted by upgrades to the core.
2. *No modifications* to SAP standard objects should be made. Any extension to SAP objects can only be done through predefined extension points. SAP also commits to keeping the extension points stable.
3. Extensions should be done using *cloud-enabled* technologies and SAP's cloud syntax.

By following these rules, as shown in Figure 3.2, enterprises can ensure that their extensions are loosely coupled and future-proof (i.e., they will continue to work even if the core SAP solution gets upgraded) and that these extensions benefit from cloud-native characteristics.

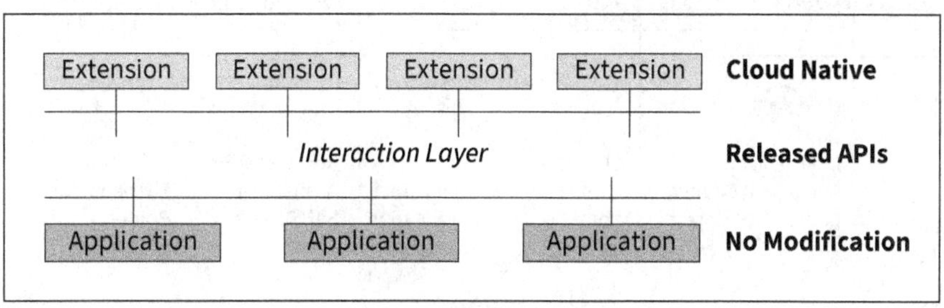

Figure 3.2: Clean Core Concept and Rules

3.2.2 Mechanism

Based on the concepts discussed in the previous section, every enterprise should try to build all their new extensions following clean core rules and convert their old extensions which don't follow the rules. SAP S/4HANA Cloud Private Edition, SAP S/4HANA Cloud Public Edition, and on-premise SAP S/4HANA offer two extensibility options:

- **On-stack**
  This is used when tight coupling with the standard solution and the need to run in the same environment are required. This is used, for example, for extending the UI, data models, etc.
- **Side-by-side**
  This is used when loosely coupled extensions running on a separate environment (SAP BTP) is feasible; for example, custom full stack applications, mobile applications, etc.

While side-by-side extensions on SAP BTP is the preferred option, as it provides maximum flexibility, in terms of upgrade impact and cloud-native capabilities, there are reasons why certain types of extensions are best done on-stack. This is why the clean core framework covers both types of extensions.

SAP has provided a number of tools, technologies, and frameworks to assess the clean core readiness of an enterprise's custom extensions and to plan for the adoption of the clean core rules. SAP's clean core extensibility framework (see Figure 3.3) allows you to assess your current extensions across four categories based on the impact of upgrades and the extent of cloud nativeness.

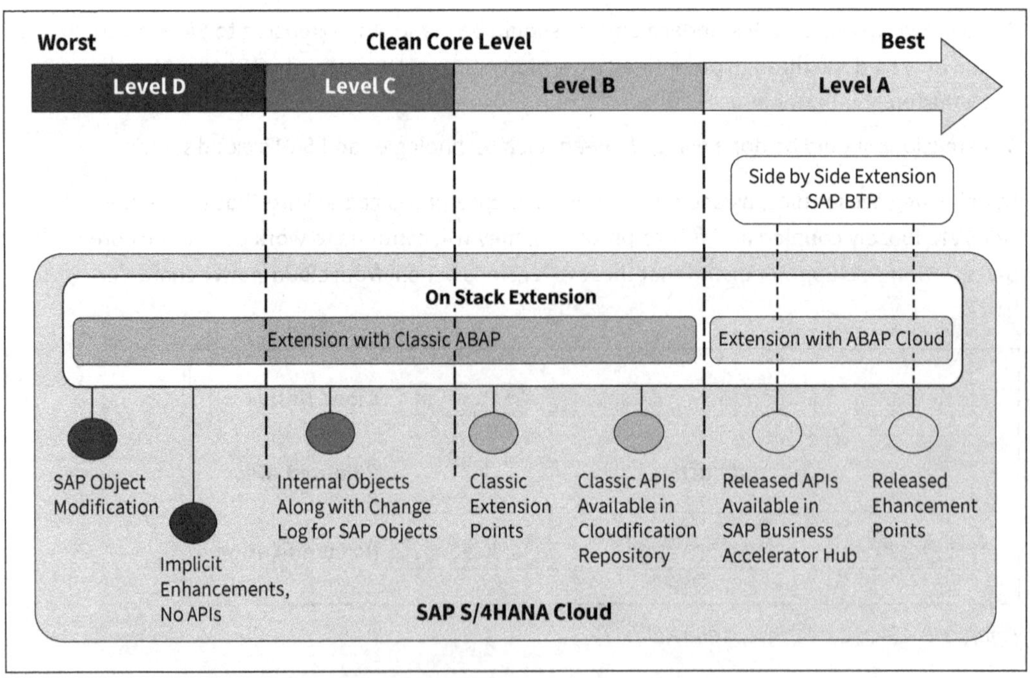

Figure 3.3: SAP's Clean Core Extensibility Framework

Let's look at each level in more detail:

- **Level A**
  This is the highest level of extension and the most preferred. These extensions interact with SAP's standard solutions through publicly released APIs and extension points. SAP ensures the stability of these APIs and extension points across versions. Hence these extensions are not impacted by upgrades. These extensions can either be:
  - Side-by-side: Developed on SAP BTP ABAP environment using SAP Build Code's low-code environment or using the pro-code environment of SAP Business Application Studio with SAP Cloud Application Programming Model.
  - On-stack: Developed in the core ERP on SAP S/4HANA Cloud ABAP environment using SAP Build tools using locally released APIs and extension points. While private cloud and on-premise customers can choose either ABAP Cloud or classic ABAP (also called standard ABAP) extensibility options, SAP S/4HANA Cloud Public Edition customers must choose ABAP Cloud extensibility.

  Level A extensions are both upgrade safe and cloud native.
- **Level B**
  These extensions leverage classic APIs published in the Cloudification Repository (*https://sap.github.io/abap-atc-cr-cv-s4hc/?states=classicAPI*). SAP ensures that these APIs remain stable across versions so Level B extensions aren't impacted by upgrades.
- **Level C**
  These extensions use internal objects of SAP and, as such, are not guaranteed by SAP to be stable across versions and might even be withdrawn. These extensions are neither upgrade safe nor cloud ready. Given that a large number of enterprises have this type of extension, SAP has provided a change log for SAP objects to help organizations understand the changes that may have occurred to these objects, allowing them to update their extensions accordingly. These change logs are delivered through simplification database accessible through SAP Support Portal.
- **Level D**
  These types of extensions are *not* recommended as they include modifications to standard SAP objects, the use of SAP objects SAP has called "unfit" (marked as "noAPI" in the Cloudification Repository), direct table writes, the use of implicit enhancement, etc.

These categories allow organizations to monitor and manage their enhancements, with the goal of having as many enhancements as possible at Level A or Level B. Check out SAP Note 3578329 for more information.

SAP provides RISE with SAP customers with dashboards (available in SAP Cloud ALM) that measure the following KPIs related to the clean core, enabling customers to monitor and govern their progress toward an improved clean core state:

- Clean core share: How many of their custom developments fall in which clean core level
- Technical debt score: Identifies objects using obsolete or soon to be obsolete technologies

- Unused code share: How many custom developments are not used
- Business modifications: Number of this type of modification which has upgrade risk

In the RISE with SAP methodology, SAP has recommended the "get clean" and "stay clean" approaches so that enterprises can convert their existing classical development to the clean core way and ensure that they do not introduce new non-clean core developments over time.

3.2.3 Benefits

While the benefits of moving to clean core extensibility are clear in terms of upgrade readiness and cloud readiness, the fact that many organizations have not attained the perfect clean core state yet points to the trade-offs they face.

While IT teams want to move to clean core for the "technical" benefits, enterprises often struggle to find a business justification for the change. The effort, cost, potential disruption, and required change management often do not clearly justify the transition. Hence, enterprises most often follow this path:

1. Ensuring that anything new that they develop is built in a clean core way so as not to increase the technical debt.
2. If they touch an existing classical development due to a business need, they will apply clean core principles and make the new version clean core compliant
3. Migrate those classical developments which, if migrated, provide business benefits such as enabling adoption of AI, automation, scale, usability transformation, process transformation, or improved data.

A combination of these paths ensures slow but steady progress toward a clean core. However, enterprises also need a clear vision of the benefits they aim to achieve by adopting a clean core. While clean core drives the dual benefits of upgrade readiness and cloud-native capabilities, some enterprises may have an upgrade event approaching. Hence, they may want to accelerate the transition to a clean core, as otherwise their upgrade event will be effort, time, and cost intensive. Given the nature of business processes, some enterprises may benefit more from cloud-native properties than others, as needs such as scaling, seasonality, the nature and intensity of customer interaction, transitioning to mobility platforms, and adopting off-the-shelf AI capabilities may vary. Accordingly, enterprises can define the speed and drivers of clean core migrations.

3.3 Microservices

Custom applications deployed in the cloud are no longer built using a monolithic architecture but rather consist of smaller functionalities called microservices. SAP BTP, Cloud Foundry environment and SAP BTP, Kyma runtime support the architecture for using microservices to build applications.

In this section you'll learn about the concept behind microservices and how SAP BTP enables microservices. We'll cover several generic concepts like design considerations for creating microservices and the twelve-factor app methodology, concepts that are equally applicable in the SAP context. We'll then show you how to search for the right microservice and how to consume them in your cloud-native applications. We'll also cover how SAP BTP, Kyma runtime is used to create microservices. Chapter 7 will discuss how Joule/SAP Build can be used to generate microservices.

3.3.1 What Are Microservices?

The concept of microservices has existed for a long time, from a service-oriented architecture (SOA) perspective. Recently, microservices have become popular among enterprise software developers because they are more lightweight and fine-grained than SOA services. Table 3.1 provides a summary of how cloud-native applications have evolved and the characteristics applications must have to mature in a particular state. This table starts with the characteristics of a cloud-ready application and moves on to show the characteristics of a cloud-friendly application, which is more mature. Then comes cloud-resilient applications, which may include additional features for managing the application, including proactive testing.

State	Characteristics
Cloud-native	▪ Microservice architecture and principles ▪ Application programming interface (API)-first design
Cloud-resilient	▪ Design for failure ▪ Apps unaffected by dependent service failure ▪ Proactive testing for failure ▪ Metrics and monitoring included ▪ Cloud-agnostic runtime implementation
Cloud-friendly	▪ Twelve-factor app methodology ▪ Horizontally scalable ▪ Leverage of the platform for high availability
Cloud-ready	▪ No file system requirements or use the Amazon Simple Storage Service (S3) API ▪ Self-contained application ▪ Platform-managed ports and addressing ▪ Consumption of off-platform services using platform semantics

Table 3.1: Cloud-Native Maturity

Finally, for cloud-native applications, the core concepts you should leverage are microservices and API-first design principles. Let's start by defining microservices.

A *microservice* provides a small piece of business functionality, runs its own processes, and is independently deployable. The microservice architectural style is an approach to developing a single application as a suite of small services, each running its own processes and communicating with lightweight mechanisms, often an HTTP resource API. These services are built around business capabilities and independently deployable by fully automated deployment machinery. A bare minimum of centralized management for these services is available, which may be written in different programming languages and may use different data storage technologies. Microservices have the following characteristics:

- Microservices are independent of each other or are loosely coupled.
- Microservices can be developed in different languages.
- Microservices communicate with each other language-neutral APIs, for instance, using the representational state transfer (REST) protocol.
- A microservice is individually scalable. We recommend following the twelve-factor app methodology for API designs (explained later in this section).

The use of DevOps principles is the recommended practice for microservice implementation and maintenance. For more information about DevOps, see Chapter 13.

From a business perspective, using microservices enables you to leverage services and features from different vendors, which can be integrated into an application, instead of building everything from scratch or using only resources from one vendor.

Individual services represent separate transactions that can be executed in separate databases. The infrastructure provides tools for overcoming challenges and failures from a system perspective as well as from a perspective of the service, which is stateless. A *stateless service* is a service that is complete by itself and does not remember the state of its data during the execution process.

[+]

Tip

You can find more details about the microservices concept at *https://martinfowler.com/microservices/*.

3.3.2 How Does SAP Business Technology Platform Enable Microservices?

Currently, SAP BTP enables the creation and consumption of microservices in two ways: via SAP BTP, Cloud Foundry environment and SAP BTP, Kyma runtime. SAP BTP, Cloud Foundry environment enables the creation and deployment of microservices. SAP HANA extended application services, advanced model (SAP HANA XSA) for on-premise applications and Cloud Foundry for cloud-based applications provide the platforms for microservices-based, polyglot web applications. The Kyma environment is also available with SAP BTP, which enables the development and deployment of microservices (and functions).

For Cloud Foundry, application containers provide an isolation layer between microservices, which can be written in separate languages, since a core part of the app server for that particular language is deployed in the corresponding app container.

As mentioned earlier, a particular function for use as a microservice can be developed using any of the available buildpacks. For Java, the Spring framework is useful and quite powerful with its dependency injection framework.

For web-based applications, a web server is required, and Tomcat is the default web server supported by Cloud Foundry as well as for the Neo platform ever since SAP BTP became available. One of its many advantages is that Tomcat requires little memory to run and is also part of the SAP Java buildpack.

Thus, to design an application and its microservices, you need SAP Business Application Studio (or SAP Web IDE), and you need access to the services you want to create your application around. Your application will be created as an *.mta* file, which is a multitarget application (MTA) file that can include a user interface (UI), business logic, and one or more workflows or an Internet of Things (IoT) service. Depending on which of these is the core feature of your application, you can use the corresponding development tabs in SAP Business Application Studio spaces.

In the following sections, we'll walk you through the architecture that enables microservices, microservice design considerations, and the twelve-factor app.

High-Level Architecture

As shown in Figure 3.4, the different components work together to deploy applications, including their microservices, in Cloud Foundry. These details will be quite transparent to your developers as well.

User account and authentication (UAA) is a service for regular authentication and for federated authentication, which is useful for authenticating technical or platform user access to an SAP Business Technology Platform instance. UAA is provided by an OAuth2 server and logon servers working together.

Extended services for user account and authentication (XSUAA), however, conducts authentication from a business perspective, checking whether a user has the authorization to access a specific application and perform a specific task. The platform application router (which is not the same as the AppRouter) routes incoming traffic to the appropriate component, either a Cloud Controller component or a hosted application running on a Diego Cell.

The router queries the Diego Bulletin Board System (BBS) to determine which cells and containers each application currently runs on. This query is performed periodically, and this information helps the platform application router recalculate new routing tables.

The request goes through the Cloud Controller, and buildpacks (including some buildpacks of applications deployed to the application code) and droplets are persisted in the binary large object (blob) store, which stores large binary files.

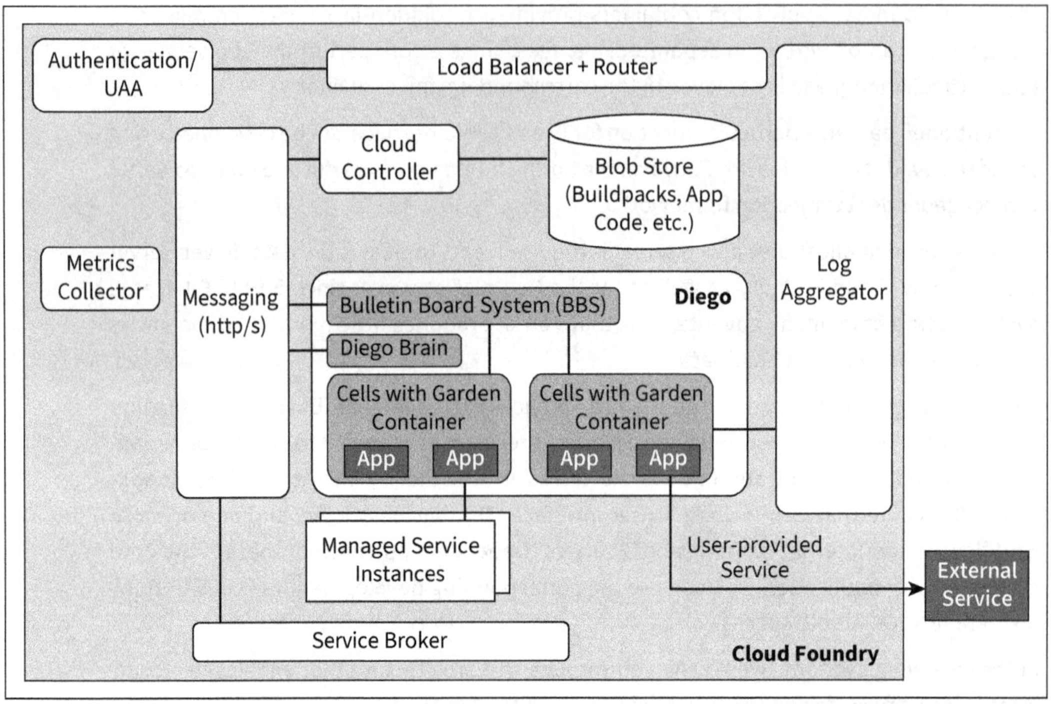

Figure 3.4: Cloud Foundry High-Level Architecture

Diego is the Cloud Foundry runtime and container management system, which consists of a house manager, which is a bulletin board system. This manager checks which applications are still running.

An application can provide a custom health endpoint to be called to check this information, but this check is optional. If an application is unhealthy, Diego will stop the application, delete the application from the container, and then re-create and restart the application, using the information found in the blob store. If an application is healthy, then a route is assigned to the application instance.

Diego basically distinguishes between tasks (like database schema migrations) and long-running processes (LRPs) and, depending on this distinction, decides which cell container the application should be deployed to. Overall, an application is deployed to one individual container. This isolation helps you scale horizontally because an application, when running in isolation, can have other instances running next to it.

Then, the load balancer can distribute the load between several application instances. Figure 3.4 also shows an application bound to an external service via the concept of user-provided services, or the application can be bound to a managed service instance.

The service broker is responsible for providing the service instance to an application, which has the service provisioned and bound to it. This service can be any service provided within the organization or the platform or via third-party software as a service (SaaS) providers.

The service broker is responsible for creating a Docker container and for deploying the service instance into that container. Furthermore, the service broker is responsible for binding this managed service instance, like MongoDB, to all instances of the application. Thus, the service broker is basically a microservice implementing just a few endpoints that are then executed by Cloud Foundry. The log aggregator streams the application log for the operators, while the metrics collector gathers metrics and statistics from the components, thereby helping users monitor a Cloud Foundry deployment. All components must provide an endpoint that offers this kind of information.

From the point of view of leveraging SAP BTP, Kyma runtime, for consuming or creating microservices, the architecture is quite transparent to developers. In this case, Kyma runs on a Kubernetes cluster and is available as a managed service on top of SAP BTP. You can create microservices in any language in Kyma.

SAP BTP, Kyma runtime supports a microservice-based architecture, event handling or consumption, authentication, logging, tracing (which in turn helps in monitoring and alerting in case of failures or issues), consumption of services provided by hyperscalers, and also API management services. All these features are enabled through the managed service option in SAP BTP.

Microservice Design Considerations

While microservices provide many advantages in terms of reuse, independent scalability, and more, some costs are associated with these benefits. These costs are incurred through interacting with each other (communication overhead) or through having a separate persistence that requires synching, which will increase overall latency. Thus, you should consider a few points when designing an application and the microservices to be used within it, such as the following:

- **Domain-driven design**
  You can apply domain-driven design, a concept invented by Eric Evans, by using a particular business context bounded to a model for that domain. An example of a domain could be insurance, finance, or manufacturing. In other words, in a particular domain, a data model cannot support all the subdomains in that domain, especially if the domain is quite large. A data model can have perspectives based on particular uses, and if all these perspectives are merged together, you may end up with a huge messy model that may not work at all. Thus, you should have domains and subdomains that are independently modeled. As shown in Figure 3.5, three types of domains can be considered: The core domain is the main area of innovation and intellectual property, and this domain will receive the business impact and answer the question of why the system is being built. Then, a generic domain is required to run any application and is non-differentiating for your business, like user management or security; you should simply buy this domain. Finally, supporting domains act as the glue code between specific parts of your system but do not really serve as the core.

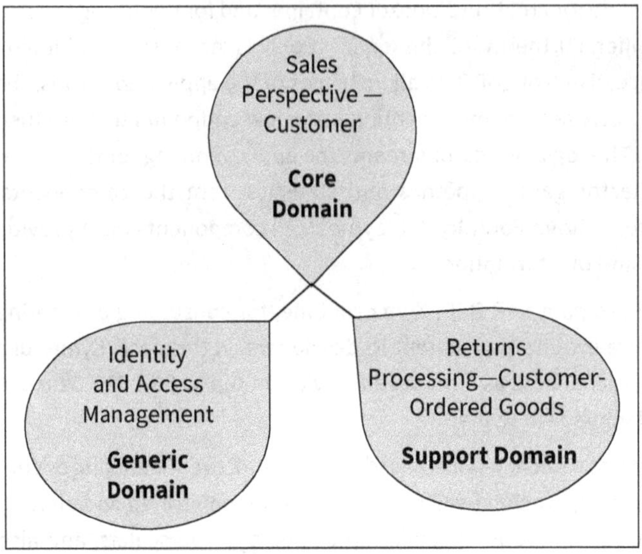

Figure 3.5: Examples of the Three Domains

- **Granularity**
  One major consideration is granularity. Not every functionality should be broken down into a minute microservice. Thus, a monolithic application should be broken down into a limited number of microservices but not randomly into hundreds of microservices.
- **Scalability**
  If a particular function must be made more scalable or resilient, then creating the function as a separate microservice makes sense. If you want to scale horizontally, design the function to have more app instances to take on the load.
- **API-first microservice design**
  Depending on your business requirements, an API can be designed first and then the other components of the microservice can be built, for instance, the UI for either a mobile app or a web app. If the API is designed well, then reusing the API will be easy, and thus, many applications can be built using this API. Sometimes, if a public API is available and suitable for a particular business need, this API can serve as the starting point for designing a microservice.
- **Type of service calls**
  If all the service calls within an application use REST APIs and/or a design decision leads to a lot of interservice calls, a high degree of latency in response time may arise, especially if all the service calls are synchronous. Also, the decoupling of services is an essential design parameter. You should avoid any transactions between microservices because calls between microservices can be thousands of times slower than calls within a microservice. For messages, performance should not be a problem, but with synchronous REST calls, you may notice performance issues. Remember that response times often add up.

- **Outside-in modeling approach**
  This approach means that the focus of the design is on business events, which are modeled with help of domain experts. The relation between events—between inputs/prerequisites and outcomes—is also mapped out. This model can then be translated into a technical solution.
- **Resiliency**
  Built-in resiliency in a microservice means that the application can have multiple functions as independently deployed units. If any unit fails to deploy or has a bug, you should be able to isolate the unit without shutting down the entire application.

Twelve-Factor App

The twelve-factor app is a modern methodology to help design SaaS solutions. This methodology can be made applicable to the SAP domain as well. The twelve-factor app principles state the following:

1. Codebase: One codebase tracked in revision control, many deploys.
2. Dependency: Explicitly declare and isolate dependencies.
3. Config: Store configuration in the environment.
4. Backing services: Treat backing services as attached resources.
5. Build, release, run: Strictly separate build and run stages.
6. Processes: Execute the app as one or more stateless processes.
7. Port binding: Export services via port binding, which is handled by the SAP platform and does not need any extra action.
8. Concurrency: Scale out via the process model.
9. Disposability: Maximize robustness with fast startups and graceful shutdowns.
10. Development/production parity: Keep development, staging, and production as similar as possible.
11. Logs: Keep logs as event streams.
12. Admin process: Run admin/management tasks as one-off processes.

> **Note**
> More information about twelve-factor app design principles is available at *https://12factor.net/*.

3.3.3 Finding the Right Microservice

The best ways to find microservices include searching for business services in SAP Business Accelerator Hub (*api.sap.com*) or SAP Discovery Center where several solutions, both from SAP and from partners, are available. Business solutions like SAP Complaint Handling or SAP Data Quality Management functions are available as microservices with set integration points with backend systems (both SAP S/4HANA and other systems). Figure 3.6 shows some examples of microservices available on SAP Business Accelerator Hub.

API Management in SAP BTP acts as a service broker in the Cloud Foundry marketplace and thus enables you to manage your digital assets on Cloud Foundry. API Management then creates an instance and binds the Cloud Foundry application to the created service instance, which automatically creates an API proxy in the API portal. These API Management services in SAP BTP play a vital role in managing the lifecycle of APIs and microservices that have been exposed. In the same way, API Management helps in the consumption of microservices by applications built on SAP BTP, Kyma runtime as well.

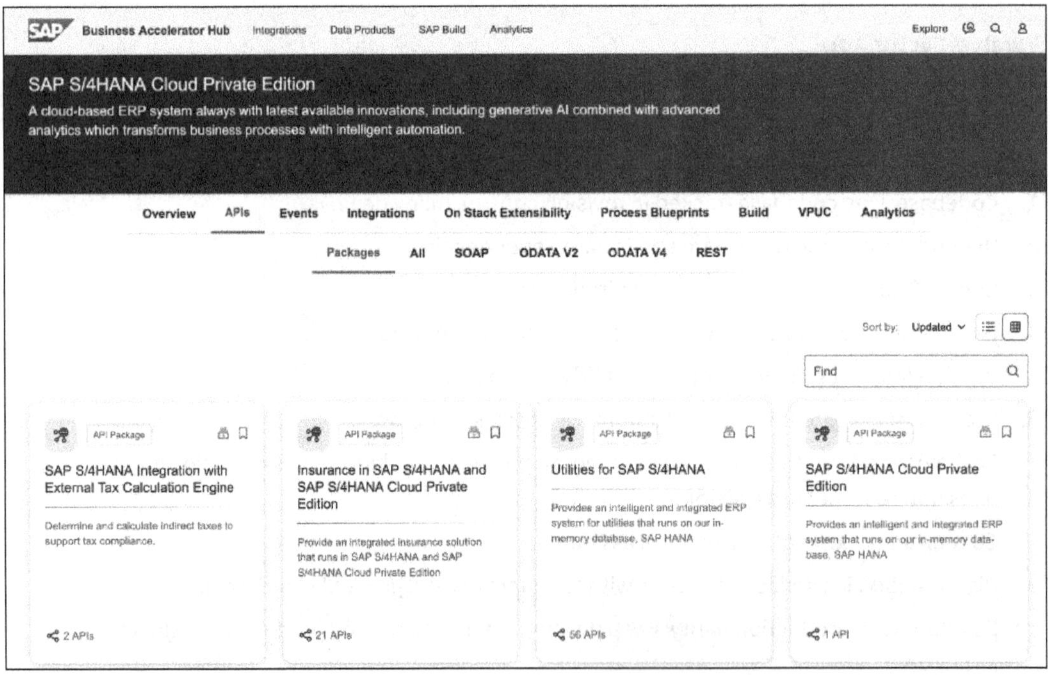

Figure 3.6: SAP Business Accelerator Hub

3.3.4 Consuming Microservices

As mentioned earlier in this chapter, microservices can be complete applications and can have their own UIs, can use multiple services themselves, and of course can use APIs.

Microservices can be consumed at the UX layer, or messaging integration may exist. Finally, while APIs are the most common way for integrating microservices, other methods can be important since powerful applications can be built using these options, with less effort and less trouble. Microservice integration with UX can occur through hyperlinks or transclusion (inclusion through hypertext).

Messaging integration is preferable to synchronous API calls, as mentioned earlier. Some tips to keep in mind while designing this integration include the following:

- The microservice API should be designed with ample resources.
- You'll need to apply a consistent uniform resource identifier (URI) convention for an intuitive understanding of what the object can do or related actions that you might take with an object.
- We do not recommend using a singular service endpoint to process different kinds of actions.
- APIs should be designed to use the right HTTP verbs and status codes and should follow the right protocols.
- Changes to an API should be backwards compatible to minimize damage to an integration, and the creation of new versions should be avoided. Otherwise, a complete reimplementation of the integration may be required.

3.3.5 Microservices in SAP BTP, Kyma Runtime

While Chapter 11, Section 11.3, provides a step-by-step approach of creating an application using SAP BTP, Kyma runtime in this section, we'll touch on some basics about Kyma's built-in mechanisms for dealing with microservices.

Since SAP BTP, Kyma runtime uses Kubernetes, your application (which can be a microservice or a solution extension) must be deployed in a container. So, after creating an application (using any environment), its Docker image must be created in Kyma, and then this image is deployed. Kyma includes functionality for capturing events from any application. These events can be configured to trigger an application in Kyma, for example. Applications using the SAP Cloud Application Programming Model, explained in detail in Chapter 6, can be deployed in SAP BTP, Kyma runtime as well. Kyma can also connect to Event Mesh.

Kyma can connect to multiple SAP systems that expose their functions as APIs. Thus, you can build a microservice in SAP BTP, Kyma runtime that consumes or extends these APIs. Kyma also supports the development of microservices that are connected to non-SAP systems or to combinations of SAP and non-SAP systems. These microservices can be standalone functionalities or extensions of functionalities provided by an existing SAP SaaS solution (e.g., SAP S/4HANA Cloud, SAP SuccessFactors, SAP Fieldglass, etc.).

3.4 Integrating Cloud-Native Applications

This section explores various tools and capabilities of SAP Integration Suite and establishes SAP Integration Suite as the integration platform to enable a seamless digital transformation for any enterprise. It also walks through using SAP Data Intelligence as an integration platform, the prebuilt integration scenarios on SAP Business Accelerator Hub, migrating SAP Process Orchestration interfaces to SAP Integration Suite, and some common integration scenarios.

3.4.1 Introduction to SAP Integration Suite

SAP Integration Suite is an enterprise-grade integration platform as a service (iPaaS) on SAP BTP. It represents how SAP has made a fundamental transformation from a traditional on-premise middleware system to a comprehensive, cloud-native integration platform that addresses today's hybrid, multicloud enterprise needs.

Based on cloud-native architecture, the strategic decision to move toward cloud integration reflects SAP's recognition that traditional on-premise middleware cannot fully meet the scalability, agility, and cost requirements of modern enterprises. This shift was initiated and driven by customer demands to reduce infrastructure overhead, accelerate deployment cycles, and enable dynamically robust integration capabilities that scale according to business needs.

When discussing business transformation, we should also consider the following:

- Ongoing market conditions are accelerating the shift from custom integration development to platform-based approaches to reduce the total cost of ownership. The rise of ecosystem business models in collaboration with clients and partners emphasizes the importance of robust, scalable, and standardized integration capabilities to minimize extensive redevelopment.
- The need to reduce build effort by incorporating visual modeling tools, template-based development, and extensive preconfigured adapters for integrating different systems. The platform also introduced integration packages that come with module- and line of business-specific integration flows, documentation, and configuration templates to accelerate common integration scenarios.
- There is a need to improve the user experience in monitoring message processing, using preconfigured templates, deploying code, and providing comprehensive testing and debugging capabilities integrated into the development environment. The platform provided role-based access controls for various capabilities and collaborative development features that help different development teams work effectively and collaboratively on integration projects while maintaining governance and quality standards.
- Hybrid cloud adoption presents unique challenges in network security, data governance and synchronization, latency, and protocol translation between various types of on-premise and cloud systems. Organizations demand secure, scalable integration platforms that support easy, seamless connectivity, real-time message processing, compliance, encryption, authentication, and audit reporting.

The latest capabilities of SAP Integration Suite include AI-assisted integration, event-driven architecture, API management, B2B integration, a clean core strategy, and agentic AI enablement. Figure 3.7 depicts SAP Integration Suite's system integration capabilities for various processes.

SAP Integration Suite is essential for organizations that need a unified platform capable of supporting real-time business processes, API-driven ecosystems, and event-based architectures to reduce operational overhead. To address these challenges SAP Integration

Suite provides core and specialized capabilities. While setting up SAP Integration Suite, we can activate different capabilities based on our requirements, which we'll look at in the following sections.

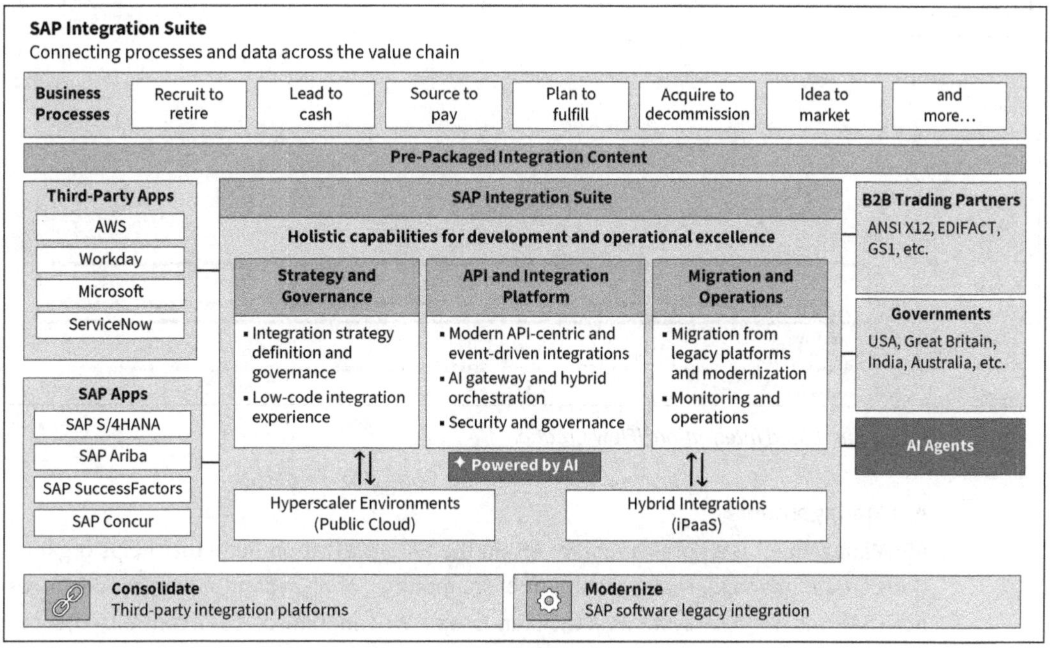

Figure 3.7: SAP Integration Suite in Q3 2026

Cloud Integration

Cloud Integration is one of the core capabilities of SAP Integration Suite. Previously known as SAP Cloud Platform Integration, Cloud Integration allows us to design, deploy, and operate various integration flows to connect SAP and non-SAP systems and applications across cloud and on-premise landscapes for A2A, B2B, and B2G scenarios.

From the middleware perspective, Cloud Integration is the core capability where any customization, content-based routing, message enrichment, and transformation can be performed based on event-driven triggers or scheduled automated flows.

Cloud Integration is part of SAP's low code/no code offerings and provides a graphical integration flow editor for coding (see Figure 3.8). It offers three types of connectivity options: standard adapters (HTTP(S), SOAP, OData, SFTP, RFC, JDBC), custom adapters (built with the adapter development kit), and connectors. These connectivity options help SAP Integration Suite to connect to external systems and protocols.

Currently, Cloud Integration supports over 80 standard adapters and more than 170 connectors for integrating third-party applications and systems. It also includes role-based security, monitoring, and error-handling features to ensure secure execution of integration flows.

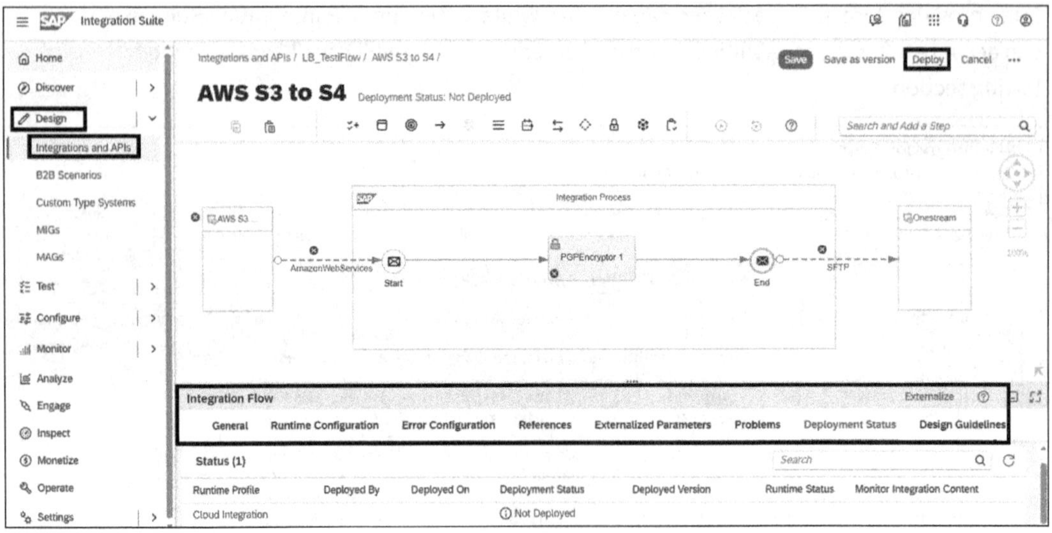

Figure 3.8: Cloud Integration: iFlow Creation

API Management

API Management is a core capability within the SAP Integration Suite that helps organizations create, manage, scale, secure, publish, monitor, analyze, and monetize APIs across both SAP and non-SAP systems in their landscape. We can think of it as a centralized, secure digital gateway that allows different applications to securely connect to the registered backend system to access business services and data through APIs. API Management has four core components, as follows:

- API gateway: It primarily provides secure routing by enforcing runtime security rules and efficiently directing incoming API traffic to backend systems or applications.
- API portal: This is used for designing an API with a set of governance policies. It provides a workspace for the development team to design, build, configure, and publish APIs. It also helps create and apply API policies.
- Developer portal: This is a self-service repository where developers can discover APIs, read API documentation, and request access tokens to use the APIs.
- API analytics: API Management also provides out-of-the-box visual dashboards to track API usage metrics, performance latency, error rates, and security threats in real time.

API Management includes key functionalities to design and create new API endpoint proxies, publish them in SAP Business Accelerator Hub, and expose existing REST, SOAP, or OData services as managed APIs. It can also create API endpoint proxies without changing backend registered applications and standardize API access across multiple systems within the organizational landscape.

Comparing API Management and Cloud Integration, you should understand that Cloud Integration is primarily used for complex data transformation and mapping logic, stateful message routing, and multi-system orchestration, especially B2B EDI routing. On the other hand, API Management exposes backend services as API endpoints to external systems, enforcing strict corporate rate limits and API keys. Users or vendors need to go to the developer portal, subscribe to the API, and obtain the API token key to access the backend service through the published API endpoint proxies. Figure 3.9 shows an example of configuring an API proxy in API Management.

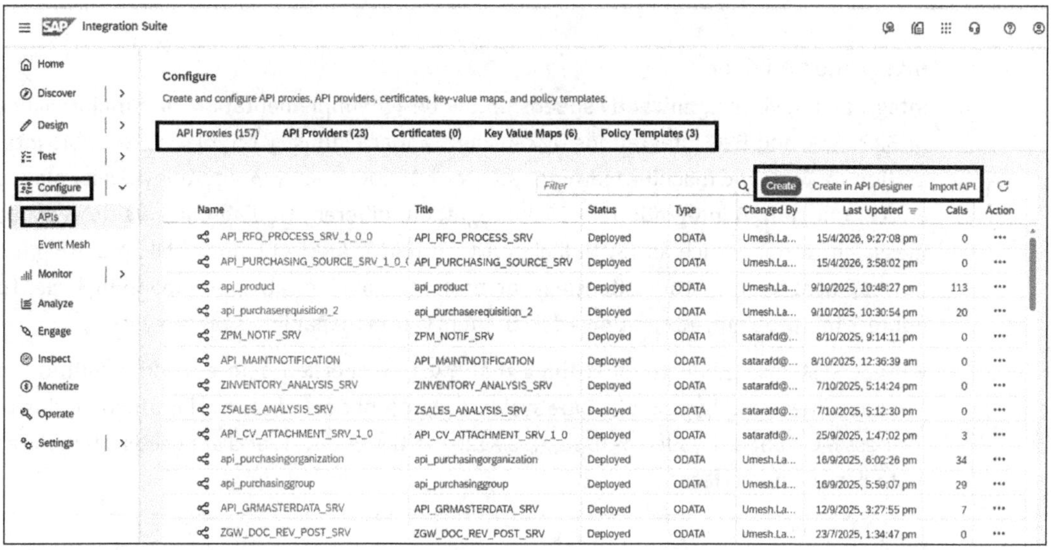

Figure 3.9: API Management

Open Connectors

Open Connectors is another core capability that offers prebuilt, feature-rich connectors to simplify and accelerate integration with non-SAP cloud applications such as Salesforce, HubSpot, Slack, SharePoint, and DocuSign. It provides a unified, normalized API layer with over 150 prebuilt, production-ready connectors using open RESTful APIs and JSON, offering decoupled integration. Open Connectors is used to integrate various non-SAP SaaS applications—such as CRM, help desk, payment, storage, and more—to support a one-to-many integration pattern through common resources.

The main components of Open Connectors are as follows:

- Prebuilt connectors: These provide instant access to a wide range of non-SAP SaaS applications for CRM, helpdesk, marketing, and finance.
- Unified APIs (hubs): These normalizes a group of similar third-party services into functional hubs so that different vendor APIs use the exact same resource names, pagination, and error handling.
- Common authentication: This reduces the cost to harmonize different third-party authentication methods (OAuth, custom tokens, basic auth) into a single, standardized approach for all SaaS applications.

- Bulk data framework: When transferring large volumes of data asynchronously, it uses standard file formats like CSV or JSON irrespective of a vendor's specific batch architecture.

Open Connectors enriches Cloud Integration's adapters by extending connectivity through a unified API layer for non-SAP SaaS applications, and can be used in iFlows or via API Management policies and proxies.

Integration Advisor

Integration Advisor significantly speeds up the design, implementation, and maintenance of B2B, A2A, and B2G integrations by reusing standard industry templates, reusable integration content, and machine learning. Let's look at how exactly Integration Advisor plays a critical role in B2B integration. In B2B integration, different trading partners use various messaging formats such as JSON, XML, IDOC, EDIFACT, ANSI X12, and cXML, requiring the development team to work extensively on message transformation and mapping logic. To address this problem, Integration Advisor offers various capabilities of its own:

- Type systems: This is the first thing that must be checked before working with other capabilities (see Figure 3.10). Type systems represent a set of central libraries of industry message standards, such as IDoc, ASC X12, and UN/EDIFACT, and are used as templates to design an interface.

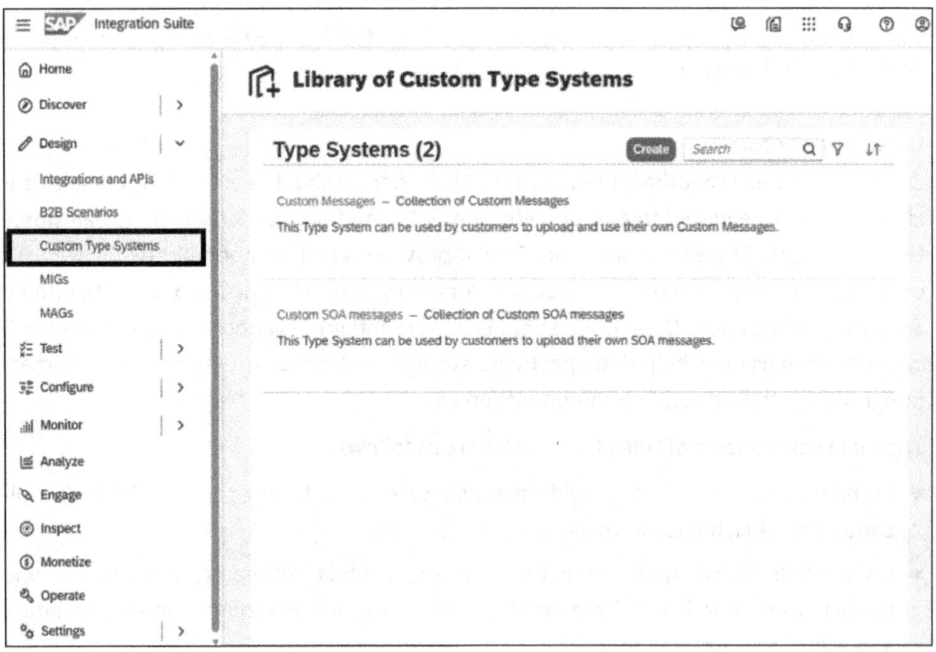

Figure 3.10: Integration Advisor Type Systems

- Message implementation guideline: This is a template for the inbound and outbound message formats. There will be one message implementation guideline created for the source type system, and another must be defined for the target type system message format. Each message implementation guideline must include the specifications for constraints, code lists, qualifiers, and business rules (see Figure 3.11).

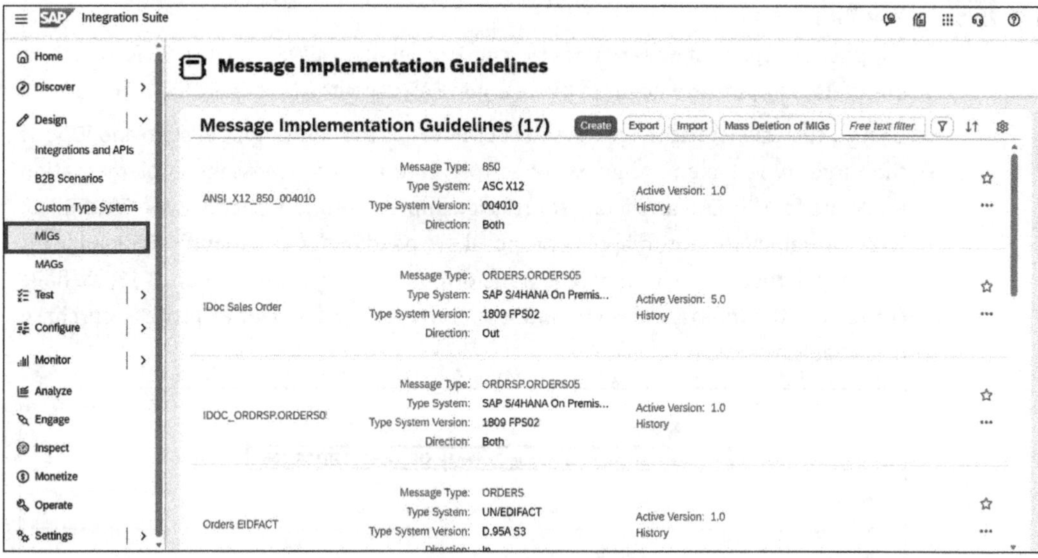

Figure 3.11: Integration Advisor Message Implementation Guideline

- Mapping guideline: This capability specifically supports mapping between source and target message implementation guidelines that are already defined. Once message implementation guidelines are selected, Integration Advisor provides intelligent automapping proposals derived from an industry knowledge graph and standards to reduce development effort (see Figure 3.12).

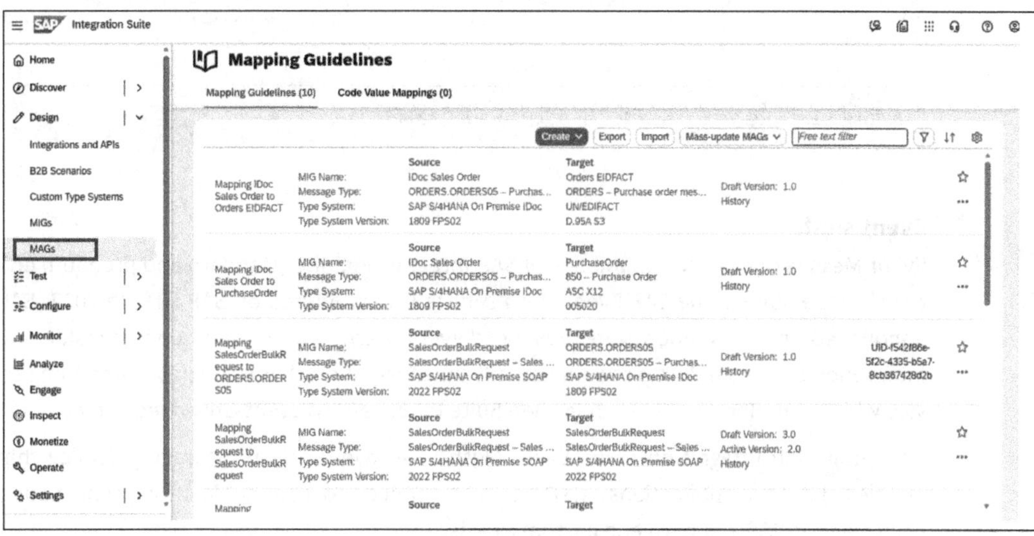

Figure 3.12: Integration Advisor Mapping Guideline

Type systems, message implementation guidelines, and mapping guidelines offer both industry standard templates and custom templates.

[zB]

Example

Suppose an organization receives a purchase order in the EDIFACT ORDERS format from one of its suppliers and needs to process it in SAP using the IDoc ORDERS05 format.

Using Integration Advisor, we will select EDIFACT as the source type system and IDoc as the target type system. Then we need to create a source message implementation guideline for ORDERS and a target message implementation guideline for ORDERS05. With the help of the automapping proposal, we need to define a mapping guideline for common industry-specific message transformations. As shown in Figure 3.13, we have completed the interface design, and it is ready for deployment to process purchase orders automatically.

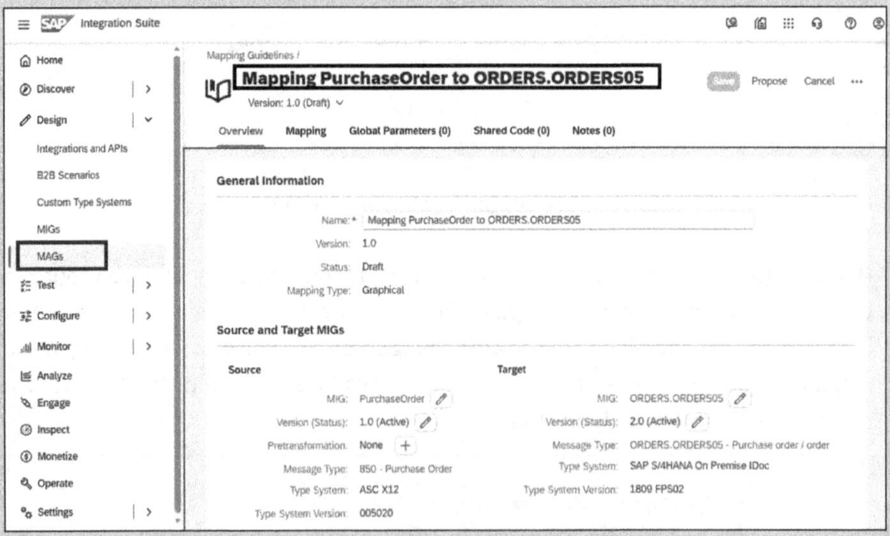

Figure 3.13: Integration Advisor Mapping Guideline Mapping

Event Mesh

Event Mesh is one of the capabilities of SAP Integration Suite (standard and premium editions), but a standalone SAP Event Mesh service is also offered on SAP BTP. In 2022, SAP announced SAP Integration Suite, advanced event mesh, which is not bundled inside SAP Integration Suite. SAP Integration Suite, advanced event mesh is a separate, premium service which complements SAP Integration Suite for advanced event-driven requirements.

So, within SAP Integration Suite, Event Mesh is an event-driven messaging service that enables different applications, systems, and services to communicate asynchronously using various SAP standard and custom events.

Instead of one system calling another synchronously, different systems publish events to a particular topic in Event Mesh, which then channels the event data to the interested applications based on their subscriptions to that topic.

Event Mesh consists of several key components, including the following:

- Business event: This represents an action which gets executed at the end of a step in a business process. In the case of SAP S/4HANA, many standard business events are generated automatically at the end of the process. Generally, for SAP S/4HANA, the standard business event naming convention is sap.s4.beh.<businessobject>.v1.<BusinessObject>.<Operation>.v1. For example, when discussing production order creation and change, business events such as BO.ProductionOrder.Created and BO.ProductionOrder.Confirmed get generated.
- Event publisher: This is a system or application which generates an event. Continuing our example, the publisher is the SAP S/4HANA system using the business event handling framework. Once the event is activated in SAP S/4HANA, the outbound topic binding/subscription communicates the event to Event Mesh.
- Topic: A topic is generally used to categorize events. The topic which we need to create in Event Mesh for production order creation is sap.s4.beh.productionorder.v1.ProductionOrder.Created.v1. For the production order confirmation, the topic would be sap.s4.beh.productionorder.v1.ProductionOrder.Changed.v1.
- Queue: In Event Mesh (see Figure 3.14), a message queue needs to be defined to receive all the messages in the queue sequentially. A webhook needs to be defined at the queue level to consume the messages from the queue (see Figure 3.15).

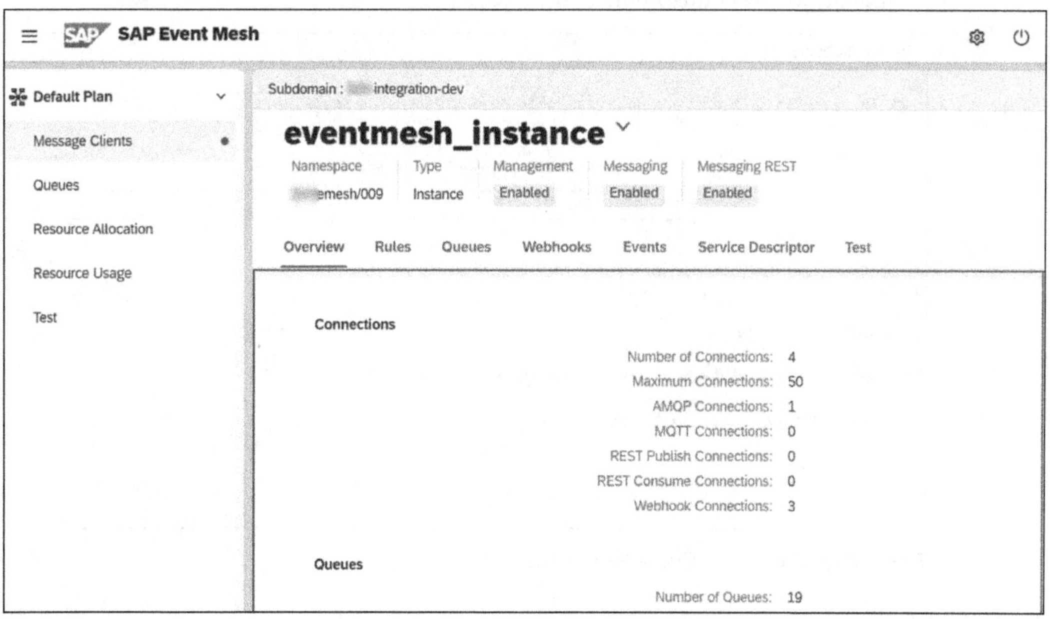

Figure 3.14: Event Mesh

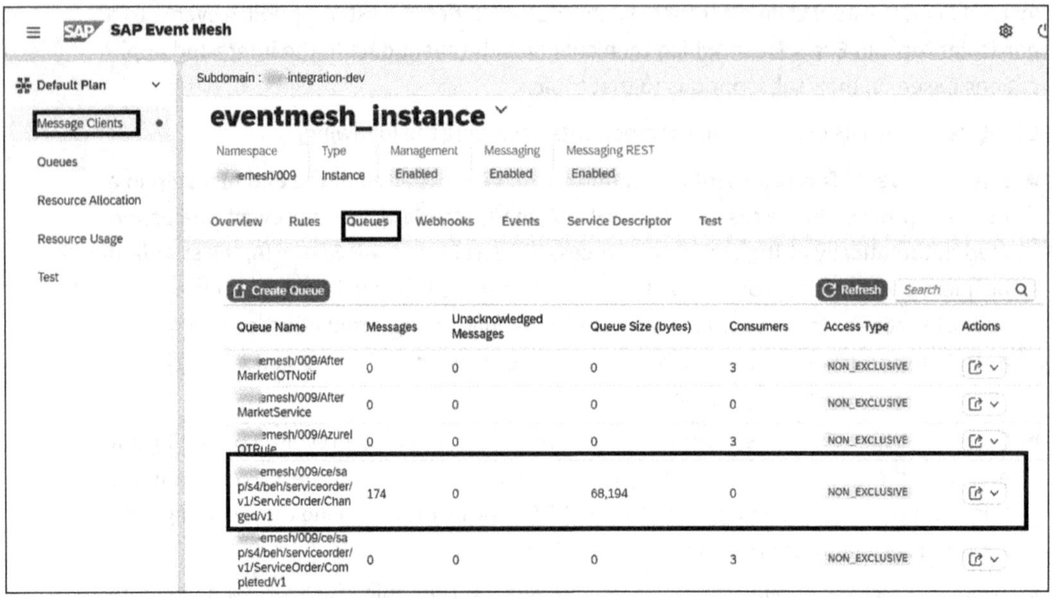

Figure 3.15: Event Mesh Message Queues

- Subscriber: This is an application that subscribes to a topic to receive information about the execution of events. Many systems can subscribe to one topic. In our example (see Figure 3.16), all production order-related information can be consumed by any custom application on the SAP BTP side to create a digital thread, providing end-to-end process visibility for the entire production process.

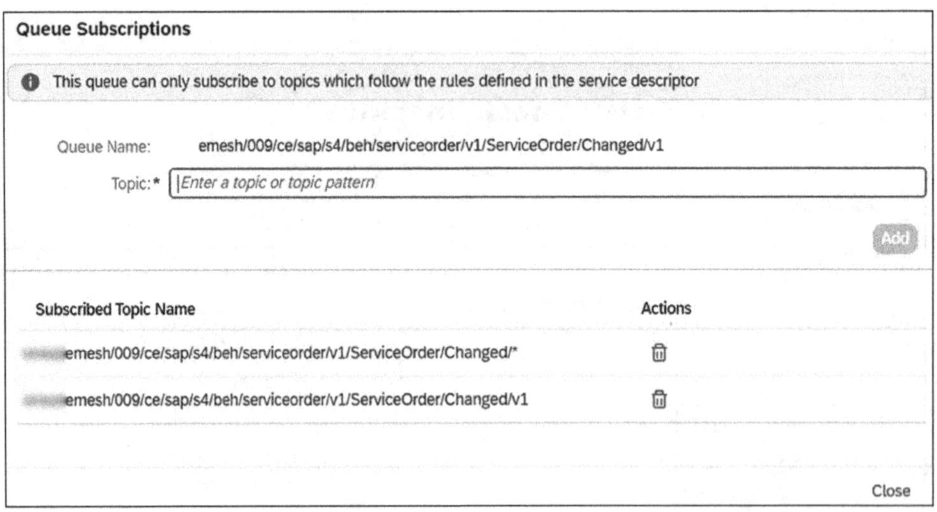

Figure 3.16: Event Mesh Queue Subscription

Trading Partner Management

Trading Partner Management is a specialized capability of SAP Integration Suite, like Integration Advisor, that simplifies the onboarding and management of B2B trading partners and helps us model, govern, and operate all B2B relationships with external business partners through centrally managed profiles, agreements, and runtime artifacts for data exchange.

An organization needs to exchange numerous business documents with various external partners, including suppliers, customers, logistics providers, financial institutions, and government agencies. Different message formats, such as EDIFACT, ANSI X12, XML, IDoc, and cXML, as well as communication protocols like AS2, SFTP, FTP, and HTTPS, will be used for different partners.

Now let's take a deep look at how exactly Trading Partner Management works and what the building blocks are. Following are the two main components that need to be configured for Trading Partner Management to work correctly:

- Partner profiles and agreement: First, the company's own profile needs to be defined, followed by the trading partners' profiles, which should include specific identifiers, communication endpoints, protocols, supported B2B standards or message types, and certificates. Once the profiles are defined, we need to create one agreement template and that template can be used to expedite the onboarding of partners. Upon successful creation of the agreement, the agreement gets added as a runtime artifact into the partner directory of Cloud Integration for dynamic, per-partner processing at runtime (see Figure 3.17).

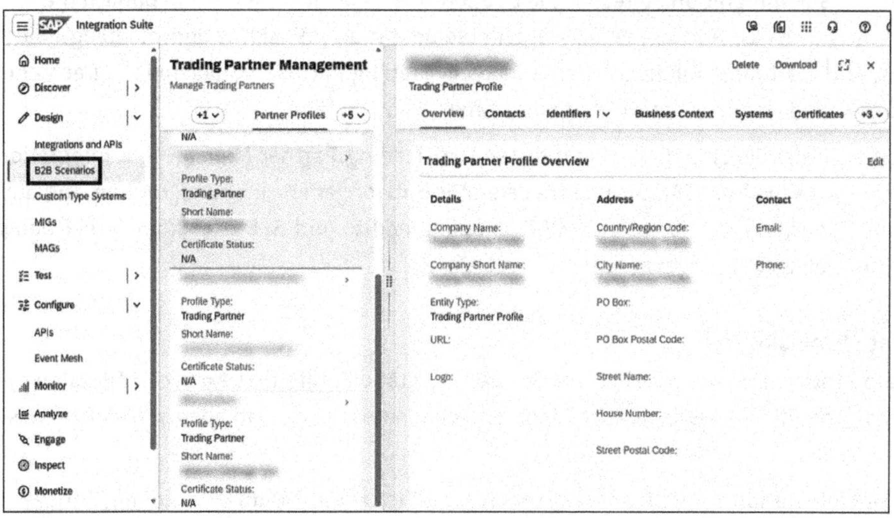

Figure 3.17: Trading Partner Management

- Partner directory: The partner directory acts as the configuration repository which gets consumed by Cloud Integration iFlows at runtime. This repository helps iFlows to get the adapter and mapping parameters details dynamically for each B2B interface (see Figure 3.18).

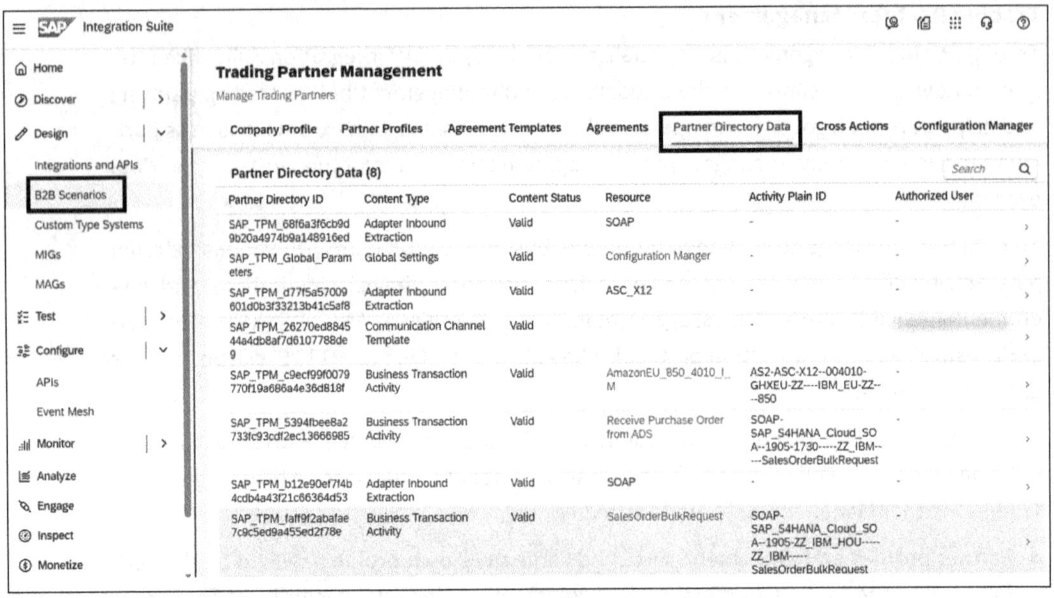

Figure 3.18: Trading Partner Management Partner Directory Data

Another interesting point is that Trading Partner Management depends on Integration Advisor, Cloud Integration, and message queues. Before activating the Trading Partner Management capability, we activated Cloud Integration, Integration Advisor, and the message queue in SAP Integration Suite. For the B2B interface, we must model and publish the corresponding source and target message implementation guidelines and mapping guidelines. These artifacts will next be referenced in Trading Partner Management agreements and executed via the generic iFlows at runtime.

Another prerequisite is deploying the standard Trading Partner Management integration content package from SAP Business Accelerator Hub for generic inbound, routing, and outbound iFlows for AS2, IDoc, and SOAP, enabling end-to-end B2B execution with Trading Partner Management data at runtime.

OData Provisioning

OData Provisioning is a capability of the SAP Integration Suite that exposes SAP Gateway–based backend services as OData endpoints. It eliminates the need to operate the on premise gateway hub.

On SAP Integration Suite's home page, there is no tile called “OData Provisioning”; once it is activated, a tile named **Access Data in SAP Business Suite** appears. To use it, the user needs to have a custom role collection assigned. This custom role collection must have the two OData Provisioning–specific roles, `ODPManage` and `ODPAPIAccess`, assigned to it. Once the role collection is assigned, the user will be able to access the tile to register OData services and expose them as cloud endpoints on the SAP BTP side, eliminating the need to develop additional gateways or hubs.

To connect to the backend ABAP, assuming it is an on-premise system, a destination specifically defined for OData Provisioning in the SAP BTP tenant cockpit is required. Through the cloud connector, the registered services publish business data and logic from SAP Business Suite or SAP S/4HANA as OData services on SAP BTP. Any user-centric applications that consume these OData services gain the following benefits: simplified exposure of gateway services in the cloud, reduced TCO, centralized management, and maintenance and security responsibilities for OData services remain with SAP.

Comparing API Management and OData Provisioning, API Management adds governance, security policies, analytics, rate limiting, and a developer portal by placing an API proxy in front of any API. On the other hand, OData Provisioning focuses solely on registering and publishing gateway OData services to SAP BTP via SAP Integration Suite.

Example

If we need to retrieve purchase order details using the standard OData service `API_PurchaseOrder_2`, OData Provisioning publishes the service for direct consumption, while API Management creates a proxy for the same service and applies policies when routing calls—such as a `GET` on the purchase order entity set to return header and item data.

Data Space Integration

Now, let's talk about Data Space Integration, which offers a dataspace connector based on Eclipse dataspace components, involving one provider company and one consumer company. Data Space Integration enables different consumer companies to discover data offers, enforce usage policies, negotiate contracts, and exchange data securely and with data sovereignty.

To access Data Space Integration, a user needs to have the appropriate Data Space Integration role collections—`DataspaceConsumer` and `DataspaceProvider`—assigned to them in SAP BTP.

Data Space Integration enables sovereign, policy-driven data sharing across ecosystems such as Catena-X with certified interoperability, centralized governance, and seamless integration with SAP and other third-party applications via Cloud Integration, all with correct access control in place.

Consider a business use case where a provider publishes a data asset—such as a purchase order–like dataset—along with policies and a contract definition in the Data Space Integration control plane—and makes it discoverable for a group of partners. The consumer's control plane then negotiates a contract through Data Space Integration APIs, which either initiates a data transfer process or returns access credentials according to the agreed policies.

Alongside Data Space Integration, Cloud Integration plays a crucial role by handling the actual data movement as a data plane, where a provider iFlow delivers the data, and a consumer iFlow receives, transforms, and routes it to target systems. For this entire process or data flow, both parties use Data Space Integration for cataloging, negotiation, and monitoring, along with Cloud Integration and optional backend systems.

Integration Assessment

Integration Assessment is primarily for enterprise architects, business domain experts, and developers. It leverages SAP Integration Solution Advisory Methodology to define, document, and govern an enterprise integration strategy through SAP guided assessments and technology selection recommendations.

To utilize this capability, we need to activate the Integration Assessment feature in the SAP Integration Suite tenant and assign the appropriate role collections to ensure that users can create and analyze assessment requests.

The tool helps capture integration requirements, design the interface by selecting suitable integration technologies and policies, and build an inventory view of the integration landscapes.

It also improves performance by detecting bottlenecks early, ensures better data consistency across systems, provides compliance assurance, and offers future readiness through a scalable solution. SAP highly recommends using Integration Assessment for any interface evaluations, as it employs a questionnaire-based approach that proposes best-fit integration technologies with appropriate rationale and justification.

[zB]

Example

To create a purchase order using SAP's standard OData services in SAP S/4HANA, we use Integration Assessment to record the business requirement (purchase order creation), capture nonfunctional and security requirements, and recommend the appropriate technology pattern—such as API/OData centric development using SAP Integration Suite capabilities like Cloud Integration, API Management, or OData Provisioning.

Edge Integration Cell

Edge Integration Cell is part of the SAP Integration Suite and is an optional hybrid integration runtime that allows us to design and monitor in the cloud when deploying and running integrations and APIs in a private landscape to meet data residency, security, and compliance requirements.

Edge Integration Cell separates the control plane from the execution plane by providing a centralized, cloud-based design, monitoring, and governance while keeping runtime processing local and avoiding the need for any public endpoints. This setup enables numerous performance improvements by placing processing near backend systems, ensuring strict data sovereignty, and using secure outbound-only communication from the local network. It is generally built on a lightweight, Kubernetes-based architecture that includes local runtime engines, persistence, caching capabilities, and lifecycle management components. Edge Integration Cell is well suited for the following:

- SAP PI/SAP PO migration scenarios
- Sensitive data processing
- Shop floor or IoT use cases that require low latency and localized execution

Migration Assessment

Migration Assessment is another important automated capability of SAP Integration Suite that analyzes and evaluates an existing SAP Process Integration/SAP Process Orchestration (SAP PI/SAP PO) landscape and provides a comprehensive fact-based estimate of the technical effort to migrate integration scenarios to SAP Integration Suite. It focuses on extracting and assessing design-time artifacts and serves as the basis for overall migration planning.

It connects to the SAP PO system and extracts metadata for integration scenarios, mappings, channels, and other artifacts supported by specific SAP PO versions.

Migration Assessment works similarly to Integration Assessment for each of the interfaces it analyzes. It recommends ideal design according to SAP Integration Solution Advisory Methodology. This tool evaluates and provides S/M/L/XL sizing and day-level estimates, focuses on analysis and scoping—not execution—and delivers dashboards and exportable reports summarizing adapters, rules, protocols, functional gaps, and recommendations.

This upfront assessment reduces project risk, accelerates planning, and supports both budgeting and phased migration strategies. It also generates the integration flow automatically for each interface. This tool is ideal for customers preparing SAP PI/SAP PO modernization or SAP S/4HANA transitions who need a data-driven view of their integration landscape. The results form the baseline for later technical conversion using migration tooling, which actually boosts the end-to-end modernization.

Model Context Protocol-Based Integration

In the context of the Autonomous Enterprise, the Model Context Protocol (MCP) is a critical component that enables AI to securely connect with external systems, data sources, APIs, tools, and documents through a consistent interface. Built on an open standard, the MCP server exposes all its capabilities, resources, and metadata in a structured format that AI agents can dynamically discover and consume. Instead of creating custom integrations for each system, MCP clearly simplifies integration and enhances interoperability across various platforms and technologies.

SAP has introduced MCP based integration as part of SAP Integration Suite, providing a secure and governed gateway to connect AI agents to enterprise systems. This will ultimately help streamline MCP-based integration, where MCP servers on one end expose APIs, data sources, and business services through standardized metadata, and SAP Integration Suite on the other end manages enterprise-grade capabilities such as protocol mediation, routing, transformation, mapping, traffic management, monitoring, and security policy enforcement for both SAP and non-SAP agents in a controlled, scalable, and compliant manner.

[»]

Integration Roadmap

As part of the SAP NetWeaver maintenance strategy, SAP Process Orchestration 7.5 will only be supported through 2030 with extended support.

However, SAP is introducing several AI-powered capabilities in SAP Integration Suite, such as AI-powered API specification and documentation generation, exposure of APIs as MCP servers for agentic AI scenarios, and AI-based automated field mapping based on contextual similarity, to name a few. A snapshot of the roadmap as of 2026 is provided in Figure 3.19. This roadmap is subject to change.

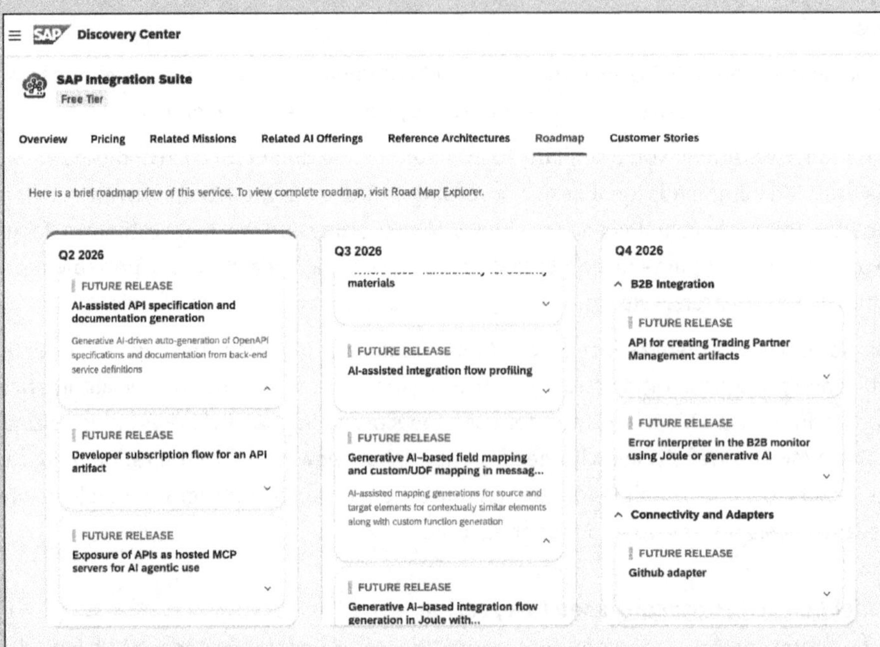

Figure 3.19: Roadmap of SAP Integration Suite 2026

3.4.2 SAP Data Intelligence

SAP Data Intelligence is a comprehensive AI platform that can help you to build intelligent enterprise solutions using a wide variety of features and services addressing the needs of multiple enterprise users and stakeholders, including data scientists, data engineers, developers, and administrators. SAP Data Intelligence also acts as an integration platform by supporting robust connectors and metadata governance. It allows you to discover, refine, and orchestrate numerous types of data from various enterprise sources. With a wide spectrum of features, SAP Data Intelligence platform ultimately helps you to:

- Integrate a wide variety of data sources and manage remote systems in a central place by supporting a robust set of connection types, including ABAP, SAP Integration Suite, Azure Data Lake, IBM DB2, Hadoop Distributed File System (HDFS), and many more.
- Govern and manage metadata assets spread across diverse systems and disparate sources, and apply a wide gamut of logical operators to dive more deeply into and extract insights out of enterprise data.
- Create data processing pipelines or graphs and provide a runtime and design-time environment for data-driven scenarios.
- Use the machine learning scenario manager to organize data science artifacts, manage related tasks through a single cockpit, and get easy access to Jupyter Notebook.
- Discover, refine, govern, and orchestrate numerous types of data from various enterprise sources and ensure business transformation through managed interfaces, open-source integration, cloud-based and on-premise deployments, and third-party interfaces.
- Use various tools and applications that are accessible through the SAP Data Intelligence launchpad and draw from its comprehensive set of features to execute an end-to-end data science execution cycle.
- Stipulate what information must be captured as part of an experiment (e.g., model parameters, model metrics, model hyperparameters, etc.) and view this information through the metric explorer within the machine learning scenario manager.
- Choose from over 250 built-in operators that can be directly used to build graphs for your business processes or as the basis for a custom operator to design powerful data pipelines.

Figure 3.20 shows the overall architecture of SAP Data Intelligence, along with its various components and capabilities.

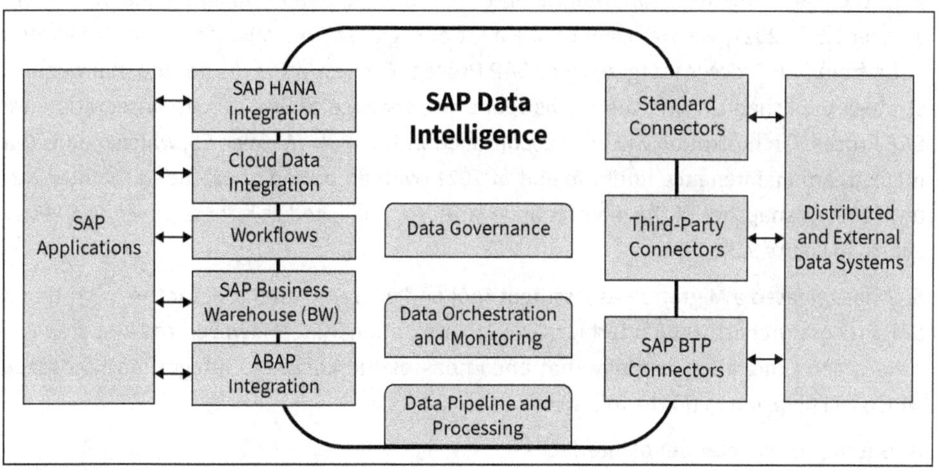

Figure 3.20: SAP Data Intelligence Architecture

3.4.3 SAP Business Accelerator Hub

As mentioned previously, SAP Business Accelerator Hub provides a single catalog to discover, explore, and test all the available SAP and partner APIs. Figure 3.21 shows the prebuilt integration scenarios on SAP Business Accelerator Hub.

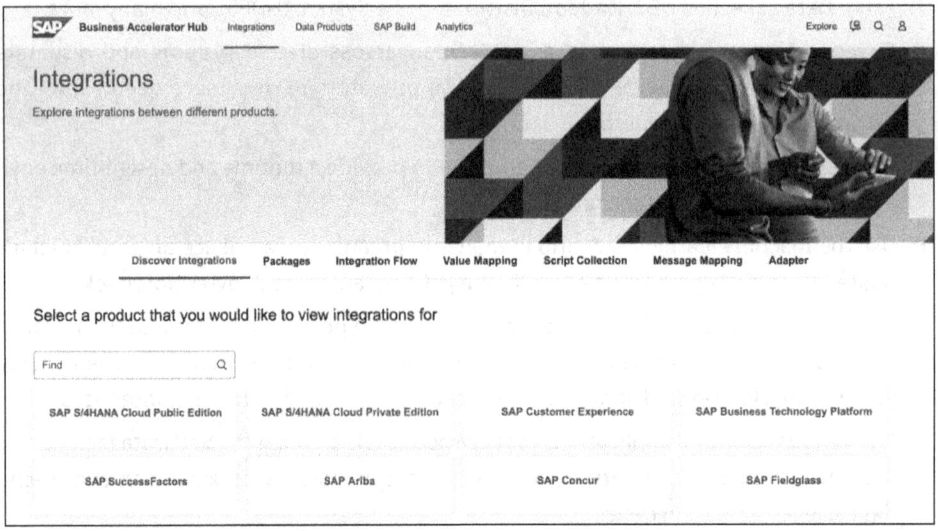

Figure 3.21: SAP Business Accelerator Hub Integration Catalog

You can explore various APIs based on your enterprise's integration needs. The API workspaces can be used to work on the content. You can then test APIs across various categories to extend your integration solutions.

3.4.4 Migrating SAP Process Integration or SAP Process Orchestration Content to SAP Integration Suite

With the rise of SAP Integration Suite and the proposed end of maintenance of SAP NetWeaver 7.5 in 2027, we are likely to see a lot of organizations migrate to SAP Integration Suite from SAP Process Integration or SAP Process Orchestration. As per the maintenance strategy published by SAP, the mainstream maintenance of SAP Process Integration and SAP Process Orchestration will be in accordance with SAP NetWeaver 7.5, which means that mainstream maintenance until the end of 2027, with an option of extended maintenance until 2030. A snapshot of the maintenance strategy, published at *http://s-prs.co/v550433*, is shown in Figure 3.22.

SAP has released a Migration Assessment tool and an automated tool for the migration of SAP Process Orchestration interfaces to SAP Integration Suite. However, the first step is to have a good migration strategy that encompasses the complete integration landscape, which will help you to do the following:

- Get the correct number of interfaces
- Evaluate the landscape spread (cloud, SAP, and non-SAP systems in the landscape)

- Assess the various systems involved with their protocols/connectivity
- Categorize interfaces based on complexity, business processes, or sender/receiver systems, as best suited to strategize the migration
- Consider the cost of moving

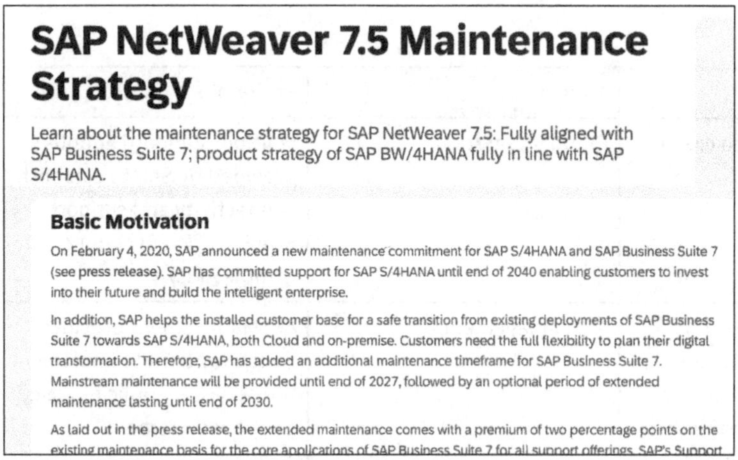
SAP NetWeaver 7.5 Maintenance Strategy

Learn about the maintenance strategy for SAP NetWeaver 7.5: Fully aligned with SAP Business Suite 7; product strategy of SAP BW/4HANA fully in line with SAP S/4HANA.

Basic Motivation

On February 4, 2020, SAP announced a new maintenance commitment for SAP S/4HANA and SAP Business Suite 7 (see press release). SAP has committed support for SAP S/4HANA until end of 2040 enabling customers to invest into their future and build the intelligent enterprise.

In addition, SAP helps the installed customer base for a safe transition from existing deployments of SAP Business Suite 7 towards SAP S/4HANA, both Cloud and on-premise. Customers need the full flexibility to plan their digital transformation. Therefore, SAP has added an additional maintenance timeframe for SAP Business Suite 7. Mainstream maintenance will be provided until end of 2027, followed by an optional period of extended maintenance lasting until end of 2030.

As laid out in the press release, the extended maintenance comes with a premium of two percentage points on the existing maintenance basis for the core applications of SAP Business Suite 7 for all support offerings. SAP's Support

Figure 3.22: SAP NetWeaver Maintenance Strategy

There can be two categories of interfaces: those that need a redesign (interfaces involving APIs, business process management, etc.), and those where some artifacts can be reused (such as message mappings or operations mappings). Many organizations are running older versions of SAP Process Integration or SAP Process Orchestration and already have matured, strategic integration platforms. Upgrading to the latest version of SAP Process Orchestration will enable you to upgrade to the latest Java-based single-stack installations and lower your total cost of ownership by 25% to 50%. SAP Process Orchestration 7.5 supports the iFlow format of SAP Integration Suite and can also be used as a solution for hybrid integration.

You can take advantage of SAP Integration Suite capabilities and prebuilt content for new integration scenarios. SAP Integration Suite provides a holistic approach for a variety of integrations, including A2A, B2B, B2G, APIs, or event-based integrations. Integrations can also be designed in SAP Integration Suite and deployed on-premise on SAP Process Orchestration. This way, you can leverage the best of both worlds and move to a hybrid architecture that is readily scalable.

You should have an incremental migration strategy to move the integrations to SAP Integration Suite. You can reuse the integration content on SAP Process Orchestration by connecting the enterprise service repository (ESR) to SAP Integration Suite and importing the message mappings and operation mappings. This gives you a great advantage and reduces a lot of effort and time for development. SAP Integration Suite is the future of integration, and it allows you to plan a path with the least disruption and fastest innovation that best suits your business needs.

3.4.5 Scenario-Based Recommendations for Integration Tools

Now that you have seen the capabilities of SAP Integration Suite, the most important step is to understand how and when to choose a tool that will not just provide a solution but is also performance-oriented and in line with your integration strategy. Table 3.2 provides a list of real-world scenarios and the SAP Integration Suite tool that will come in handy in that situation.

Scenario	Tool	Rationale
Integration between cloud applications	Cloud Integration	Multiple standard adapters provided by SAP to establish connectivity and support SOAP/HTTP/API connectivity (simple APIs)
API integration involving additional functionalities, such as mediation, security (basic, keys, and certificates), and traffic management (spike arrest, concurrent rate, and quota)	API Management	API Management provides various options to implement the necessary policies based on the requirement
Integration of AI to address business challenges by acquiring data from SAP and non-SAP sources	SAP Data Intelligence	To deliver data-driven innovation and intelligence across an enterprise, unifying scalable enterprise AI and intelligent information management, and make the most out of different data assets through seamless integration
Integration between on-premise SAP and legacy applications	▪ SAP Process Orchestration ▪ Alternatively, you can build the interface in SAP Integration Suite and deploy it on the SAP Process Orchestration runtime if appropriate product versions are available	▪ Designing integrations is easier with SAP Integration Suite. SAP Process Orchestration should be used if there is a lot of data traffic involved in the interfaces between on-premise systems instead of transferring the traffic to SAP Integration Suite ▪ Matured monitoring and error handling tools in SAP Process Orchestration

Table 3.2: Sample Scenarios and the Appropriate Integration Tool

Scenario	Tool	Rationale
		▪ Stable B2B integration solutions in SAP Process Orchestration, e.g., through EDIs
Integration scenarios that involve event-based messaging and publish-subscribe scenarios	Event Mesh	Event Mesh uses events to communicate, can decouple the event generator and consumer systems, and acts as a message broker. Cloud-native applications on SAP BTP can use these events to trigger further actions
For B2B scenarios involving primarily cloud-based applications; to accelerate designing and building processes on SAP Integration Suite	Integration Advisor	Integration Advisor offers standard content libraries for various B2B message types, is easy to map, and reduces effort and time by up to 60%
Integration scenarios involving banking via APIs, e.g., sending payment files	▪ Enable SOAP services on SAP S/4HANA Cloud (SAP API Business Hub) and Cloud Integration ▪ If a customer has a disparate non-SAP middleware for connecting with external systems, use SAP API Business Hub for the SOAP services on SAP S/4HANA Cloud and the non-SAP middleware tool to establish connectivity with the external application ▪ Use SAP Multi-Bank Connectivity (an add-on for SAP S/4HANA, out of the scope of this chapter)	▪ SAP Business Accelerator Hub provides standard SOAP services for bank communication ▪ Less manual work ▪ The middleware tool (either SAP or non-SAP) must be present to provide the necessary quality of service for the integrations
Migrating SAP Process Orchestration interfaces to Cloud Integration involving B2B scenarios	Integration Advisor	By leveraging the content library of Integration Advisor, including the structure and mapping, the migration is accelerated

Table 3.2: Sample Scenarios and the Appropriate Integration Tool (Cont.)

Note that this is not an exhaustive list and is intended only to be a sample of potential scenarios. The examples shown here may change, subject to the given landscape and integration scenarios as well as SAP's tooling strategy and the evolution of SAP Integration Suite. There can also be complex scenarios where a combination of integration tools might be needed to suit the landscape and requirements.

3.5 Summary

In this chapter, we covered the most important building blocks in Cloud Foundry and cloud-native applications—microservices and integrations of cloud-native applications. We explored important design principles for microservices and discussed how you can expose microservices as external services for consumption. Finally, we emphasized how microservices should be created from a business perspective—they contain specific functionalities that can then require one or more components, whether an API or a UI or more. The advantage for your business is increased agility in terms of getting a function up and running quickly, independent of other dependent services. Thus, a microservice-based architecture can bring about huge change in the business architecture for your custom non-SAP application space as well as in the SAP space.

You also learned about the capabilities of SAP Integration Suite that make it a perfect choice for a digital transformation tool. We explored how Cloud Integration can connect to a variety of applications using adapters. We discussed how API Management and SAP Business Accelerator Hub enable you to explore and use various APIs to suit your individual business needs. We walked through the Open Connectors that enable SAP Integration Suite to connect to a wide variety of applications, and we discussed how the Integration Advisor provides a library from which you can choose prebuilt, industry-standard integration content that suits your scenario. We also saw how SAP Data Intelligence provides a platform based on AI that enables you to deliver intelligent and data-driven solutions. We then saw the important points to be noted when planning a migration from SAP Process Integration and SAP Process Orchestration to SAP Integration Suite.

Chapter 4
SAP Business Application Studio

In this chapter, we'll provide an overview of SAP Business Application Studio, a cloud service provided by SAP Business Technology Platform (SAP BTP) for the development of business applications. We'll walk you through the steps for setting up SAP Business Application Studio dev spaces for various business scenarios, and you'll learn about integration with GitHub and migration scenarios for moving applications from SAP Web IDE. Additionally, we are also going to introduce SAP Build Code and describe how we can take advantage of it to accelerate the enterprise scale application development process.

SAP Business Application Studio is a cloud service available on SAP BTP that offers you a development environment for developing business applications serving an intelligent enterprise. It accelerates the development of applications as it provides runtimes, desktop-like user interfaces (UIs), wizards, code editors, command-line interfaces (CLI)/terminals, local test runs, and deployment features without installing any software on local machines.

SAP Business Application Studio is based on the open-source integrated development environment (IDE) Eclipse Theia, which embraces the Microsoft Visual Studio Code (VS Code) experience.

At the core of SAP Business Application Studio is the dev space, which provides tools and preinstalled runtimes for business scenarios, such as SAP Fiori, SAP S/4HANA extensions, workflows, and more. This approach simplifies the setup of a development environment since tools are installed automatically depending on the business scenario and allows for efficient development, testing, building, debugging, local runs and deployment of business applications.

In this chapter, we'll cover the capabilities found in SAP Business Application Studio, setting up SAP Business Application Studio with different dev space types to support application development for various business scenarios, and developing first a cloud application using SAP Business Application Studio. We are also going to introduce SAP Build Code and walk through how you can leverage the Joule-based code assistant to generate various artifacts inside SAP Business Application Studio. Finally, at the very end, we'll cover moving artifacts or migrating projects from SAP Web IDE to SAP Business Application Studio.

4.1 Capabilities

SAP Business Application Studio is a cloud service in SAP BTP that provides dev spaces to cater to the development of business applications for various scenarios. Each scenario requires a different set of tools and capabilities. This flexibility is achieved through dev spaces. Additional tools can be installed in a dev space for extending applications using predefined extensions. SAP Business Application Studio provides the flexibility to test your applications in dev spaces without deploying them in the SAP BTP, Cloud Foundry environment. Figure 4.1 depicts the range of dev spaces offered by SAP Business Application Studio that are most essential for enterprise application development.

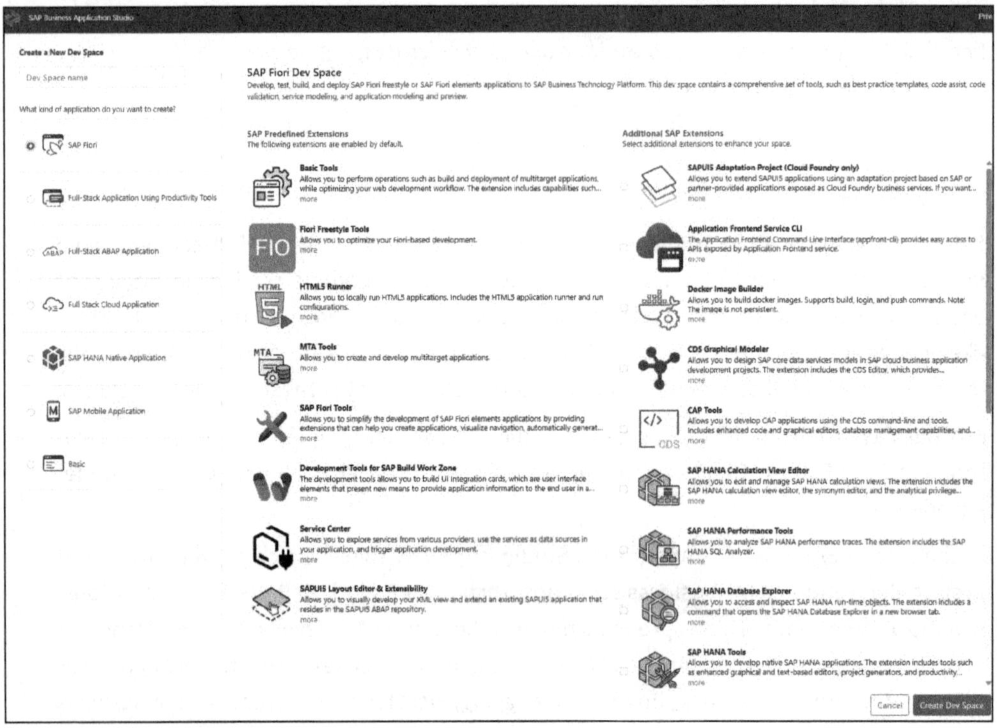

Figure 4.1: Dev Spaces

SAP Business Application Studio facilitates end-to-end development of applications in SAP ecosystems in the following ways:

- It integrates with a Git client for better lifecycle management of code. Projects can be cloned from Git client.
- It provides editors for SAP-specific technologies like ABAP and SAP HANA databases.
- It allows application testing to be performed locally (without deployment) even if applications consume services from remote sources.
- It enables building and deploying applications as multitarget applications (MTAs). Logically, an MTA is a single application with components built using multiple technologies.

The MTA approach helps you build and deploy applications developed using multiple components simply by maintaining the necessary dependencies in the *mta.yaml* file.

- Keeping pace with the AI revolution, especially in the field of code and application generation, SAP Business Application Studio now contains Joule inside of it, which accelerates code and application artifact generation including test scripts, data, etc. through simple prompts.

4.2 Setting Up SAP Business Application Studio

In this section, we'll look at the various kinds of dev spaces, describe the steps for creating dev spaces and projects within a dev space, and show you how to integrate a workspace within a dev space with a Git client, namely, GitHub.

4.2.1 Plans

Over the last few years, SAP Business Application Studio has gone through a transformation, especially in terms of features and components. Before configuring SAP Business Application Studio in your enterprise landscape, it is important to understand the service plans that SAP offers so that you can select the most suitable components. In terms of service plan, SAP currently offers subscription-based and consumption-based plans which include those listed in Table 4.1.

Plan Name/Discovery Plan	Description
standard/standardedition	Suitable for productive development in preconfigured dev spaces.
free/Free	The free offering is only for evaluation purposes and contains restrictions on the development environments. Support is limited to community channels and this plan is not covered by SLAs.
build-code/Build Code	The **Build-code** plan is essential if you wish to work with AI-assisted source code and the associated artifact generation. This plan is only available to subaccounts subscribed to SAP Build Code.
Trial/Trial	The trial offering is intended for evaluation purposes for those looking to take a sneak peek at the available functionalities within a limited period (30 days) in the SAP BTP trial environment. The trial plan is also limited to one runtime and two dev spaces per user; after the trial period, all artifacts automatically get deleted.

Table 4.1: SAP Business Application Studio Subscription Plans

4.2.2 Setup Instructions

SAP Business Application Studio runs on SAP BTP, Cloud Foundry environment. Therefore, you must have or create an account in SAP BTP. An enterprise SAP BTP account is preferred, but a trial account will suffice for training and research. Regardless of the SAP BTP account subscription level, you'll need a subaccount in your SAP BTP global account with administrator authorization as a prerequisite. When you create a trial account, a dev subaccount is also created, and you'll automatically become an administrator.

Once you have administrative privileges for your subaccount, you can now subscribe to the SAP Business Application Studio service. To initiate the setup of SAP Business Application Studio, all you need to do is to navigate to **Services • Service Marketplace** and search for "SAP Business Application Studio" in your trial subaccount and then open the SAP Business Application Studio application and click **Create**. This step will create an instance of SAP Business Application Studio in that subaccount. Before you can use this SAP Business Application Studio instance for development, you'll need to manage authorizations and roles to grant access to yourself or to other developers. SAP BTP provides three distinct role collections specific to SAP Business Application Studio:

- `Business_Application_Studio_Administrator`: This role template uses the role to manage user data.
- `Business_Application_Studio_Developer`: This role collection is the required role template for SAP Fiori developers and for other types of development (e.g., extensions).
- `Business_Application_Studio_Extension_Deployer`: This role template consists of a role to allow the deployment of simple extensions.

As an administrator of your subaccount, you can now assign one or more of these roles to your SAP Business Application Studio users. The process of assigning a role is simple: From your subaccount, navigate to **Security • Role Collections** and click on the desired role collection (e.g., **Business_Application_Studio_Developer**). Then, click **Edit** to maintain the users with their email addresses and identity provider(s) and then click **Save**. Once you assign a developer role to a user, that user can then open SAP Business Application Studio by navigating to **Instances and Subscriptions** and then selecting the **Go to Application** option.

Next, we'll set up a dev space based on the type of application that you are going to build, but before that, let's try to understand the types of dev spaces offered by SAP Business Applications Studio.

4.2.3 Types of Dev Spaces

A dev space in SAP Business Application Studio is created by first selecting a dev space type. In the following sections, we'll provide an overview of the various dev space types provided in SAP Business Application Studio, as well as their corresponding extensions.

Overview

A dev space type provides all the tools that you may require for the creation of business applications in the form of predefined SAP extensions and create additional SAP extensions. You can select additional extensions or create your own extensions and add them to your dev space.

[«]

> **Note**
>
> You can also install additional extensions/tools (beyond SAP extensions) and software components for developing applications using the terminal, using a dedicated UI, or directly in the code. These extensions are not supported or endorsed by SAP.

The extensions provided by SAP across all dev space types include the following:

- **Core data services (CDS) graphical modeler**
  Allows you to design CDS models in your application development projects. This extension includes the CDS editor, which provides capabilities for developers to visually design CDS models. Model artifacts include entities, types, unstructured types, enums, arrayed types, events, and associations, as well as compositions, projects, and services, all of which can accelerate business application development using the low-code/no-code programming paradigm.
- **SAP Cloud Application Programming Model tools**
  Allows you to develop SAP Cloud Application Programming Model applications using the CDS command line and tools. Tools include enhanced code and graphical editors, database management capabilities, and tools for running your applications in local environments
- **SAP HANA calculation view editor**
  Allows you to edit and manage SAP HANA calculation views. The extension includes the SAP HANA calculation view editor, the synonym editor, and the analytical privilege editor.
- **SAP HANA performance tools**
  Allows you to analyze SAP HANA performance traces. The extension includes the SAP HANA SQL analyzer.
- **SAP HANA database explorer**
  Allows you to access SAP HANA runtime objects. The SAP HANA database explorer is web browser-based tool that allows you to connect to SAP HANA databases and SAP HANA containers for exploring, viewing, and creating SAP HANA catalog objects.
- **SAP HANA tools**
  It allows you to develop native SAP HANA applications. The extension includes tools such as enhanced graphical and text-based editors, project generators, and other productivity tools.
- **HTML5 application templates (deprecated)**
  Allows you to create an HTML5 application using a template wizard.

- **Java tools**
  Allows you to develop and run Java applications. It includes debugging capabilities, enhanced code editors, and Spring tools. The following tools are installed as part of this extension:
  - Language support Java in VS Code
  - Java Debugger
  - Spring Tools 4
  - Java test support
  - Project Manager for Java
- **Python tools**
  You can take advantage of Python-related tools to deliver Python-based projects including IntelliSense, formatting, linting, and debugging support for Python files and Jupyter Notebooks.
- **Development tools for SAP Build Work Zone**
  You can use these development tools to create, edit, bundle, and deploy UI integration cards.
- **SAP Business Application Studio extension development**
  It allows you to use create, deploy, and manage your own SAP Business Application Studio extensions. You can then include these extensions in your SAP Business Application Studio environment. The SAP Business Application Studio extension management CLI tool will be installed as part of the extension.
- **Workflow module**
  Allows you to create workflow applications and process templates.
- **SAPUI5 adaptation projects (Cloud Foundry)**
  Allows you to extend SAPUI5 applications using adaptation projects and a visual editor.
- **Basic tools**
  Allows you to optimize your web development workflow. The extension includes capabilities such as a code editor, file management capabilities, split view and window management, preferences, and setting.
- **SAP Fiori freestyle tools**
  This allows you to optimize your SAP Fiori-based development. This extension provides SAP Fiori templates with the following tools:
  - SAPUI5 CLI
  - Grunt CLI
  - abap-deploy CLI
- **HTML5 runner**
  It allows you to locally run HTML5 applications. This tool includes an HTML5 application runner and predefined run configuration.

- **MTA tools**
  Allows you to perform some operations, such as building, deploying, and validating MTAs. The following tools are installed as part of this extension:
  - Cloud Foundry CLI
  - SAP BTP, Cloud Foundry environment deployment plugin
  - Cloud MTA Build Tool
  - MTA module runner (a VS Code extension)
- **SAP Fiori tools**
  Allows you to simplify the development of SAP Fiori elements applications by providing extensions to help you create applications, visualize navigation, automatically generate code, and more. The following extensions are installed:
  - SAP Fiori elements Yeoman generator
  - Application modeler
  - Guided development
  - Service modeler
  - XML annotation language server
- **Mobile services app development extended tools**
  Allows you to build end-to-end mobile applications. Includes modeling and generating OData services, creating native mobile apps from metadata with the mobile development kit (MDK), and building content for SAP Mobile Cards.
- **SAP HANA smart data integration tools**
  Allows you to manage data federation, replication, and transformation in SAP HANA. The extension includes graphical editors such as the flowgraph, replication task, and file format editors.
- **Docker image builder**
  Allows you to build Docker images and includes support to build, log in, and push commands.
- **Headless testing framework**
  Allows you to run cross-platform end-to-end tests on SAPUI5 applications in the background through various tools like OPA5, WebdriverIO, etc.
- **Launchpad module**
  Allows you to add a launchpad to your multitarget application.
- **Application Frontend service command line interface (appfront-cli)**
  Allows you to easily access APIs exposed by the Application Frontend service.
- **Full stack ABAP tools**
  Allows you to develop full stack applications for modeling business objects using the ABAP Cloud development model, supporting remote ABAP modeling and local UI development. This extension includes capabilities such as project creation, opening remote

ABAP CSN files, graphical modeling and visualization, ABAP artifact generation, service publication, and UI creation for ABAP services.

SAP Fiori Dev Space

This dev space, shown in Figure 4.2, provides tools for developing, testing, and deploying SAPUI5 freestyle or SAP Fiori elements applications to SAP BTP, Cloud Foundry environment; ABAP on-premise; or SAP BTP ABAP environment. SAP Fiori elements provide predefined floorplans for common application use cases. Developers can use SAP Fiori elements to create SAP Fiori applications using OData services and annotations without any JavaScript UI coding. The SAPUI5 runtime interprets the metadata and annotations from the OData service and maps this information to the corresponding views of the application. The SAPUI5 freestyle option provides templates for generating SAP Fiori applications. Developers can select templates and OData service endpoints to start application development. Flexibility is provided to developers to create applications based on OData services and using JavaScript UI coding for the UI logic.

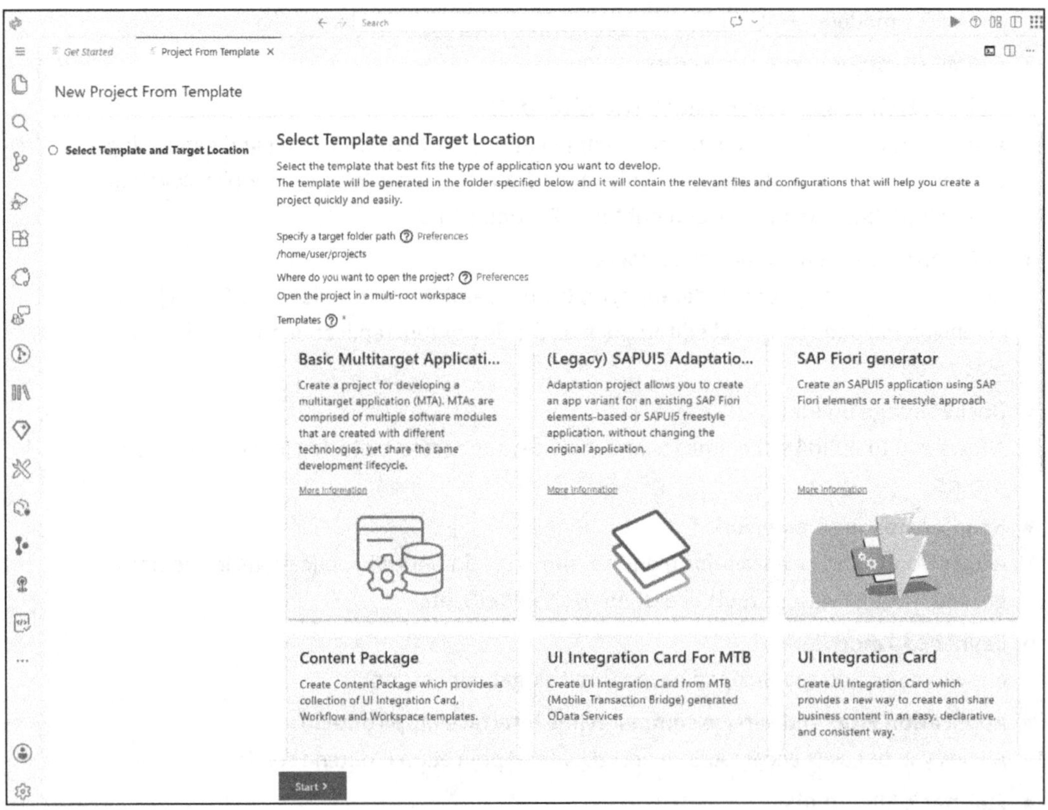

Figure 4.2: SAP Business Application Studio: Dev Space Based on the SAP Fiori Dev Space Type

Table 4.2 compares the SAP Fiori elements and SAPUI5 freestyle approaches.

Feature	SAPUI5 Freestyle	SAP Fiori Elements
Design	Freestyle designs based on JavaScript coding	SAP template based
Developer skills	Web development	Annotations
Ownership	Your own coding and UI logic	You own annotations and SAP's UI logic
Maintenance	High	Low

Table 4.2: Differences Between SAPUI5 Freestyle and SAP Fiori Elements Development

Table 4.3 provides a list of extensions that are installed by default or which can be installed based on need.

SAP Business Application Studio Extension	Type
Development tools for SAP Build Work Zone	Default extension
Basic tools	Default extension
SAP Fiori freestyle tools	Default extension
HTML5 runner	Default extension
MTA tools	Default extension
SAP Fiori tools	Default extension
Service center	Default extension
SAPUI5 layout editor and extensibility	Default extension
SAPUI5 adaptation project (Cloud Foundry only)	Additional extension
Application Frontend service CLI	Additional extension
Docker image builder	Additional extension
CDS graphical modeler	Additional extension
SAP Cloud Application Programming Model tools	Additional extension
SAP HANA calculation view editor	Additional extension
SAP HANA performance tools	Additional extension
SAP HANA database explorer	Additional extension
SAP HANA tools	Additional extension

Table 4.3: SAP Fiori Dev Space Extensions

SAP Business Application Studio Extension	Type
SAP HANA XSA tools	Additional extension
HTML5 application template	Additional extension
Java tools	Additional extension
Launchpad module	Additional extension
Python tools	Additional extension
SAP HANA smart data integration tools	Additional extension
Headless testing framework	Additional extension
SAP Business Application Studio extension development	Additional extension
SAP Workflow Management	Additional extension

Table 4.3: SAP Fiori Dev Space Extensions (Cont.)

Full Stack Application Using Productivity Tools Dev Space

This dev space provides tools for developing, testing, building, and deploying applications using high-productivity tools and an interactive UI, thereby reducing complexity and accelerating the development lifecycle, as shown in Figure 4.3.

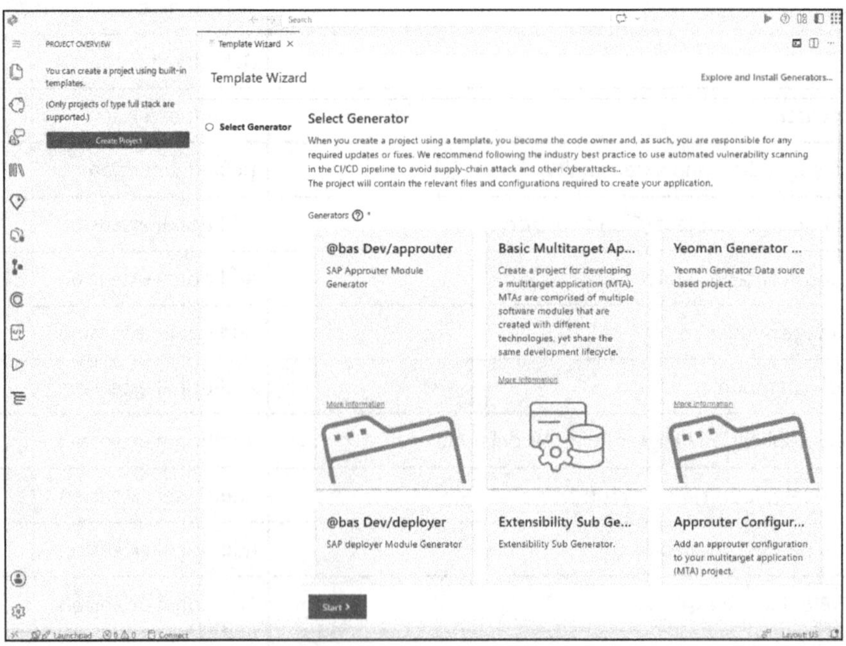

Figure 4.3: SAP Business Application Studio: Dev Space Based on the Full Stack Application Using Productivity Tools

Table 4.4 provides a list of extensions that are installed by default or which can be installed based on need.

SAP Business Application Studio Extension	Type
Basic tools	Default extension
CDS graphical modeler	Default extension
SAP Cloud Application Programming Model tools	Default extension
Productivity toolkit: release build 1.38.0	Default extension
SAP Mobile Services app development tools	Default extension
SAP Fiori freestyle tools	Default extension
HTML5 runner	Default extension
Java tools	Default extension
MTA tools	Default extension
SAP Fiori tools	Default extension
Service center	Default extension
SAPUI5 layout editor and extensibility	Default extension
SAPUI5 application project (Cloud Foundry only)	Additional extension
Application Frontend service (CLI)	Additional extension
Docker image builder	Additional extension
SAP HANA calculation view editor	Additional extension
SAP HANA performance tools	Additional extension
SAP HANA database explorer	Additional extension
SAP HANA tools	Additional extension
SAP HANA XSA tools	Additional extension
Launchpad module	Additional extension
Python tools	Additional extension
SAP HANA smart data integration tools	Additional extension
Development tools for SAP Build Work Zone	Additional extension

Table 4.4: Full Stack Application Using Productivity Tools Dev Space Extensions

SAP Business Application Studio Extension	Type
Headless testing framework	Additional extension
SAP Business Application Studio extension development	Additional extension
SAP Workflow Management	Additional extension

Table 4.4: Full Stack Application Using Productivity Tools Dev Space Extensions (Cont.)

Full Stack Cloud Application Dev Space

This dev space provides tools for building applications and services using SAP Cloud Application Programming Model, SAP Fiori, and Java or Node.js, as shown in Figure 4.4.

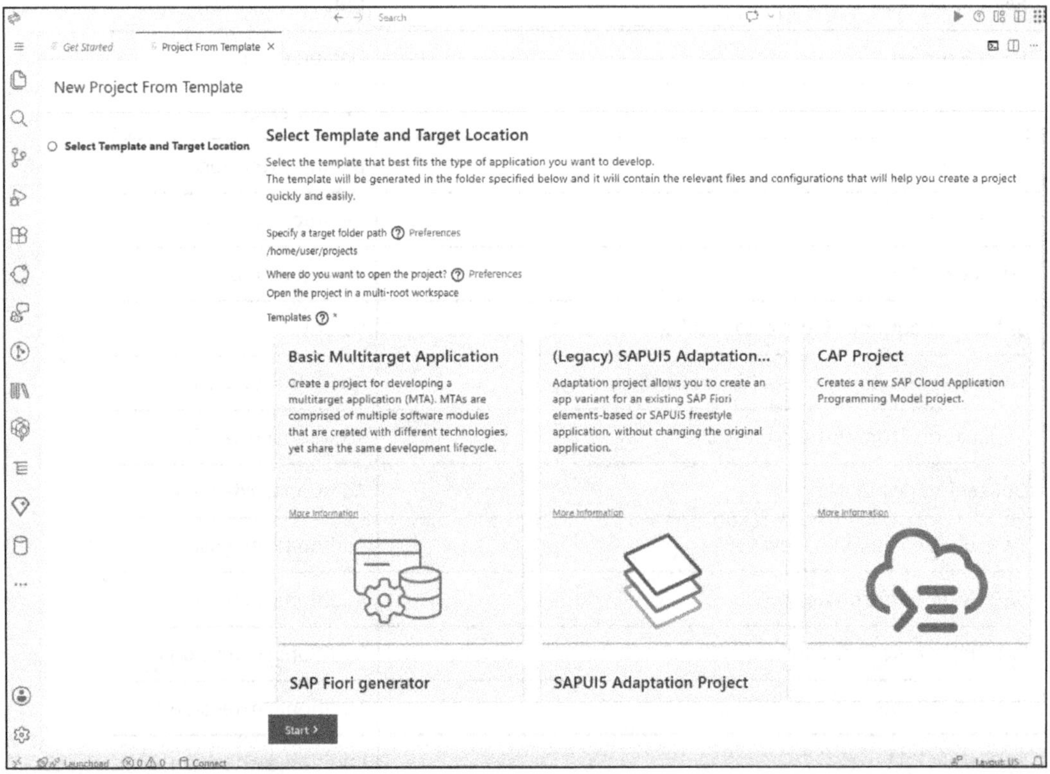

Figure 4.4: SAP Business Application Studio: Dev Space Based on Full Stack Cloud Application

Table 4.5 provides a list of extensions that are installed by default or which can be installed based on need.

SAP Business Application Studio Extension	Type
Basic tools	Default extension
CDS graphical modeler	Default extension
SAP Cloud Application Programming Model tools	Default extension
SAP Fiori freestyle tools	Default extension
SAP HANA database explorer	Default extension
Java tools	Default extension
MTA tools	Default extension
SAP Fiori tools	Default extension
Service center	Default extension
SAPUI5 layout editor and extensibility	Default extension
Application Frontend service (CLI)	Additional extension
Docker image builder	Additional extension
SAP HANA calculation view editor	Additional extension
SAP HANA performance tools	Additional extension
SAP HANA tools	Additional extension
SAP HANA XSA tools	Additional extension
HTML5 runner	Additional extension
Launchpad module	Additional extension
Python tools	Additional extension
SAP HANA smart data integration tools	Additional extension
Development tools for SAP Build Work Zone	Additional extension
Headless testing framework	Additional extension
SAP Business Application Studio extension development	Additional extension
SAP Workflow Management	Additional extension

Table 4.5: Full Stack Cloud Application Dev Space extensions

SAP HANA Native Application Dev Space

This dev space provides editors for the development of SAP HANA database artifacts, like calculation views, tables, stored procedures, synonyms, and table functions, as well as tools for building, testing, and deploying native SAP HANA business applications, as shown in Figure 4.5.

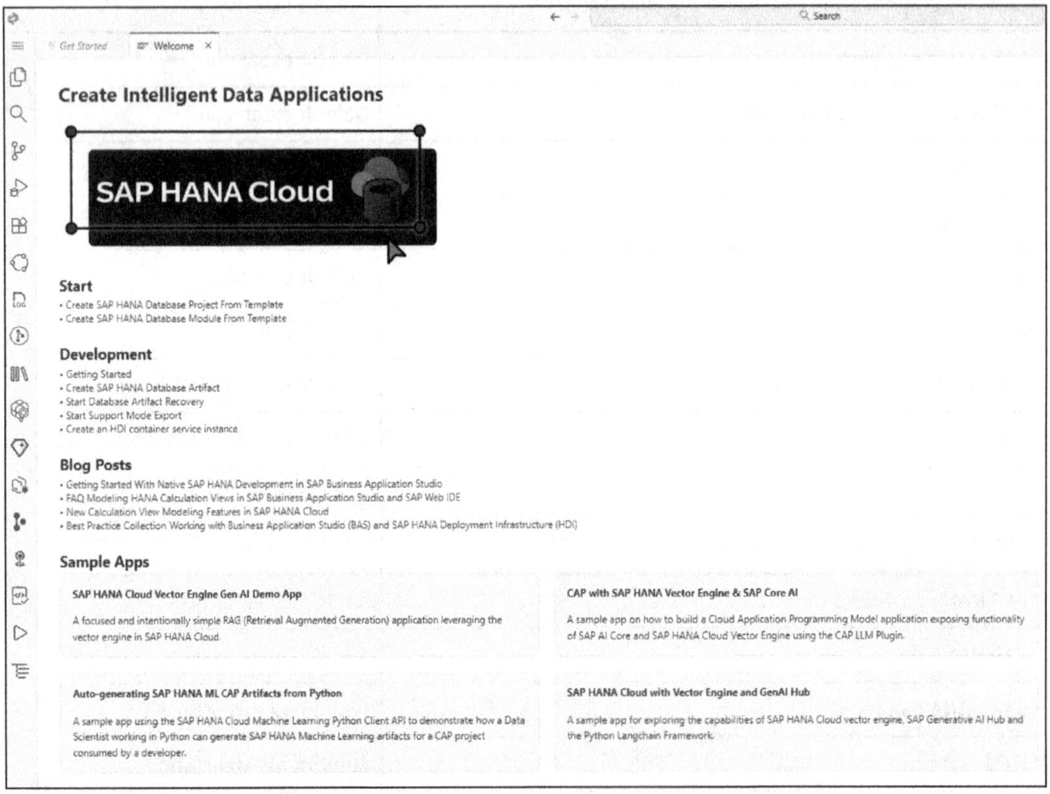

Figure 4.5: SAP Business Application Studio: Dev Space Based on the SAP HANA Native Application Dev Space Type

Table 4.6 provides a list of extensions that are installed by default or which can be installed based on need.

SAP Business Application Studio Extension	Type
SAP HANA calculation view editor	Default extension
SAP HANA database explorer	Default extension
SAP HANA tools	Default extension

Table 4.6: SAP HANA Native Application Dev Space Extensions

SAP Business Application Studio Extension	Type
Basic tools	Default extension
MTA tools	Default extension
Service center	Default extension
SAP HANA smart data integration tools	Default extension
SAPUI5 application project (Cloud Foundry only)	Additional extension
Application Frontend service CLI	Additional extension
Docker image builder	Additional extension
CDS graphical modeler	Additional extension
SAP Cloud Application Programming Model tools	Additional extension
Fiori Freestyle Tools	Additional extension
SAP HANA performance tools	Additional extension
SAP HANA XSA tools	Additional extension
HTML5 runner	Additional extension
Java tools	Additional extension
Launchpad module	Additional extension
Python tools	Additional extension
SAP Fiori tools	Additional extension
SAPUI5 layout editor and extensibility	Additional extension
Development tools for SAP Build Work Zone	Additional extension
Headless testing framework	Additional extension
SAP Business Application Studio extension development	Additional extension
SAP Workflow Management	Additional extension

Table 4.6: SAP HANA Native Application Dev Space Extensions (Cont.)

SAP Mobile Application Dev Space

This dev space provides the MDK for developing and deploying customized iOS and Android apps to SAP Mobile Services, as shown in Figure 4.6.

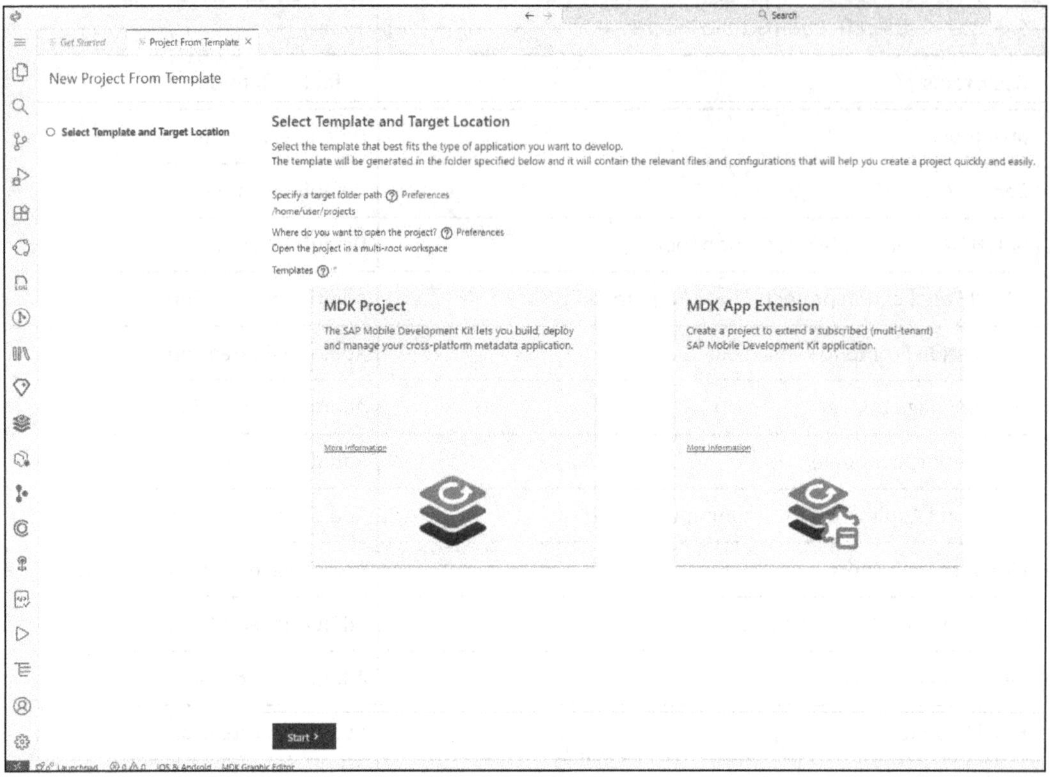

Figure 4.6: SAP Mobile Application Dev Space Extensions

Table 4.7 provides a list of extensions that are installed by default or which can be installed based on need.

SAP Business Application Studio Extension	Type
Basic tools	Default extension
SAP Mobile Services app development tools	Default extension
HTML5 runner	Default extension
Java tools	Default extension
SAP Mobile Services app development extended tools	Default extension
MTA tools	Default extension

Table 4.7: SAP Mobile Application Dev Space Extensions

SAP Business Application Studio Extension	Type
Service center	Default extension
SAP Fiori freestyle tools	Additional extension
SAP Fiori tools	Additional extension
CDS graphical modeler	Additional extension
SAP Cloud Application Programming Model tools	Additional extension
SAPUI5 layout editor and extensibility	Additional extension
SAPUI5 application project (Cloud Foundry only)	Additional extension
Application Frontend service (CLI)	Additional extension
Docker image builder	Additional extension
SAP HANA calculation view editor	Additional extension
SAP HANA performance tools	Additional extension
SAP HANA database explorer	Additional extension
SAP HANA tools	Additional extension
SAP HANA XSA tools	Additional extension
Launchpad module	Additional extension
Python tools	Additional extension
SAP HANA smart data integration tools	Additional extension
Development tools for SAP Build Work Zone	Additional extension
Headless testing framework	Additional extension
SAP Business Application Studio extension development	Additional extension
SAP Workflow Management	Additional extension

Table 4.7: SAP Mobile Application Dev Space Extensions (Cont.)

SAP Basic Tool Extensions Dev Space

This dev space provides the required basic tool extensions, as shown in Figure 4.7.

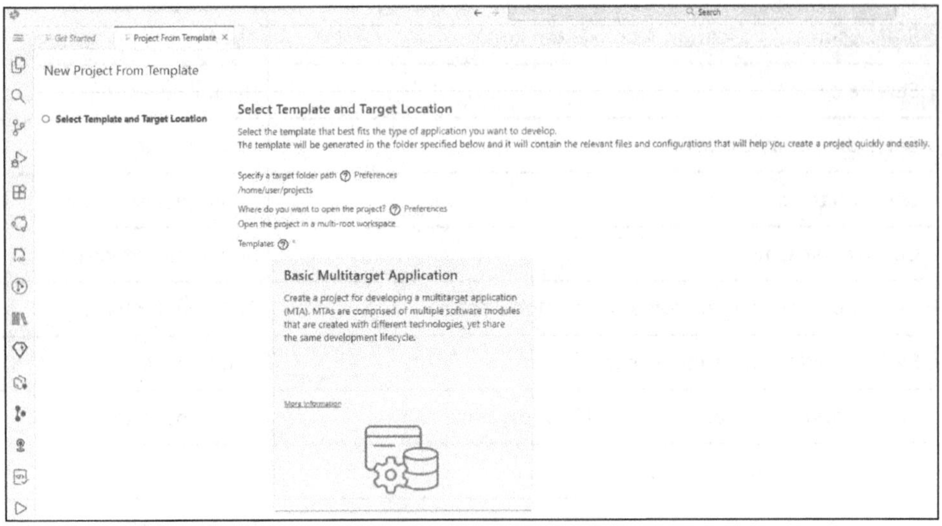

Figure 4.7: SAP Basic Tool Extensions Dev Space

Table 4.8 provides a list of extensions that are installed by default or which can be installed based on need.

SAP Business Application Studio Extension	Type
Basic tools	Default extension
MTA tools	Default extension
Service center	Default extension
CDS graphical modeler	Additional extension
SAP Cloud Application Programming Model tools	Additional extension
SAP Fiori freestyle tools	Additional extension
SAP HANA database explorer	Additional extension
Java tools	Additional extension
SAP Fiori tools	Additional extension
SAPUI5 layout editor and extensibility	Additional extension
Application Frontend service (CLI)	Additional extension
Docker image builder	Additional extension
SAP HANA calculation view editor	Additional extension
SAP HANA performance tools	Additional extension

Table 4.8: Full Stack Cloud Application Dev Space Extensions

SAP Business Application Studio Extension	Type
SAP HANA tools	Additional extension
SAP HANA XSA tools	Additional extension
HTML5 runner	Additional extension
Launchpad module	Additional extension
Python tools	Additional extension
SAP HANA smart data integration tools	Additional extension
Development tools for SAP Build Work Zone	Additional extension
Headless testing framework	Additional extension
SAP Business Application Studio extension development	Additional extension
SAP Workflow Management	Additional extension

Table 4.8: Full Stack Cloud Application Dev Space Extensions (Cont.)

Full Stack ABAP Application Dev Space

The newly introduced ABAP application dev space on SAP Business Application Studio is tailored to support the creation of full stack ABAP applications for modeling business objects using the ABAP Cloud development model and offers a comprehensive set of tools, including project templates, graphical modeling and visualization, service publishing, and UI development to enrich the development process, as shown in Figure 4.8.

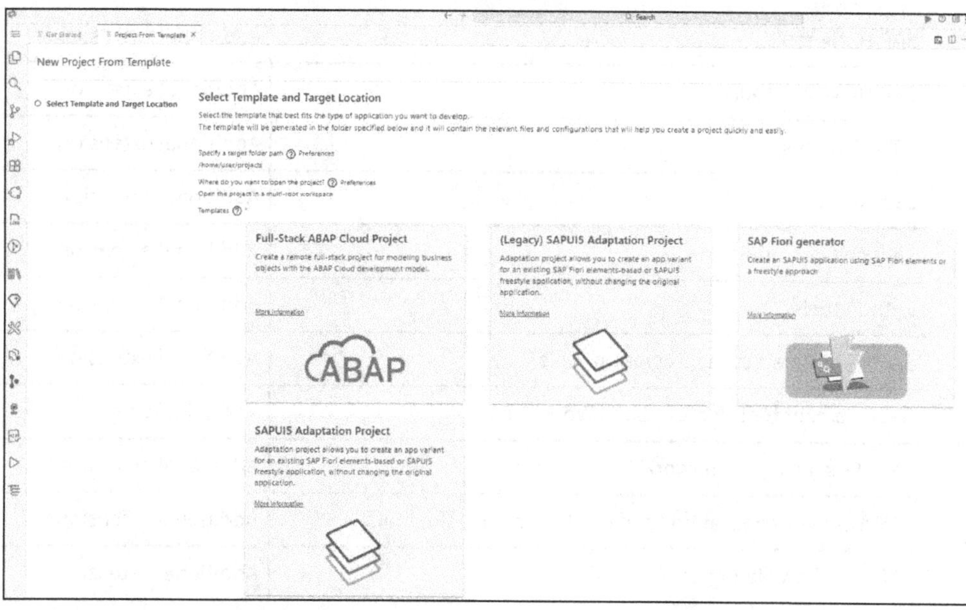

Figure 4.8: ABAP Full Stack Application Dev Space

Table 4.9 provides a list of extensions that are installed by default or which can be installed based on need.

SAP Business Application Studio Extension	Type
Basic tools	Default extension
CDS graphical modeler	Default extension
MTA tools	Default extension
SAP Fiori tools	Default extension
Full stack ABAP tools	Default extension
Service center	Default extension
SAPUI5 layout editor and extensibility	Additional extension
Application Frontend service (CLI)	Additional extension
Docker image builder	Additional extension
SAP Cloud Application Programming Model tools	Additional extension
SAP Fiori freestyle tools	Additional extension
SAP HANA calculation view editor	Additional extension
SAP HANA performance tools	Additional extension
SAP HANA database explorer	Additional extension
SAP HANA tools	Additional extension
SAP HANA XSA tools	Additional extension
HTML5 runner	Additional extension
Java tools	Additional extension
Launchpad module	Additional extension
Python tools	Additional extension
SAP HANA smart data integration tools	Additional extension
Development tools for SAP Build Work Zone	Additional extension
Headless testing framework	Additional extension
SAP Business Application Studio extension development	Additional extension
SAP Workflow Management	Additional extension

Table 4.9: ABAP Full Stack Application Dev Space Extensions

4.2.4 Creating Dev Spaces in SAP Business Application Studio

In this section, we'll focus on creating an SAP Business Application Studio dev space in an SAP BTP trial account. The following steps are similar to those you'd follow when creating an SAP Business Application Studio dev space in an enterprise account:

[+]

Tip

You must create your trial account before proceeding with the next steps. During the creation of a trial account, a subaccount called "trial" is created; SAP BTP, Cloud Foundry environment is enabled; and a space called "dev" is created.

1. Log on to your SAP BTP trial account via the URL *https://cockpit.hanatrial.onde mand.com/trial/#/home/trial*, as shown in Figure 4.9.

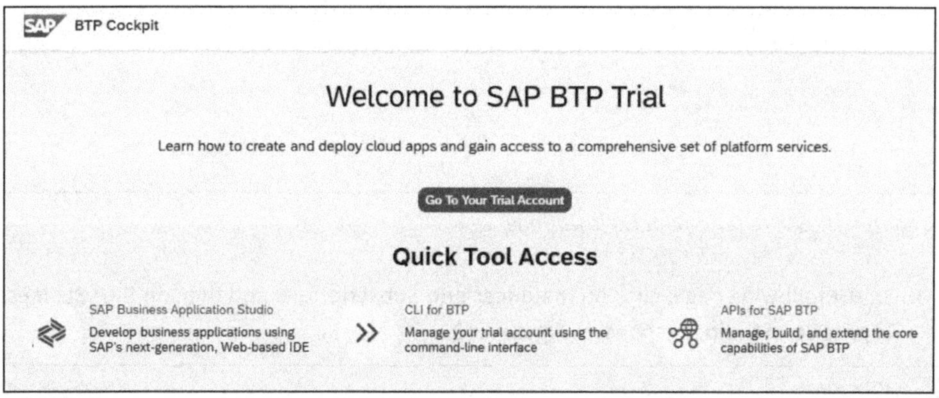

Figure 4.9: SAP BTP: Trial Account Logon Screen

2. SAP Business Application Studio can be launched by clicking on **SAP Business Application Studio**, which will open the solution in a new browser window, as shown in Figure 4.10.

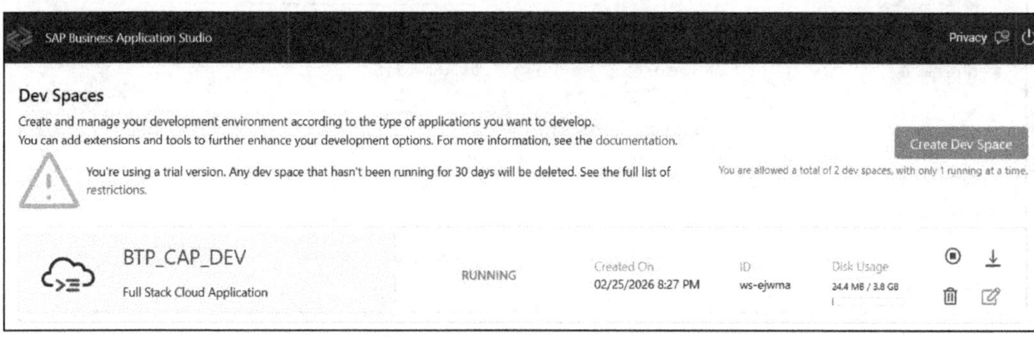

Figure 4.10: Launching SAP Business Application Studio

3. SAP Business Application Studio can also be launched by clicking on the **Go To Your Trial Account** button (shown earlier in Figure 4.9) and then clicking on **trial** under the **Subaccounts** tab, as shown in Figure 4.11.

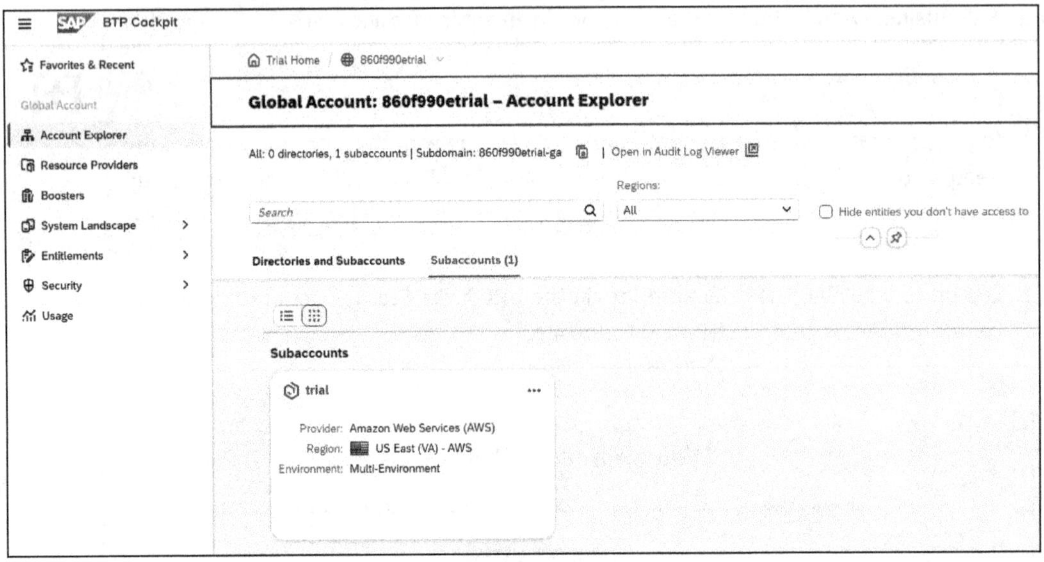

Figure 4.11: SAP BTP: Account Explorer Page

4. On the following page, click on **Instances and Subscriptions** and then on **SAP Business Application Studio**, as shown in Figure 4.12.

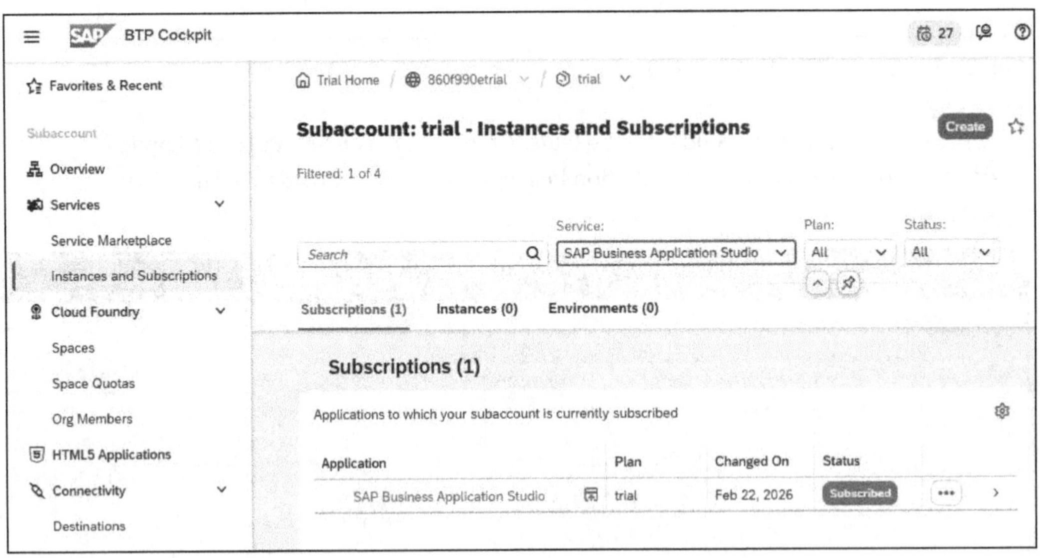

Figure 4.12: Subaccount Page Displaying Instances and Subscriptions

5. By clicking **SAP Business Application Studio**, you'll be taken to SAP Business Application Studio landing page, which we saw in Figure 4.10.

[+]

Tip

Bookmark the URL of your SAP Business Application Studio web page so that you can launch SAP Business Application Studio directly through the URL.

If your subaccount lacks a subscription to the SAP Business Application Studio service, follow these steps:

[«]

Note

Since the trial subaccount we are using already has a subscription to the SAP Business Application Studio service, for our examples, we created another subaccount called "SAPBAS."

1. Navigate to the **Subaccount** page in the SAP BTP cockpit, as shown in Figure 4.13.

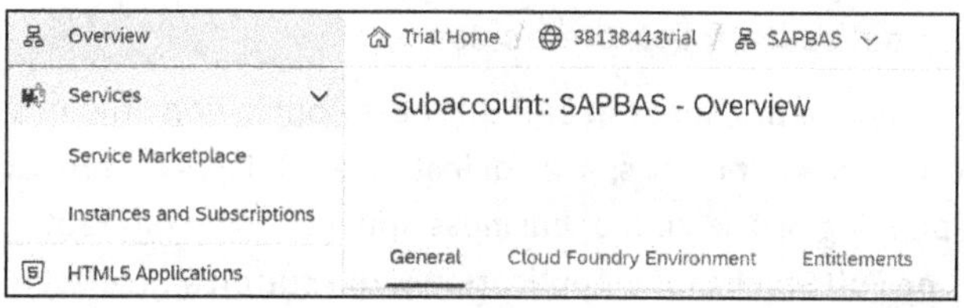

Figure 4.13: SAP Business Application Studio: Subaccount Page

2. Click on **Service Marketplace** in the pane on the left and then enter "STUDIO" into the search box. Click on **Create** as shown in Figure 4.14.

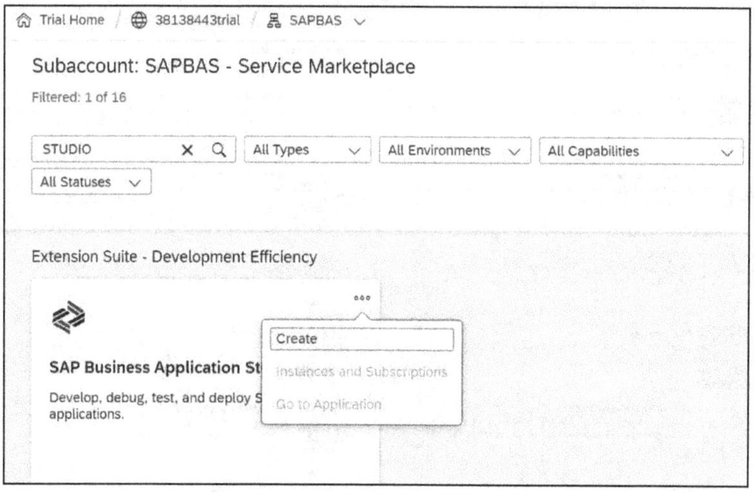

Figure 4.14: Creating an SAP Business Application Studio Service Instance

3. From the **Service** dropdown list, select **SAP Business Application Studio**, and from the **Plan** dropdown list, select **trial**, as shown in Figure 4.15, and then click **Create**.

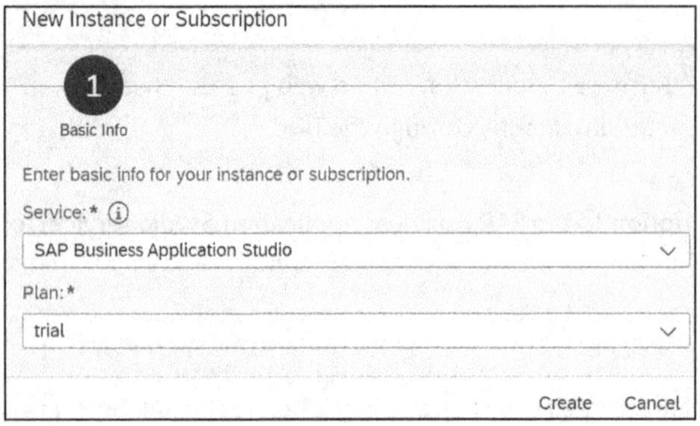

Figure 4.15: Creating an SAP Business Application Studio Service Instance with a Plan

4. The confirmation of an SAP Business Application Studio subscription in your subaccount is then displayed.

Returning to the task of creating a dev space, follow these steps:

1. Click on **Create Dev Space** to create a dev space in SAP Business Application Studio, as shown in Figure 4.16. Provide a name for the dev space, for instance, "BTP_CAP". You can select the dev space type depending on the kind of business application being created. For this exercise, we'll select the **Full Stack Cloud Application** radio button and select all the checkboxes under **Additional SAP Extensions**. Click on the **Create Dev Space** button.

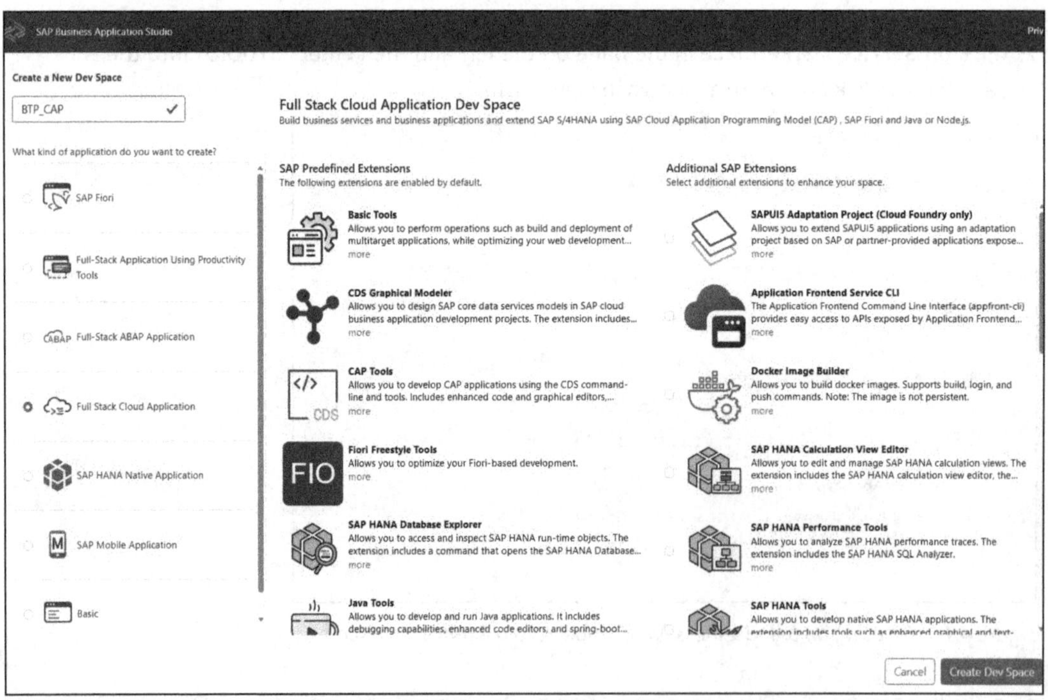

Figure 4.16: SAP Business Application Studio: Creating a Dev Space

Tip

Extensions cannot be removed after a dev space has been created.

2. After a dev space has been created, it will have a status of **Running**, as shown in Figure 4.17.

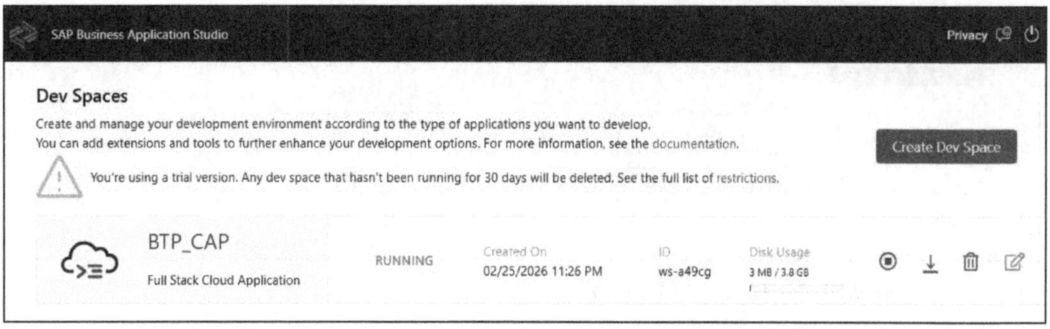

Figure 4.17: SAP Business Application Studio: Dev Space Running

Note

Following are some of the icons that you can use in the dev space you have created:

- starts a dev space.
- stops a dev space.
- downloads the dev space's content.
- deletes a dev space (along with its content).
- enables you to edit a dev space (i.e., adding extensions).

3. Click on the dev space name (i.e., BTP_CAP) to navigate to the welcome page for the dev space, as shown in Figure 4.18.

Warning

The space allocated for a dev space in an enterprise account is 10 GB, but for a trial account, the allocation is only 4 GB. If the content of a dev space exceeds its limit, you may encounter issues when working in the dev space.

Note

Each time you open the dev space, the installation of the relevant extensions begins. This may take some time, delaying when you can start working on application development.

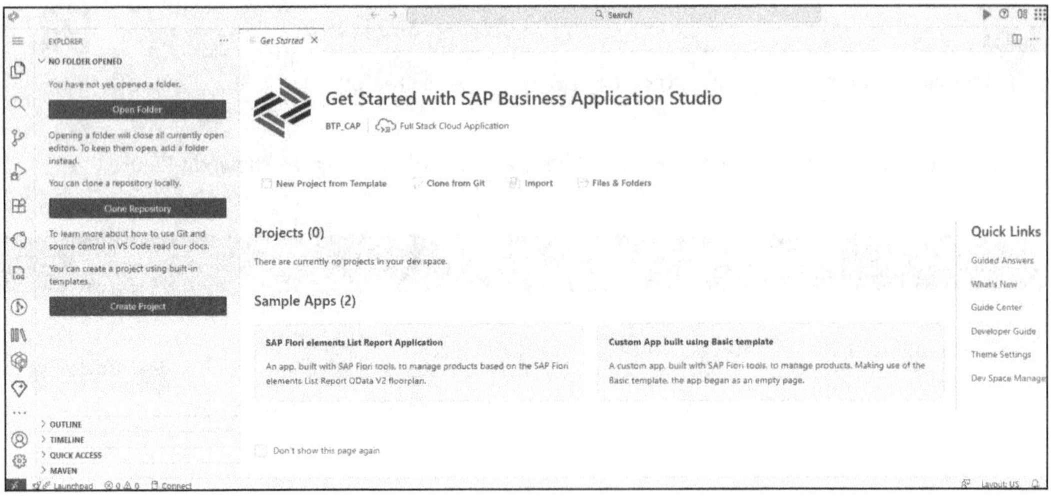

Figure 4.18: Dev Space Welcome Page

4.2.5 Workspaces Inside Dev Spaces

A workspace is a collection of one or more projects that can be opened in the file explorer in SAP Business Application Studio. When a workspace has multiple projects, potentially saved in different file locations or even in different workspaces, this kind of workspace is called a multiroot workspace. Workspaces provide isolation in terms of configuration settings, debugging configurations, and task configurations and can store and restore UI states. A new workspace in a dev space can be created by selecting **File • Save Workspace As...**, as shown in Figure 4.19.

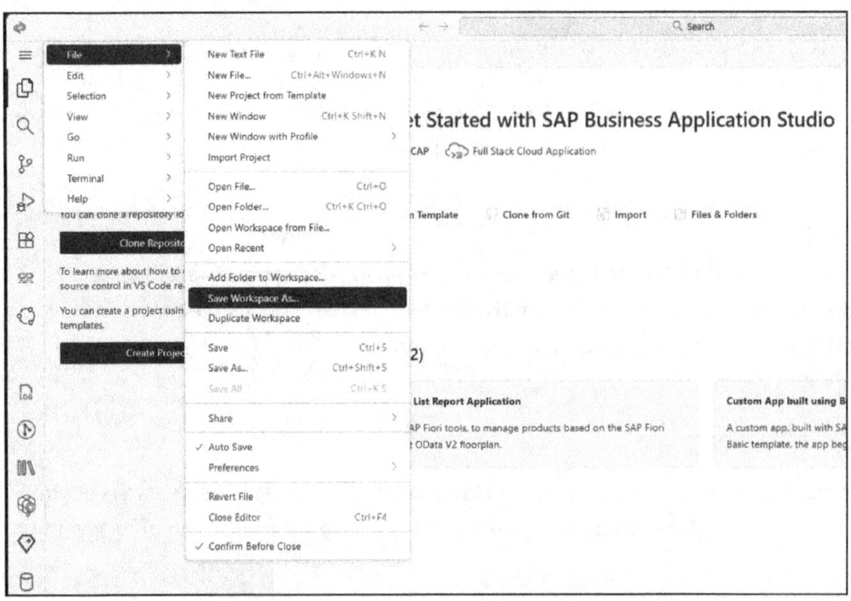

Figure 4.19: SAP Business Application Studio: Creating a New Workspace in a Dev Space

Next, make a selection in the **Format** dropdown list and provide a name for your workspace, as shown in Figure 4.20.

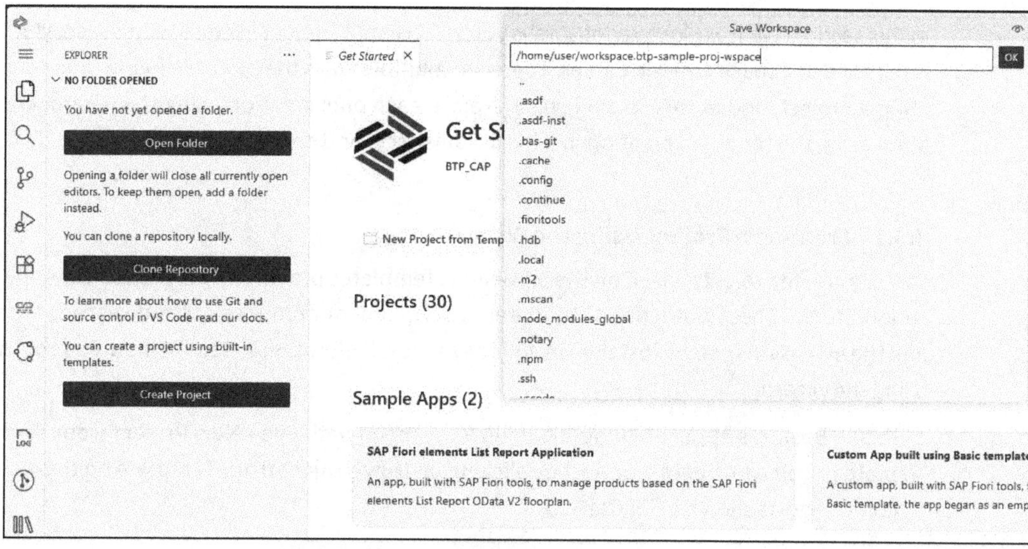

Figure 4.20: Saving a New Workspace: Providing Details

As soon as the workspace is saved, you will be automatically placed inside your newly created workspace and you can find the workspace info at the top of the screen, as shown in Figure 4.21.

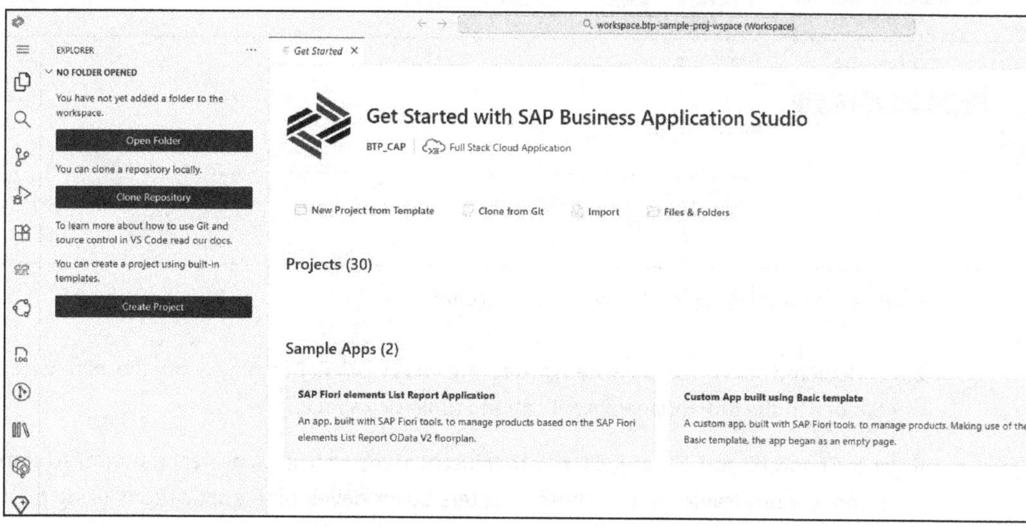

Figure 4.21: Inside Workspace

Now that we have established our dev space, let's step forward to understand the relevant artifacts inside SAP Business Application Studio and the process of setting up your first SAP BTP project!

4.3 Developing Your First Cloud Application Using SAP Business Application Studio

In this section, we'll explore various ways you can create a project inside a workspace which is part of a dev space created earlier. There are multiple ways that you can follow while creating a project, and in this section, we'll explore each one: creation using the wizard, creation using the terminal, creation in a local workspace, and creation via import.

4.3.1 Creating a Project Using the Wizard

You can create projects based on the predefined templates provided by SAP Business Application Studio. These predefined templates are dependent on the selected dev space type and the extensions you've installed in your dev space. Follow these steps to create a project using the wizard:

1. In SAP Business Application Studio, follow the menu path **File • New Project from Template** or you can create a project by clicking on **New Project from Template** on the welcome page as shown in Figure 4.22.

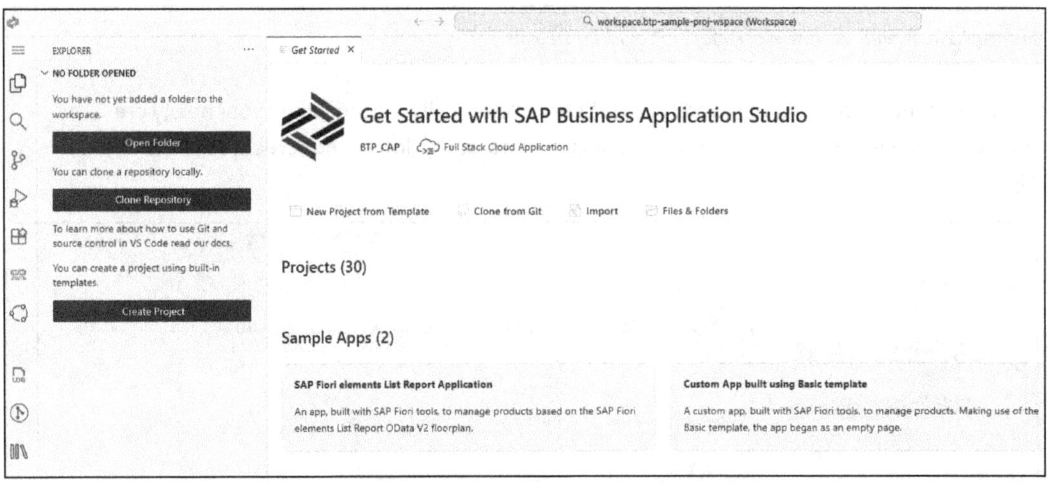

Figure 4.22: Creating Project Using via the Welcome Page

2. As depicted in Figure 4.23, templates are made available based on the extensions installed in the SAP Business Application Studio dev space.

Refer to Chapter 7 and Chapter 8 for further instructions on using a project template to start your application development. These chapters cover developing applications using a full stack application dev space and Node.js and Java runtimes extensively, so we won't go into further details in this chapter.

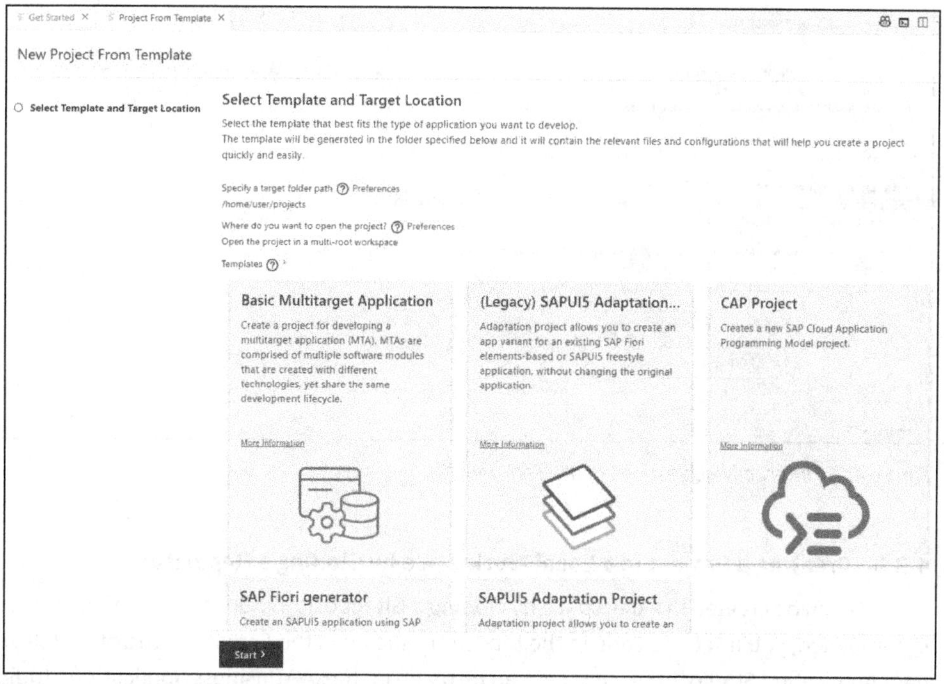

Figure 4.23: Template and Target Location for the Project Created Using Wizard

4.3.2 Creating a Project Using the Terminal

You can also create projects based on the development scenario and extensions installed in your dev space. To create a project using terminal, follow these steps:

1. Open a new terminal by following the menu path **Terminal • New Terminal**.
2. In the terminal window, enter "yo" and select a project type by pressing the Up and Down arrows, as shown in Figure 4.24.

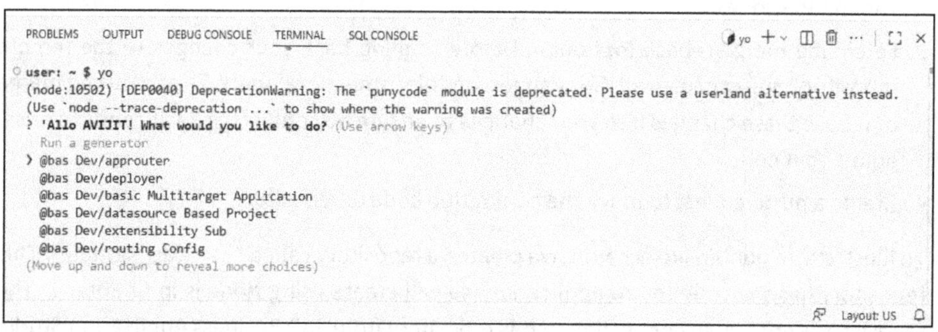

Figure 4.24: Project Types During the Creation of a Project via the Terminal

3. Answer the interactive questions asked by the terminal. An example is shown in Figure 4.25. We will cover this process in more detail in Chapter 7 and Chapter 8.

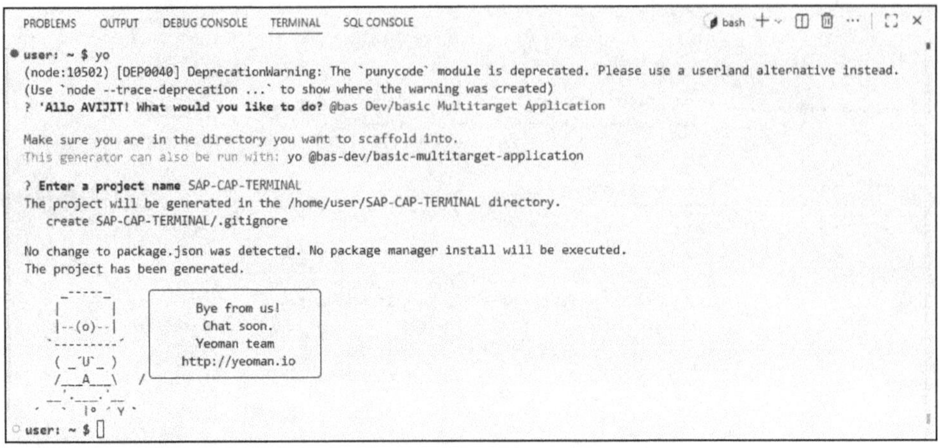

Figure 4.25: Interactive Questions in the Terminal

4.3.3 Creating a Project in a Local Workspace by Cloning a Repository

Projects can be created in a dev space by cloning a Git repository. Once the cloning process is completed, you'll see the code in the repository shown in the **Explorer** pane of SAP Business Application Studio. Code changes can be made in the SAP Business Application Studio explorer, and these changes will be tracked through source control.

Changes can be committed to the repository from your local project or workspace by following these steps:

1. Clone the repository via the Git source control system. A local master branch is created and is visible in the workspace. If you have additional branches, you can create local branches for each remote branch.
2. Add, modify, or delete the code in the cloned project. The changes made are visible in the **Source Control:Git page** in SAP Business Application Studio. Stage your changes and commit the code.
3. Fetch and merge/rebase (optional): Before sending back your changes to the remote repository, you can fetch all the changes made by other developers. Then, you can merge or rebase these changes into your changes to ensure no conflicts arise. If conflicts arise, adjust your code.
4. Create a push request to move the committed code to repository.

To illustrate, in our GitHub account, we created a repository called "btp_cap_sample". This repository has the code for an application we will create using Node.js in Chapter 7. The repository is *private*, so connecting to this repository from SAP Business Application Studio will require a password or access token. Any GitHub user can be provided access to this repository, and each user can generate their own access token for connecting to the repository. Access tokens are preferred to passwords so that access to the repository is not exposed.

Note

In this example, we assume that the user cloning the repository in SAP Business Application Studio has a user created in GitHub and has access to the repository.

Create a project by cloning a Git repository by following these steps:

1. Log on to GitHub with the URL for the repository, as shown in Figure 4.26.

Figure 4.26: Logging on to the GitHub Repository

2. Click on the user profile icon and select **Settings • Developer Settings**, as shown in Figure 4.27.

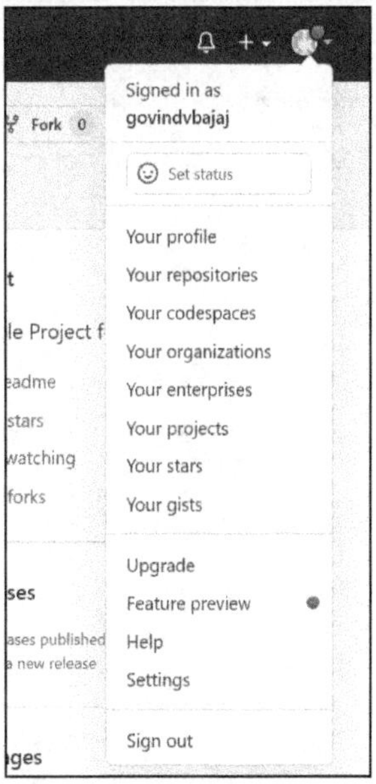

Figure 4.27: User Profile in GitHub

3. Next, click on **Personal access tokens**, as shown in Figure 4.28.

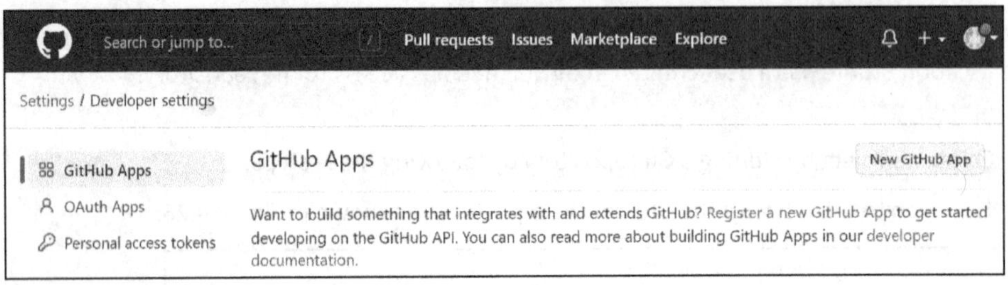

Figure 4.28: Navigation Path to Personal Access Tokens

4. Click on **Generate new token**, as shown in Figure 4.29.

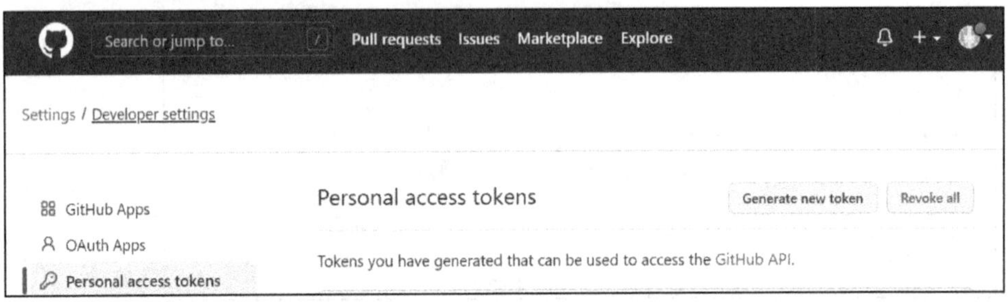

Figure 4.29: Generate a New Token

5. Provide the name of the token in the **Note** field and, in the **Select scopes** section, select the checkboxes for **repo, gist**, and **write:discussion**, as shown in Figure 4.30.

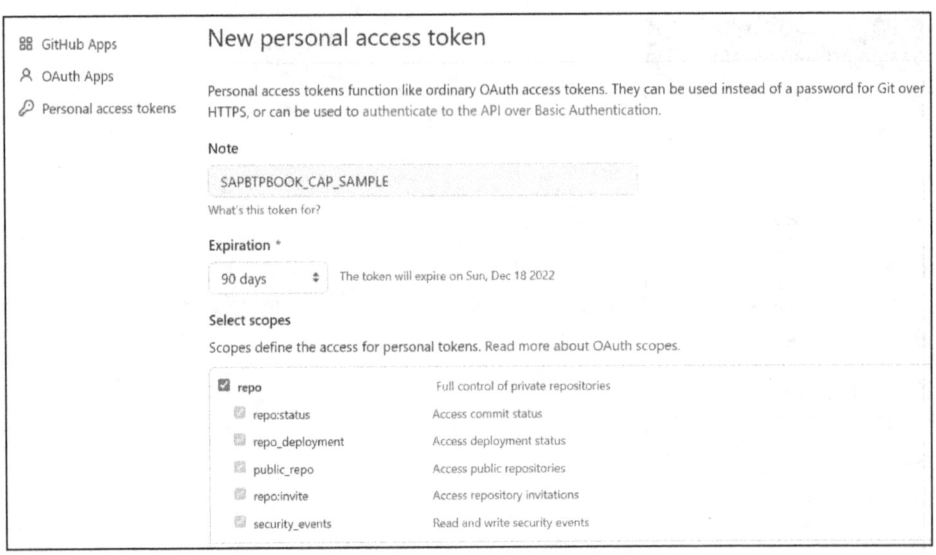

Figure 4.30: Creating a Personal Access Token

6. Click on **Generate token**, as shown in Figure 4.31.

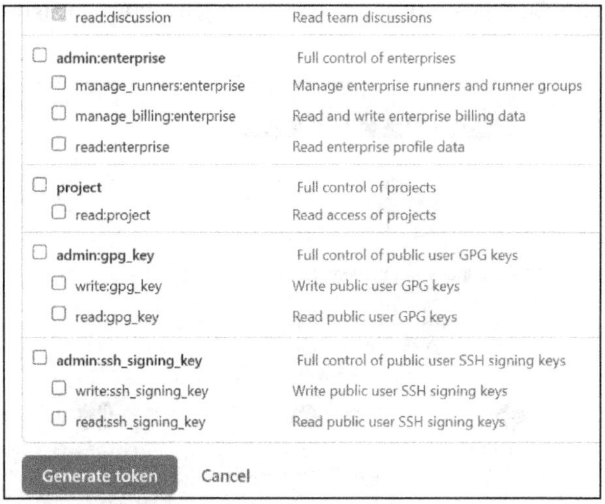

Figure 4.31: Generating an Access Token

7. Copy the access token by clicking the icon, as shown in Figure 4.32.

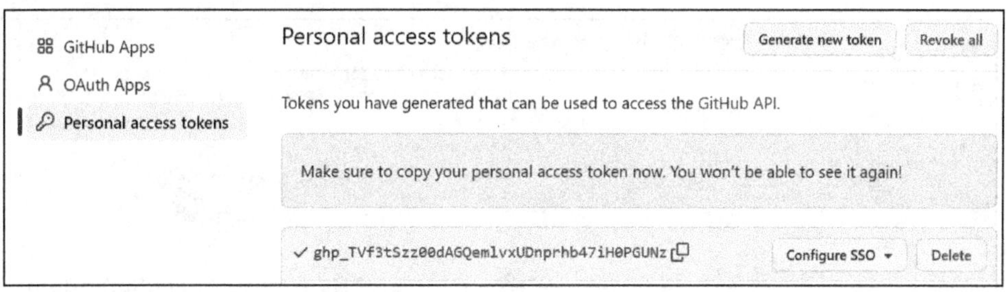

Figure 4.32: Copying a Personal Access Token

8. In the next step, we'll connect to the repository from SAP Business Application Studio. Start by clicking on the **Clone from Git** tile, as shown in Figure 4.33.

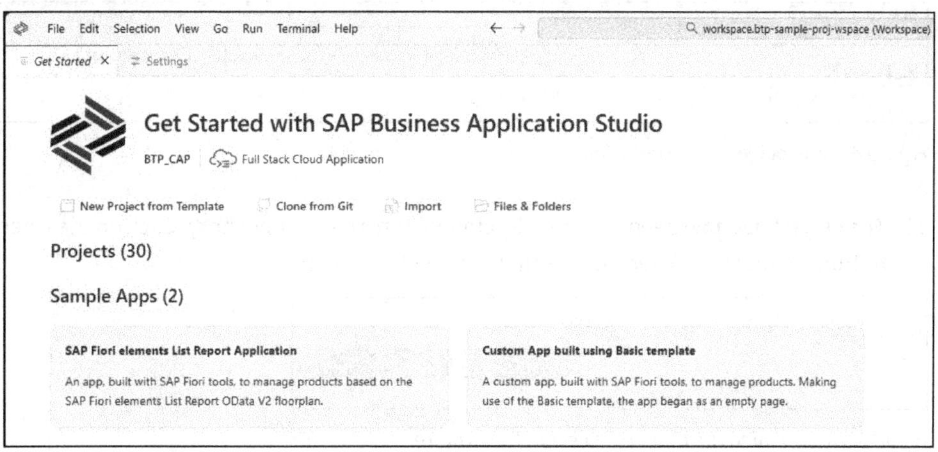

Figure 4.33: Clone from Git Tile on the Welcome Page

9. Enter the URL associated with the Git repository and press Enter, as shown in Figure 4.34.

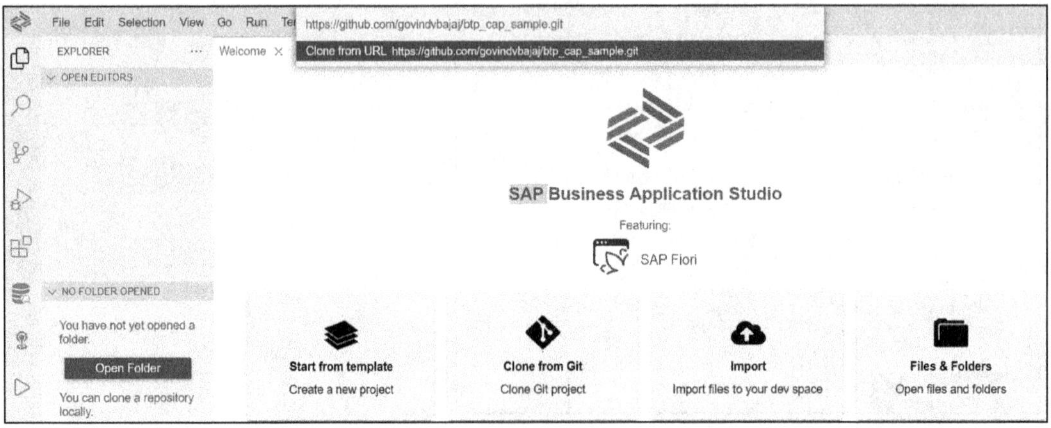

Figure 4.34: Providing the URL for the Git Repository

10. Provide your Git username and press Enter, as shown in Figure 4.35.

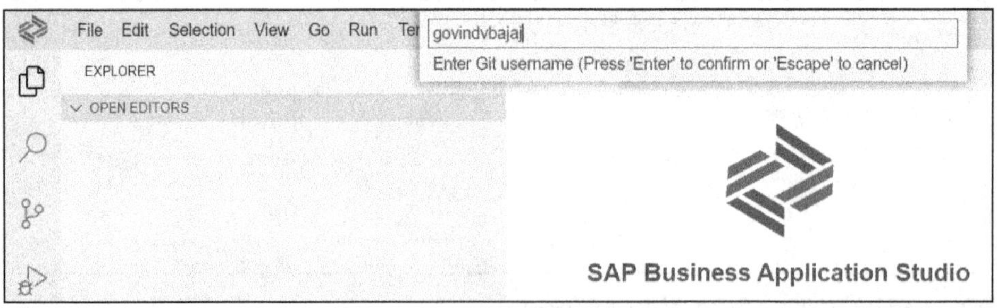

Figure 4.35: Providing a Git Username

11. Provide the access token copied earlier and press Enter, as shown in Figure 4.36.

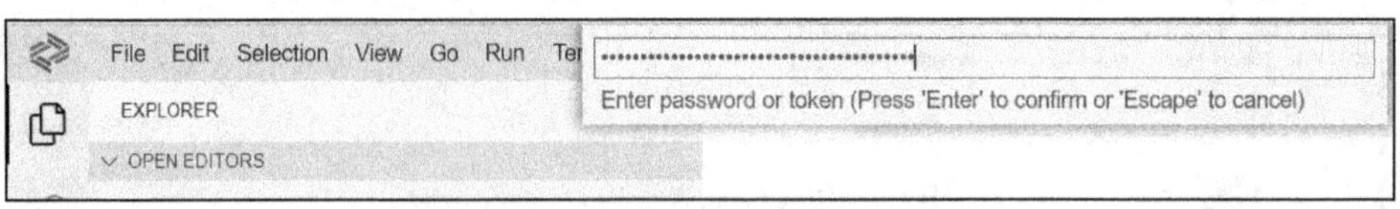

Figure 4.36: Providing an Access Token

12. The project has now been successfully cloned from the Git repository. Click on the **Open** button, as shown in Figure 4.37, and the project will open.

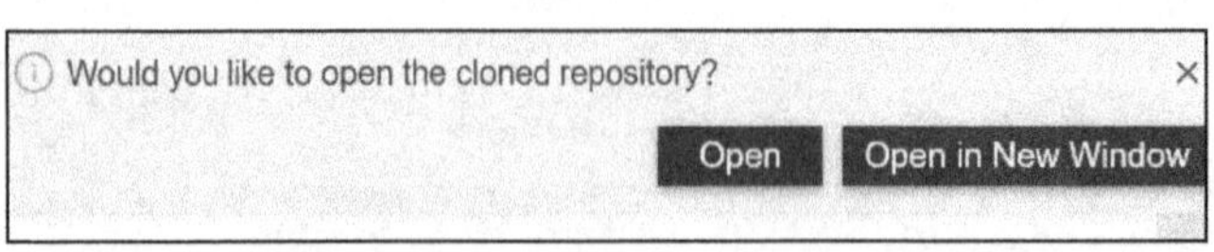

Figure 4.37: Opening the Cloned Project in the Explorer

4.3.4 Creating a Project by Importing a Project

A project can be downloaded from a dev space as a TAR zip file that can be imported into any other dev space. To create a project by importing a project, follow these steps:

1. From the welcome page, click on the **Import** tile, as shown in Figure 4.38. Alternatively, you can follow the menu path **View • Command Palette**, enter "Import" and then select **SAP Business Application Studio:Import**.

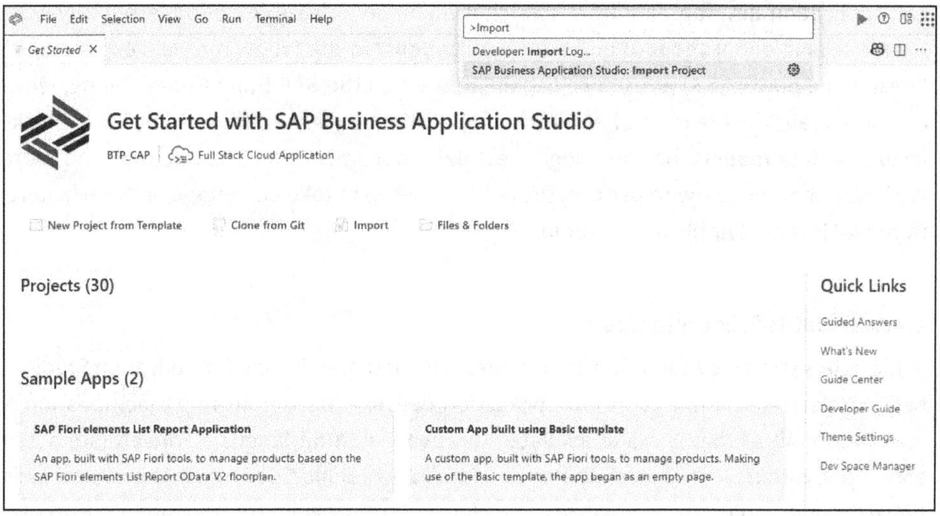

Figure 4.38: Import Through the Command Palette

2. Navigate to the folder that contains the file you wish to import and click **Open**.
3. Now that the project is successfully imported, click on **Open in New Workspace**, as shown in Figure 4.39.

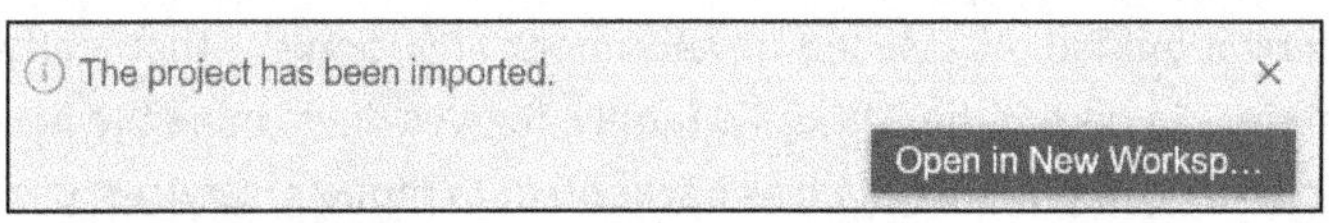

Figure 4.39: Confirmation of Project Import

4. The explorer shows the project created by the import, as shown in Figure 4.40.

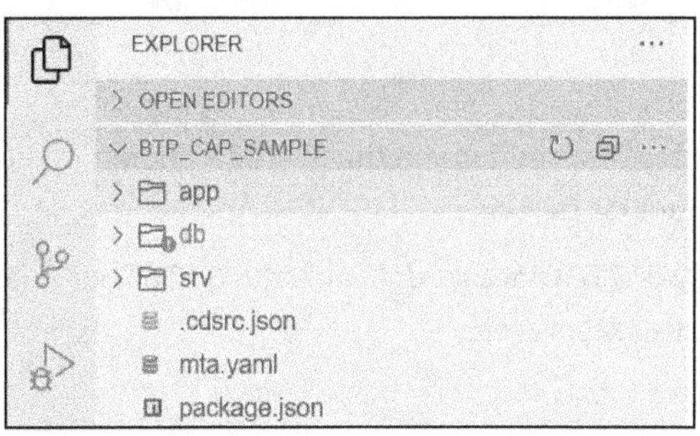

Figure 4.40: Project Created by Import

4.4 Leveraging SAP Build Code and Joule

In the era of generative AI, project creation within SAP Business Application Studio is no longer limited to templates and wizards. AI solutions are now working together to assist developers in creating all potential artifacts, including project setup, data model creation, code, scripts, and test data generation at various stages of the project management lifecycle. With increasing demand, shortened go-to-market times, and a lack of skilled developers, application development from scratch is no longer a feasible option, especially when we are talking about modern and edge applications. In the following, we'll explore SAP's latest Joule powered AI assistance capabilities within the SAP Build Code offering, which can accelerate the creation of enterprise applications by generating key building blocks, including data models, backend logic, test data, documentation, test scripts, and more. We'll also show you how to perform the required setup to take advantage of these features from SAP Business Application Studio.

4.4.1 What Is SAP Build Code?

Joule, SAP's strategic AI solution, is also integrated into SAP Business Application Studio to help accelerate different stages of application delivery, starting from project creation to integration. All of this is made available through a natural language–driven interaction embedded within SAP Business Application Studio. SAP Build Code offers AI-powered assistance through Joule to increase developer efficiency by providing code generation, accelerated API access, CI/CD integration, security functionality, and automation of collaboration and enterprise application deployment.

With SAP Build Code, building custom developments and extensions based on SAP's clean core strategy can be accomplished through simple natural language statements. This approach brings together business users and professional developers to collaborate and gain productivity by enabling a unified, AI-assisted development environment that works seamlessly across SAP systems. As part of the SAP Build family, it gives developers the flexibility to generate various critical components, such as backend data models, services, user interfaces, test data, complex business logic, and documentation, using a blend of AI, low-code, and pro-code tools—requiring only a simple prompt explaining the application requirements end to end.

4.4.2 Set Up SAP Build Code on SAP BTP

To get started with SAP Build Code, we need to first set it up on the SAP BTP tenant; in this case, we are going to leverage our trial account, as follows:

1. As shown in Figure 4.41, navigate to your SAP BTP trial account using the **Go To Your Trial Account** option and then navigate to the **Boosters** area.

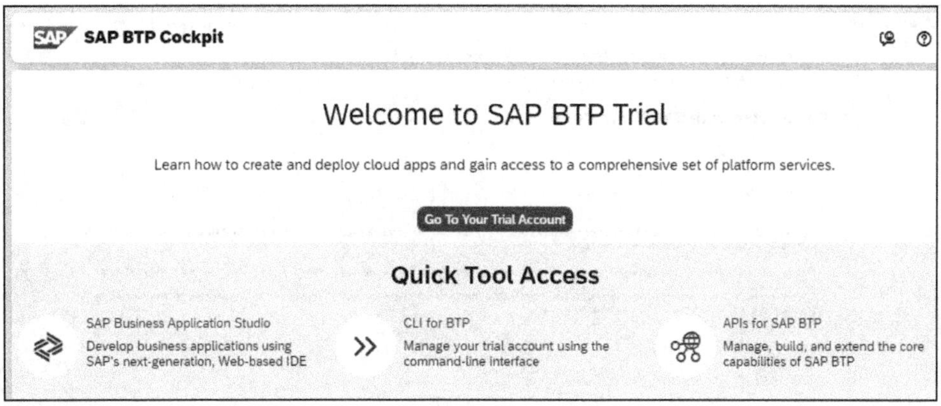

Figure 4.41: Login to Trial Account

2. Search for "Build Code" as shown in Figure 4.42, and you will see the booster for SAP Build Code setup called **Get Started with SAP Build Code.**

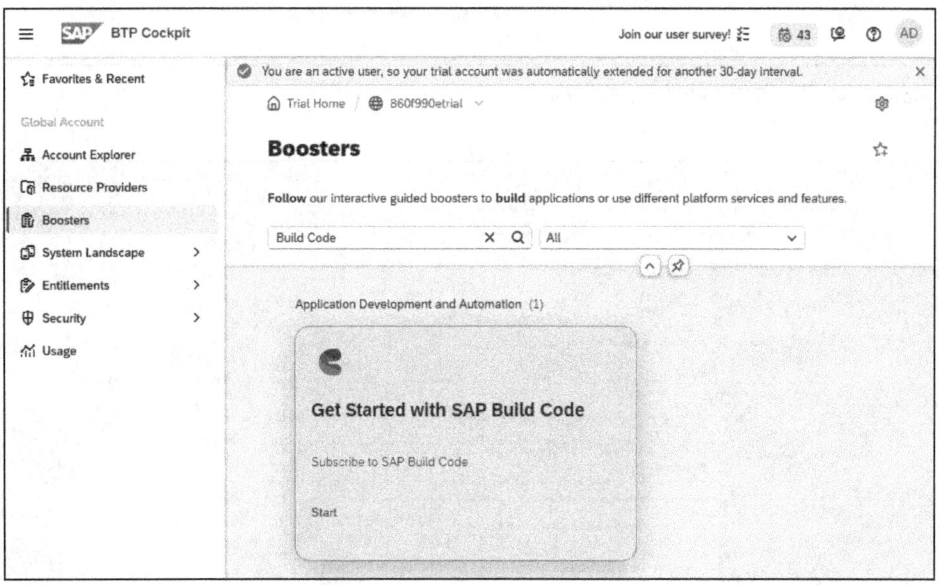

Figure 4.42: SAP Build Code Booster

3. Clicking the tile will take you to next screen, shown in Figure 4.43, where you can get to know SAP Build Code, its components, and the resources required for it to run.
4. Click the **Start** button to initiate the booster. A **Booster executed successfully** message will appear once it has run, as shown in Figure 4.44.

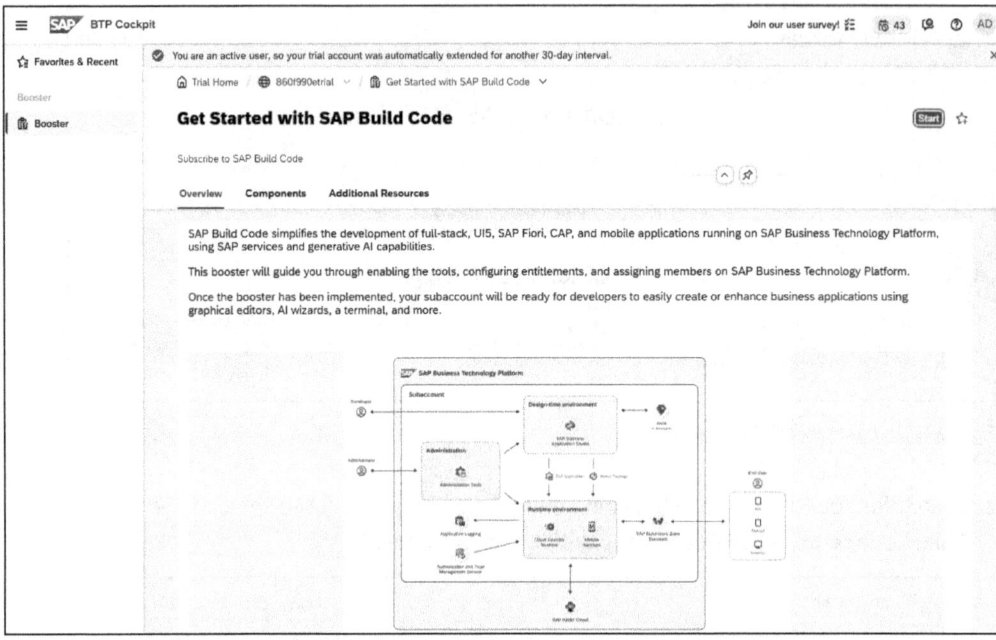

Figure 4.43: Run SAP Build Code Booster

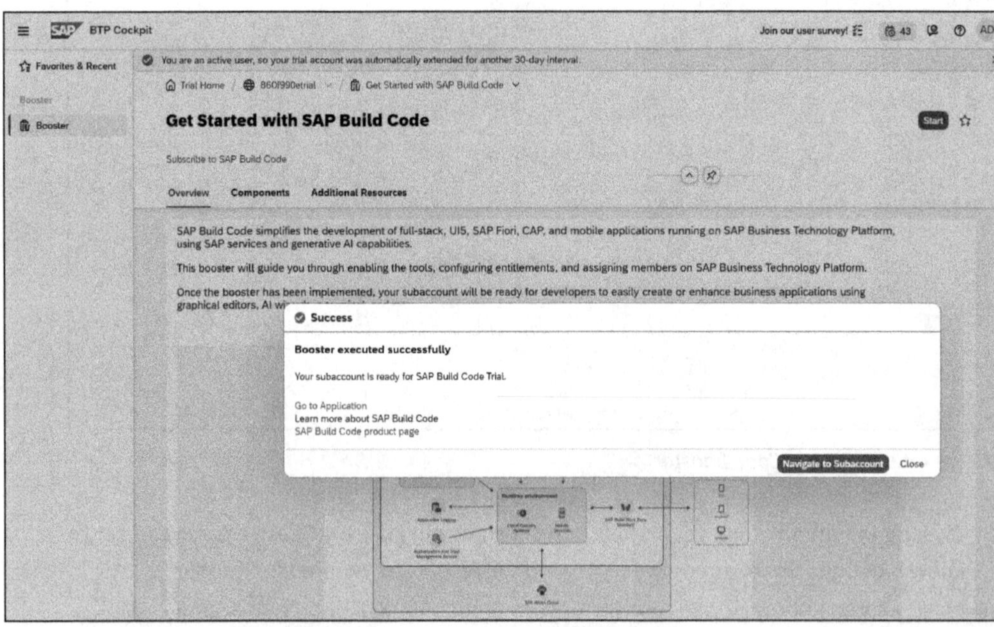

Figure 4.44: Successful Booster Execution

You can now choose the **Navigate to Subaccount** option to directly navigate to your subaccount and validate the setup. As shown in Figure 4.45, inside the **Subscriptions** tab, it shows **SAP Build Code** as a subscribed application and clicking on it launches the SAP Build lobby.

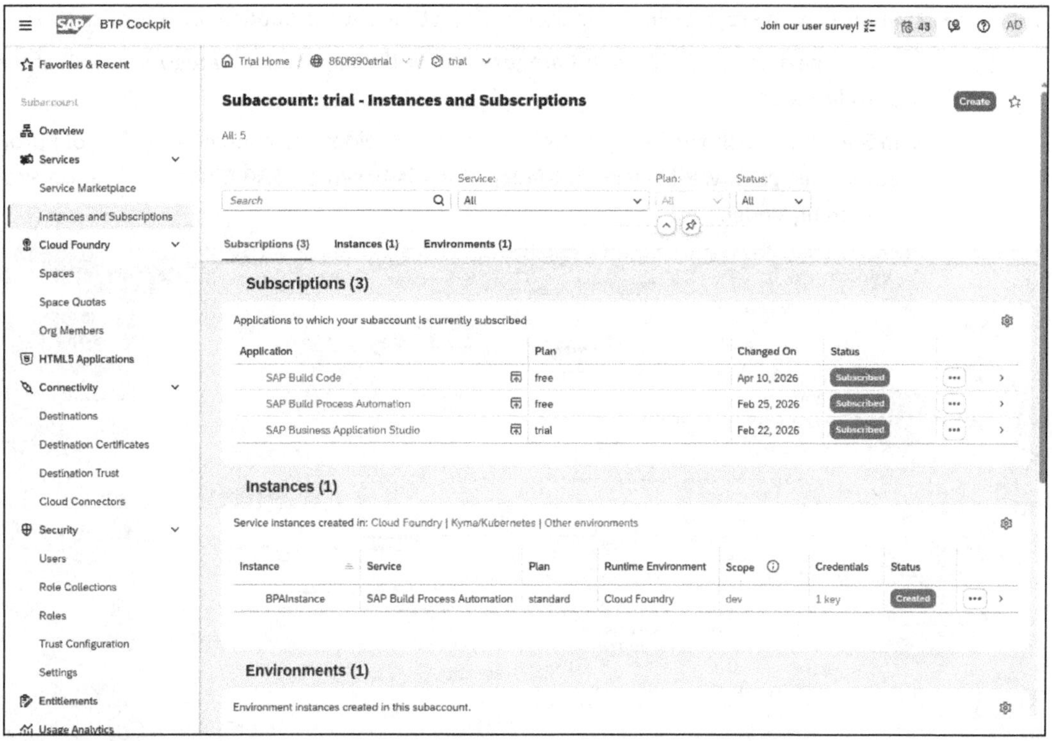

Figure 4.45: SAP Build Code Subscription

4.4.3 Create a Project Using SAP Build Code

To create a project, follow these steps:

1. Click on the **Create** option in the top right corner of the **Items** table, as shown in Figure 4.46.

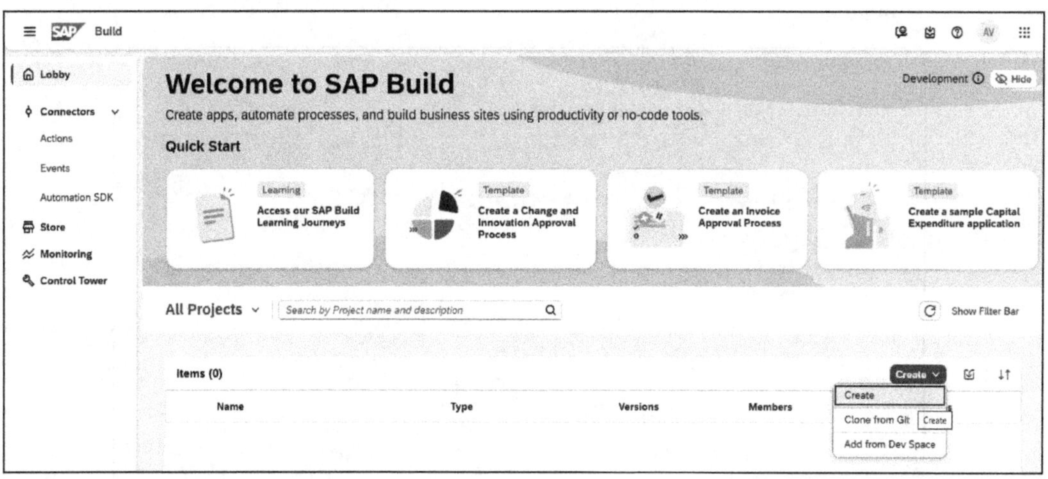

Figure 4.46: Create Project Using SAP Build Code

2. In the **Create Project** window in **Step 1: Objective**, choose **Application** and click **Next.**
3. On the next screen for **Step 2: Category**, select the application **Category** as **Full-stack** and click **Next.**
4. In **Step 3: Type**, shown in Figure 4.47, select the development configuration for your project. In this case, we'll select **Full-Stack Node.js** based on SAP Cloud Application Programming Model.

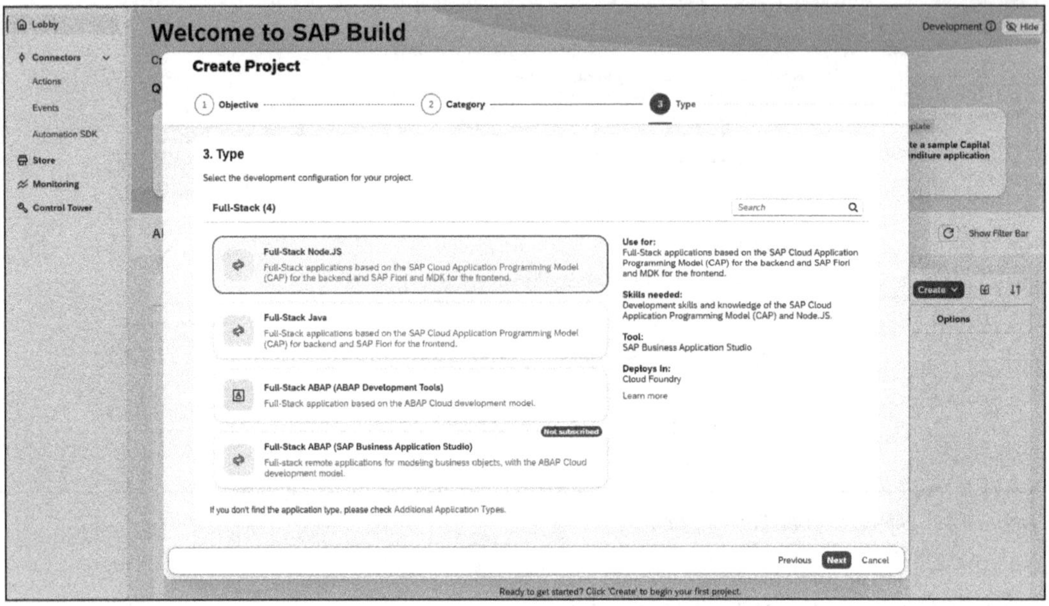

Figure 4.47: Select Development Configuration

5. Figure 4.48 shows the screen where you need to enter the **Name** of your application, its **Description,** and its **Dev Space**. You can either create a **New** or choose from a list of **Existing** ones.

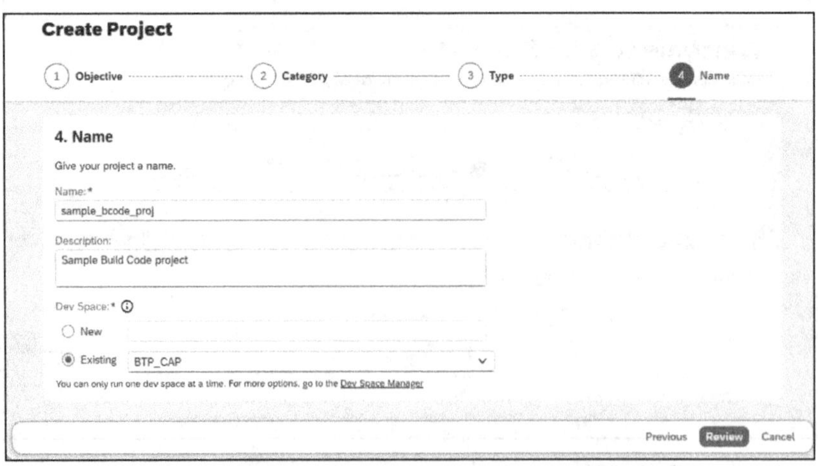

Figure 4.48: Enter Application Name and Dev Space

6. Finally, the project gets created and appears inside the SAP Build lobby, as shown in Figure 4.49. Its initial status will be **Pending Creation** as long as the project creation is still in progress.

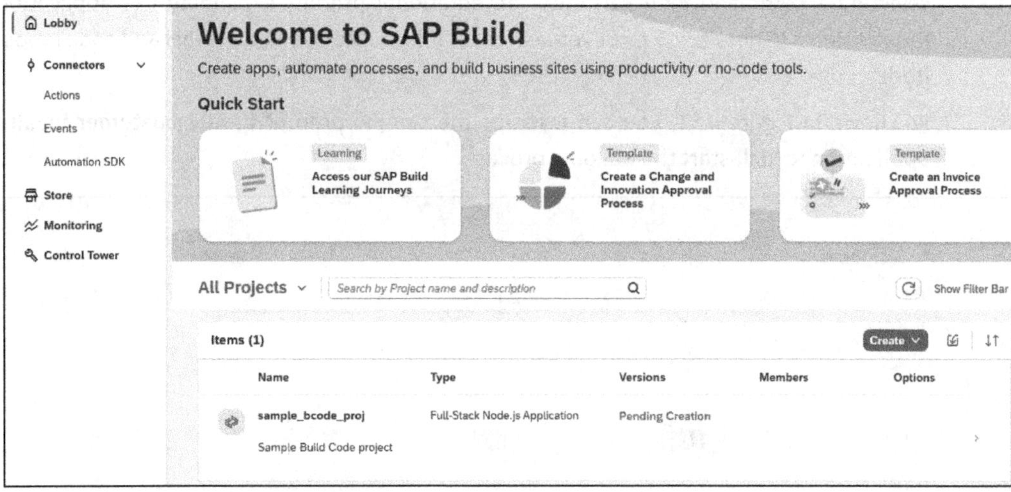

Figure 4.49: Project Creation Successful

Clicking the project name launches the storyboard inside SAP Business Application Studio with the Joule interaction window turned on. You can open the Joule editor by clicking the icon.

Inside the storyboard shown in Figure 4.50, the left side panel shows the Joule editor with multiple assistive commands that you can choose from to start building.

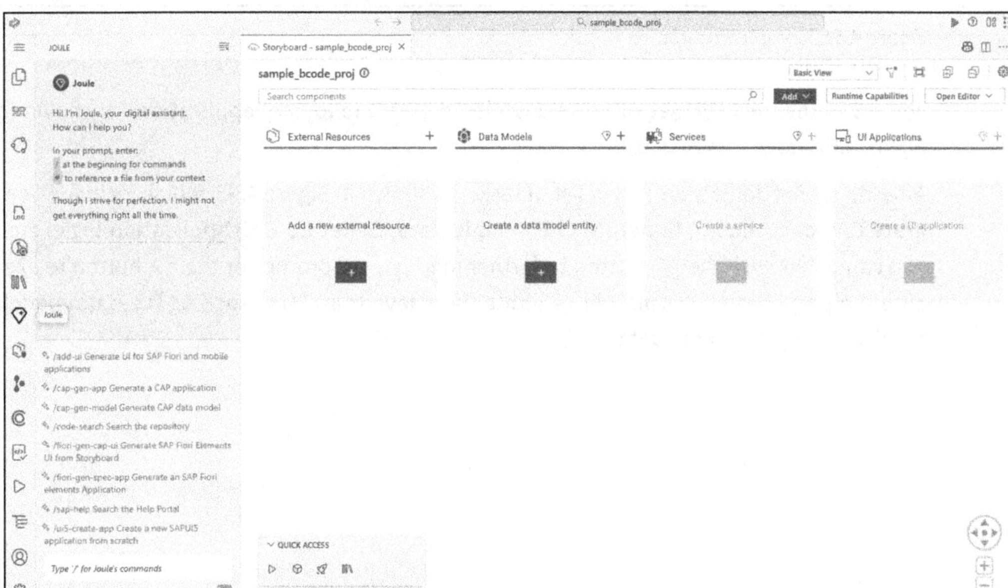

Figure 4.50: Storyboard

For example, the **/cap-gen-model Generate a CAP application** command instructs Joule to create an SAP Cloud Application Programming Model application, while the **/cap-gen-model Generate CAP data model** command is used to create a data model. In addition to selecting a command, you can enter an additional prompt explaining the application requirements in detail. You can even search for information in help portals and code repositories.

As shown in Figure 4.51, you can leverage the sample prompt **Create customer loyalty application** to kick-start the building process.

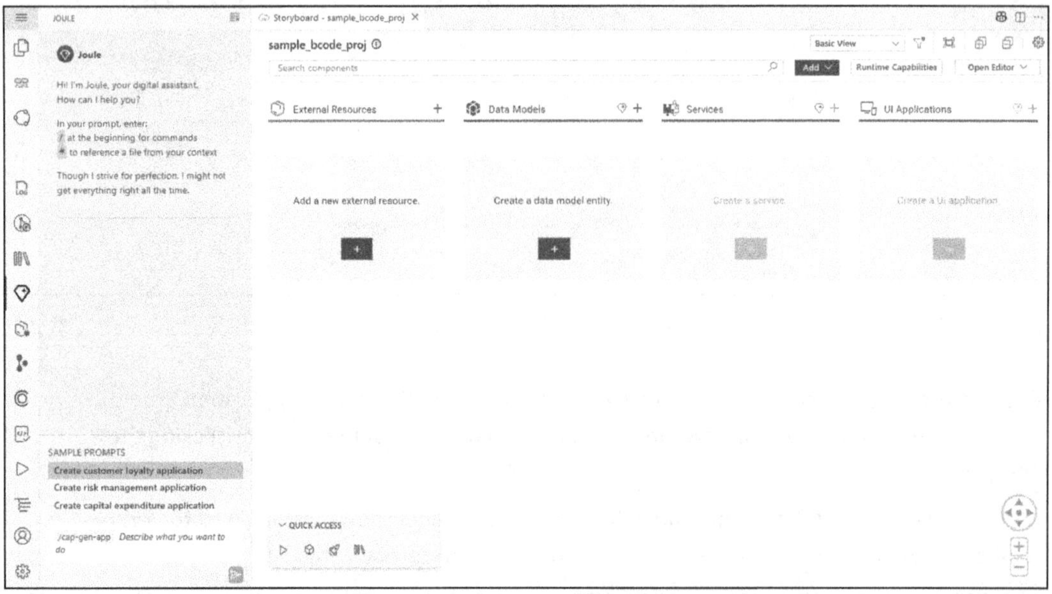

Figure 4.51: Sample Prompt

Joule will then list the set of tasks associated with the loyalty application, as shown in Figure 4.52, for you to review.

After reviewing Joule's proposed list of tasks, as shown in Figure 4.53, which includes **Generate CAP data model, Generate CAP sample data, Generate CAP application logic**, etc., you can either click the button to further modify your prompt or the button to execute the proposed task. This will generate the relevant artifacts such as the data model, application logic, sample data, etc.

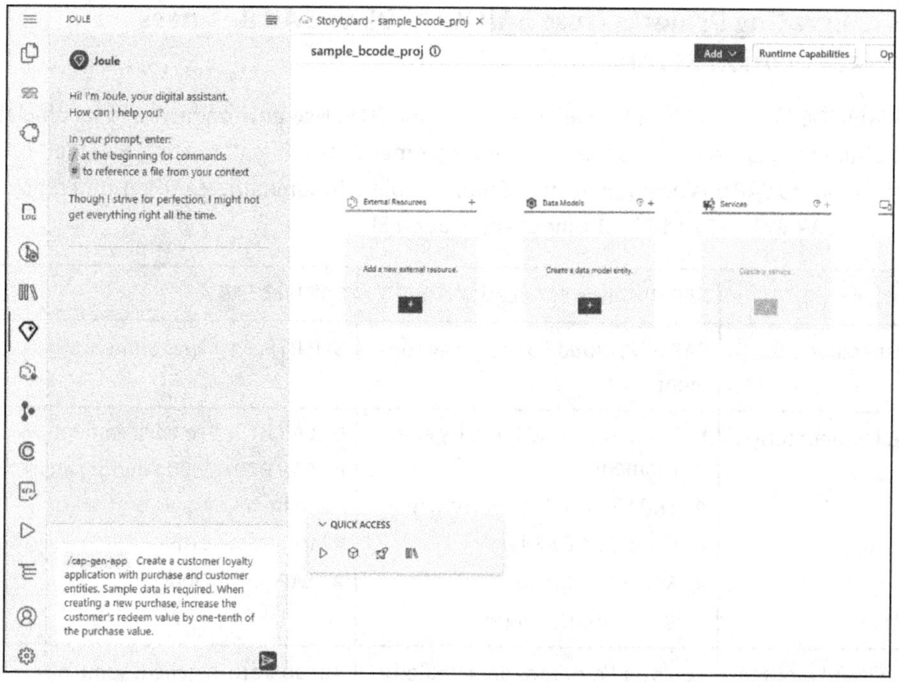

Figure 4.52: Review Joule Tasks

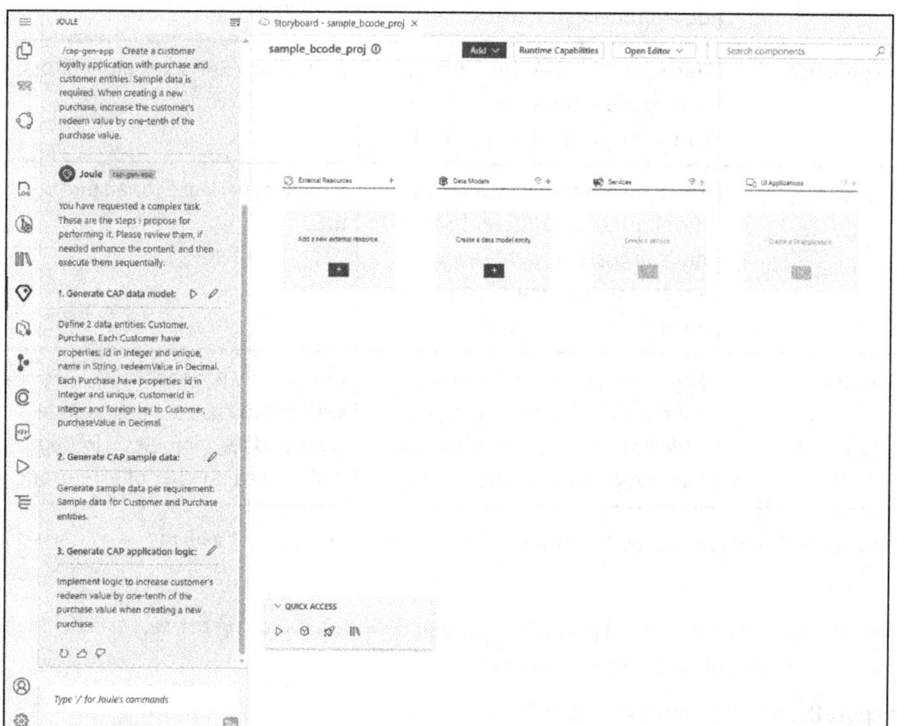

Figure 4.53: Execute Task

4.5 Migrating Projects from SAP Web IDE to SAP Business Application Studio

SAP Web IDE (full stack) is only available on the SAP BTP, Neo environment and its official end-of-life date is December 31, 2028. SAP recommends moving all existing SAP Web IDE applications to SAP Business Application Studio. Table 4.10 summarizes the key differences between SAP Web IDE and SAP Business Application Studio.

Feature	SAP Business Application Studio	SAP Web IDE
Availability	SAP BTP, Cloud Foundry environment	SAP BTP, Neo environment
Deployment target	■ SAP BTP, Cloud Foundry environment ■ SAP BTP ABAP environment ■ On-premise ABAP ■ SAP HANA Cloud ■ SAP S/4HANA Cloud	■ SAP BTP, Neo environment ■ SAP BTP, Cloud Foundry environment ■ On-premise ABAP ■ SAP HANA Cloud
Tools and runtimes	Tools and runtimes are enabled based on the dev space type and extensions to cater to the specific development scenario.	All tools are installed, some of which may not be required for all development scenarios.
UI experience	Based on VS Code and other open-source tools. Superior code editors with guided development.	Proprietary SAP tools and editors.
Personas	To cater to different scenarios, developers with different personas can customize their dev spaces for more enhanced experience.	Similar experience for all personas.
Orientation	More oriented toward open-source tools and technologies, which makes it scalable from an integration perspective.	Completely proprietary to SAP, with limited use of open-source tools and technologies. Limited with a fixed number of extensions.

Table 4.10: Differences Between SAP Business Application Studio and SAP Web IDE

In this example, we'll migrate a project created in SAP Web IDE using the server-side JavaScript (XS-JS) compatibility layer as an example:

1. Export the project from SAP Web IDE.
2. The project is downloaded as a ZIP file. Unzip this file.

3. In SAP Business Application Studio, open the project folder using the menu path **File • Open**. Select the project folder and then click **Open**.
4. Drag and drop the exported and unzipped folder into the **PROJECTS** area, as shown in Figure 4.54.

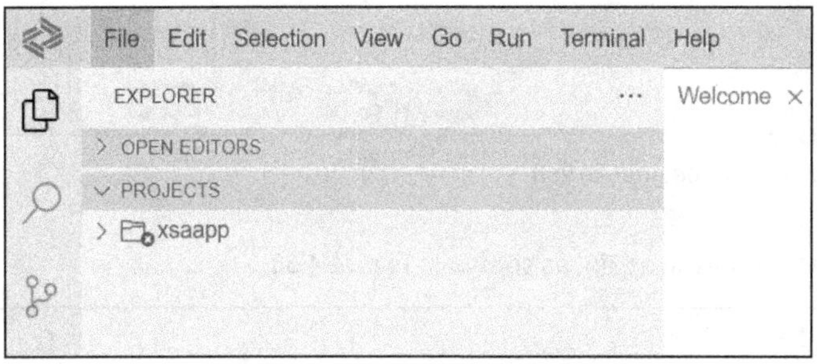

Figure 4.54: Uploading an SAP Web IDE Project in the SAP Business Application Studio Explorer

5. Open a new terminal by following the menu path **Terminal • New Terminal**. A new terminal opens with a path pointing to the **PROJECTS** folder.
6. Initialize package.json through a terminal window by entering "npm init," as shown in Figure 4.55.

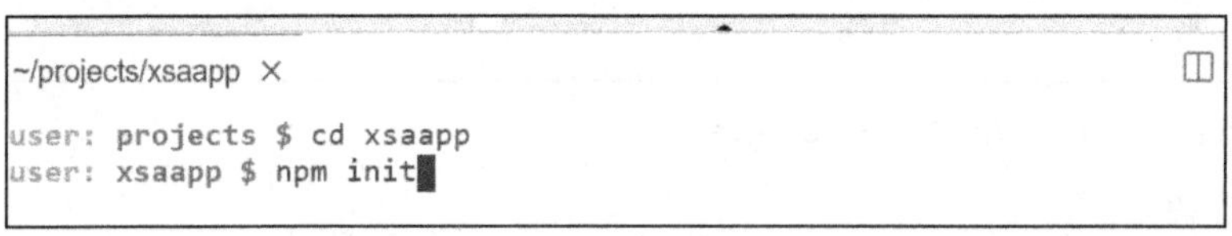

```
~/projects/xsaapp ×
user: projects $ cd xsaapp
user: xsaapp $ npm init
```

Figure 4.55: New Terminal Window

7. Enter the information prompted for in the terminal, as shown in Figure 4.56.

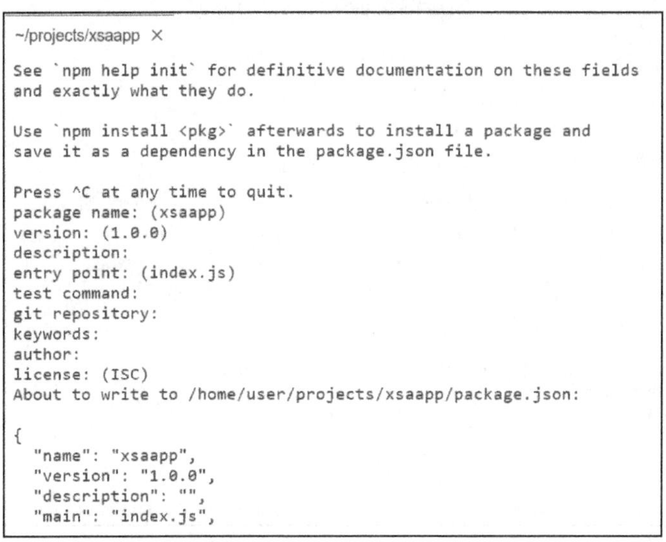

```
~/projects/xsaapp ×
See `npm help init` for definitive documentation on these fields
and exactly what they do.

Use `npm install <pkg>` afterwards to install a package and
save it as a dependency in the package.json file.

Press ^C at any time to quit.
package name: (xsaapp)
version: (1.0.0)
description:
entry point: (index.js)
test command:
git repository:
keywords:
author:
license: (ISC)
About to write to /home/user/projects/xsaapp/package.json:

{
  "name": "xsaapp",
  "version": "1.0.0",
  "description": "",
  "main": "index.js",
```

Figure 4.56: Terminal Inputs for npm init

8. Delete the file *ui5.yaml* under the webapp package. In the terminal window, enter "ui5init." Once initialized, the migration prompt appears. Click on **Start Migration**, as shown in Figure 4.57.

Figure 4.57: Starting the Migration of the Application

9. Click on the **Start Migration** button again, as shown in Figure 4.58.

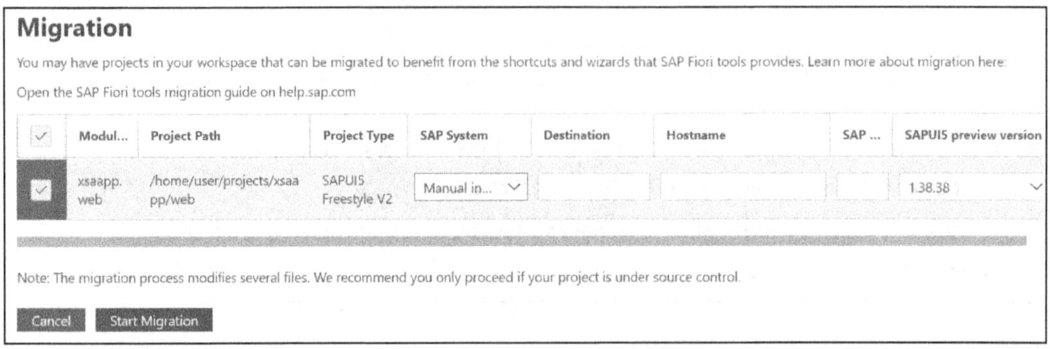

Figure 4.58: Migration of the UI to SAPUI5 Freestyle

10. The results of the migration process are displayed automatically by the system, as shown in Figure 4.59.

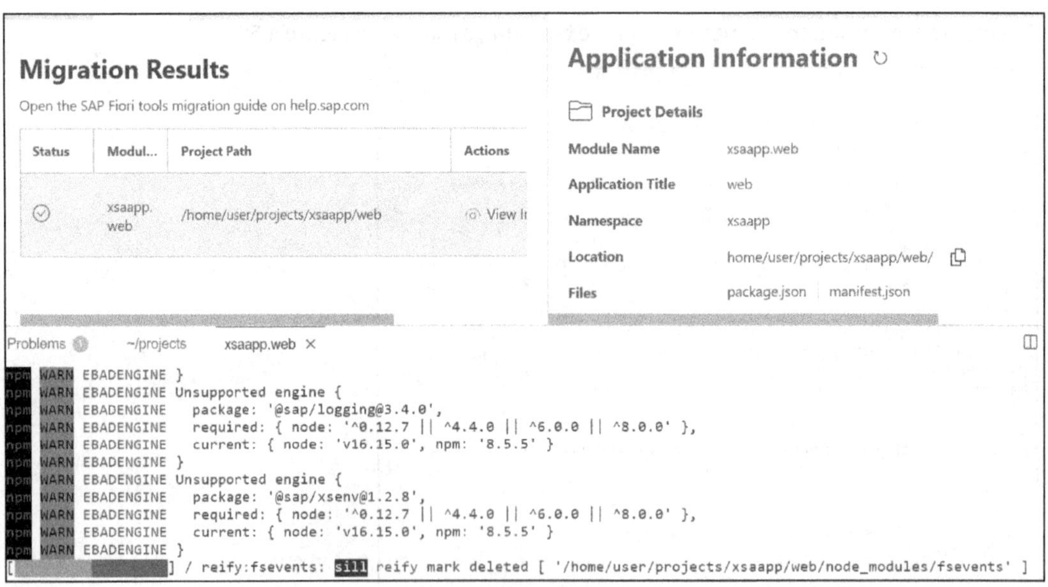

Figure 4.59: Migration Log

11. The project structure changes as additional files are automatically added by the previous steps, as shown in Figure 4.60.

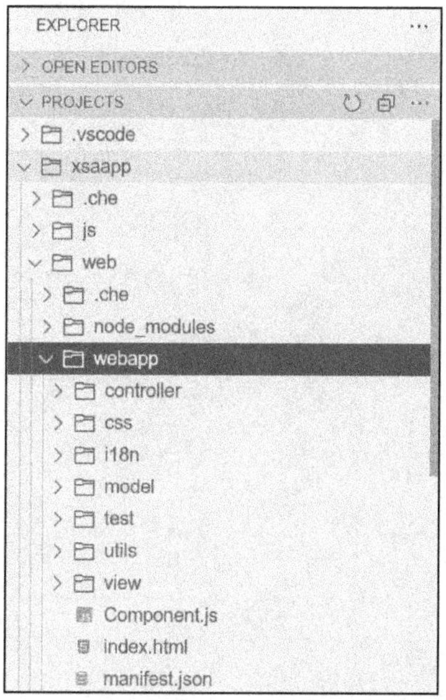

Figure 4.60: Project Structure after Migration

You're now ready to build and run the project in SAP Business Application Studio.

4.6 Summary

In this chapter, you learned all about SAP Business Application Studio, including its capabilities, the different types of dev spaces available, the extensions associated with these dev space types, and how to create application projects through various approaches. We described in detail how to integrate SAP Business Application Studio with GitHub and clone a project from a repository in Git. We also introduced SAP Build Code and how to use it to create applications on SAP BTP. At the very end of the chapter, we went through the process of migrating a project created in SAP Web IDE to SAP Business Application Studio. These lessons will be used extensively in Chapter 7 and Chapter 8 when we discuss application development

Chapter 5
SAP HANA Cloud

SAP HANA Cloud is a modern database as a service (DBaaS) for analytics, machine learning (ML), and artificial intelligence (AI). It is based on data lakehouse architecture for scalable processing and storage of structured, semistructured and unstructured data through a relational database and data lake files. It has multimodel engines for processing spatial, graph, vector, relational and AI/ML workloads.

SAP HANA Cloud is a DBaaS that powers mission-critical applications and real-time analytics with petabyte-scale storage and in-memory computing. It processes, stores, and queries multimodal data using various multimodel data engines, making SAP HANA Cloud the data backbone for AI-infused business applications—both transactional and analytical—as well as AI agents.

In this chapter, we will cover the features of SAP HANA Cloud, its medallion data architecture based on data lakehouses, the SAP HANA Cloud central interface, the SAP HANA Cloud knowledge graph engine, and the SAP HANA Cloud vector engine.

5.1 Features of SAP HANA Cloud

SAP HANA Cloud is a data platform built on cloud infrastructure that can scale both storage and compute power, seamlessly blending multimodal data for analytics and AI/ML. The features and capabilities of SAP HANA Cloud are illustrated in Figure 5.1.

The key features of SAP HANA Cloud are as follows:

- A data platform that stores, processes, and federates data to run intelligent applications in a single cloud solution.
- In-memory processing through SAP HANA Cloud, SAP HANA database for both analytical and transactional relational data loads.
- Store and access structured, semistructured, and unstructured data in SAP HANA Cloud, data lake files and analyze a petabyte relational data using the data lake relational engine.
- Smart integration with cloud and on-premise systems to access data virtually or ingest data into SAP HANA Cloud. This smart data federation simplifies the data architecture by eliminating need for unnecessary data redundancy

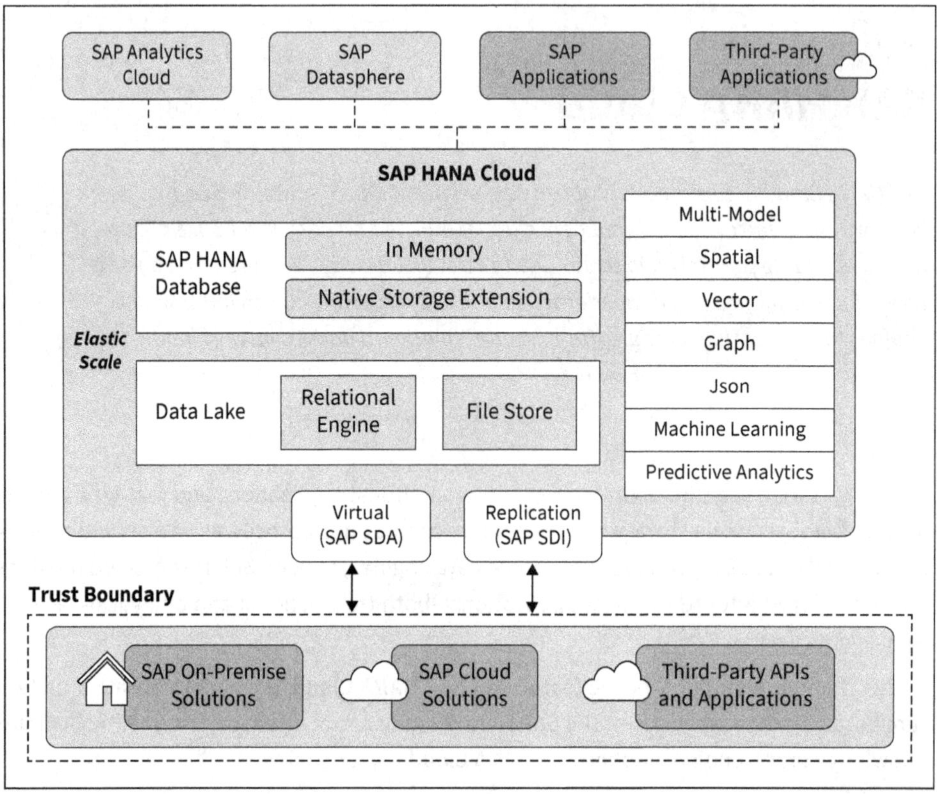

Figure 5.1: Features of SAP HANA Cloud

- SAP HANA Cloud has multimodel engines to process all types of business data to generate business insights and provide business context for generative AI use cases. In addition to the relational engine, SAP HANA Cloud has engines like spatial, graph, document store, and vector for advanced analytical capabilities. It seamlessly blends data processed by these engines with relational data using SQL queries.
- SAP HANA Cloud provides interfaces and APIs for custom applications built in SAP BTP or other platforms. Analytical applications can consume curated data, while transactional applications can create data in SAP HANA Cloud.
- SAP HANA Cloud provides an elastic scale of both compute power and storage resources to manage workloads in a performance optimized and cost-effective way.
- SAP HANA Cloud has embedded machine learning and predictive libraries for advanced analytics.
- SAP HANA Cloud supports data tiering through native storage extension (NSE) for warm data and data lake files for cold data. Data tiering in SAP HANA Cloud is shown in Figure 5.2.

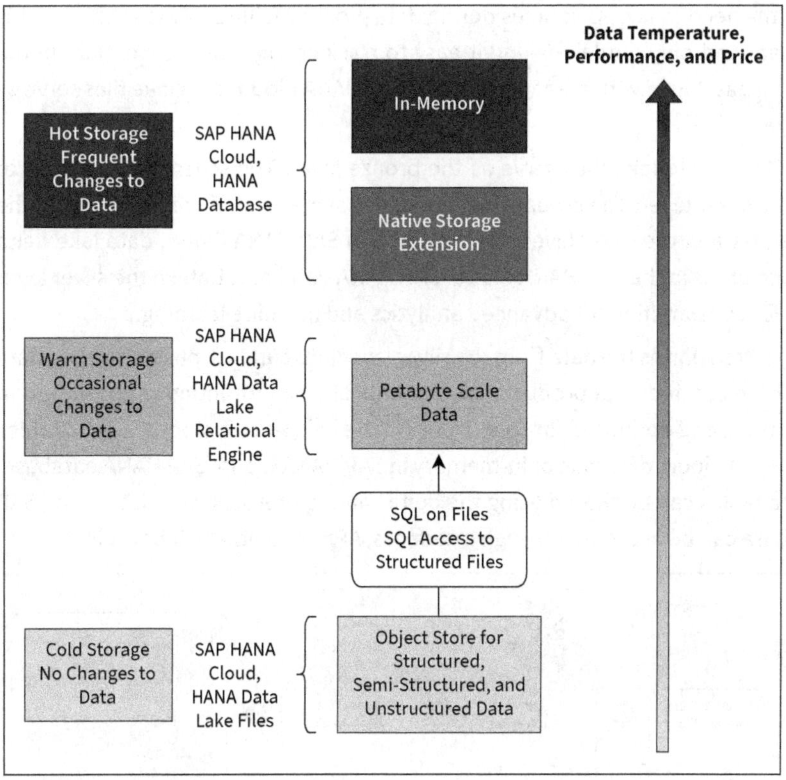

Figure 5.2: Data Tiering in SAP HANA Cloud

5.2 Medallion Data Architecture

One of the challenges of modern data management is ensuring data quality—cleansed, transformed, semantically rich, and well governed data. Medallion data architecture's multilayered approach implements data quality gates that ensure data is progressively cleansed, transformed, and trustworthy as it converts from its raw state to being ready for analytics and advanced analytics consumption.

The focus on quality in medallion data architecture solves one of the critical challenges faced by organizations: The need to be data-driven. It logically organizes data in a data lake into multiple layers—bronze, silver and gold—that progressively improve the structure and quality.

Medallion data architecture in SAP HANA Cloud is shown in Figure 5.3. Before getting into the process details for converting raw data to consumable data, let's look at the layers of medallion data architecture:

- The bronze layer stores ingested, unprocessed raw data in its original format. The purpose of this layer is to serve as a single, immutable source of truth for all historical data, eliminating the need to access sources during data pipeline failures. Data in this layer is updated in append-only mode, making it easy to track changes over time. Data in the bronze layer is cataloged with managed access. SAP HANA Cloud, data lake files serve as a bronze layer.

- SAP HANA Cloud, data lake files serve as the bronze layer. The schema is enforced or evolves in the silver layer. The silver layer retains the same level of granularity as in the bronze layer. Data in the silver layer can be stored in SAP HANA Cloud, data lake delta tables or in columns in the SAP HANA Cloud, SAP HANA database. Data in the silver layer is designed for consumption for advanced analytics and machine learning.
- The gold layer transforms the data from the silver layer into curated, business ready data products. These curated data products are semantically rich for intuitive consumption by business users and optimized for speed. Data in the gold layer is stored in delta tables within SAP HANA Cloud, data lake or in-memory in SAP HANA Cloud, SAP HANA database. Data in delta tables can be shared using the delta sharing protocol, and data in the SAP HANA database can be accessed through interfaces, APIs, and other client tools.

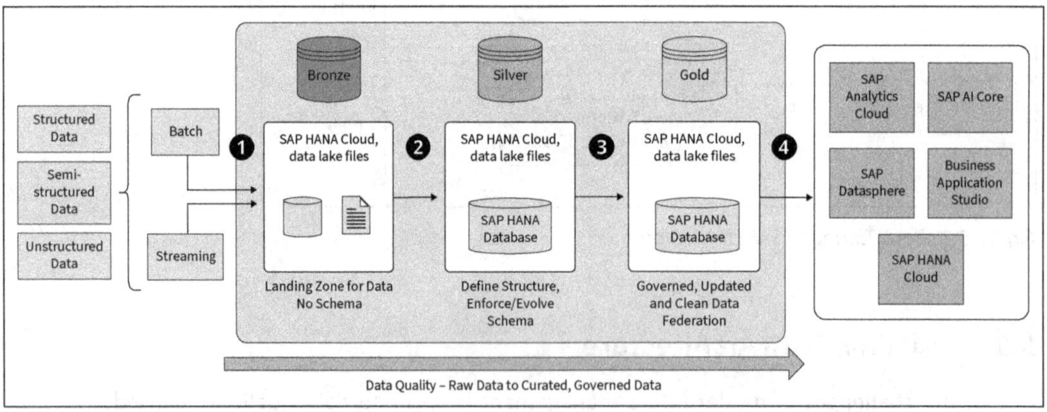

Figure 5.3: Medallion Data Architecture in SAP HANA Cloud

The steps in the end-to-end data pipeline for the medallion data architecture in SAP HANA Cloud are as follows:

❶ Data ingestion: Structured, semistructured, and unstructured data from various sources is ingested using different approaches like streaming or batch. The ingestion approach depends on factors like data sources, acceptable data latency, and performance.

❷ Process, curate, and store: Raw data is cleansed, validated, semantically enriched, and transformed into structures (through enforcement of schema or evolution of schema).

❸ Model and share: Data in the silver layer is modeled into data products and shared using the delta sharing protocol or other interfaces, APIs, and client tools

❹ Consume: Data in the gold layer is shared with business intelligence (BI) reporting tools (SAP and third party), SAP BTP services like SAP Datasphere, SAP HANA Cloud, SAP AI Core, and machine learning tools.

5.3 SAP HANA Cloud Central

SAP HANA Cloud Central is the primary interface for administration and management of SAP HANA Cloud. It serves as a centralized dashboard for database administrators to provision, manage, and monitor SAP HANA Cloud, SAP HANA database and SAP HANA Cloud, data lake. The core capabilities of SAP HANA Cloud Central are as follows:

- Provisioning of SAP HANA Cloud, SAP HANA database or data lake instances manually or using JSON files containing all the instance creation parameters
- Instance management activities like create, stop, start, recover, delete, and resize
- Monitoring the instance health through alert configurations and monitoring CPU, memory, and disk usage
- Tools for self-service migration of an on-premise SAP HANA database to an SAP HANA Cloud instance

[+]

Tip
Subscribe to the SAP HANA Cloud - Tools service in SAP BTP to enable the SAP HANA Cloud Central interface.

The following administration tasks can be performed in SAP HANA Cloud Central:

- Get an overview of the status of all SAP HANA Cloud instances that you can access.
- Review notifications and alerts.
- Check the configured memory, compute power, and storage consumption of an instance.
- Check the high availability configuration of an instance.
- Deploy new SAP HANA Cloud instances.
- Change the configuration of an instance.
- Add a data lake to an existing instance.
- Start, stop, and delete instances.
- Recover or recreate an instance.
- Open the SAP HANA cockpit for an instance.
- Open the database explorer for an instance.

The SAP HANA cockpit enables SAP HANA Cloud administrators to perform administration and monitoring tasks at the database instance level. Administrators can use the SAP HANA cockpit to monitor system resources (memory, CPU, and storage), analyze SQL statements, configure system settings, and manage users and authorizations.

The SAP HANA database explorer is a client tool to access database objects and make changes to the data. The SAP HANA database explorer allows developers to query information about the database using SQL, explore catalog objects, and view the database diagnostic files.

5.4 SAP HANA Cloud Knowledge Graph Engine

SAP HANA Cloud knowledge graph engine is one of the multimodel engines in SAP HANA Cloud. It processes structured, semistructured, and unstructured data into knowledge graphs and allows you to analyze graph data. Before we get into the details of the SAP HANA Cloud knowledge graph engine, let's understand what knowledge graphs are.

5.4.1 Knowledge Graphs

A knowledge graph is a structured, semantically rich representation of information. It depicts the entities and their relationships in a graph format. Unlike relational databases that store data in rigid schema-on-write tables, knowledge graphs model information as interconnected objects with meaningful relationships and store them in graph databases.

A knowledge graph can become a foundational asset for AI, as the structured and contextual information it provides can improve response accuracy by enabling grounding of large language models (LLMs) through retrieval-augmented generation (RAG), enhancing reasoning by allowing traversal and logical inference, and increasing explainability through explicit relationships. Knowledge graphs are built using the following three components:

1. Nodes: Nodes represent entities like people, places, concepts, and things.
2. Edges: Edges represent relationships or connections between entities.
3. Labels and properties: These describe characteristics/attributes of nodes and edges.

Figure 5.4 shows an example of sales order information represented as a knowledge graph using nodes, edges, labels, and properties.

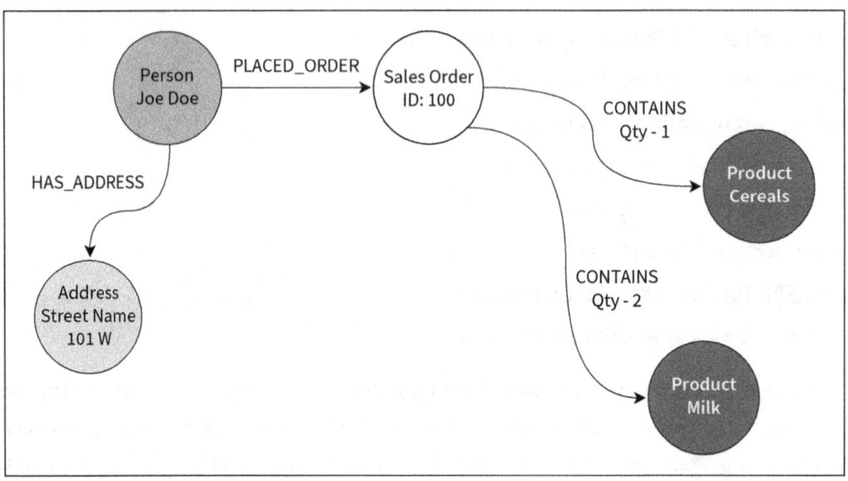

Figure 5.4: Example of Sales Order as a Knowledge Graph

The basic unit of data in a knowledge graph is a *triple* and it consists of a subject, a predicate, and an object. While subjects and objects are entities, a predicate is an edge. A triple from our sales order example is shown in Figure 5.5.

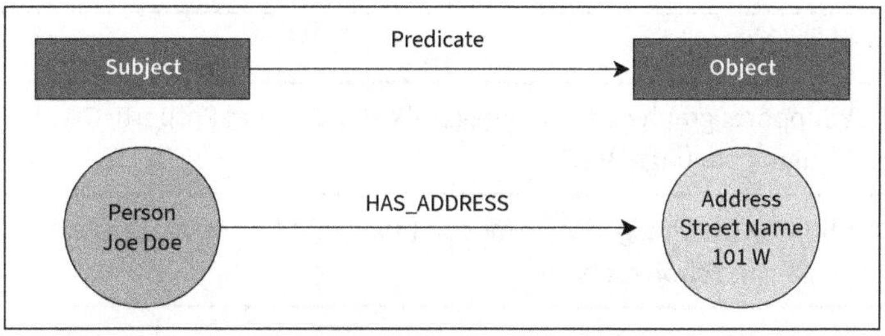

Figure 5.5: Example of a Triple in a Knowledge Graph

Triples in knowledge graphs are defined by the Resource Description Framework (RDF). The RDF is a standard model of data interchange on the web. Each part of the triple is uniquely identified by an Internationalized Resource Identifier (IRI).

Contextual or semantic information in a knowledge graph is based on standardized vocabularies or ontologies. Ontologies are a structured way to describe entities and their relationships in a domain of interest. This creates a common vocabulary, understanding for sharing of information across applications, data integration by mapping source data to common ontology, and semantic interoperability.

5.4.2 Features and Capabilities of SAP HANA Cloud Knowledge Graph Engine

The knowledge graph engine in SAP HANA Cloud helps with modeling, querying, and analyzing graph data efficiently. It uses the in-memory capabilities of SAP HANA Cloud, SAP HANA database for high performance. The knowledge graph engine supports SQL extensions that allow integration of graph data with relational data, making it a very powerful engine for contextual analysis of relational data. The features of the knowledge graph engine are listed in Table 5.1

Feature	Description
Graph data processing	The SAP HANA Cloud knowledge graph engine helps to model complex relationships, query and transverse through these models, and help analyze the graph data using contextual information.
In-memory processing	The SAP HANA Cloud knowledge graph engine uses the SAP HANA Cloud, SAP HANA database in-memory capabilities to provide fast query performance.
Graph format	Graph data is stored in schema-less structures.
Integration with SQL	Seamless integration with the SQL engine allows users to combine graph queries with traditional SQL queries for simplified access to relational and graph data.

Table 5.1: Key Features of SAP HANA Cloud Knowledge Graph Engine

Feature	Description
Query languages	Supports graph query languages like Cypher and Property Graph Query Language (PGQL).
Graph algorithms	Built-in graph algorithms for common tasks like centrality measures and shortest path.

Table 5.1: Key Features of SAP HANA Cloud Knowledge Graph Engine (Cont.)

The knowledge graph engine provides interfaces and APIs that facilitate interaction with systems for data ingestion, querying, and management. Interfaces enable seamless integration with other systems for graph-based data management and analytics within an organization.

Interfaces such as SPARQL (the primary interface for processing, querying, and analyzing graph data), SQL, and event streaming are provided for seamless analysis, along with SAP HANA client APIs supporting various runtimes like JavaScript, APIs for data management, RESTful APIs for RDF data loads, and service brokers for Cloud Foundry and Kubernetes environments. These interfaces and APIs allow users to connect with the knowledge graph engine in a variety of ways, including custom applications, real-time analytics, and AI agents. The key capabilities of the knowledge graph engine are listed in Table 5.2.

Capability	Description
Data ingestion and storage	The knowledge graph engine supports ingestion of structured, semistructured and unstructured data. Processed data is stored in a graph-based storage system for efficient read and write operations.
Data modeling with RDF or OWL	The knowledge graph engine uses the RDF or Web Ontology Language (OWL) for modeling data that is semantically rich and easy to traverse through relationships.
Querying using SPARQL	The knowledge graph engine supports SPARQL, a query language for RDF data. SPARQL can handle complex queries that can traverse the graph data.
Integration and APIs	The knowledge graph engine is integrated with other engines in SAP HANA Cloud, allowing for seamless graph data movement and processing. APIs provide interaction with external systems.

Table 5.2: Capabilities of SAP HANA Cloud Knowledge Graph Engine

Capability	Description
Security and governance of graph data	The knowledge graph engine enforces data security and privacy policies to ensure data privacy and compliance. Tools and frameworks are provided to support data governance, including data lineage, metadata management, and quality checks.
Visualization	The knowledge graph engine supports visualization of graph data using SAP Analytics Cloud or external tools.

Table 5.2: Capabilities of SAP HANA Cloud Knowledge Graph Engine (Cont.)

[+]

Getting Started with the Knowledge Graph Engine

Installation of SAP HANA Cloud knowledge graph engine is done through SAP HANA Cloud Central. A prerequisite for installation of the SAP HANA Cloud knowledge graph engine is to have an active SAP HANA Cloud instance in your SAP BTP subaccount. Steps for installing the knowledge graph engine in the SAP HANA Cloud instance are as follows:

1. Log on to SAP HANA Cloud Central.
2. Select the SAP HANA Cloud instance where you wish to install the knowledge graph engine.
3. Enable the knowledge graph engine service in the **Services** section of SAP HANA Cloud Central.

5.5 SAP HANA Cloud Vector Engine

SAP HANA Cloud vector engine is a multimodel engine which processes unstructured data into vector embeddings, stores vector embeddings in a vector database, and queries these vector embeddings with seamless integration with other multimodel engines. Vector embedding provides additional context to the prompts through RAG, thereby increasing the accuracy of responses from LLMs.

[«]

What is RAG?

Retrieval-augmented generation (RAG) is an architecture that optimizes the performance of LLMs by connecting them with external, up-to-date knowledge bases. RAG retrieves documents from these external knowledge bases at query time, incorporates new and context-aware data and then feeds it to LLMs for more accurate, current, and domain-specific responses.

Before we get into the details of the SAP HANA Cloud vector engine, let us understand core concepts of vector engines:

- Vectors embeddings: Also called vectors, these are long arrays of numbers used to mathematically represent the semantic meaning and context of unstructured data like text, images, audio, or video. Unstructured data is translated into a multidimensional array, with each number in the array representing a feature or characteristic of the data. Data with similar semantics and characteristics are assigned vectors that are mathematically close to each other. This allows machine learning models and AI applications to understand and search data based on semantic meaning
- Similarity search: This is the process of finding vectors that are semantically similar. This is mathematically achieved by using metrics that measure how closely two vectors align in multidimensional space. Common methods include cosine similarity, which measures the angle between vectors, and Euclidean distance, which calculates the straight-line distance separating vectors. These metrics help the vector engine identify vector embeddings in the database that are similar to the user's natural language query. Vector-based similarity search is heuristic rather than being based on exact matches, making it more effective for natural language processing.
- Indexing techniques: These are algorithms that organize vector embeddings in vector databases for faster and scalable similarity search. Some of the indexing algorithms are as follows:
  - Hierarchical navigable small world (HNSW): Also known as *approximate nearest neighbor search*, this is a graph-based algorithm that enables quick navigation between similar vectors.
  - Locality-sensitive hashing (LSH): This uses a hash-based indexing technique to group similar vectors into buckets using hash functions for faster comparisons.
  - Product quantization (PQ): This uses a quantization-based compression technique to compress vectors into smaller representations to reduce memory usage while maintaining search quality.

[+]

Note

The terms vectors and vector embeddings are used interchangeably throughout this chapter.

5.5.1 Journey from Unstructured Data to Vectors

Conversion of raw unstructured data into vector embeddings broadly involves the following steps, as illustrated in Figure 5.6:

1. Data chunking is the process of breaking large pieces of unstructured data into smaller, semantically meaningful segments, enabling foundational models to process information within their manageable context limits and memory capacities. Data chunking helps

retain the semantic and contextual meaning of multimodal data (data with multiple types such as audio, video, and text), enables targeted and accurate information retrieval from LLMs, and improves performance by using data chunks instead of large multimodal datasets.

Note

Data chunking in SAP HANA Cloud can be achieved using PAL_TEXTSPLIT in the Application Function Library (AFL) and can be called from SQL and Textsplitter in the Python ML API.

2. Once the data has been chunked, it is encoded into vectors using embedding models. Vector embeddings represent these chunks as arrays of numbers, making it mathematically possible to compare them based on their semantic and contextual meanings.
3. After the embeddings model converts chunks into vector embeddings, these vectors are stored in the vector database of SAP HANA Cloud. Data in a vector database is indexed for the quick search and retrieval of vector embeddings.
4. Vector embeddings stored in vector database are ready to be consumed by RAG-based AI applications. User queries, such as natural language sentences, images, or prompts, are converted into vector embeddings and compared with vector embeddings in the database. Retrieved vector embeddings are used to provide the latest semantic context to user queries before they are sent to LLMs.

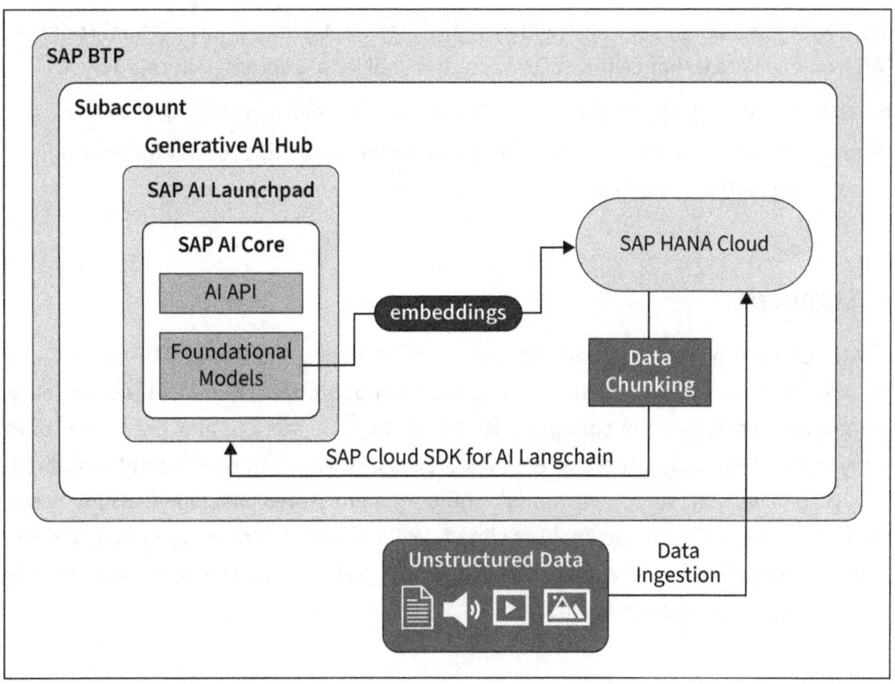

Figure 5.6: Solution Diagram for Vector Embedding

5.5.2 Benefits and Use Cases of SAP HANA Cloud Vector Engine

The SAP HANA Cloud vector engine helps organizations working with AI and large volumes of unstructured data. Some of the benefits of the SAP HANA Cloud vector engine are as follows:

- Processes unstructured multimodal data into vector embeddings for semantic, context-aware, and up-to-date responses to user queries.
- Fast similarity search using index techniques to find semantically similar items across huge volumes of vector embeddings.
- The SAP HANA Cloud vector engine integrates with AI pipelines through LLMs, RAG applications, and recommendation engines.
- The SAP HANA Cloud vector engine combines vector similarity with traditional filters like tags, categories, and timestamps to refine search results.

The SAP HANA Cloud vector engine is transforming how we search, recommend, generate, and interpret content and is powering a growing number of use cases across industries. Some of these use cases are as follows:

- Enables semantic search based on meaning rather than exact keywords ensuring the usefulness of knowledge bases, organizational internal tools, and support centers
- Facilitates conversational AI by retrieving similar responses or content for user queries
- Recommendation of similar products or content based on user preferences for an interactive engagement with platforms
- Helps with grounding user query with semantic, context-aware information through the RAG framework to enhance the accuracy and reliability of content generated by AI
- Image and video searches through the comparison of vector embeddings
- Storing vector embeddings of all previous interactions making it easier for continuity in longer conversations and tasks

5.6 Summary

SAP HANA Cloud is a modern DBaaS for advanced analytics and AI/ML. SAP HANA Cloud, SAP HANA database, and SAP HANA Cloud, data lake support data lakehouse and medallion data architectures. It has the capability for petabyte-scale storage and the power of in-memory computing. Its multimodel engines process, store, and analyze multmodal data such as graphs, spatial, vector, relational, and document stores. Machine learning is supported by the built-in Predictive Analysis Library (PAL). Curated, semantically rich, and context-aware data products make SAP HANA Cloud an AI database and serve as a comprehensive data platform for all business insights.

Chapter 6

SAP Cloud Application Programming Model

SAP Cloud Application Programming Model is an application development approach for creating cloud applications. The model provides developers with tools, languages, and libraries that can help reduce development overhead by simplifying the development process. This chapter will introduce you to the model and its use in detail.

In this chapter, we'll explain the concepts behind SAP Cloud Application Programming Model and its unique capabilities. We'll describe the technical components available in SAP Cloud Application Programming Model and then go deeper into various implementation concepts through simple examples. At the end of this chapter, you should have a fair understanding of this application development approach and its related steps for building enterprise-grade applications. Let's begin our journey by understanding the concepts behind SAP Cloud Application Programming Model.

6.1 Overview of SAP Cloud Application Programming Model

A *programming model* is one level higher than programming languages and describes best practices when building applications in a standardized way. A programming model comprises of one or more programming languages, tools, libraries, templates, and more. A well-defined programming model will help you develop your applications quickly and help you test, debug, and troubleshoot. SAP Cloud Application Programming Model guides the developer down a golden path to developing applications based on best practices through the use of out-of-the-box content. The key differentiator of SAP Cloud Application Programming Model is its *domain models* that focus on domains instead of technical disciplines. Before we describe domain models in detail in subsequent sections, let's first explore the various technical components of SAP Cloud Application Programming Model, as shown in Figure 6.1.

As shown in Figure 6.1, SAP Cloud Application Programming Model is based on many open-source technologies (e.g., Java, Node.js, etc.). However, it also contains many SAP specific technologies, such as core data services (CDS), SAP Fiori, SAP HANA, and more. CDS is the definition language for building domain models and for defining services. SAP Fiori is the set of design principles for achieving a better user experience (UX). SAP HANA is the database where all database artifacts are created behind the scenes. You can also use any other UX technology and other databases as well; however, building SAP Cloud Application Programming Model-based applications always starts with domain modeling using CDS.

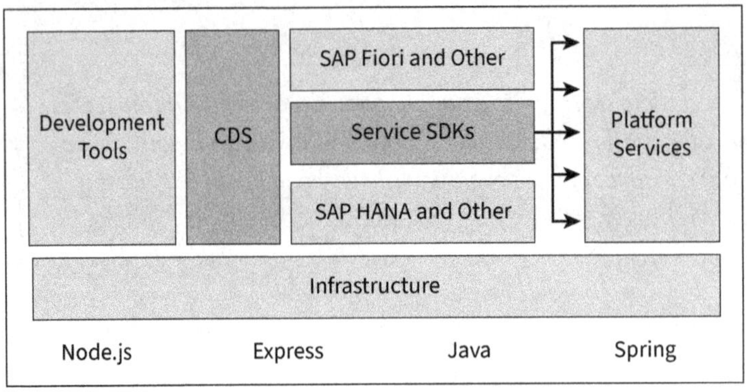

Figure 6.1: SAP Cloud Application Programming Model: Technical Components

SAP Cloud Application Programming Model provides two runtime choices (Java and Node.js), and its service software development kits (SDKs) provide various platform services to handle repetitive tasks in a platform-agnostic way. As mentioned earlier, SAP Cloud Application Programming Model also contains best practices out of the box. You also have multiple options when selecting development tools as it has dedicated tools support for SAP Business Application Studio, Visual Studio Code (VS Code), and Eclipse. All behavioral aspects of SAP Cloud Application Programming Model are based on services and events. Services are stateless, declared using CDS, and can be accessed by other services using application programming interfaces (APIs). Services react to various events.

Now, you have some basic ideas about SAP Cloud Application Programming Model and its technical components. Before we dive into further details and build an application, we'll shift our focus to development tools. Even though multiple tool choices are available, we'll primarily focus on SAP Business Application Studio. Since Chapter 4 already covered the setup of SAP Business Application Studio and creating your project, we'll only cover a few SAP Cloud Application Programming Model-specific pieces of project setup in the next section.

6.2 Project Setup

As soon as your project is created, it's visible in SAP Business Application Studio, as shown in Figure 6.2.

You can now open the project workspace by clicking on the project's name. Once the workspace is open, you'll find three primary folders (**app**, **db**, and **srv**), as shown in Figure 6.3.

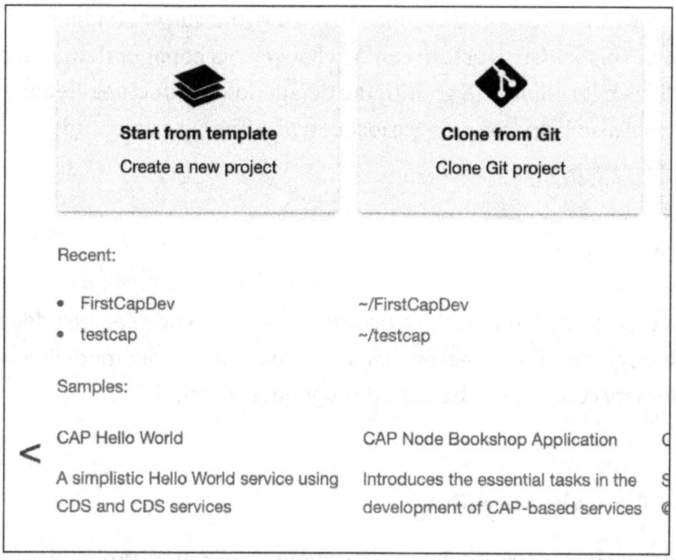

Figure 6.2: SAP Cloud Application Programming Model Project Available in SAP Business Application Studio

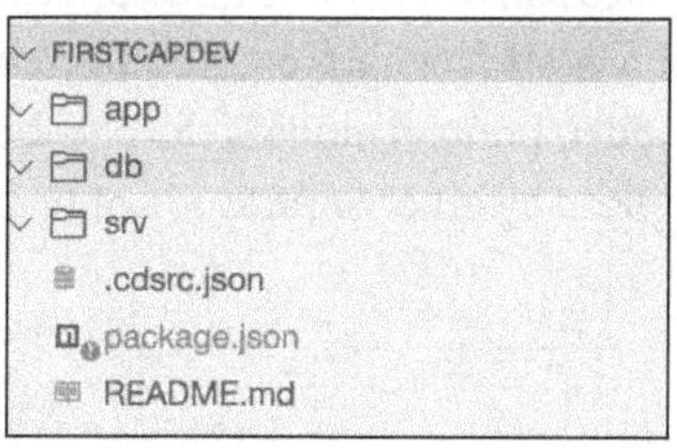

Figure 6.3: Folder Structure of an SAP Cloud Application Programming Model Project

You user interface (UI) content should be stored in the **app** folder; the domain model and database-related artifacts should go to the **db** folder, and the service definition and implementation should be maintained in the **srv** folder. The *package.json* file is your project's description in JavaScript Object Notation (JSON). We'll explore the significance and usage of all these folders and files in the following sections.

While you're in your project workspace, you can open the terminal window by following the menu path **Terminal • New terminal**. You should see the current folder is your project name, as shown in Figure 6.4.

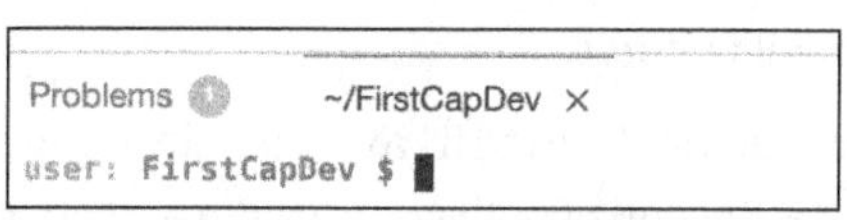

Figure 6.4: Terminal Showing Project Name as the Current Folder

Now, you can enter the command `cds env get folders` to review your current folder structure, as shown in Figure 6.5. This folder structure can be changed via configuration in the *package.json* file, but we'll keep it simple and go with the default folder structure. Keeping this default project layout will also help in the easy execution of other `cds` commands.

```
user: FirstCapDev $ cds env get folders
{ db: 'db/', srv: 'srv/', app: 'app/' }
```

Figure 6.5: Command to View the Folders

Our project has now been created with the basic structure, and further code can be added under various folders. Although mentioned earlier, let's now explain domain modeling in greater detail since this primary concept will be used throughout the rest of this chapter.

6.3 Understanding Domain Modeling

A domain model is a key element when building SAP Cloud Application Programming Model-based applications, where the focus is on the domain level, defined in a business-centric way, rather than defining the finer details of a technical implementation. The domain model is the starting point when building any application using SAP Cloud Application Programming Model. As shown in Figure 6.6, database artifacts are created from a domain model either directly or through a persistence model. Domain models can also be exposed as services for UIs. Domain models can also be reused by other SAP Cloud Application Programming Model projects.

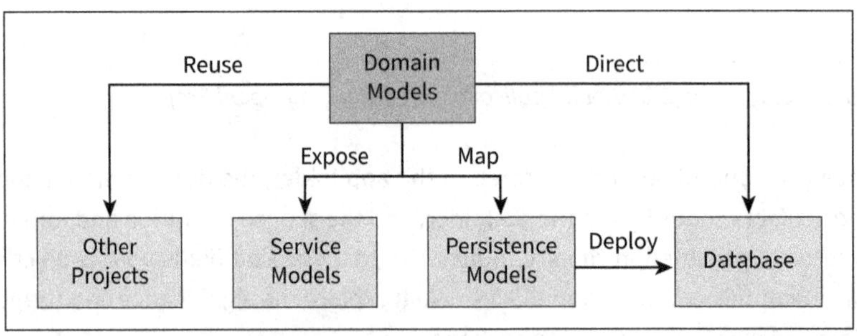

Figure 6.6: Central Role of Domain Model

The purpose of domain modeling is to enhance collaboration between developers and domain experts by providing a shared language. Technically, a domain model is like an entity relationship model, with many more additional features.

Let's now create our first domain model and add various features to it. Your domain models must be created in your defined **db** folder. Open SAP Business Application Studio and open your previously created project workspace, right-click on the project workspace, and select **New File** from the context menu. Create a file with the name "schema.cds" under the **db** folder, as shown in Figure 6.7.

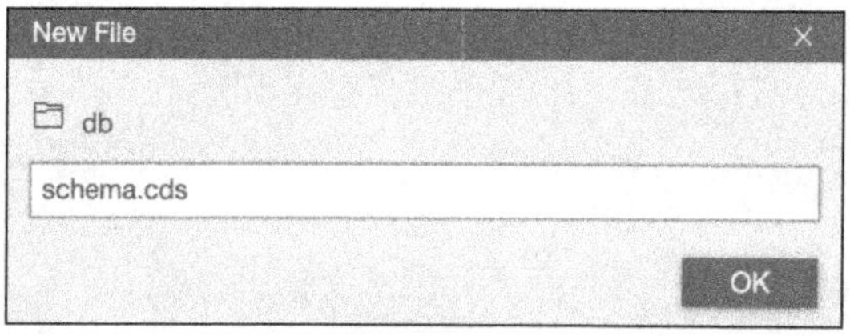

Figure 6.7: Creating a schema.cds File to Define the Domain Model

Once the file is created, you can add the code for the domain model. Listing 6.1 shows some basic CDS code that describes one entity for a sales order with a bunch of attributes.

```
namespace company.orders;

entity Header {
  key ordernum : Integer;
  customer: String(10);
  salesorg: String(4);
  distchan: String(2);
  division: String(2);

}
```

Listing 6.1: Initial Code for Building a Header Entity

Listing 6.1 shows just one entity with the name Header, which is the order's header information with five different attributes. The key for this entity is ordernum. The namespace is optional, but normally, you would add a namespace directive on top of your domain model (e.g., the entity Header can be referred to as company.orders.Header outside of your domain model).

Let's now try to understand entity association with the enhanced source code shown in Listing 6.2.

```
namespace company.orders;

entity Header {
  key ordernum : Integer;
  itemno: Association to many Items;
  customer: Association to Customers on customer.customerID =
CustomerID;
  salesorg: String(4);
  distchan: String(2);
  division: String(2);
  CustomerID: String;
```

```
}

entity Customers {
    key customerID: String(10);
    name: String(40);
    parent: Association to Header;
}

entity Items {
    itemno: Integer;
    product: Association to Products;
    qty: Integer;
    price: Decimal(15,2);
    parent: Association to Header;
}

entity Products {
  productno: String(40);
  name: String(50);

}
```

Listing 6.2: Entities with Associations

Associations define the relationships between entities. As shown in Listing 6.2, the `Header` entity is associated with the entities `Items` and `Customers`. Similarly, the `Items` entity is further associated with products and so on. When you define a service on top of this model, the service can generate complex join conditions based on these defined associations to fulfill any query operation. These associations can also be used to create or update complex hierarchical data. If you carefully look into the nature of the associations for the `Header` entities, you'll find two different patterns of association: one with the `on` clause and another without the `on` clause. Any association with the `on` clause is called an *unmanaged association* because you can use any arbitrary elements to join two entities. *Managed associations*, on the other hand, always use the primary key of the associated entity and thus don't need the `on` condition. Additionally, as shown in Listing 6.2, some associations have one-to-many relationships. Maintaining many-to-many relationship is also possible but is not shown in Listing 6.2.

After defining the domain model, let's define a service to expose this model as an OData service. Our service should ideally be created in the **srv** folder, as shown in Figure 6.8.

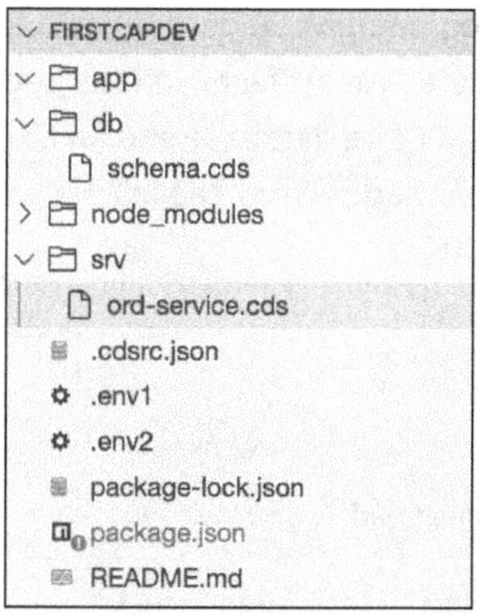

Figure 6.8: Service File ord-service.cds

Listing 6.3 shows the source code of the ord-service.cds file, where we've exposed all four entities as a part of the OData service.

```
using { company.orders as ord } from '../db/schema';

service CatalogService {

   entity Orders as SELECT from ord.Header {*,
     customer.name as name
    } ;
   entity Orderitems as select from ord.Items{*};
   entity Customers as select from ord.Customers;
   entity Products as select from ord.Products;

}
```

Listing 6.3: Service Exposure of the Domain Model

We'll now open a new terminal window and run the command `cds watch`, as shown in Figure 6.9.

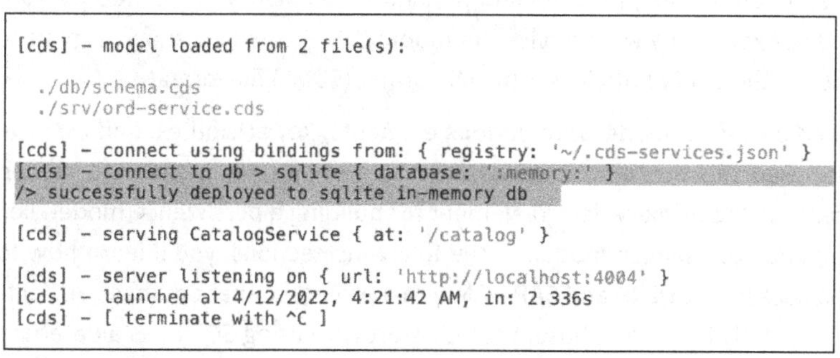

```
[cds] - model loaded from 2 file(s):

  ./db/schema.cds
  ./srv/ord-service.cds

[cds] - connect using bindings from: { registry: '~/.cds-services.json' }
[cds] - connect to db > sqlite { database: ':memory:' }
/> successfully deployed to sqlite in-memory db

[cds] - serving CatalogService { at: '/catalog' }

[cds] - server listening on { url: 'http://localhost:4004' }
[cds] - launched at 4/12/2022, 4:21:42 AM, in: 2.336s
[cds] - [ terminate with ^C ]
```

Figure 6.9: cds watch

As shown in Figure 6.9, by default, SAP Cloud Application Programming Model connects to an in-memory SQLite database to create all the necessary database artifacts. In later sections, you'll learn how to switch this connection to an actual SQLite database and an SAP HANA database separately. However, let's now explore the exposed service. The `cds watch` command will display the popup window shown in Figure 6.10.

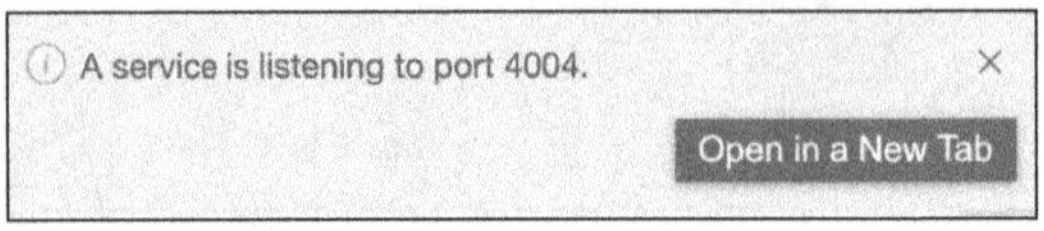

Figure 6.10: Popup Window for Testing Your Service for the Domain Model

Clicking on **Open in a New Tab** will open a new browser tab where you can see your service details, as shown in Figure 6.11.

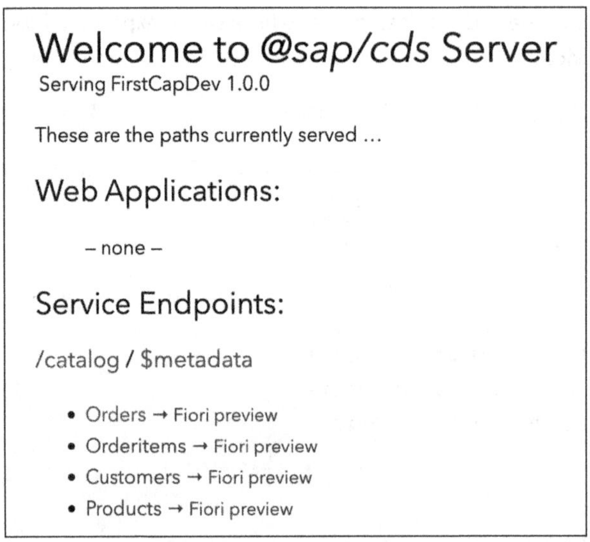

Figure 6.11: Exposing Service and Entity Set Details

Clicking on **$metadata** will open the OData metadata that you can explore further to understand your service, its entity sets, its associations, and its elements as defined by your domain model and as exposed by your service definition CDS. Figure 6.12 shows a glimpse of the metadata file, in the Entity Data Model in XML format (EDMX) file format.

Now, you've created a domain model with various elements, key attributes, and associations and have exposed this domain model as an OData service. As mentioned earlier, domain models serve as the primary design element for building a persistence model (i.e., database artifacts) and your service model. In the following sections, you'll learn how to add different databases (e.g., SQLite and SAP HANA) to a project so that database artifacts can be created persistently in the database. So, let's start with using SQLite as a database as the persistence medium for our project.

```
▼<edmx:Edmx xmlns:edmx="http://docs.oasis-open.org/odata/ns/edmx" Version="4.0">
  ▼<edmx:DataServices>
    ▼<Schema xmlns="http://docs.oasis-open.org/odata/ns/edm" Namespace="CatalogService">
      ▼<EntityContainer Name="EntityContainer">
        ▼<EntitySet Name="Orders" EntityType="CatalogService.Orders">
            <NavigationPropertyBinding Path="itemno" Target="Orderitems"/>
            <NavigationPropertyBinding Path="customer" Target="Customers"/>
          </EntitySet>
        ▼<EntitySet Name="Orderitems" EntityType="CatalogService.Orderitems">
            <NavigationPropertyBinding Path="product" Target="Products"/>
            <NavigationPropertyBinding Path="parent" Target="Orders"/>
          </EntitySet>
        ▼<EntitySet Name="Customers" EntityType="CatalogService.Customers">
            <NavigationPropertyBinding Path="parent" Target="Orders"/>
          </EntitySet>
          <EntitySet Name="Products" EntityType="CatalogService.Products"/>
        </EntityContainer>
      ▼<EntityType Name="Orders">
        ▼<Key>
            <PropertyRef Name="ordernum"/>
          </Key>
          <Property Name="ordernum" Type="Edm.Int32" Nullable="false"/>
          <NavigationProperty Name="itemno" Type="Collection(CatalogService.Orderitems)"/>
        ▼<NavigationProperty Name="customer" Type="CatalogService.Customers">
            <ReferentialConstraint Property="CustomerID" ReferencedProperty="customerID"/>
          </NavigationProperty>
          <Property Name="salesorg" Type="Edm.String" MaxLength="4"/>
```

Figure 6.12: *OData Metadata in an EDMX File*

6.4 Using the Default SQLite Database

Earlier, you saw how an SAP Cloud Application Programming Model project by default provides an SQLite in-memory database if you do not want to persist the data while restarting the service. Thus, every time you load the service or a related app, the data will be reset to its initial state. This feature is a good way to check and verify your initial service or application without connecting to a database. If you want this database connection to change to an SQLite database, run the command `cds deploy --to sqlite`, as shown in Figure 6.13.

```
user: FirstCapDev $ cds deploy --to sqlite
/> successfully deployed to ./sqlite.db

/> updated ./package.json
```

Figure 6.13: Deploying CDS Artifacts to an SQLite Database

As shown in Figure 6.13, the CDS command also updated the *package.json* file. Your project's *package.json* should now contain the code shown in Listing 6.4.

```
"cds": {
    "requires": {
      "db": {
        "kind": "sqlite",
        "credentials": {
          "database": "sqlite.db"
        }
```

```
        }
    }
}
```

Listing 6.4: package.json Content for Deployment to the SQLite Database

You can now run the command `cds deploy --to sqlite: memory`, and you'll see your *package.json* file has been updated, as shown in Listing 6.5.

```
"cds": {
    "requires": {
      "db": {
        "kind": "sqlite",
        "credentials": {
          "database": "memory"
        }
      }
    }
}
```

Listing 6.5: package.json Content for Deployment to In-Memory SQLite

So, now you understand how to select an SQLite in-memory database using the `cds` command. You may also need some default data for your database so that you can quickly test your application. SAP Cloud Application Programming Model provides an easy way to update your database from CSV files. The first step is to create a folder called **data** under the **db** folder, as shown in Figure 6.14.

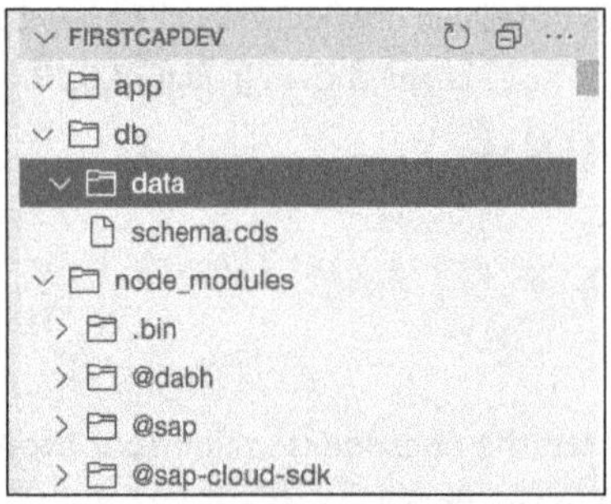

Figure 6.14: Test Data Folder Under the db Folder

Create a CSV file with the correct element name as the column header, and your test data in the rows shown in Figure 6.15. Then, drop the CSV file into the data folder. The name of this file must be in the format *<namespace>-<Entity Name>.csv*.

```
company.orders-Header.csv
orderno,salesorg,distchan,division,CustomerID
12,1000,10,10,C1234
13,1000,10,10,C1234
14,1000,10,10,C1234
```

Figure 6.15: Sample Test Data Content for the Header Entity

Once your file is loaded, run the command `cds deploy --to sqlite` again to check for the successful loading of the data, as shown in Figure 6.16. In our case, we deployed the data for just one entity, but you can create other CSV files for other entities and load them together.

```
user: FirstCapDev $ cds deploy --to sqlite
 > filling company.orders.Header from ./db/data/company.orders-Header.csv
/> successfully deployed to ./sqlite.db

/> updated ./package.json
```

Figure 6.16: Deploying an SQLite Database with Test Data

Now, use the `cds watch` command again and then open the `Orders` entity to review its test data, as shown in Figure 6.17.

```
{
    "@odata.context": "$metadata#Orders",
    "value": [
        {
            "ordernum": 12,
            "salesorg": "1000",
            "distchan": "10",
            "division": "10",
            "CustomerID": "C1234"
        },
        {
            "ordernum": 13,
            "salesorg": "1000",
            "distchan": "10",
            "division": "10",
            "CustomerID": "C1234"
        },
        {
            "ordernum": 14,
            "salesorg": "1000",
            "distchan": "10",
            "division": "10",
            "CustomerID": "C1234"
        }
    ]
}
```

Figure 6.17: Test Data for the Order Entityset of Service

You should now be comfortable adding SQLite databases to your applications and loading test data. Test data can also be loaded in a similar way for an in-memory SQLite database.

However, we generally use an SAP HANA database as the persistence layer for enterprise-grade applications, and thus, you might be curious about switching this SQLite database for an SAP HANA database. Let's tackle that scenario next!

6.5 Using the SAP HANA Cloud Database

Since the option of adding an SQLite database is provided by default with any SAP Cloud Application Programming Model project, getting set up doesn't require much effort. But remember that the SQLite database is part of your project and can be seen in the project structure. The database will be deployed to SAP BTP, Cloud Foundry environment together with the other artifacts. If you want to develop an enterprise-grade application, however, you should use an SAP HANA Cloud database provisioned in SAP BTP, Cloud Foundry environment. Chapter 5 already covered SAP HANA Cloud in some detail, so we'll only walk through the steps specific to using SAP HANA as your database for your SAP Cloud Application Programming Model application here.

Before you can add SAP HANA as the database to your application, you'll need to log on to SAP BTP, Cloud Foundry environment, via SAP Business Application Studio by following the menu path **View • Find Command**. Once logged on to SAP BTP, Cloud Foundry environment, click the **Cloud Foundry** link in the toolbar on the left in SAP Business Application Studio, which should display details about your SAP BTP, Cloud Foundry environment, along with your SAP HANA Cloud instance, as shown in Figure 6.18.

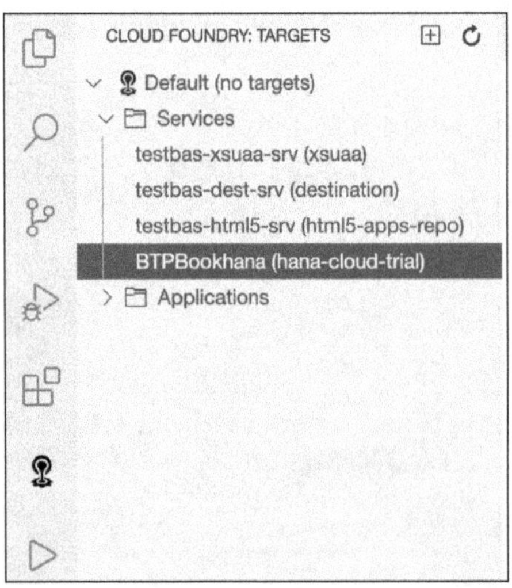

Figure 6.18: SAP HANA Cloud Instance in SAP BTP, Cloud Foundry Environment from SAP Business Application Studio

Now, go back to the project explorer and open the terminal window again. If you created your project previously using the terminal window, you may be lacking the required *multi-target application (MTA)*-specific features in your project. In this case, use the command cds add mta. This step is not required if you created a SAP Cloud Application Programming Model project using the wizard and selected the **MTA Based SAP Business Technology Platform Deployment** option. After optionally adding *mta* features, you can now run the command cds add hana in your terminal window, which will perform two primary changes to your *package.json* file:

- The command will add an hdb driver for SAP HANA as a dependency.
- A data source of type sql will be added to the cds.requires.db block.

Use the command cds deploy --to hana to deploy the database artifacts to your SAP HANA Cloud instance, as shown in Figure 6.19.

```
user: FirstCapDev $ cds deploy --to hana
[cds.deploy] - Starting deploy to SAP HANA ...

[cds.deploy] - Creating build tasks
[cds.deploy] - Running build

[cds.deploy] - Using container FirstCapDev-db

[cds.deploy] - Running 'cf create-service hana hdi-shared FirstCapDev-db' wit
Creating service instance FirstCapDev-db in org 69fcc01ctrial / space dev as
OK

Create in progress. Use 'cf services' or 'cf service FirstCapDev-db' to check
|
Showing info of service FirstCapDev-db in org 69fcc01ctrial / space dev as ga
```

Figure 6.19: Deploying the Project's Database Artifacts to SAP HANA

Deployment triggers the creation of an instance of **SAP HANA Schemas & HDI Containers**, as shown in Figure 6.20. Service keys will be created, and then all the SAP HANA database tables/views in the SAP HANA deployment infrastructure (HDI) container will be deployed.

Instance	Service	Plan
BTPBookhana	SAP HANA Cloud	hana
FirstCapDev-db	SAP HANA Schemas & HDI Containers	hdi-shared

Figure 6.20: SAP HANA Deployment Infrastructure Schemas in SAP BTP, Cloud Foundry Environment

Note

In our example, we added an extra record to our test data file(s) before deploying to SAP HANA, as shown in Figure 6.21.

```
db > data > company.orders-Header.csv
ordernum,salesorg,distchan,division,CustomerID
12,1000,10,10,C1234
13,1000,10,10,C1234
14,1000,10,10,C1234
15,1000,10,10,C1234
```

Figure 6.21: Extra Record in the Test Data File Before Its Deployment to SAP HANA

If you want to run your application from SAP Business Application Studio but need to see data from your SAP HANA database, you must change the `kind` property of the *package.json* file to `hana`, as shown in Figure 6.22, and then run `cds watch`.

```
},
"cds": {
  "requires": {
    "db": {
      "kind": "hana",
      "credentials": {
        "database": "sqlite.db"
      }
```

Figure 6.22: Changing the kind Property to hana

Now, when viewing your order data, you'll see the extra record is now available because the data is coming from the SAP HANA database, as shown in Figure 6.23.

```
{
  "@odata.context": "$metadata#Orders",
  "value": [
    {
      "ordernum": 12,
      "salesorg": "1000",
      "distchan": "10",
      "division": "10",
      "CustomerID": "C1234"
    },
    {
      "ordernum": 13,
      "salesorg": "1000",
      "distchan": "10",
      "division": "10",
      "CustomerID": "C1234"
    },
    {
      "ordernum": 14,
      "salesorg": "1000",
      "distchan": "10",
      "division": "10",
      "CustomerID": "C1234"
    },
    {
      "ordernum": 15,
      "salesorg": "1000",
      "distchan": "10",
      "division": "10",
      "CustomerID": "C1234"
    }
  ]
}
```

Figure 6.23: Additional Record Coming from the SAP HANA Database

Switching the `kind` property to `sql` will show only three records, which means your SQLite database has three records, but SAP HANA has four records for the `Header` entity (i.e., `Order Entityset`) of the OData service.

6.6 Generating SAP Fiori User Interfaces

SAP Cloud Application Programming Model provides out-of-the-box support for SAP Fiori elements applications, driven by annotations. Figure 6.24 shows links to several SAP Fiori elements applications, listed under **Service Endpoints**.

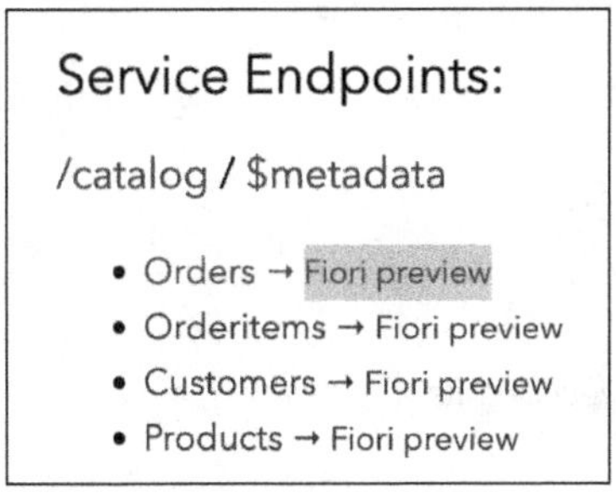

Figure 6.24: SAP Fiori Preview

These links are for previews of possible SAP Fiori applications, as shown in Figure 6.25.

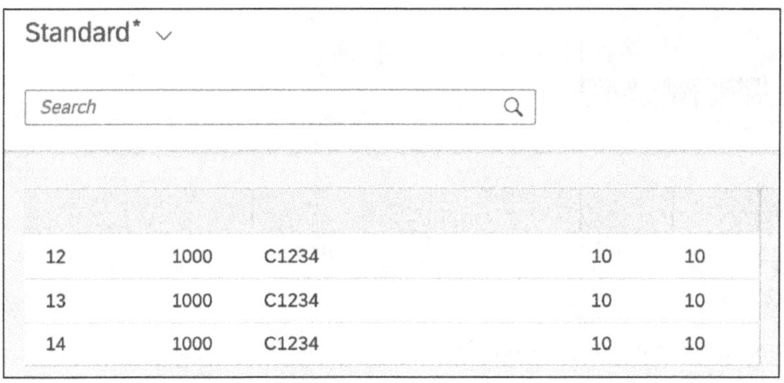

Figure 6.25: SAP Fiori App in Preview Mode

To build a full-fledged SAP Fiori app, you'll need to add annotations for your domain model or your service model. Annotations are additional metadata that you can use to instruct the SAPUI5 framework how the UI should look. Based on your defined annotations at runtime, the SAPUI5 framework will derive the necessary UIs for you. Thus, through annotations, you can create low-code/no-code SAP Fiori applications. We recommend adding separate annotation files to the **app** folder, but you can technically put annotations anywhere in a model.

SAP Business Application Studio provides another powerful feature, SAP Fiori tools, for building SAP Fiori applications on top of your SAP Cloud Application Programming Model service. To create an SAP Fiori app, follow the menu path **File • New Project From Template** and then select **SAP Fiori Application**. On the next screen, select **SAP Fiori Elements** and **List report object page**. Then, you'll select an SAP Cloud Application Programming Model project by its project path and select a service, as shown in Figure 6.26. Once you complete these steps in the project, your SAP Fiori-related files should be visible in the **app** folder, as shown in Figure 6.27.

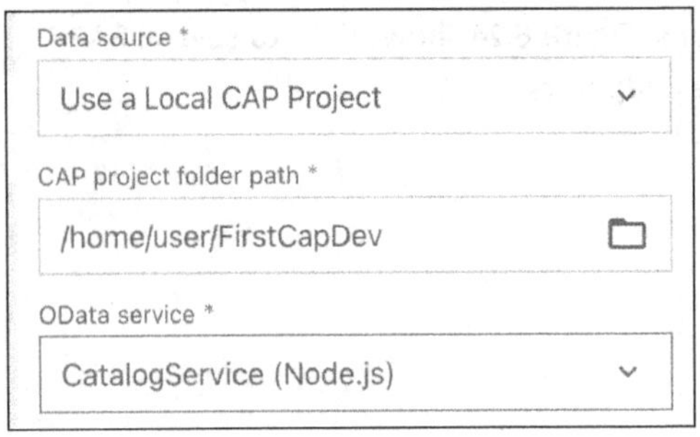

Figure 6.26: SAP Cloud Application Programming Model Project Details

Figure 6.27 shows a separate annotation file added to the **app • orders** folder.

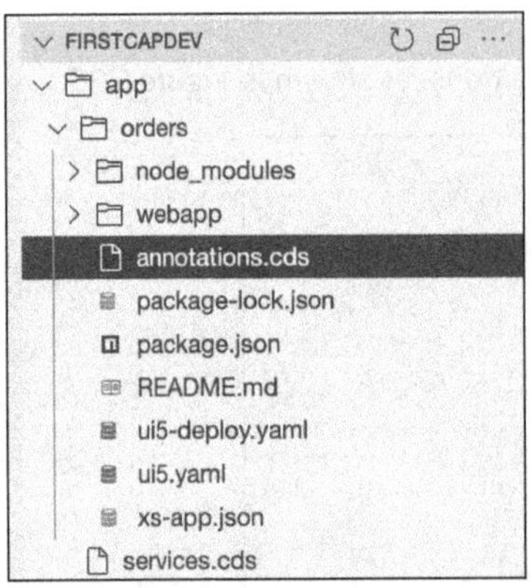

Figure 6.27: Annotation File in the app/orders Folder

Some annotation features include the following:

- Annotations start with @; for instance, `@UI.LineItem` is for table fields.
- You can annotate service entities, entity elements, and more.
- Annotations support internationalization (i.e., i18n for supporting multiple languages).
- Through annotations, you can define the draft functionality for editable entities.

SAP Business Application Studio supports code completion for annotations so that you can create annotations more quickly. Listing 6.6 shows a generated annotation file with several different types of annotations created.

```
using CatalogService as service from '../../srv/ord-service';
annotate service.Orders with @(
    UI.LineItem : [
        {
            $Type : 'UI.DataField',
```

```
            Label : 'ordernum',
            Value : ordernum,
        },
        {
            $Type : 'UI.DataField',
            Label : 'salesorg',
            Value : salesorg,
        },
        {
            $Type : 'UI.DataField',
            Label : 'distchan',
            Value : distchan,
        },
        {
            $Type : 'UI.DataField',
            Label : 'division',
            Value : division,
        },
        {
            $Type : 'UI.DataField',
            Label : 'CustomerID',
            Value : CustomerID,
        },
    ]
);
annotate service.Orders with @(
    UI.FieldGroup #GeneratedGroup1 : {
        $Type : 'UI.FieldGroupType',
        Data : [
            {
                $Type : 'UI.DataField',
                Label : 'ordernum',
                Value : ordernum,
            },
            {
                $Type : 'UI.DataField',
                Label : 'salesorg',
                Value : salesorg,
            },
            {
                $Type : 'UI.DataField',
                Label : 'distchan',
                Value : distchan,
            },
```

```
                {
                    $Type : 'UI.DataField',
                    Label : 'division',
                    Value : division,
                },
                {
                    $Type : 'UI.DataField',
                    Label : 'CustomerID',
                    Value : CustomerID,
                },
            ],
        },
        UI.Facets : [
            {
                $Type : 'UI.ReferenceFacet',
                ID : 'GeneratedFacet1',
                Label : 'General Information',
                Target : '@UI.FieldGroup#GeneratedGroup1',
            },
        ]
);
```

Listing 6.6: Annotations for SAP Fiori Apps

The `UI.LineItem` annotation is responsible for creating table fields in the list report, while `UI.FieldGroups` is responsible for displaying fields on the object page, and `UI.Facets` is responsible for organizing object page fields. Many other different UI annotations are available, which you can explore at *cap.cloud.sap*. If you again enter the `cds watch` command, you'll see that your test screen has been changed to display the newly created app, as shown in Figure 6.28.

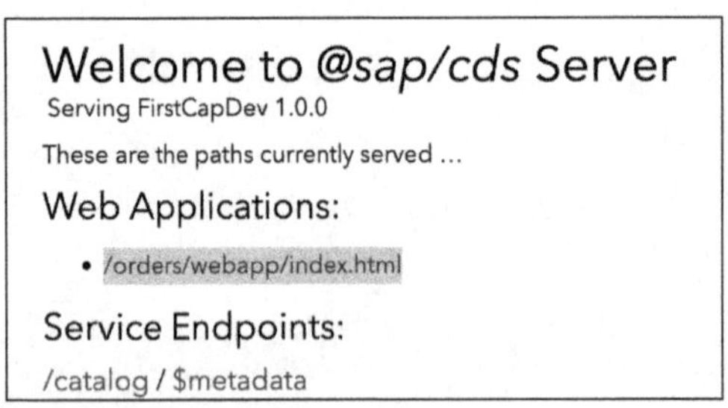

Figure 6.28: Test Screen Updated to Test the Newly Created SAP Fiori App

At this point, our SAP HANA database is available, the domain model is complete, a service has been exposed, and a UI will be generated through SAP Fiori elements. Now, everything can be tested via SAP Business Application Studio, but we'll deploy our application to SAP BTP, Cloud Foundry environment.

6.7 Deploying Applications to SAP BTP, Cloud Foundry Environment

An SAP Cloud Application Programming Model application can be deployed in various ways, depending on your use case, for instance, in the following ways:

- Deploy to SAP BTP, Cloud Foundry environment: For simple use cases.
- Deploy to SAP BTP, Cloud Foundry environment as a multitenant software as a service (SaaS) application: For applications where a cloud service provider will deploy the app, and multiple customers will use the app with their own data.
- Deploy to the continuous integration (CI)/continuous delivery (CD) pipeline: For deployment automation.
- Simple SAP BTP, Cloud Foundry environment deployment: Using Cloud Foundry's native `cf push` command.

In this book, we'll focus on the first option, deploying to SAP BTP, Cloud Foundry environment. First, however, let's explore some key tools essential for deployment:

- Cloud MTA Build Tool: This tool creates the *.mtar* file, which is an MTA archive file, from the deployment descriptor (i.e., *mtad.yaml*) of the SAP Cloud Application Programming Model application.
- The Cloud Foundry CLI plugin
- The MTA plugin for the Cloud Foundry CLI plugin: This tool is used to deploy the *.mtar* file created by the Cloud MTA Build Tool to SAP BTP, Cloud Foundry environment.

With these tools in place, to execute the deployment process, follow these steps:

1. If you created the project using the wizard in SAP Business Application Studio, then the Cloud MTA Build Tool and the Cloud Foundry CLI with the MTA plugin have already been installed for your project. Run the commands `mbt` and `cf` to ensure these tools are available in your project.
2. If you used an SQLite database for testing and now you want to switch to an SAP HANA database for deployment, run the command `cds add hana --for production`. In this context, `production` means the production landscape that has the SAP HANA database as a persistence.
3. Now, add extended services for user account and authentication (XSUAA)-based authentication using the command `cds add xsuaa --for production`. You'll learn more about XSUAA authentication in Section 6.9.2, but for now, simply run the command.
4. Add the *AppRouter* as a gateway. You can make your app available to users in two ways after deployment: using the AppRouter as the entry point or using the SAP Fiori launchpad. We'll use the AppRouter option in this scenario, so we'll run the command `cds add approuter --for production`.

5. If you previously used the `cds add mta` command, then your project folder already has the required deployment descriptor; otherwise, run this command to add the *mta.yaml* file to your project structure.
6. You should now freeze dependencies using the command `npm update --package-lock-only`. Freezing dependencies will create a file called *package-lock.json*. This file may also be present in your project hierarchy depending on how you created the project or how you added some other tools.
7. Use the Cloud MTA Build Tool to build and assemble everything into an *mta.tar* archive. Use the command `mbt build -t gen --mtar mta.tar` to create the *mta.tar* file under the **gen** folder.
8. Make sure your SAP HANA database instance is up and running, which is particularly important if you're using a trial account because this instance is stopped automatically when not used.
9. The final step is to deploy the *mta.tar* file to SAP BTP, Cloud Foundry environment. You can now run the command `cf deploy gen/mta.tar` to deploy the *mta.tar* file. Make sure you are logged on to SAP BTP, Cloud Foundry environment, using the command **CF:Login to Cloud Foundry** via the menu path **View • Find Command** in SAP Business Application Studio prior to running this command.

Once you complete the deployment process, your app should be visible under the **Routes** menu of the SAP BTP, Cloud Foundry environment space you selected before deployment, as shown in Figure 6.29.

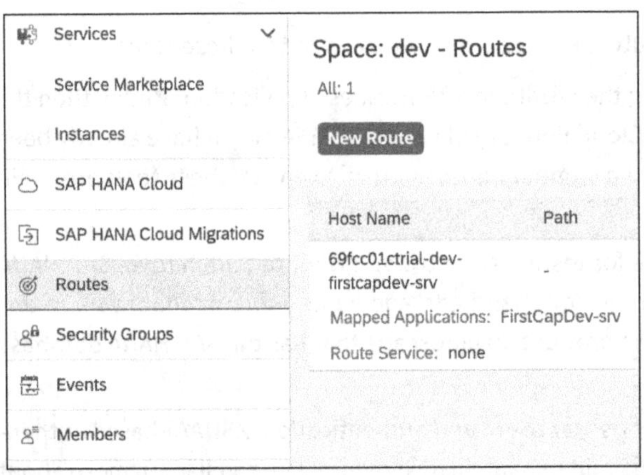

Figure 6.29: Deployed App in an SAP BTP, Cloud Foundry Environment Space

At this point, you've seen how a complete full-stack SAP Cloud Application Programming Model application, with all its database artifacts, service components, and SAP Fiori UIs, can be built and deployed to SAP BTP, Cloud Foundry environment. However, you may also encounter a situation where you need to consume external services for your application (e.g., consuming SAP S/4HANA services to build a side-by-side application for your core ERP system). We'll cover the steps for consuming external services next.

6.8 Consuming External Services

So far, we've described the steps for building an application from scratch and running its database artifacts either on local SQLite database or on SAP BTP/SAP HANA Cloud. However, in many situations, you may need to consume external services from one or more on-premise and/or cloud-based systems, for instance, from SAP S/4HANA, SAP S/4HANA Cloud, SAP Sales Cloud, SAP SuccessFactors, and others. Your requirement might be a simple consumption step or might be complex enough to extend a service with additional fields for calculation, aggregation, etc. In this section, we'll walk you through the steps for consuming an external service, starting from design to deployment, as shown in Figure 6.30.

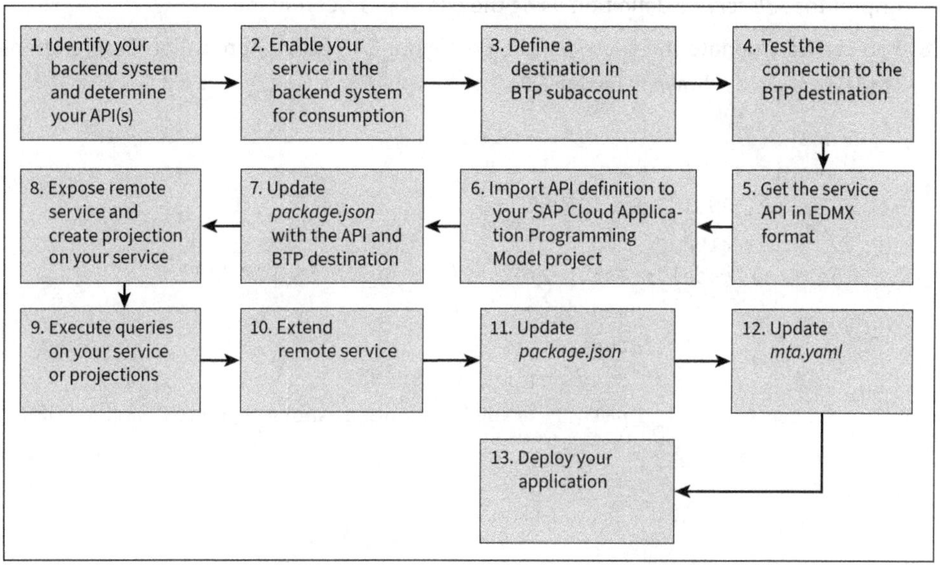

Figure 6.30: Steps for Consuming External Services

Let's briefly go over each step next:

1. The first step is simple: identify the related backend system(s) that will provide the required services. Note that your backend service can be from an on-premise SAP system (e.g., SAP S/4HANA); an on-premise non-SAP system; an SAP cloud system like SAP Ariba or SAP Sales Cloud; or even a non-SAP cloud system like Salesforce. Depending on the type of system, your configuration will be slightly different. For example, you can use the cloud connector to consume any services from SAP S/4HANA on-premise.
2. The next step is to identify one or more APIs and related services from the backend system(s). You may need to consume different services from multiple systems and mash up the data in your SAP Cloud Application Programming Model application. If your backend system is SAP S/4HANA, then you can explore the available APIs from SAP Business Accelerator Hub at *https://api.sap.com*.

3. Enable this service in your backend system. For example, you must define a communication arrangement for that specific communication scenario for SAP S/4HANA Cloud, and for SAP S/4HANA, you'll need to activate the OData service (OData API).
4. In the next step, define an SAP BTP destination in your predefined subaccount to connect to the previously enabled backend system APIs.
5. Test the connection to the SAP BTP destination and make sure it's reachable.
6. Get the API's service definition in EDMX format. For SAP APIs, you can export the EDMX file from SAP Business Accelerator Hub. SAP Cloud Application Programming Model also works with OData v4 export.
7. Import the API service definition using the `cds import` command.
8. You can now update the *package.json* file of your SAP Cloud Application Programming Model project, as shown in Listing 6.7.

```
    "cds": {
    "requires": {
        "API_BUSINESS_PARTNER": {
            "kind": "odata",
            "model": "srv/external/API_BUSINESS_PARTNER",
                "[production]": {
              "credentials": {
                 "destination": "S4HANA",
                 "path": "/sap/opu/odata/sap/API_BUSINESS_PARTNER"
                }
             }

        }
    }
}
```

Listing 6.7: Updating the package.json File with External Service Details

As shown in Listing 6.7, you can use destination for production testing after deployment or use the local testing option without the destination.

9. Next, you'll expose, extend, and project your service, as shown in Listing 6.8, which is similar to exposing a local service.

```
using {  API_BUSINESS_PARTNER as bupa } from '../srv/external/API_
BUSINESS_PARTNER';

    extend service RiskService with {
    entity BusinessPartners as projection on bupa.A_BusinessPartner;
}
```

Listing 6.8: Exposing, Extending, and Projecting Your Service

10. You can now use Node.js or Java code to query your external service for calculating additional fields and logic.
11. Update your *mta.yaml* file to add details under the `resources` section, as shown in Listing 6.9.

```
  - name: cpapp-uaa
  type: org.cloudfoundry.managed-service
  parameters:
    service: xsuaa
    service-plan: application
    path: ./xs-security.json

- name: cpapp-destination
  type: org.cloudfoundry.managed-service
  parameters:
    service: destination
          service-plan: lite
```

Listing 6.9: Resources in the mta.yaml File

12. If you're connecting to an on-premise system, you must add a connectivity service as well, as shown in Listing 6.10.

```
- name: cpapp-connectivity
  type: org.cloudfoundry.managed-service
  parameters:
    service: connectivity
    service-plan: lite
```

Listing 6.10: Connectivity Service for On-Premise Systems

13. Now, you can deploy your application in the usual way and test your remote service or additional query logic.

We've now covered both the provisioning for and the service consumption of an SAP Cloud Application Programming Model application. Before we wrap up this chapter, we'd like to explore four additional topics that are critical for developing enterprise-grade SAP Cloud Application Programming Model applications—localization, authentication and authorization, annotations, and the SAP Cloud Application Programming Model LLM plugin for building SAP Fiori applications.

6.9 Additional Topics

We're almost at the end of this chapter explaining how to build SAP Cloud Application Programming Model applications and running them on SAP BTP, Cloud Foundry environment. To enhance this chapter, we'll go further, starting with localization and showing you how to make an app available in multiple languages. Then, we'll describe XSUAA-based authentication, followed by discussion about some common annotations required for building fully functional SAP Cloud Application Programming Model applications. Last, but not least, we'll discuss the SAP Cloud Application Programming Model LLM plugin that lets you create advanced SAP Cloud Application Programming Model applications powered by AI/ML services.

6.9.1 Localization of SAP Fiori Apps

The localization of an app is nothing but the translation of various texts in your application into target languages. In the localization process, you must indicate which texts need to be translated. This step can be achieved by removing hard-coded texts and externalizing your texts as i18n texts, as shown in Figure 6.31.

```
app > orders > annotations.cds
using CatalogService as service from '
annotate service.Orders with @(
    UI.LineItem : [
        {
            $Type : 'UI.DataField',
            Label : '{i18n>ordernum}',
            Value : ordernum,
        },
        {
            $Type : 'UI.DataField',
            Label : '{i18n>salesorg}',
            Value : salesorg,
        },
        {
            $Type : 'UI.DataField',
            Label : '{i18n>distchan}',
            Value : distchan,
        },
        {
```

Figure 6.31: i18n Texts for Translations

As shown, these texts are indicated by the `i18n>` prefix in curly braces, and each word next to that prefix serves as a key to uniquely identify that text. Next, you'll create i18n files to contain the translations for each key. A typical i18n file is shown in Figure 6.32, with default language (English) texts.

```
app > orders > webapp > i18n > i18n.properties
# This is the resource bundle for orders

#Texts for manifest.json

#XTIT: Application name
appTitle=App Title

#YDES: Application description
appDescription=A Fiori application.
ordernum = Order Number
salesorg = Sales Org.
distchan = Distribution Channel
division = Division
```

Figure 6.32: i18n File Content (Default)

When you created the SAP Fiori elements module for your SAP Cloud Application Programming Model project, a default i18n file might have already been created with the name *i18n.properties*, in the *app/<app name>/webapp/i18n* folder. You can edit this file further to add additional texts, as shown in Figure 6.32.

Other options for storing i18n files include in the *_i18n* folder or in the *_assets/i18_* folder. Using the property `cds.i18n.folders` in the *package.json* file, you can change the folder path as needed. In this way, you can create multiple files for different languages, as shown in Figure 6.33.

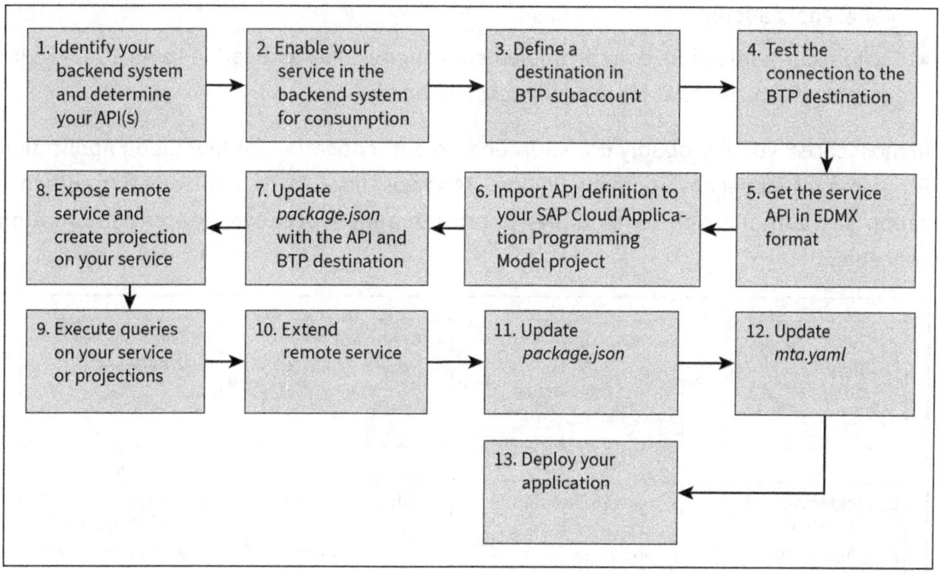

Figure 6.33: i18n Files for Different Languages

Figure 6.33 shows three variations of i18n files: The file without any language specification will be applied first and then language-specific i18n files will be applied depending on the user's preferred language. The user's language is determined in multiple ways (e.g., browser settings, `sap-language` URL parameter, HTTP request header's `accept-language` parameter, default language configured at app level, and so on).

We hope you have a good sense of how the localization concept can be applied for your SAP Cloud Application Programming Model projects. Let's now shift our focus to another area—XSUAA-based authentication and authorization.

6.9.2 Authentication and Authorization

SAP provides various authentication strategies for use with SAP Cloud Application Programming Model applications. By default, SAP Cloud Application Programming Model uses `mocked-auth` in development (in SAP Business Application Studio) and `jwt-auth` in production (in SAP BTP). You can override these settings by specifying a different authentication strategy in the *package.json* file. Some options include the following:

- Dummy authentication: This type of authentication creates a user that passes all authorization checks.
- Basic authentication: Authentication with user ID and password that can be used for testing.
- JSON Web Token (JWT)-based authentication: Authentication is performed using tokens (JWT) in the authorization header of the incoming HTTP request.
- XSUAA-based authentication: This option is also based on JWT and additionally offers SAML attributes. This choice is commonly used for SAP Cloud Application Programming Model applications.
- Custom authentication: You can implement your own authentication using a JavaScript file and referencing that file in your *package.json* file.

In most cases, you'll probably use XSUAA-based authentication for SAP Cloud Application Programming Model development. Figure 6.34 shows the steps for implementing authentication and authorization using XSUAA to provide a high-level overview of authorization concepts.

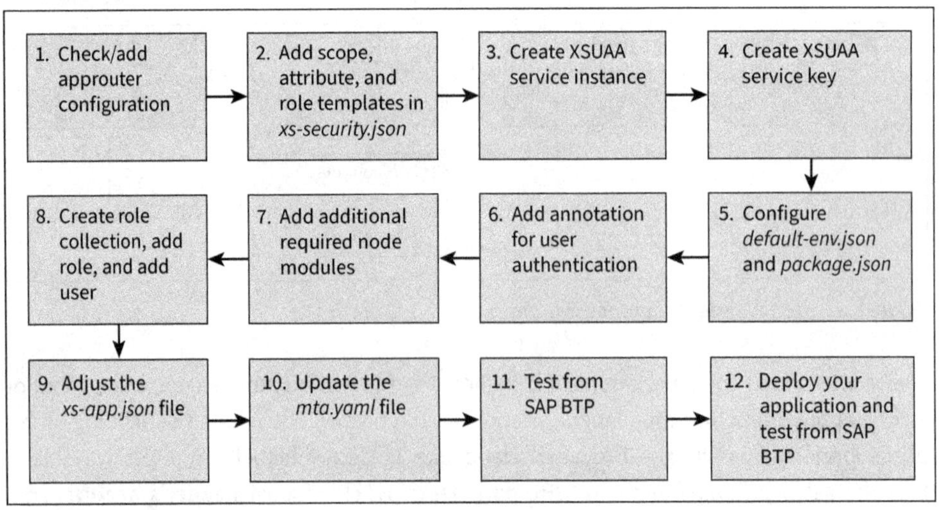

Figure 6.34: Steps for Setting Up Authentication and Authorization

Let's go over each step shown in Figure 6.34 in sequence:

1. Add the AppRouter, which is the central entry point through which a calling component can access your SAP Cloud Application Programming Model application. If you haven't added an AppRouter to your project, doing so is the first step you should perform. The AppRouter can be added either using the command `cds add approuter --for production` or using the wizard available via the menu path **View • Find Command** and selecting the option **MTA Module from Template.** Opening the wizard will display the AppRouter option, as shown in Figure 6.35.

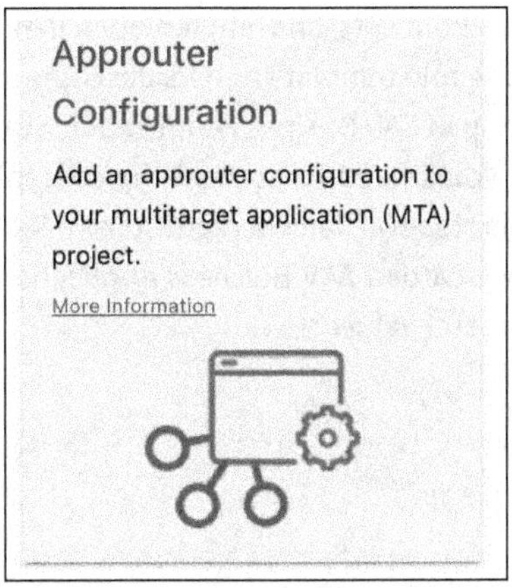

Figure 6.35: Adding an AppRouter Using a Wizard

Using the wizard will help you create the AppRouter for your SAP Cloud Application Programming Model project.

2. When you add an AppRouter in the first step, the *xs-security.json* file is created in the project's root folder. Using the wizard will also add some sample content to this file, as shown in Listing 6.11.

```
{
  "xsappname": "firstcapdev",
  "tenant-mode": "dedicated",
  "description": "Security profile of called application",
  "scopes": [
    {
      "name": "uaa.user",
      "description": "UAA"
    }
  ],
  "role-templates": [
    {
      "name": "Token_Exchange",
      "description": "UAA",
      "scope-references": [
        "uaa.user"
      ]
    }
  ]
}
```

Listing 6.11: Default Content of the xs-security.json File

The `scopes` element enables various authorization concepts and implements them using Node.js or Java. You can update the name of the role template and its description to meet your requirements. To test your security setup in SAP Business Application Studio, before deploying the application to SAP BTP, you'll need to append the content shown in Listing 6.12 to your *xs-security.json* file. This step will allow a request from SAP Business Application Studio to be processed as a wildcarded SAP Business Application Studio URL maintained as the `redirect-uris` parameter value.

```
"oauth2-configuration": {
    "redirect-uris": [
        "https://*.applicationstudio.cloud.sap/**"
    ]
}
```

Listing 6.12: Redirect URI to Test from SAP Business Application Studio

3. In the next step, create an XSUAA service instance by using the command `cf create-service xsuaa application firstcapdev-uaa -c xs-security.json`.
4. Now, create the service key for the XSUAA service instance using the command `cf create-service-key firstcapdev-uaa firstcapdev-key`. Here, the `firstcapdev-key` is the service key of the service instance `firstcapdev-uaa`, created in the previous step. After successful creation of your service instance and service key, you can view them from your SAP BTP cockpit.
5. Next, add the content shown in Listing 6.13 to your *default-env.json* file.

```
"xsuaa": [{
          "name": "firstcapdev-uaa",
          "label": "xsuaa",
          "tags": ["xsuaa"],
        "credentials": {
            [...]
        }
   }],
```

Listing 6.13: Initial Content for default-env.json

6. Then, go to the SAP BTP cockpit and open the service key details of your XSUAA service instance and copy the entire content, as shown in Figure 6.36.
7. Paste this content to the *default-env.json* file for your project in the specified path, as shown in Figure 6.37.

dev-key

Form | JSON


```
{
    "clientid": "sb-firstcapdev-uaa!t54983",
    "clientsecret": "rYYIB0lPsG8rXuTkgwtdE8eQHGo=",
    "url": "https://69fcc01ctrial.authentication.us10.hana.ondemand.com",
    "identityzone": "69fcc01ctrial",
    "identityzoneid": "f88f2a73-0e86-4e52-ae42-78bd84d9f2de",
    "tenantid": "f88f2a73-0e86-4e52-ae42-78bd84d9f2de",
    "tenantmode": "shared",
    "sburl": "https://internal-xsuaa.authentication.us10.hana.ondemand.com",
    "apiurl": "https://api.authentication.us10.hana.ondemand.com",
    "verificationkey": "-----BEGIN PUBLIC KEY
        -----MIIBIjANBgkqhkiG9w0BAQEFAAOCAQ8AMIIBCgKCAQEAqazN3pqcRasZ40zWFQHBMRy0I5loAq
        pw3m9IBIkaA8C1BdrV9jZazQGXENMHTnG0at5OqzTPvYh5mf0QwZh5veHxIteanMOOpDrOcdg0FUA
        +JqqSCqWM7vBwnHToL+GwaAIceieFQsDgJL/+JgTntQ37b0MAdTCGW0zm
        +n75zVl9K2VCcfgcM2RRWnN88Z1N33TTPDJqF8T3lRf1eUhXPsPoLhiNWSH8YDkx
        +gW9nk8XhdhASJfqPu4/ovB1MiPi0Hzboc8O/mnMUq7UwlRwwd2sLf3fgdr7lgeb6PqnojmBJ1C8eew
        /SrsxZ3ui/sI9BkU0tAN0FjX4QKBI04DJwIDAQAB-----END PUBLIC KEY-----",
    "xsappname": "firstcapdev-uaa!t54983",
```


Figure 6.36: JSON Content for the XSUAA Service Key


```
"VCAP_SERVICES": {
  "xsuaa": [{
      "name": "firstcapdev-uaa",
      "label": "xsuaa",
      "tags": ["xsuaa"],
      "credentials": {
              "tenantmode": "shared",
              "sburl": "https://internal-xsuaa.authentication.us10.hana.ondemand.com",
              "subaccountid": "f88f2a73-0e86-4e52-ae42-78bd84d9f2de",
              "credential-type": "instance-secret",
              "clientid": "sb-firstcapdev-uaa!t54983",
              "xsappname": "firstcapdev-uaa!t54983",
              "clientsecret": "rYYIB0lPsG8rXuTkgwtdE8eQHGo=",
              "url": "https://69fcc01ctrial.authentication.us10.hana.ondemand.com",
              "uaadomain": "authentication.us10.hana.ondemand.com",
              "verificationkey": "-----BEGIN PUBLIC KEY-----MIIBIjANBgkqhkiG9w0BAQEFAA
              "apiurl": "https://api.authentication.us10.hana.ondemand.com",
              "identityzone": "69fcc01ctrial",
              "identityzoneid": "f88f2a73-0e86-4e52-ae42-78bd84d9f2de",
              "tenantid": "f88f2a73-0e86-4e52-ae42-78bd84d9f2de",
              "zoneid": "f88f2a73-0e86-4e52-ae42-78bd84d9f2de"
            }
      }
  ],
  "hana": [
    {
      "name": "FirstCapDev-db",
      "tags": [
```


Figure 6.37: Final Content of the default-env.json File

8. You should also update the *package.json* file, under the `cds: requires` section, with the content shown in Listing 6.14.

```
"uaa": {
          "kind": "xsuaa"
        }
```

Listing 6.14: Content for package.json Under cds.requires

9. Add `@requires: 'authenticated-user'` to your *ord-service.cds* file, as shown in Figure 6.38.

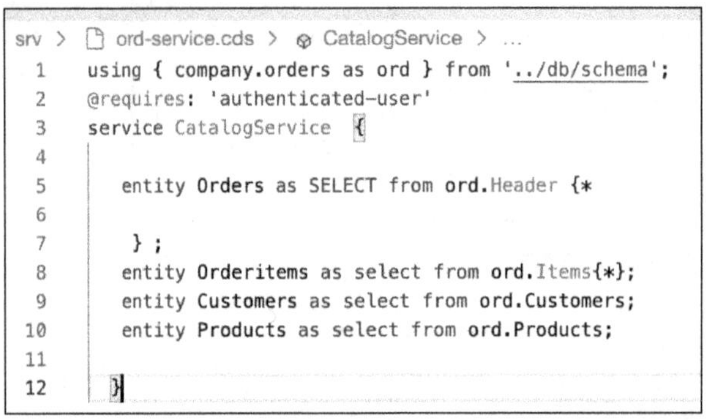

Figure 6.38: Adding the Annotation to Authenticate the Service

10. Now, add the additional node modules to your *package.json* by running the command `npm install --save passport @sap/xssec @sap/xsenv @sap/audit-logging` from the project's root folder.
11. You can now go to the SAP BTP cockpit and create a role collection. Then, add the roles created through *xs-security.json* and assign the role collection to the required users.
12. Adjust *xs-app.json* to add the statement `"authenticationMethod": "route"`.
13. Adjust the *mta.yaml* file to bind the XSUAA module to your service module.
14. You can now test your SAP Cloud Application Programming Model application's authentication by using `cds watch`, or you can deploy the application to SAP BTP and test from there.

So, you now understand how authentication works with an SAP Cloud Application Programming Model project and how to set up authentication using SAP Business Application Studio.

We'll now briefly explain how you can add authorization concepts along with authentication for your SAP Cloud Application Programming Model application. An authorization concept can be implemented using various annotations in your service definition. Let's explore some key authorization-specific annotations next:

- `@requires` adds authenticated users or adds specific custom roles for any entities.
- `@restrict` restricts or allows certain actions (e.g., read/write access) to specific custom roles. You can add various other conditions as well.
- `@readonly` restricts certain entities as read-only for all users.

Once your role and authorization design is complete, run the command `cds compile service.cds --to xsuaa > xs-security.json`, which will update the *xs-security.json* file with additional scopes and attributes.

With this brief look at authentication and authorization, let's move on to the next topic: annotations.

6.9.3 Annotations

As discussed in Section 6.6, SAP Fiori UIs are generated automatically through annotations. You may already be familiar with the annotation concept in SAP Fiori, but let's now look at annotations in more detail.

An annotation is additional metadata that enriches a service; annotations are understood by consumers that interpret the annotations in a meaningful way. Various annotation consumers exist. For instance, the ABAP platform interprets ABAP-related annotations, the SAPUI5 framework consumes SAP Fiori UI-related annotations, and the SAP S/4HANA analytics engine consumes embedded analytics annotations, among others. SAP Cloud Application Programming Model supports annotations, but these annotations are interpreted by different runtime engines. Annotations start with an @ symbol and can have multiple levels, (e.g., `@UI.LineItem`). In this section, we'll focus on the most common UI- and OData-related annotations. SAP Cloud Application Programming Model uses annotation vocabularies following the OASIS standard as well as some SAP-specific annotation vocabularies. Some examples of annotations in SAP-specific vocabularies include the following:

- `@Common`: This annotation is common for all types of vocabularies.
- `@Communication`: This annotation carries communication related information (e.g., contact, address, email, phone, etc.).
- `@PersonalData`: These annotations represent personal and/or sensitive data.
- `@Analytics`: These annotations represent various analytical resources (e.g., dimensions, measures, aggregations, etc.).
- `@UI`: These annotations are related to SAP Fiori UIs. Some key UI annotations include the following:
  - `HeaderInfo`: This annotation helps you define the various UI properties of the header section of an object page. This annotation can have further subannotations (e.g., `Title`, `Description`, `ImageUrl`, etc.).

- `LineItem`: Line-item annotations are responsible for defining various UI properties for the fields of a list report (e.g., `Labels` for column headings, `Criticality` to represent values in different colors, and `IconURL` if the field is an icon, etc.).
- `Identification`: Identification annotations similar to `LineItems` but responsible for defining various properties of an object page's fields.
- `SelectionFields`: To define the selection/filter fields for list report applications.
- `SelectionVariant`: To bundle all your filter selections together, similar to ABAP's selection-screen variant.
- `PresentatationVariant`: To bundle the result of the queried data along with their UI positions, sorting criteria, and more. This annotation is similar to an SAP List Viewer (ALV) variant.
- `SelectionPresentationVariant`: If you want to bundle `SelectionVariant` and `PresentationVariant` together using a single variant name, then you need to define the `SelectionPresentationVariant`.
- `MultiLineText`: This annotation will identify the fields that can have unlimited number of characters (string) represented by multiple lines of texts on the screen.
- `Facets`: To organize the header area of an object page into various sections and subsections.

Refer to *http://s-prs.co/v550408* for more information on the available annotations—the possibilities are endless!

6.9.4 SAP Cloud Application Programming Model LLM Plugin

So far, we have covered the process of creating SAP Cloud Application Programming Model applications using SAP Business Application Studio. However, SAP Cloud Application Programming Model applications, in the context of today's autonomous enterprise, can become significantly more sophisticated and advanced—especially when addressing complex AI/ML or enterprise integration scenarios. The SAP Cloud Application Programming Model LLM plugin is a major step toward enabling these applications to connect to the generative AI hub and SAP HANA Cloud vector engine to perform tasks such as anonymization, embeddings, similarity search, chat completion, and end-to-end retrieval-augmented generation (RAG) flows. The plugin lets you connect to LLMs via SAP AI Core/generative AI hub using SAP Cloud Application Programming Model configurations (for example, model settings in *.cdsrc.json*) and SAP BTP destinations. However, it remains a separate SAP Cloud Application Programming Model–side component rather than an SDK module of SAP AI Core. As depicted in Figure 6.39, the SAP Cloud Application Programming Model LLM plugin offers key features, including the following:

- Data anonymization: The plugin enables you to leverage SAP HANA Cloud's advanced data anonymization techniques to seamlessly protect sensitive data using simple annotations such as `@anonymize`, as well as algorithms like k-anonymity and l-diversity, ensuring data integrity and compliance.

- Enhanced AI-powered functionalities: The plugin lets you integrate SAP Cloud Application Programming Model applications with SAP AI Core, facilitating the generation of vector embeddings and performing similarity searches, which are especially crucial for applications that require precise and insightful data analysis.
- Simplified access to LLMs: The plugin also offers streamlined access to LLMs for tasks such as chat completion, which deliver contextually appropriate AI-generated responses, enhancing user interaction.
- Effective RAG: The plugin optimizes the RAG process by integrating the functionalities of both SAP AI Core and the SAP HANA Cloud vector engine into a single, streamlined method. This integration enables dynamic data fetching and utilization in AI operations, significantly boosting efficiency.

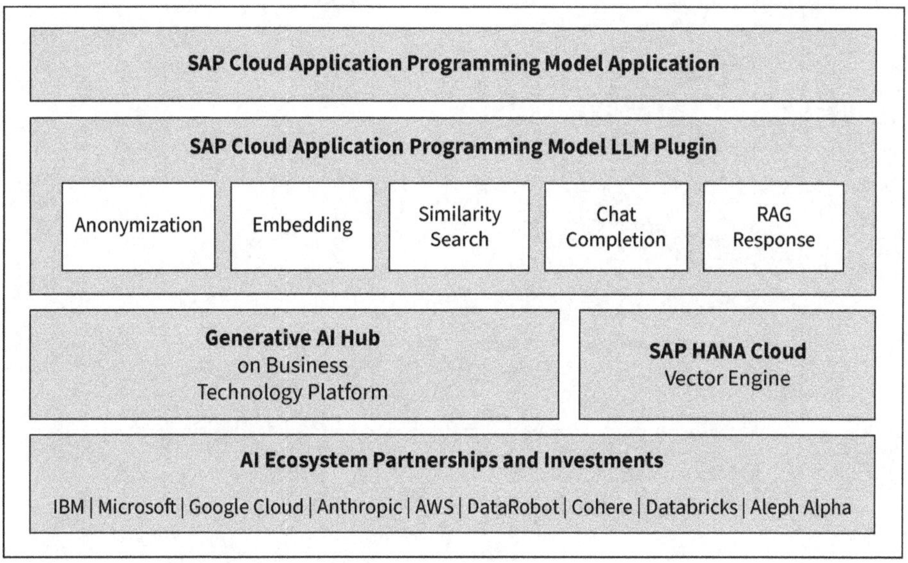

Figure 6.39: SAP Cloud Application Programming Model LLM Plugin

For more information related to the SAP Cloud Application Programming Model LLM plugin and APIs, refer to *https://github.com/SAP-samples/cap-llm-plugin-samples/blob/main/docs/api-documentation.md.*

6.10 Summary

Building a simple SAP Cloud Application Programming Model application is easy, but building an enterprise-grade one is not as simple because you'll need to handle all aspects of building a service, including service consumption from external systems, local persistence, complex UI development, authentication and authorization, and so on. SAP provides extensive ready-to-use content for faster application development, and more and more businesses are implementing this content to quickly see the benefits of using domain modeling and wizard-driven application development features. We hope to see more customers adopt SAP Cloud Application Programming Model as SAP continuously improves its features and functionalities. Concluding this chapter on SAP Cloud Application Programming Model, let's now turn to building Node.js applications in SAP BTP.

Chapter 7
Developing Node.js Applications

This chapter describes how to develop a business application using SAP Cloud Application Programming Model for Node.js. We'll be using SAP Business Application Studio for the development of backend objects using core data services (CDS), domain models, CDS service definitions, and Node.js services. Our example business application will be deployed on SAP Business Technology Platform (SAP BTP), Cloud Foundry environment, as a multitarget application (MTA).

This chapter focuses on developing a business application using the SAP Cloud Application Programming Model for Node.js. We'll be using SAP Business Application Studio, available via the SAP BTP platform, for the development of backend database objects and the CDS service definitions consumed by Node.js services in an HTML application, which we'll deploy to SAP BTP, Cloud Foundry environment.

For beginners, SAP Cloud Application Programming Model is a framework of languages, libraries, and tools for building enterprise-grade services and applications. It leverages open-source technologies like CDS and a runtime for Node.js along with SAP technologies like SAP HANA Cloud, SAP Fiori, and SAPUI5. SAP Cloud Application Programming Model based applications can be either created directly inside SAP Business Application Studio or accelerated further using the Joule-based coding assistant called SAP Build Code through simple natural language prompts.

7.1 Creating a Project Using SAP Business Application Studio

In Chapter 4, we discussed SAP Business Application Studio's capabilities and functionalities in detail. In this section, we'll focus on creating a full-stack cloud application using SAP Business Application Studio. Our simple project involves calculating the survival rate of a region, calculated as the number of births minus the number of deaths in the region. To create a new project using SAP Business Application Studio, follow these steps:

1. Log on via your SAP Business Application Studio URL and then click on **Create Dev Space**, as shown in Figure 7.1.

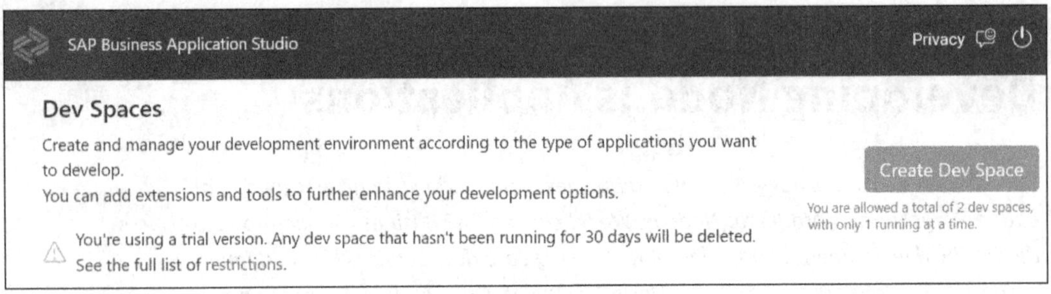

Figure 7.1: Logging on to SAP Business Application Studio

2. Create a dev space by providing the following information, as shown in Figure 7.2:
   - Provide a name for your dev space ❶.
   - Select the **Full Stack Cloud Application** radio button ❷.
   - Select additional SAP extensions ❸, which will allow SAP HANA database artifacts to be created in the space.
   - Click on **Create Dev Space** ❹.

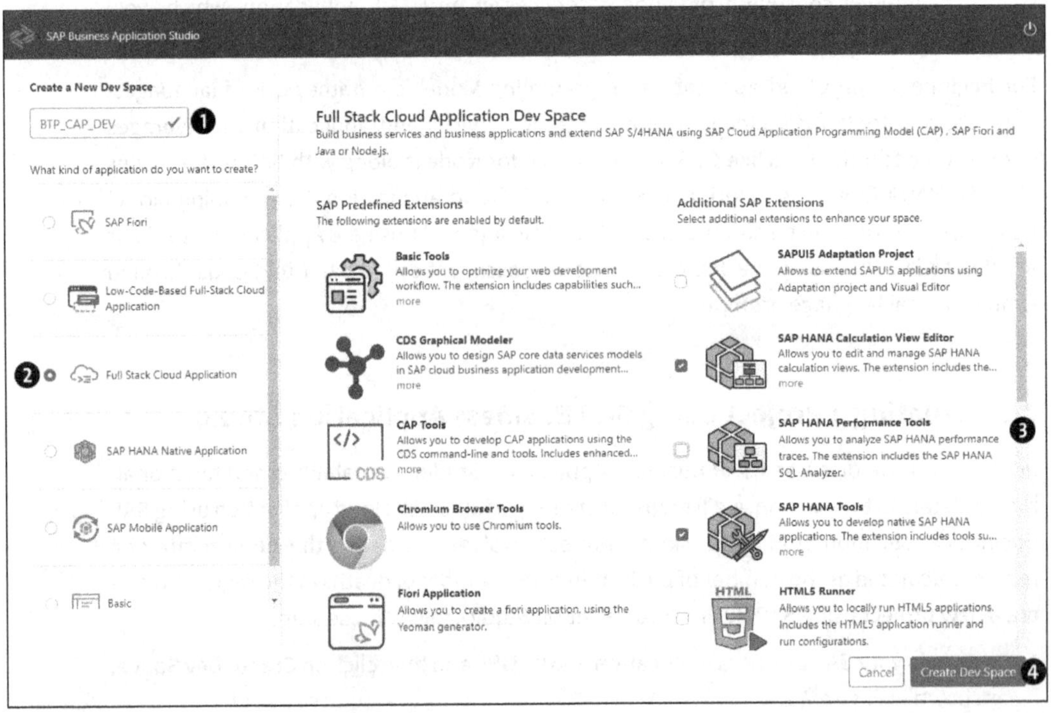

Figure 7.2: SAP Business Application Studio: Creating a Dev Space

3. Your SAP Business Application Studio dev space is now up and running, as shown in Figure 7.3.

Figure 7.3: SAP Business Application Studio: Dev Space Running

4. Click on **BTP_CAP_DEV** to launch SAP Business Application Studio, as shown in Figure 7.4.

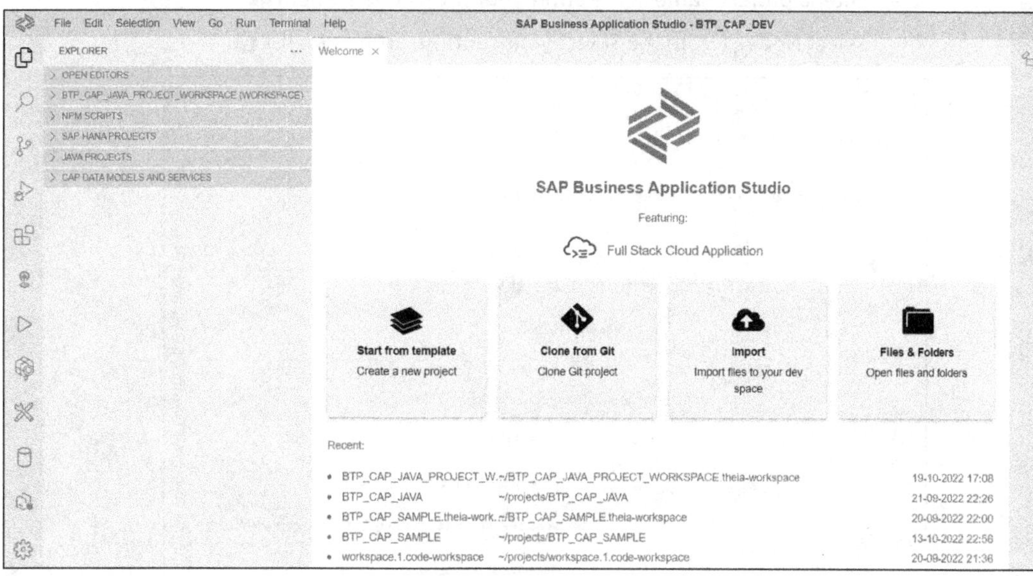

Figure 7.4: SAP Business Application Studio Launch Page

5. Let's start creating our project. Click on **Start from template**, as shown in Figure 7.4. SAP Business Application Studio provides numerous templates you can use, as shown in Figure 7.5. Select **CAP Project** and then click the **Start** button.

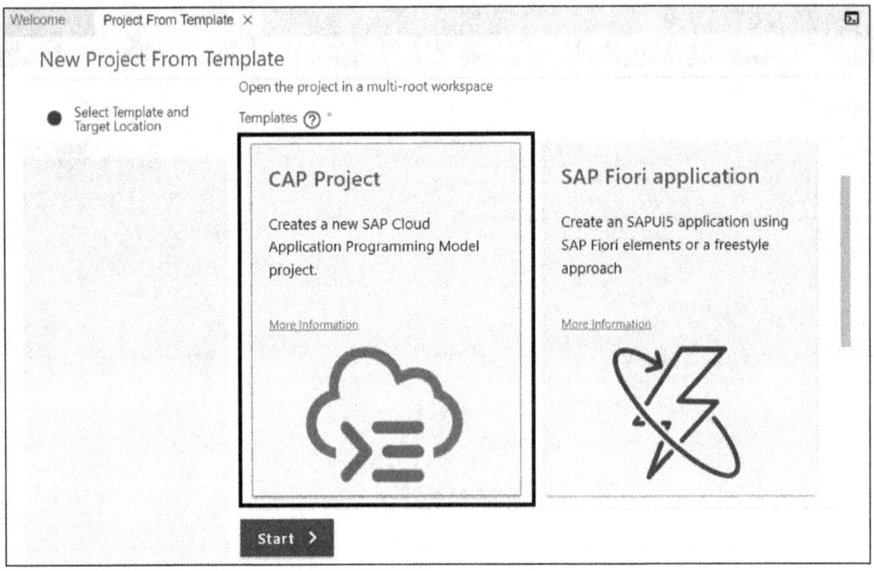

Figure 7.5: New Project from Template Screen

6. Provide the following information, as shown in Figure 7.6:
   - Enter a project name in the **Enter your project name** field ❶.
   - Select **Node.js** from the **Select your runtime** dropdown list ❷.

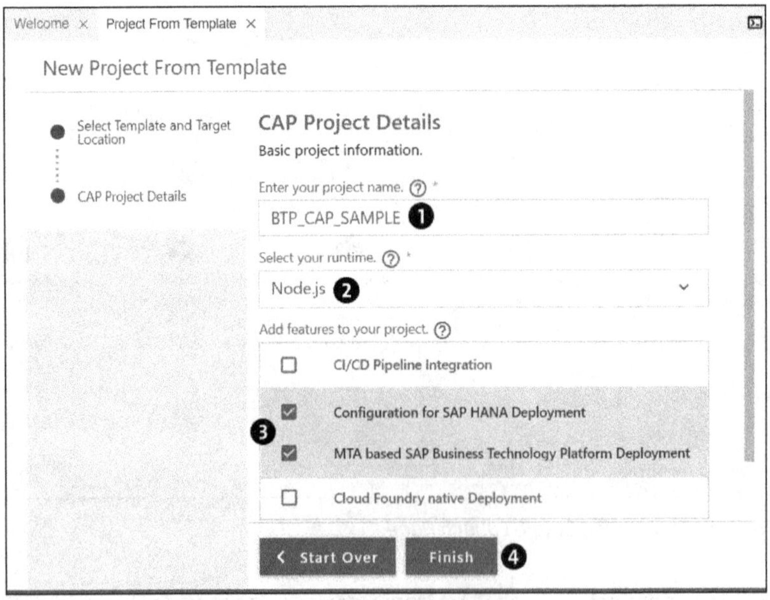

Figure 7.6: Project Details

7. Select the following checkboxes under **Add features to your project** ❸:
   - **Configuration for SAP HANA Deployment**
   - **MTA based SAP Business Technology Platform Deployment**
8. Click on **Finish** ❹.
9. Our project, called **BTP_CAP_SAMPLE**, has been created and includes the **db** and **app** modules, as shown in Figure 7.7.

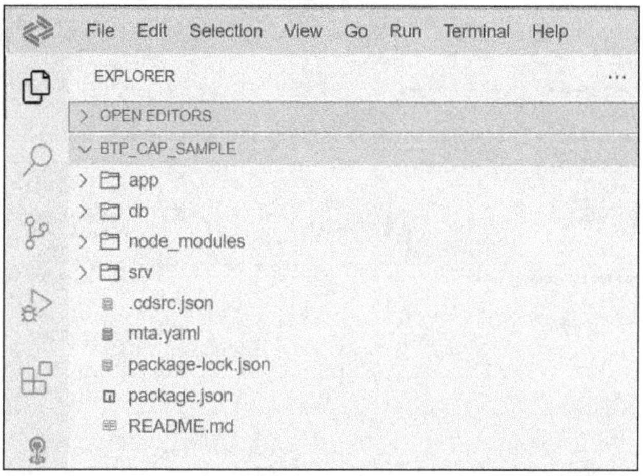

Figure 7.7: SAP Cloud Application Programming Model: Project Created

Tip

You can install various tools for business application development like the Cloud Foundry command-line interface (CLI), Visual Studio Code (VS Code), along with their extensions and user interfaces (UI), as standalone solutions, but SAP Business Application Studio provides all these tools, depending on types of business application, without local installation and also includes best practices.

7.2 Connecting to SAP BTP, Cloud Foundry Environment

Your project is created with db (database), srv (CDS service definition) and app (application) modules. We'll now proceed with connecting to SAP BTP, Cloud Foundry environment, to deploy database artifacts using SAP HANA deployment infrastructure (HDI) containers to deploy the business application. Some prerequisites for connecting to SAP BTP, Cloud Foundry environment, include the following:

- SAP BTP, Cloud Foundry environment, must be activated at the subaccount level in SAP BTP.
- A space must be created in an SAP BTP, Cloud Foundry environment, organization.

The SAP BTP, Cloud Foundry environment, API endpoint, organization name, and space are all required for connecting to the environment; this information can be found from the subaccount overview, as shown in Figure 7.8.

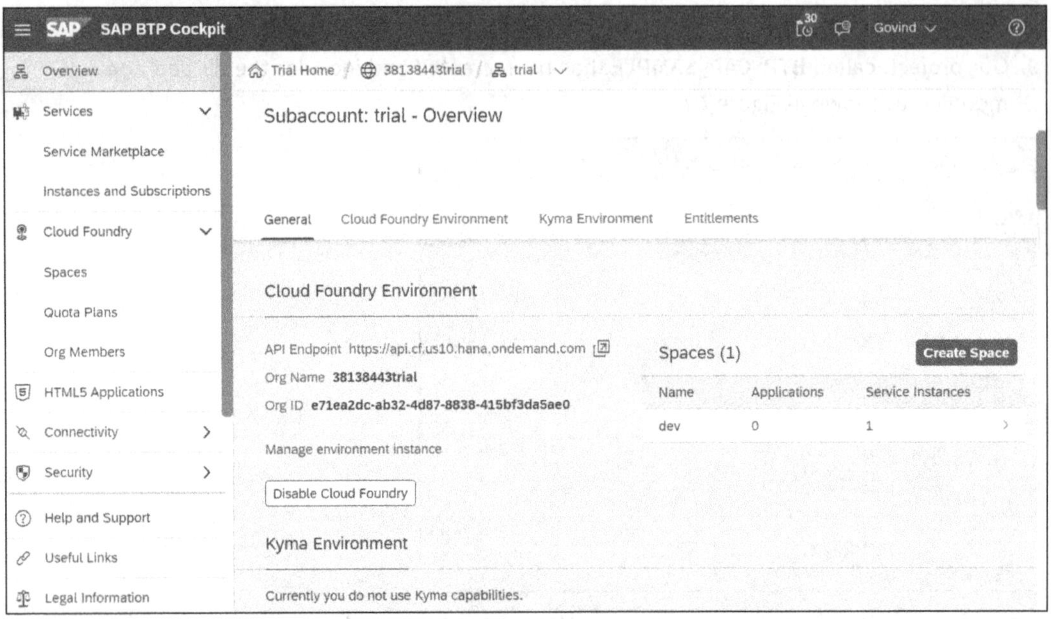

Figure 7.8: SAP BTP, Cloud Foundry Environment: Details

To connect to SAP BTP, Cloud Foundry environment, from SAP Business Application Studio, follow these steps:

1. From the menu bar, click **View • Find Command • CF: Login to Cloud Foundry**, as shown in Figure 7.9.

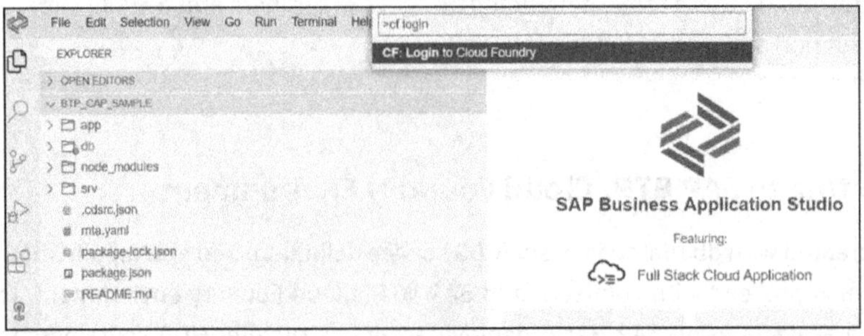

Figure 7.9: Logging on to SAP BTP, Cloud Foundry Environment, Using Commands in SAP Business Application Studio

2. Provide the API endpoint for your SAP BTP, Cloud Foundry environment, instance, as shown in Figure 7.10, and press `Enter`.

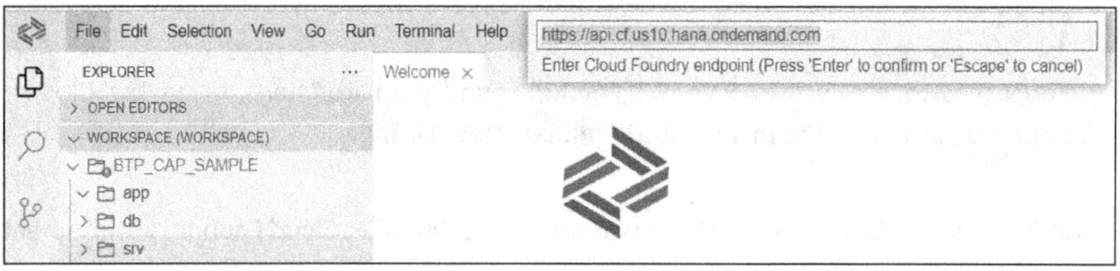

Figure 7.10: SAP BTP, Cloud Foundry Environment, API Endpoint

3. Provide the email used for logging on to SAP BTP, as shown in Figure 7.11.

Figure 7.11: Providing Logon Details

4. Provide the organization name in SAP BTP, Cloud Foundry environment, as shown in Figure 7.12.

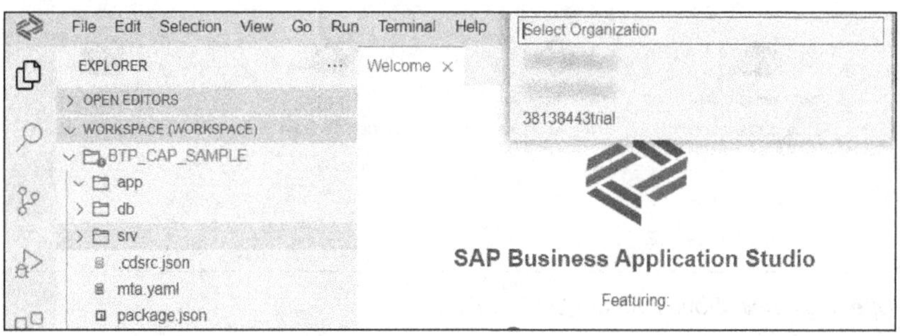

Figure 7.12: Selecting an Organization from the List

5. Provide a name for the space, as shown in Figure 7.13. Now, you're connected to SAP BTP, Cloud Foundry environment.

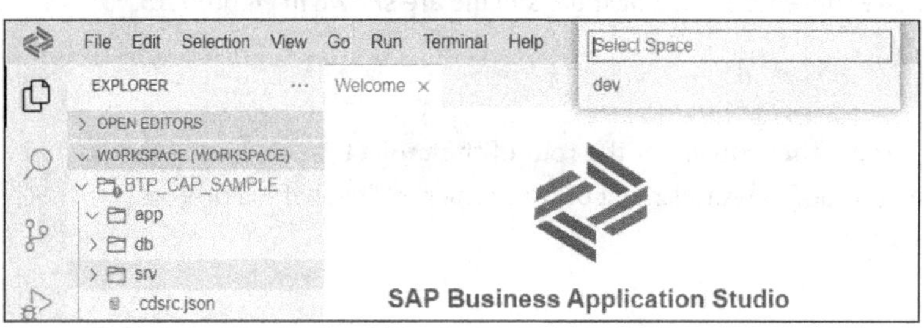

Figure 7.13: Selecting a Space from the List

[+]

Tip

You can also connect to SAP BTP, Cloud Foundry environment, using the Cloud Foundry CLI. Follow the menu path **Terminal • New Terminal.**

SAP Business Application Studio is now connected to SAP BTP, Cloud Foundry environment.

7.3 Installing Node.js Modules

To install the necessary Node.js modules, open a new Cloud Foundry CLI terminal and follow the menu path **Terminal • New Terminal**, as shown in Figure 7.14.

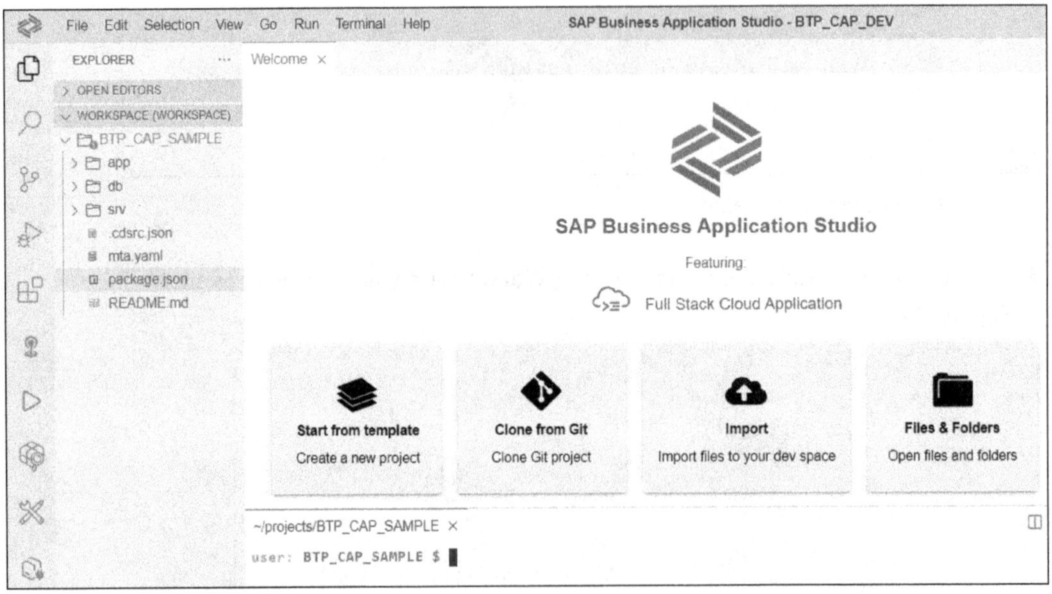

Figure 7.14: Opening a New Cloud Foundry CLI Terminal

In the terminal, run the command `npm install`. By default, this command will install all the modules listed as `dependencies` and `devDependencies` in the *package.json* file in the project root folder. A new folder called *node_modules* will be created during the installation of Node.js modules. The contents of the *package.json* file are shown in Figure 7.15.

[!]

Warning

You should make sure that you are at the root of the project in the Cloud Foundry CLI because installing Node.js libraries must occur at the root level of the project.

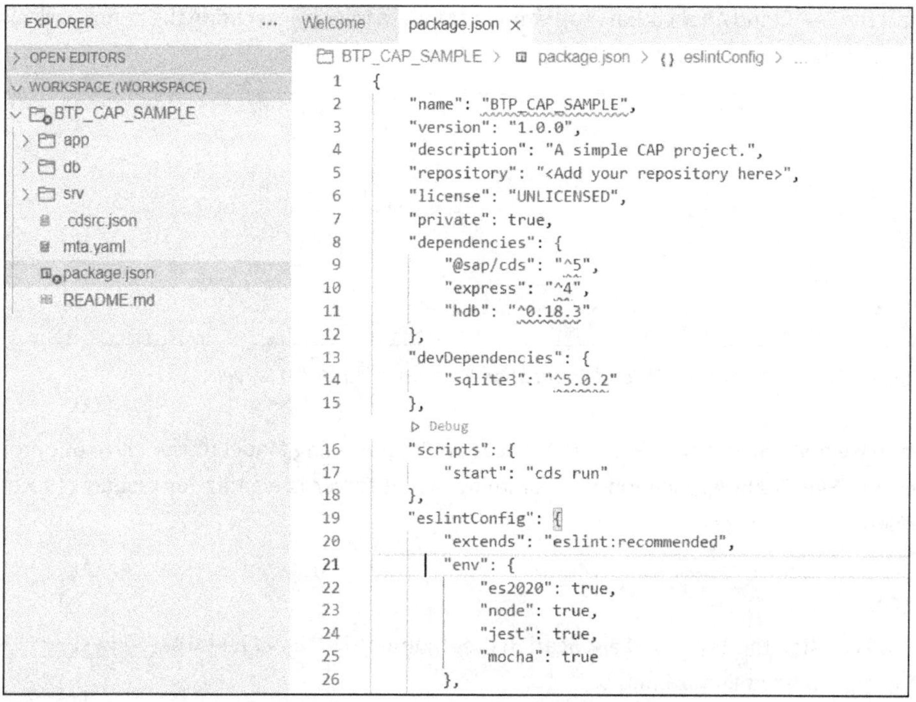

Figure 7.15: package.json File in the Root Folder of the Project

After executing the `npm install` command, the *node_modules* folder is created in the root directory of the project, and this folder includes all the required modules, along with their `dependencies` and `devDependencies`, as shown in Figure 7.16.

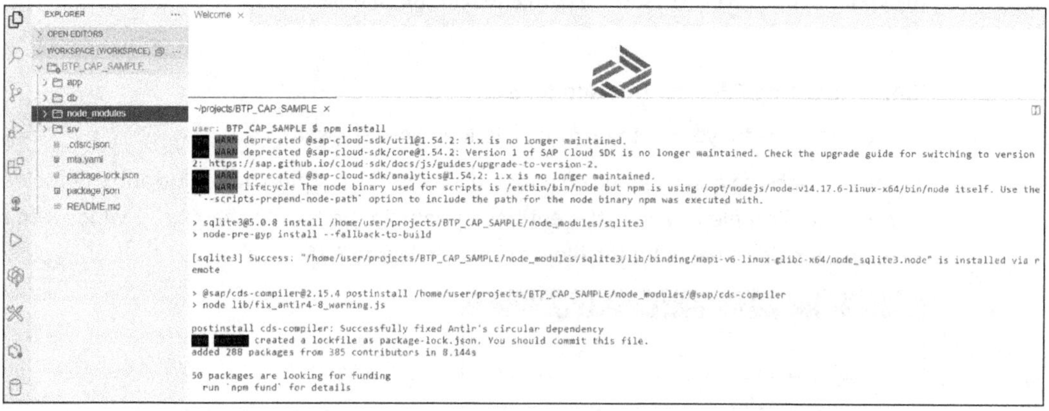

Figure 7.16: Executing the npm install Command

The SAP Cloud Application Programming Model server can provision all the related resources in your project. This server also watches all the files in your project and watches for new additions in files. It restarts automatically whenever a file is saved.

Start the SAP Cloud Application Programming Model server by executing the command `cds watch` in the terminal, as shown in Figure 7.17.

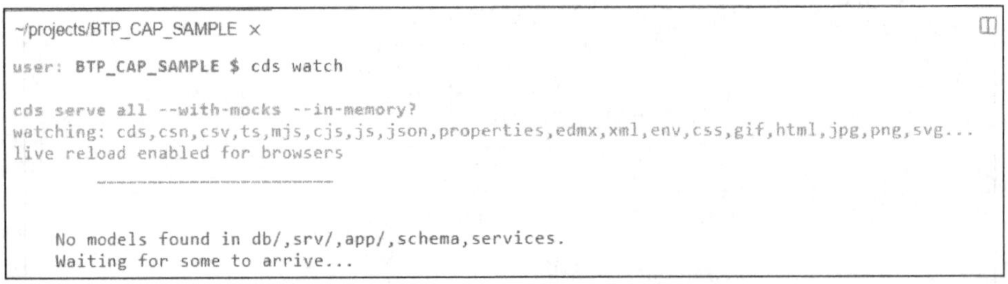

Figure 7.17: Starting the SAP Cloud Application Programming Model Server

Since we don't have any SAP Cloud Application Programming Model resources in our project, the SAP Cloud Application Programming Model server is waiting for resources to be added.

[+]

Tip

You can stop the SAP Cloud Application Programming Model server by pressing Ctrl + C in the terminal window.

7.4 Creating Core Data Services Resources for Our Sample Project

In this section, we'll show you how to create CDS domain models and services followed by the creation of data file required to run and test the solution.

7.4.1 Creating Files for Domain Models

To create the file for your domain model, follow these steps:

1. Create a *schema.cds* file in the *db* folder of your project. Right-click on **db** and then select **New File**. This file provides the entities in domain models. In the popup window that opens, provide a name for the file, as shown in Figure 7.18.

Figure 7.18: Creation of the schema.cds File

2. Open the file *db/schema.cds* and paste in the code in Listing 7.1.

```
namespace sample;

entity birth_death {
    period : String(4);
    birth_death : String(30);
    region : String(300);
    qty : Integer;
}
```

Listing 7.1: Add Code to db/schema.cds

Note

Once the *schema.cds* file is saved in the project folder, in the terminal, entering the command `cds watch` displays the message **No service definitions found in loaded models.** This message indicates that the SAP Cloud Application Programming Model server has recognized that a domain model has been created through the *schema.cds* file but you still need to create a service definition.

7.4.2 Creating Files for Service Definitions

The service definitions for entities created through the *schema.cds* file are also contained in files. The purpose of a service definition is to expose an entity as an OData service. To create the service definition file, follow these steps:

1. Create the *data-services.cds* file in the *srv* folder of your project by right-clicking on **srv** and then selecting **New File**. In the popup window, provide a name for the file, as shown in Figure 7.19.

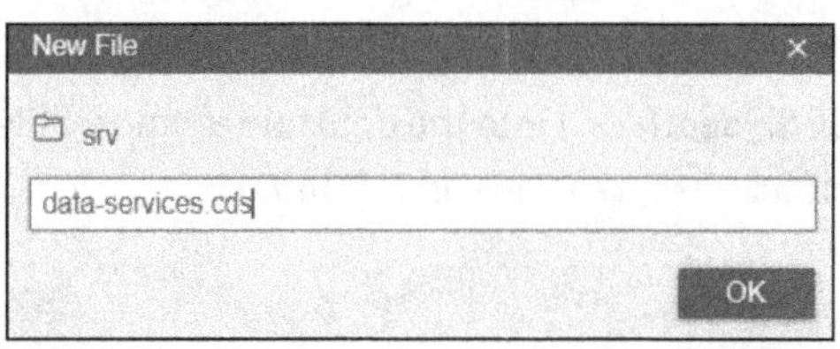

Figure 7.19: Creation of the data-services.cds File

2. Open the *srv/data-services.cds* file and paste in the code in Listing 7.2.

```
using sample from '../db/schema';

service dataservices {
    @readonly
    entity BirthDeathTable as projection on sample.birth_death;
}
```

Listing 7.2: Add Code to srv/data-services.cds

3. After the file is saved, the SAP Cloud Application Programming Model server will automatically identify the service definition file and make the file ready to serve at *http://localhost:4004*. Clicking this link will open the page shown in Figure 7.20 in a web browser.

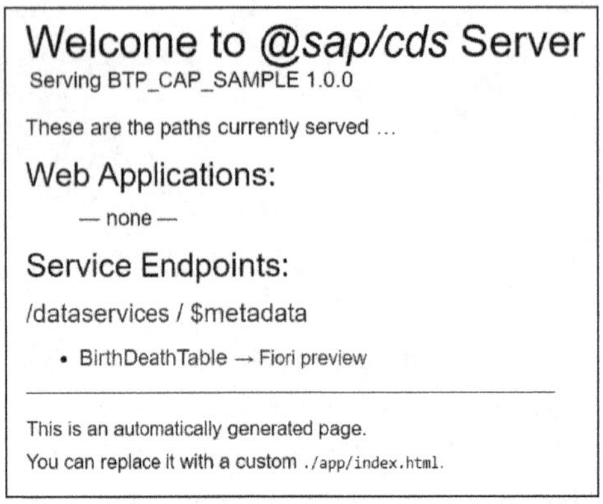

Figure 7.20: SAP Cloud Application Programming Model Server Webpage

4. Click on the **BirthDeathTable** link. No content will be displayed since no data exists in our domain model.

7.4.3 Creating Data Files

At this point in our example, no data exists in the entity **BirthDeathTable**. In this section, we'll create data files to upload data into the SQLite in-memory database by following these steps:

1. Create the subfolder *data* in the *db* folder of your project by right-clicking on **db** and selecting **New Folder**.
2. Import the CSV file by name (i.e., *sample-birth_death.csv*) into the *data* folder. Import file by right-clicking on **data** and selecting **Upload Files**, as shown in Figure 7.21.

[+] **Tip**

The naming convention of the CSV file is *<namespace>-<entity name>*, as defined in the *schema.cds* file. This naming convention ensures that data is uploaded into the associated entity.

3. Data can be previewed in your local server by clicking on **BirthDeathTable** (shown earlier in Figure 7.20).

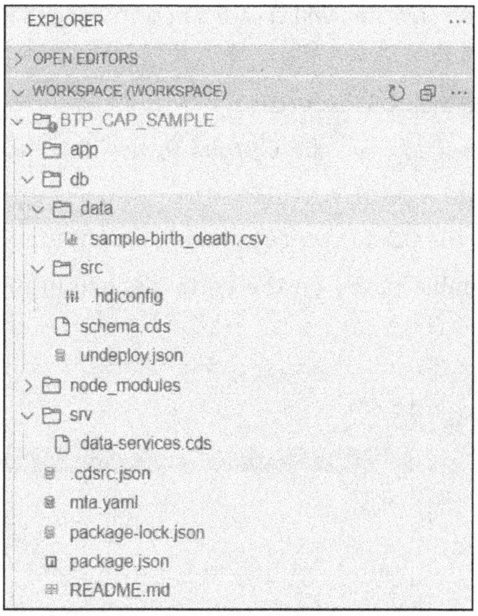

Figure 7.21: Importing a CSV File

7.5 Creating SAP HANA Cloud Artifacts

In the previous section, we created CDS resources and could view entities and services locally created in the SQLite in-memory database using the SAP Cloud Application Programming Model server. This database was listed in the `devDependencies` section of the *package.json* file in the root folder and was created when we ran the command `npm install`. We have not deployed any resources in the backend SAP HANA Cloud system, which we'll do now by following these steps:

1. Modify the *package.json* file in the root folder by replacing the code shown in Listing 7.3 with the code shown in Figure 7.22.

```
"cds": {
        "requires": {
            "db": {
                "kind": "sql"
            }
        },
```

Listing 7.3: Changes in the package.json File: Code to Replace

```
"cds": {
        "requires": {
            "db": {
                "kind": "hana"
            }
        },
```

Listing 7.4: Replacement Code

2. Run the command `cds build` in a new terminal. This step will create a new folder *gen* in the project root directory, as shown in Figure 7.22:
   - The new folder *gen* ❶ is created by building the CDS resources.
   - The *.hdbtabledata* file ❷ will initiate data load from the CSV file to the database object during deployment.
   - The *.hdbview* file ❸ creates an SQL view on the OData service exposing the entity.
   - The *.hdbtable* file ❹ creates a columnar table based on the entity defined in the *schema.cds* file.

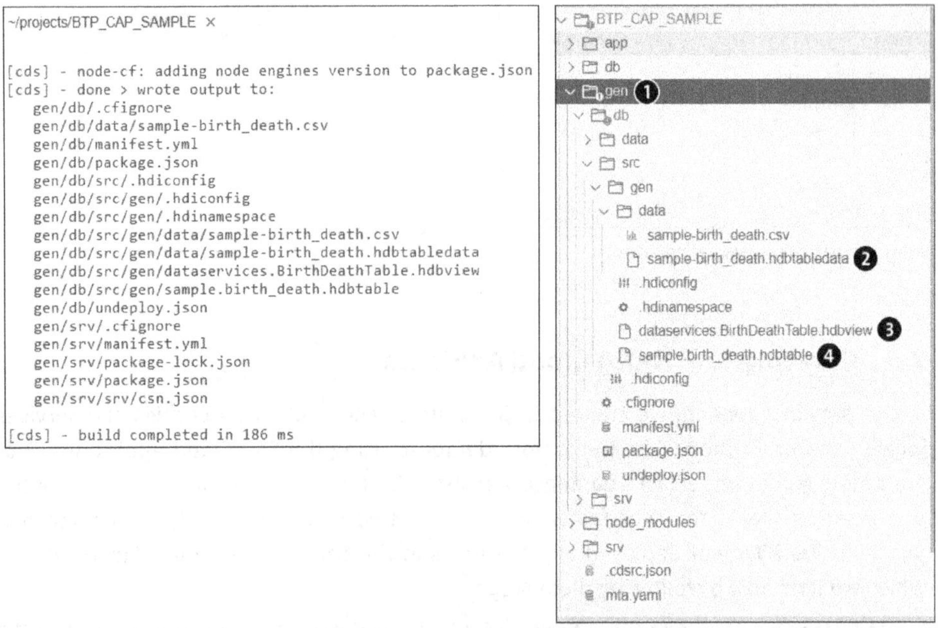

Figure 7.22: Building SAP HANA Cloud Backend Objects Based on CDS Resources

In the following sections, we are going to focus on binding the application to the cloud followed by deployment and viewing of the generated SAP HANA Cloud artifacts.

7.5.1 Binding Resources to the HDI Container Service

Now, we'll bind the resource **BTP_CAP_SAMPLE-db** to an HDI container service instance. Create an HDI container service instance by following the steps shown in Figure 7.23:

❶ In the **EXPLORER** pane, expand the **SAP HANA PROJECTS** section.

❷ Click on the bind icon.

❸ Select **Create a new service instance**.

❹ Provide the name for the service instance (i.e., BTP_CAP_SAMPLE) and press `Enter`.

❺ A message will appear in SAP Business Application Studio confirming the creation of the service instance.

❻ View the service instance in the SAP BTP cockpit.

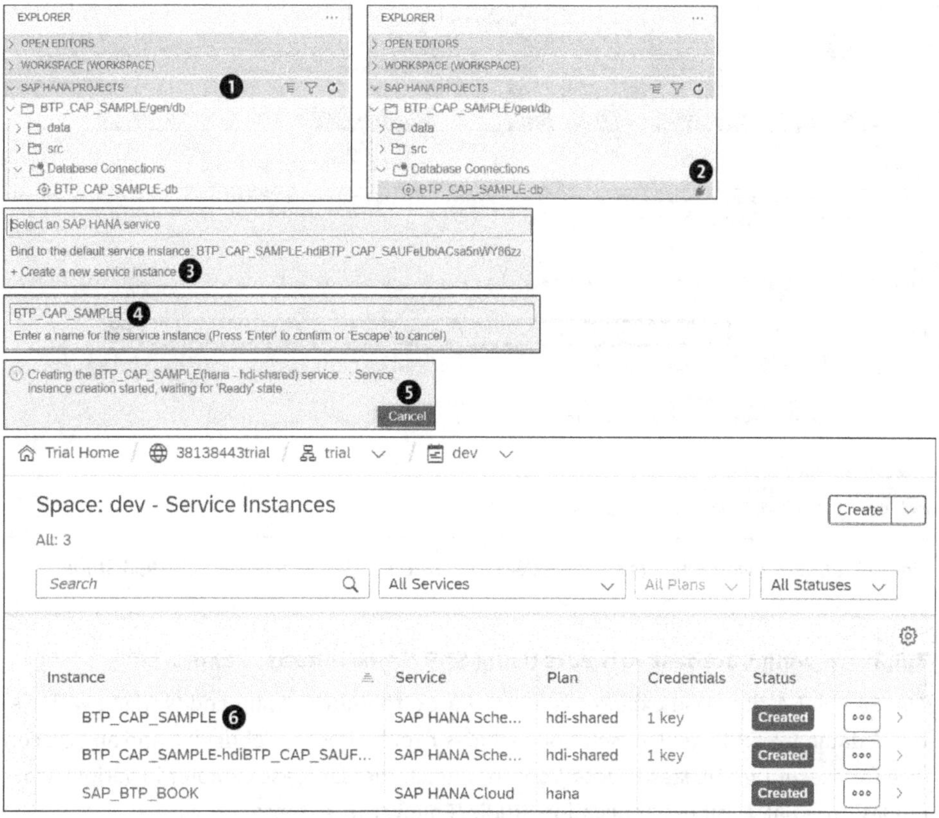

Figure 7.23: Binding a Resource to an SAP HANA Cloud Service Instance

7.5.2 Deploying Artifacts in SAP Business Application Studio

Now that we've bound the SAP HANA Cloud service instance, we're now ready to deploy the objects to SAP HANA Cloud through SAP Business Application Studio.

> **Note**
>
> Deployment using SAP Business Application Studio in non-production environments is preferred. For production environments, the preferred method is through multitarget application archive (MTAR) files.

As shown in Figure 7.24, to deploy database artifacts to SAP HANA Cloud, follow these steps:

❶ In the **EXPLORER** pane, expand the **SAP HANA PROJECTS** section.

❷ Click on the deploy icon.

❸ You can monitor the status of the deployment. The container schema name is generated by hdi-deploy, which is the schema where all objects are deployed.

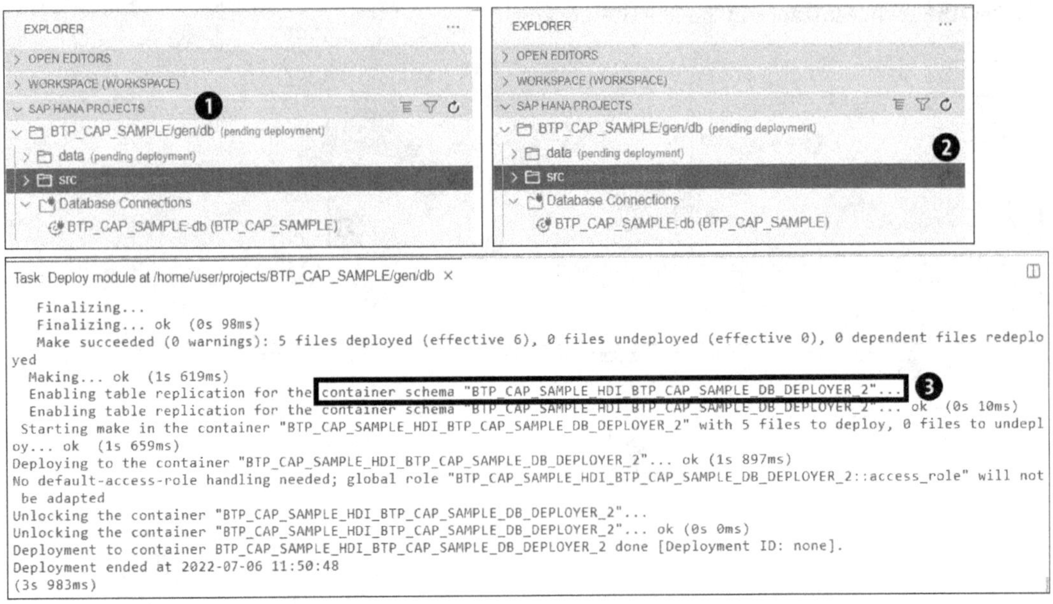

Figure 7.24: Deploying Artifacts to SAP HANA Cloud through SAP Business Application Studio

7.5.3 Viewing Database Artifacts Using SAP HANA Database Explorer

The SAP HANA database explorer is a web-based tool for viewing and working with SAP HANA database artifacts. This solution includes capabilities for debugging and an SQL console for executing SQL statements. You can launch the database explorer in various ways, but let's use SAP Business Application Studio. Follow these steps:

1. Follow the menu path **View • Find Commands**. In the search box, enter "SAP HANA: Open Database Explorer." This step will launch the database explorer in a web browser, as shown in Figure 7.25.

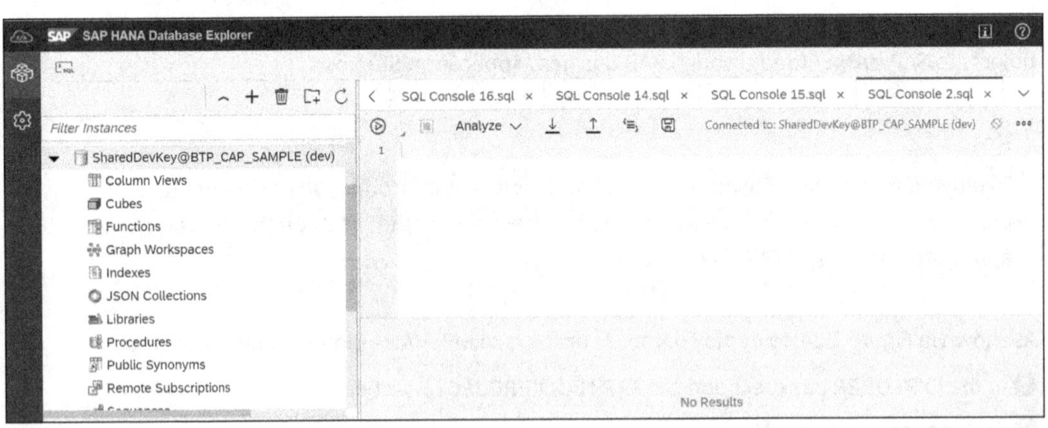

Figure 7.25: Launching the SAP HANA Database Explorer Through SAP Business Application Studio

Note

The database explorer automatically connects to the container created by the SAP HANA container service instance used during the deployment of the database artifacts.

2. To explore the database artifacts created by building the CDS models, as shown in Figure 7.26, follow these steps:
   - Double-click on **Tables** ❶.
   - Enter the table name or part of a table name ❷.
   - Right-click on a table and then select **Open Data** to view the data in the table ❸.

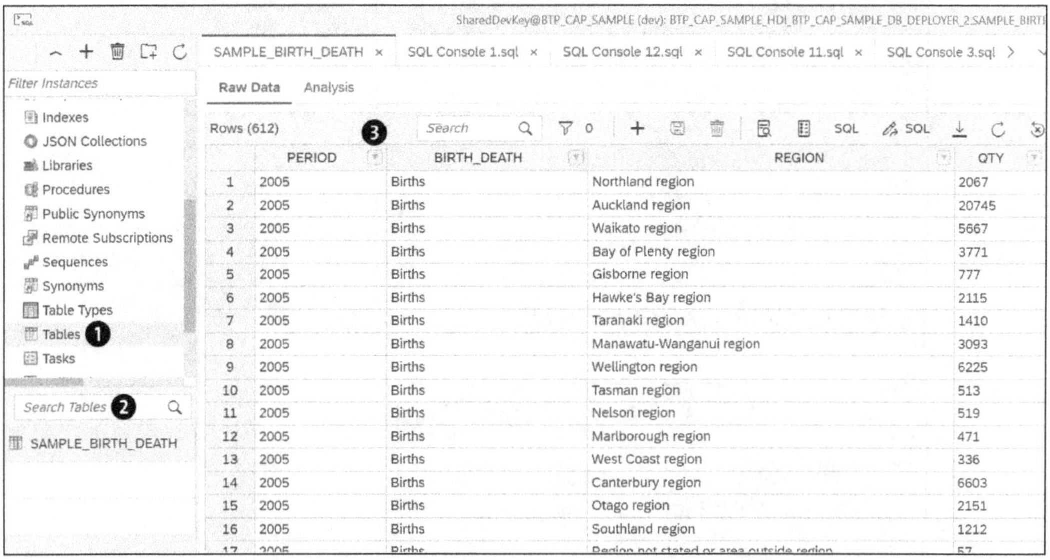

Figure 7.26: Displaying the Contents of hdbtable

7.6 Deploying Applications to SAP BTP, Cloud Foundry

Deploying an SAP Cloud Application Programming Model application built using Node.js to SAP BTP, Cloud Foundry environment, can be achieved by building an MTA project. Start by right-clicking on the **mta.yaml** file in the root folder of the **BTP_CAP_SAMPLE** project and then select **Build MTA Project**, as shown in Figure 7.27.

A file with the extension **.mtar** is created in the **mta_archives** folder, as shown in Figure 7.28.

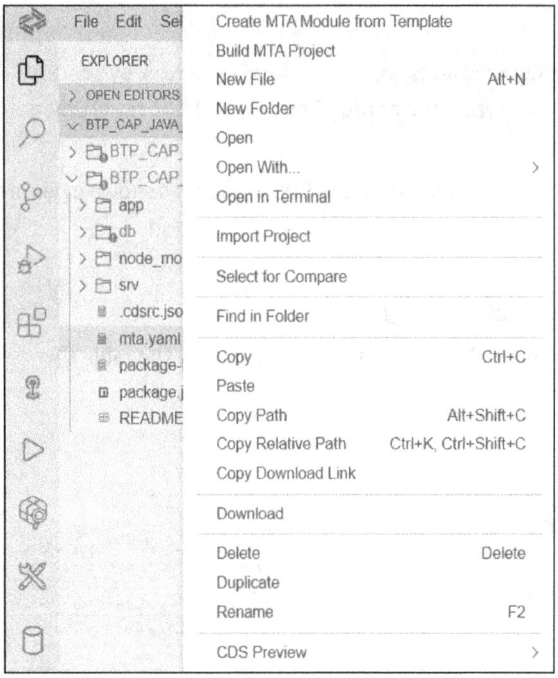

Figure 7.27: Building MTA for Deployment

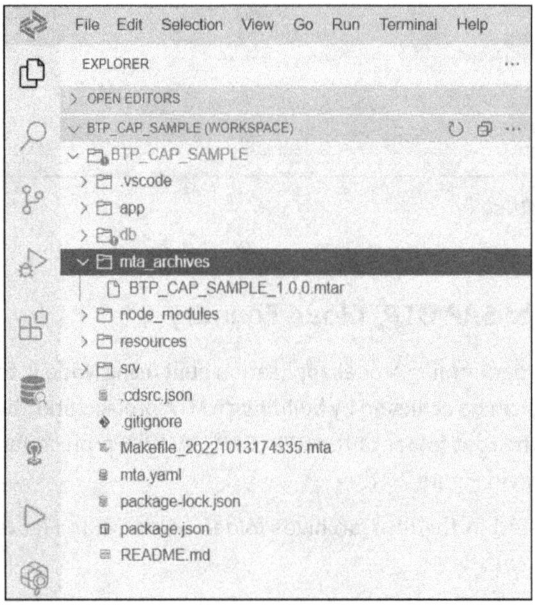

Figure 7.28: MTA Archive File Generated

Deployment can be performed by right-clicking on the MTA archive file and then selecting **Deploy MTA Archive**, as shown in Figure 7.29. The deployment log is shown in Figure 7.30.

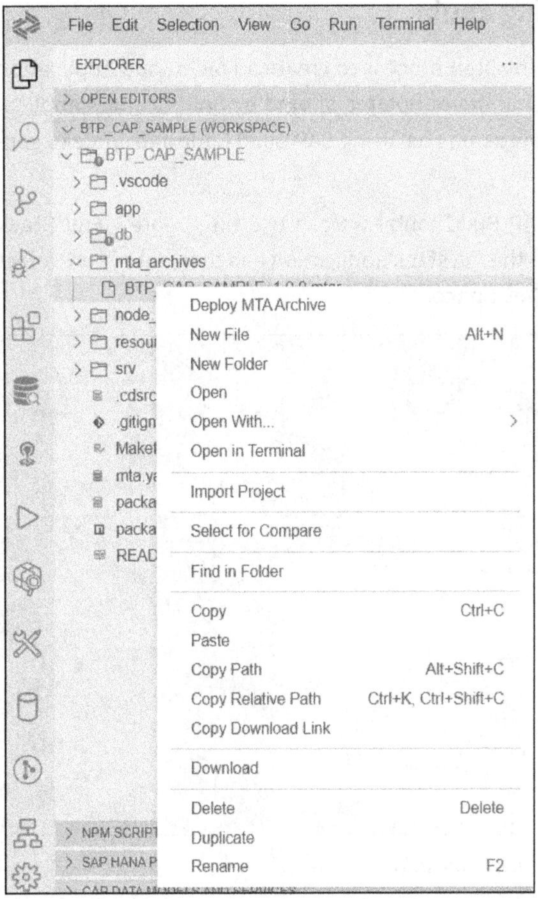

Figure 7.29: Deploying MTA Archive File

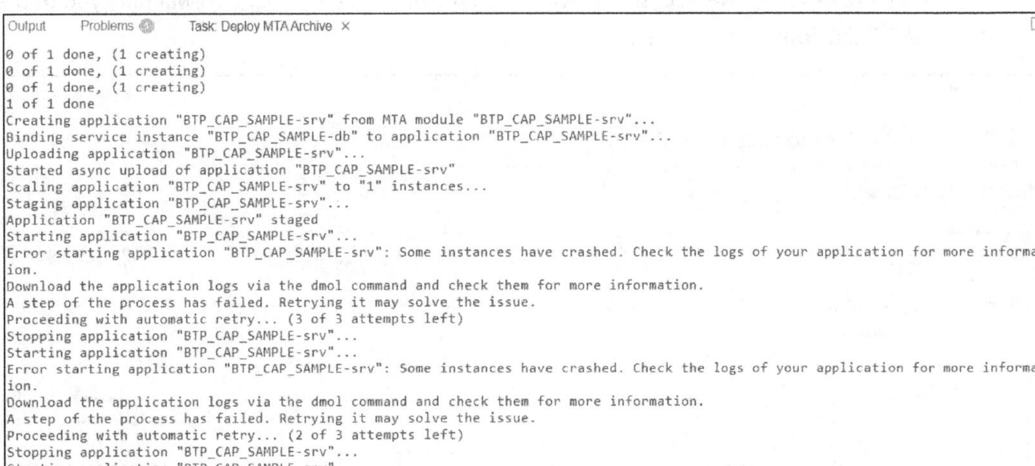

Figure 7.30: Deployment Logs and Messages in SAP Business Application Studio

7.7 Generating Artifacts Using Joule

So far, we have gone through the step-by-step process to create an SAP Cloud Application Programming Model application by using the templates offered by SAP Business Application Studio. However, we can achieve the same goal using SAP Build Code and Joule to generate the required artifacts and code.

As shown in Figure 7.31, inside the SAP Build lobby we can use the **Create • Full Stack Node.js Application** option to generate the full stack application and then enter the **Project Name** and **Description** and select the **Dev Space**.

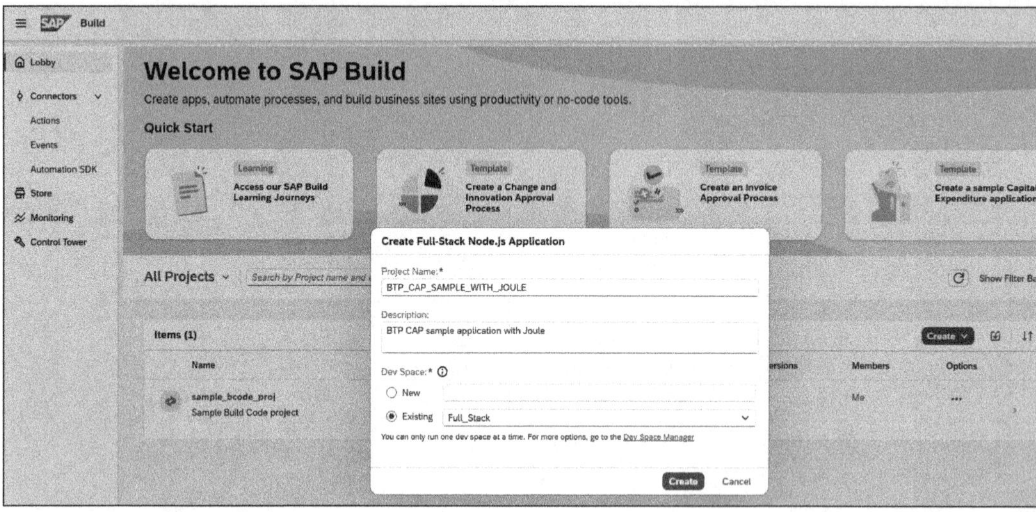

Figure 7.31: SAP Build Code: Create Full Stack Application

As soon as the project is created, you can proceed to the next step by clicking the project name (**BTP_CAP_SAMPLE_WITH_JOULE**) as shown in Figure 7.32. This will take you to the SAP Build Code launch screen.

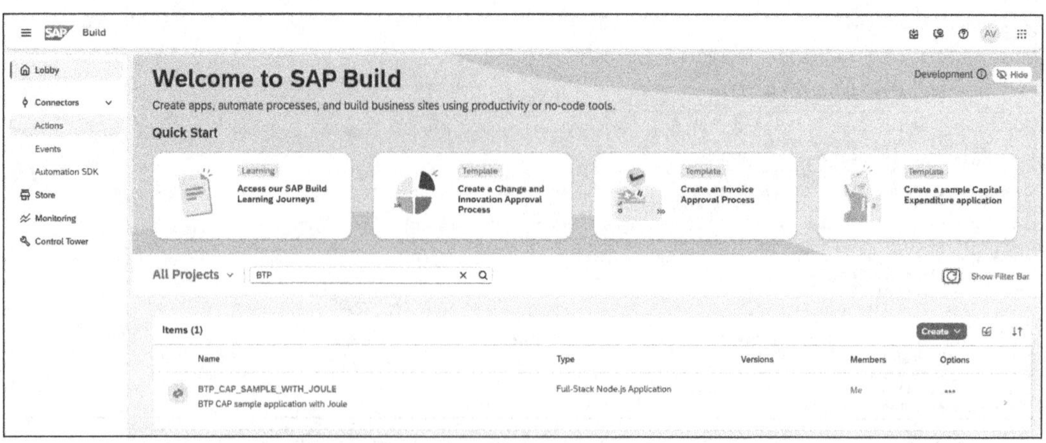

Figure 7.32: All Projects

Launching SAP Build Code ultimately takes you to the SAP Business Application Studio storyboard. Joule is already turned on, so all you need to do is to click the icon from the left side panel. Clicking the icon immediately launches the Joule editor, as shown in Figure 7.33. You will be shown details about its capabilities and a list of commands that you can perform with it.

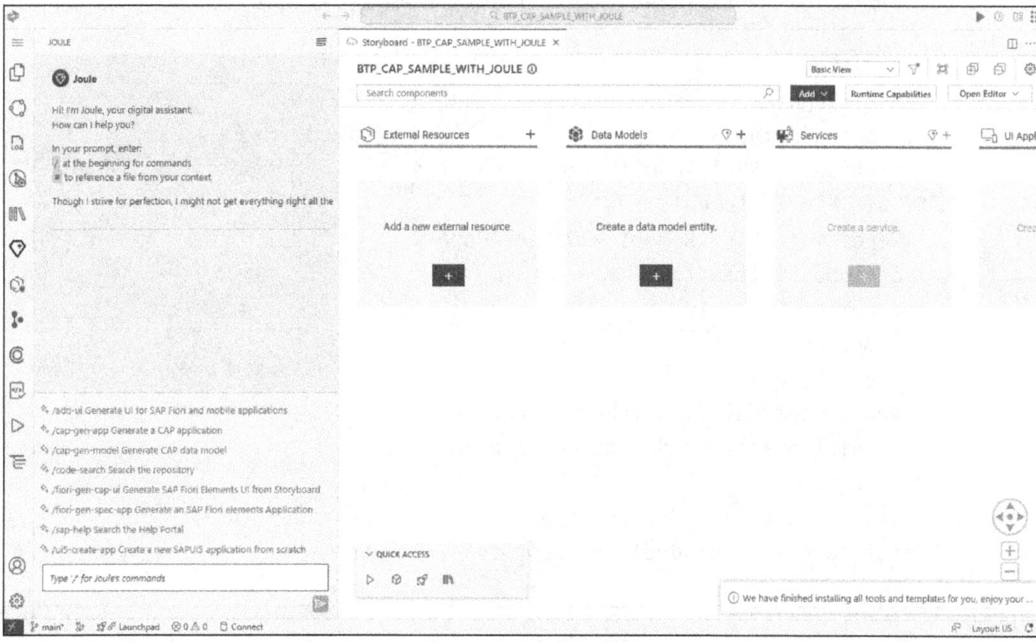

Figure 7.33: Joule Editor

To engage Joule, we need to first select one of the potential commands like **/cap-gen-app Generate a CAP application** or **/cap-gen-model Generate a CAP data model.**

Figure 7.34 shows the selected Joule command followed by the actual prompt that we are going to use to instruct Joule. A sample prompt to create an application looks as follows:

```
Create a complete SAP CAP application for managing and tracking birth
and death records.
Requirements:
1. Data Model:
Define an entity "BirthDeath" with the following fields:
- period: String(4) (represents year)
- birth_death: String(30) (values: "Birth" or "Death")
- region: String(300)
- qty: Integer
2. Add Validations:
- period must be a valid 4-digit year
- birth_death must only allow "Birth" or "Death"
- qty must be a non-negative integer
```

```
3. Service Layer:
- Create a service exposing CRUD operations for the entity
- Add a custom action to aggregate total births and deaths by region and
period
4. Business Logic:
- Implement logic to calculate totals (sum of qty) grouped by:
  - region
  - period
  - birth_death
5. Sample Data:
- Provide a CSV file with sample entries
6. UI (Fiori Elements):
- Generate a List Report + Object Page
- Include filters for:
  - period
  - region
  - birth_death
- Add analytical view (chart) showing:
  - total births vs deaths per region
7. Deployment:
- Include steps to deploy on SAP BTP Cloud Foundry
Generate clean, production-ready code with comments.
```

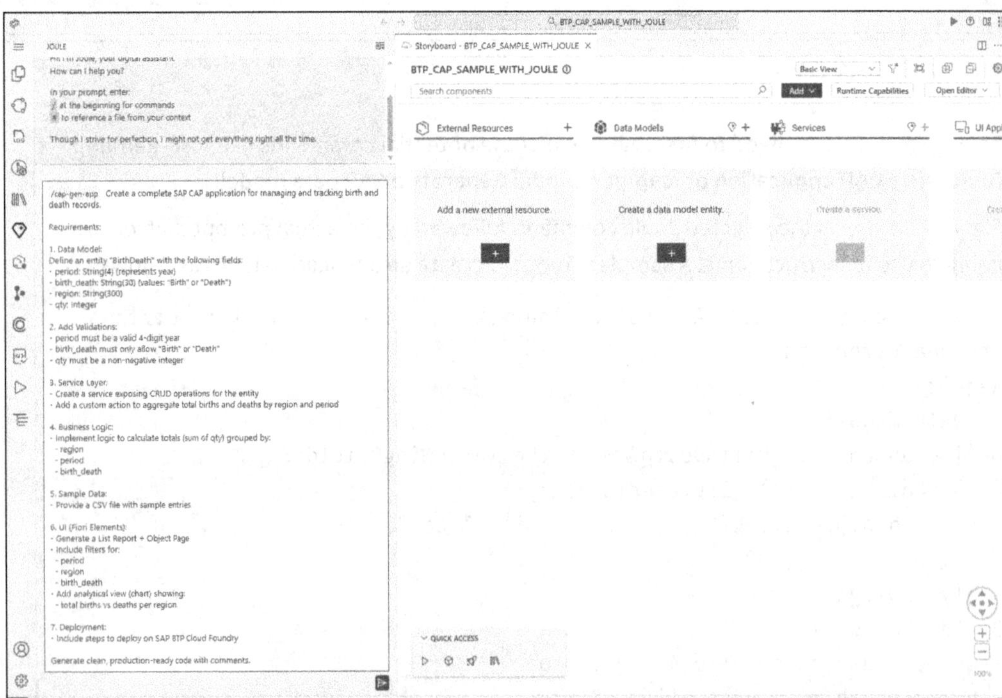

Figure 7.34: Prompt-Based Instruction

The intention of this prompt is to create an SAP Cloud Application Programming Model application that models and manages birth and death records using validations, create, read, update, and delete (CRUD) services, and aggregation logic.

Based on the prompt it is given, Joule provides a sequential list of tasks that it is planning on performing, as shown in Figure 7.35.

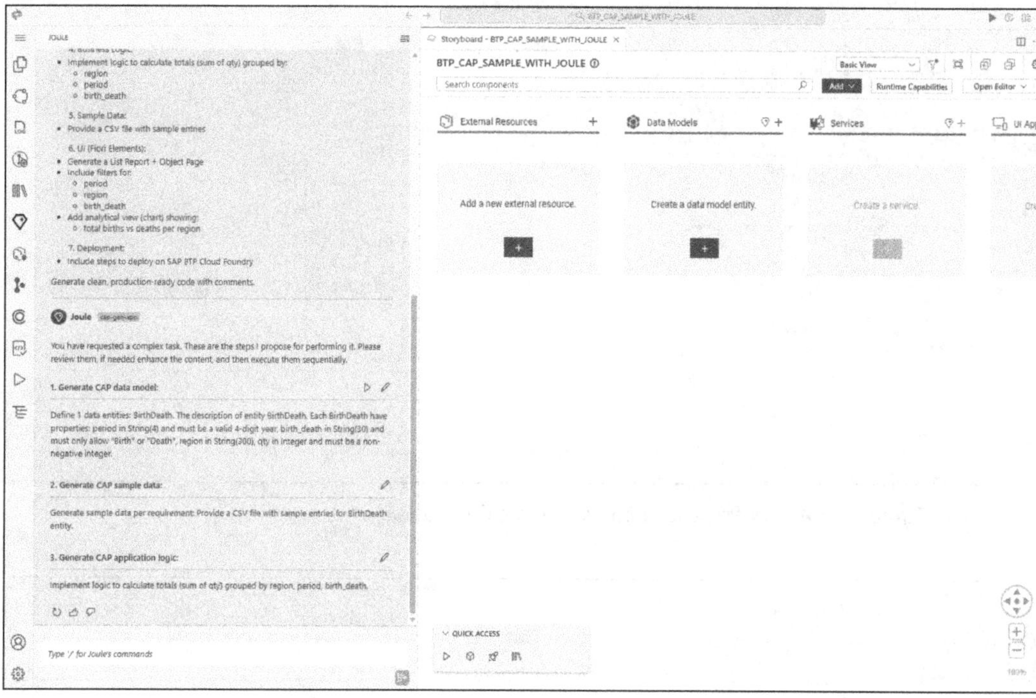

Figure 7.35: Joule Tasks: User Review

After reviewing the proposed list of tasks, you can either click the ✎ button to further enhance your prompt or click the play button ▷ to execute a task. Clicking the play button next to each task ultimately generates the required artifacts, as shown in Figure 7.36. You now just need to review the artifacts and click **Accept** to reflect the changes in the explorer.

Post acceptance, Joule generates and provides all the required artifacts, as shown in Figure 7.37, including the data model, services, test data, annotations, etc.

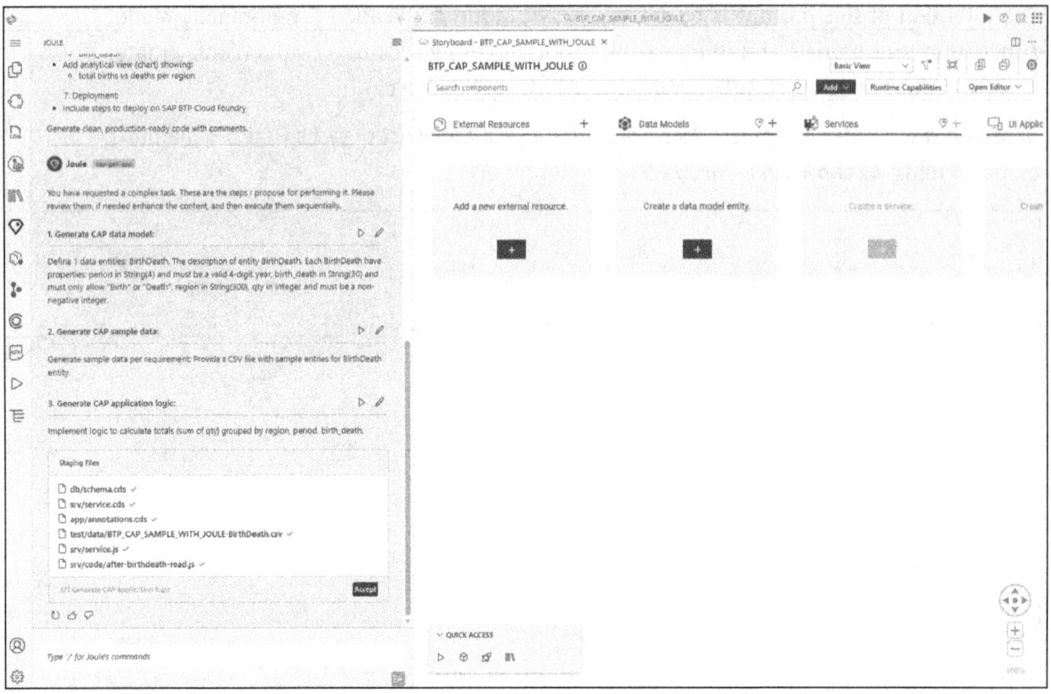

Figure 7.36: Accept Proposed Changes Before Generation

Figure 7.37: Joule-Based Artifact Generation

Finally, Figure 7.38 shows a quick sneak peek of the *schema.cds* file generated by Joule as part of this project.

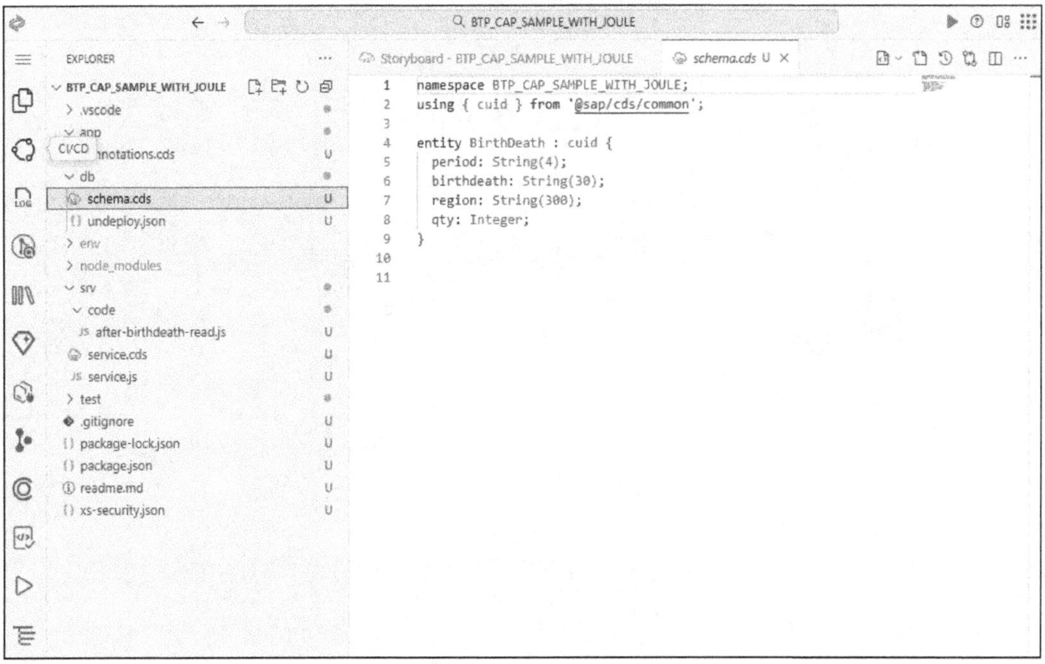

Figure 7.38: schema.cds File Generated by Joule

7.8 Summary

In this chapter, we covered various aspects of an end-to-end development lifecycle for application development using SAP Cloud Application Programming Model for Node.js. We created a CDS domain model and a service definition, exposed database artifacts (like the domain model and service definition), and deployed the application to SAP BTP, Cloud Foundry environment. At the very end, we have also explained how we can take advantage of SAP Build Code and Joule to generate an application and its associated artifacts. We hope this chapter has prepared you for application development using SAP Cloud Application Programming Model for Node.js by covering the most important functionalities and tools. In the next chapter, we'll discuss creating an SAP Cloud Application Programming Model application using the Java software development kit (SDK).

Chapter 8
Developing Java Applications

This chapter focuses on developing a business application using SAP Cloud Application Programming Model for Java. We'll be using SAP Business Application Studio for the development of backend objects using core data services (CDS) domain models, CDS service definitions, and Java services. Business applications will be deployed on SAP Business Technology Platform (SAP BTP), Cloud Foundry environment, as multi-target applications (MTAs).

In this chapter, we'll look at different aspects of developing, testing, and deploying a business application using the SAP Cloud Application Programming Model for Java. The SAP Cloud Application Programming Model Java software development kit (SDK) enables developing SAP Cloud Application Programming Model applications written in Java and supports picking the required features and adding them as part of dependencies. The SAP Cloud Application Programming Model Java SDK includes the following features:

- It enables local development using SQLite databases but can also switch to production environment, for instance, an SAP HANA Cloud database, by switching the application deployment configuration.
- Spring Boot is tightly integrated with the SAP Cloud Application Programming Model Java SDK and works as a runtime container.
- The SDK supports extended services for user account and authentication (XSUAA) authentication and the authorization service in SAP BTP.
- It supports the OData v4 protocol adapter.

In this chapter, you'll learn how to create CDS domain models and CDS service definitions; how to expose database artifacts as domain models and service definitions; and how to deploy applications to SAP BTP, Cloud Foundry environment.

Note

In Chapter 4, we covered in detail topics related to setting up SAP Business Application Studio-related activities like creating dev spaces for full-stack application development and creating new projects in a dev space. We advise you to review this chapter before proceeding with the next steps.

We'll leverage the same dev space that we created in Chapter 7, Section 7.1, for application development, as shown in Figure 8.1.

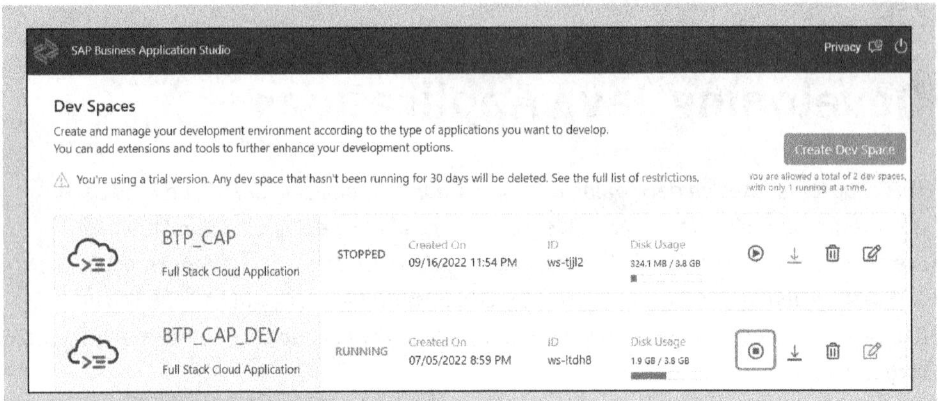

Figure 8.1: Dev Space Created in Chapter 7

8.1 Creating a Project from a Template

Recall that we created a dev space of the full-stack cloud application type in Chapter 7. In this dev space, let's now create a new project using an SAP Cloud Application Programming Model project template. Follow these steps:

1. Open the dev space by clicking on **BTP_CAP_DEV** and create a project by clicking the **Start from Template** tile on the welcome page. Select the **CAP Project** tile and click on the **Start** button, as shown in Figure 8.2.

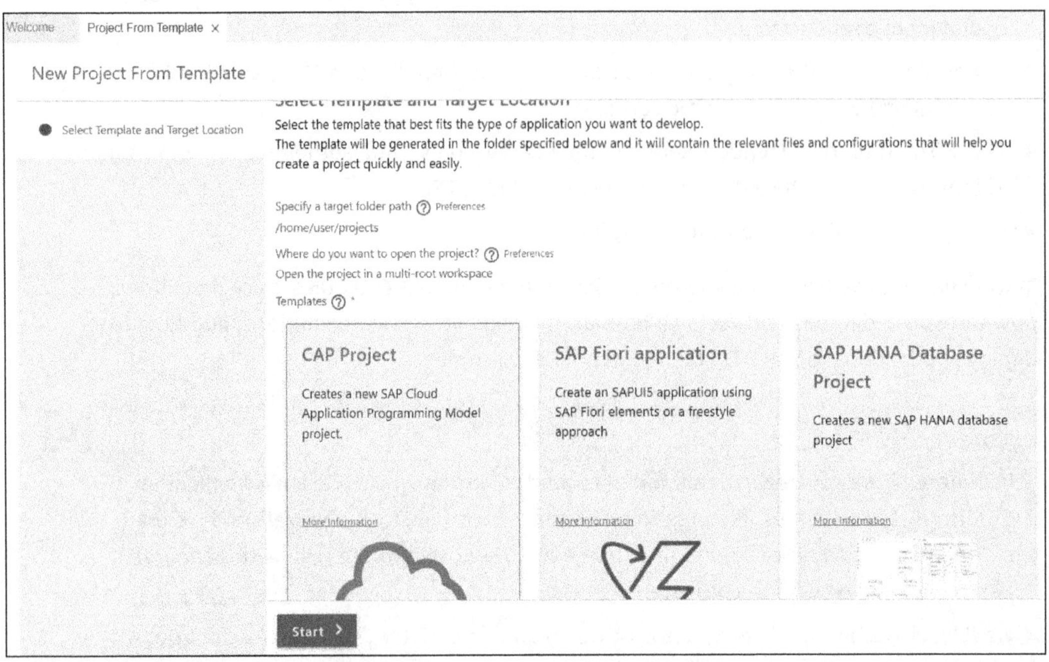

Figure 8.2: Creating a Project from a Template

2. As shown in Figure 8.3, provide a project name in the **Enter your project name** field and select the **Java** option from the **Select your runtime** dropdown list. Then, make some selections under **Add features to your project.** Click on **Finish.**

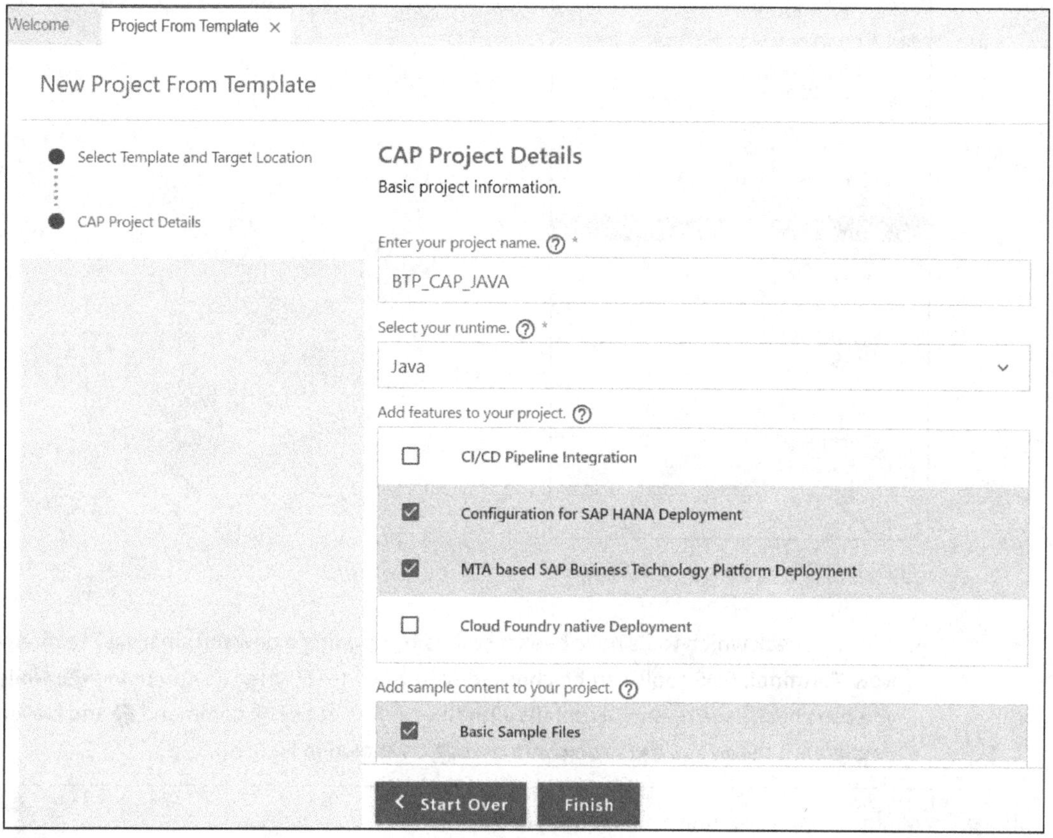

Figure 8.3: Project Details Screen

3. As shown in Figure 8.4, when the project is created, five folders are generated. Table 8.1 describes the purpose of each of these folders.

Folder	Purpose
db	Contains database artifacts and domain CDS models
srv	Contains CDS service definitions and Java backend code
srv/src/main/java	Contains Java application logic for handlers
node_modules	Contains dependencies for CDS tools
integration-tests	Maven module to perform integration tests

Table 8.1: Project Folders and Their Purposes

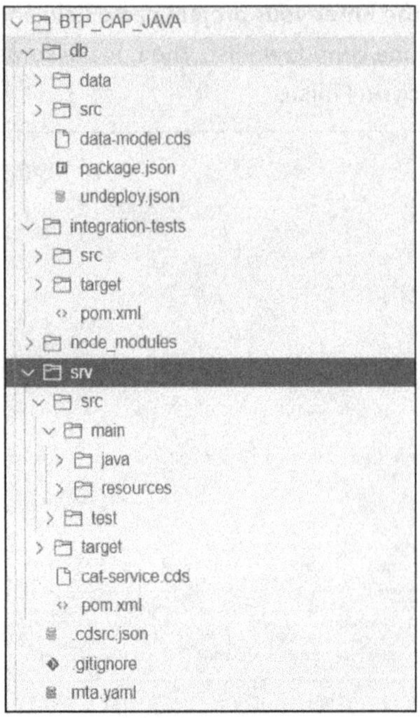

Figure 8.4: Project Structure in Explorer

4. You can check which tools have been installed by opening a new terminal via **Terminal • New Terminal**. CDS tools can be checked with the `cds --version` command ❸, while you can check Java open SDK details with the `java --version` command ❷ and Maven details with the `mvn --version` command ❶, as shown in Figure 8.5.

```
user: BTP_CAP_JAVA $ mvn --version                                                                    ❶
Apache Maven 3.8.1 (05c21c65bdfed0f71a2f2ada8b84da59348c4c5d)
Maven home: /usr/lib/maven
Java version: 11.0.6.0.1, vendor: SAP SE, runtime: /usr/lib/sapmachine-jdk/sapmachine-jdk-11.0.6.0.1
Default locale: en_US, platform encoding: ANSI_X3.4-1968
OS name: "linux", version: "5.10.114-garden-cloud-amd64", arch: "amd64", family: "unix"
```

```
user: BTP_CAP_JAVA $ java --version                                          ❷
openjdk 11.0.6.0.1 2020-02-12 LTS
OpenJDK Runtime Environment (build 11.0.6.0.1+10-LTS-sapmachine)
OpenJDK 64-Bit Server VM (build 11.0.6.0.1+10-LTS-sapmachine, mixed mode)
```

```
user: BTP_CAP_JAVA $ cds --version                      ❸
@sap/cds: 6.1.3
@sap/cds-compiler: 3.1.2
@sap/cds-dk: 6.1.4
@sap/cds-foss: 4.0.0
@sap/cds-mtx: -- missing --
@sap/eslint-plugin-cds: 2.5.0
BTP_CAP_JAVA-cds: 1.0.0
Node.js: v16.15.0
home: /extbin/npm/globals/lib/node_modules/@sap/cds
user: BTP_CAP_JAVA $ 
```

Figure 8.5: Versions of Java, CDS, and Maven installed

8.2 Creating a Java Application

In Section 8.1, we created a "shell" project in the SAP Cloud Application Programming Model Java SDK. In this section, we'll focus on building up the project by creating domain models, service definitions, and database artifacts and then consuming these services. Follow these steps:

1. The db module is not added by default in the *mta.yaml* file in the root folder of the project. You need this module for MTAs to be built with all required dependencies and related database files (artifacts). For this step, add the code shown in Listing 8.1.

```
- name: db
  type: hdb
  path: db
  requires:
  - name: hdi_db
resources:
- name: hdi_db
  type: com.sap.xs.hdi-container
  properties:
    hdi-container-name: ${service-name}
```

Listing 8.1: Adding the db Module to the mta.yaml File

2. Compile the CDS model definition added to the project with the **Basic Sample File** option during the creation of a project by running the command `mvn clean install` in a new terminal. Then, follow the menu path **Terminal • New Terminal**. The CDS domain model is located in the **db** folder, as shown in Figure 8.6.

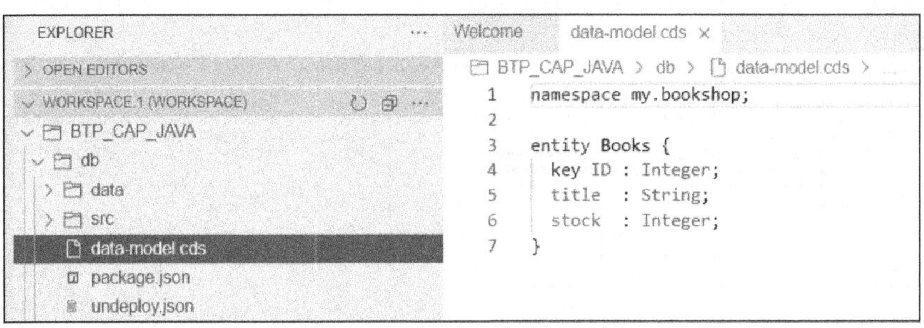

Figure 8.6: Sample Domain Model Provided During Project Creation

3. The CDS service definition based on the domain model is located in the **srv** folder, as shown in Figure 8.7.

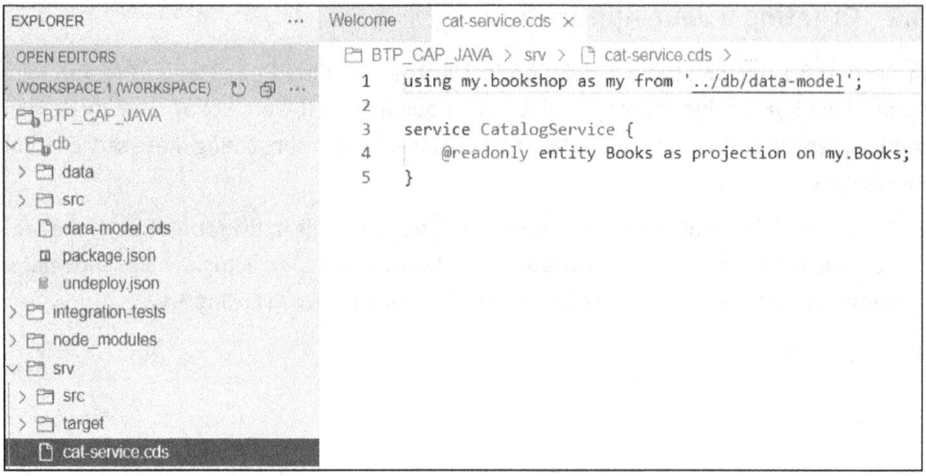

Figure 8.7: Sample CDS Service Provided During Project Creation

4. Compile the CDS model by executing the command `mvn clean install`, as shown in Figure 8.8.

Figure 8.8: Compiling the CDS Model Definition

5. Figure 8.9 shows the logs after executing the mvn clean install command.

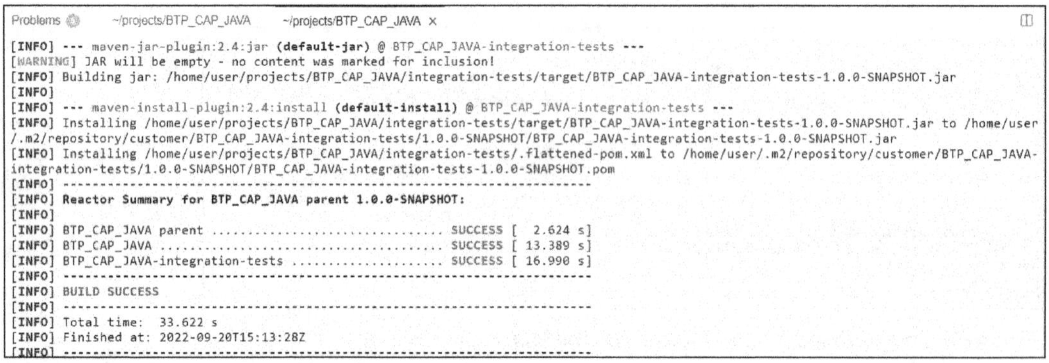

Figure 8.9: Success Message for Compilation of CDS Model Definition

6. After successfully compiling CDS model definitions, the files are added to the *srv/src/main/resources/edmx* folder, as shown in Figure 8.10. The Java runtime looks for service definitions in this folder.

Figure 8.10: Output Compilation of the CDS Model Definition

7. Let's now run the application in our local environment by running the `mvn clean spring-boot:run` command in a new terminal, as shown in Figure 8.11.

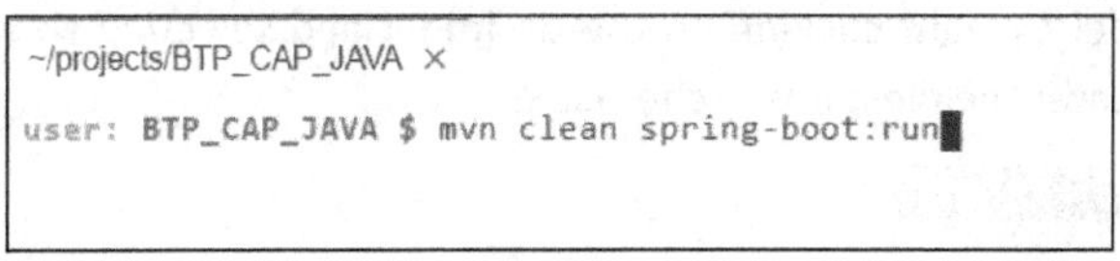

Figure 8.11: Running the Application in a New Terminal

8. A popup window appears. Click on **Open in a New Tab**, as shown in Figure 8.12.

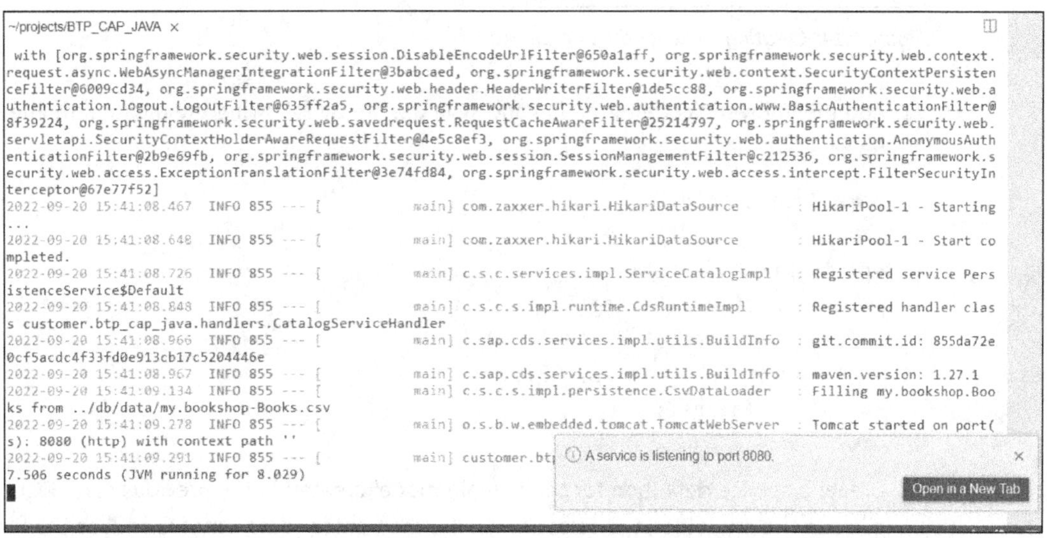

Figure 8.12: Terminal with Port Listening Message

9. A window opens in your web browser displaying details about OData v4 service endpoints, as shown in Figure 8.13.

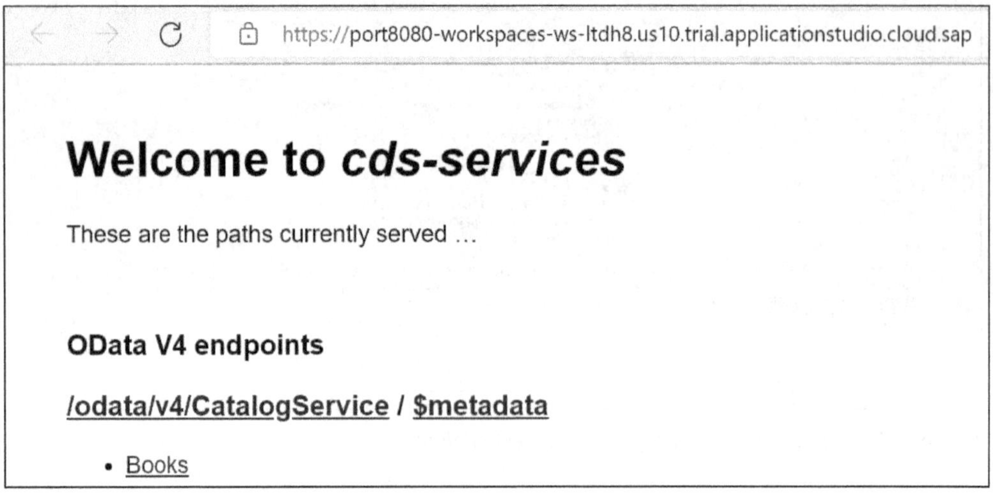

Figure 8.13: Application Start Page

10. Create a custom domain model by right-clicking on the **db** folder and selecting **New File**, as shown in Figure 8.14. Enter the file name "schema.cds."

Figure 8.14: Creating a File for the Domain Model

11. Add the following code to the file and save the file by following the menu path **File • Save**:

```
namespace sample;

entity birth_death {
    key period : String(4);
    key birth_death : String(30);
    key region : String(300);
    qty : Integer;
}
```

12. Create a service definition for the domain model created in the previous step. Right-click on the **srv** folder and select **New File,** then enter "data-services.cds" as the file name, as shown in Figure 8.15.

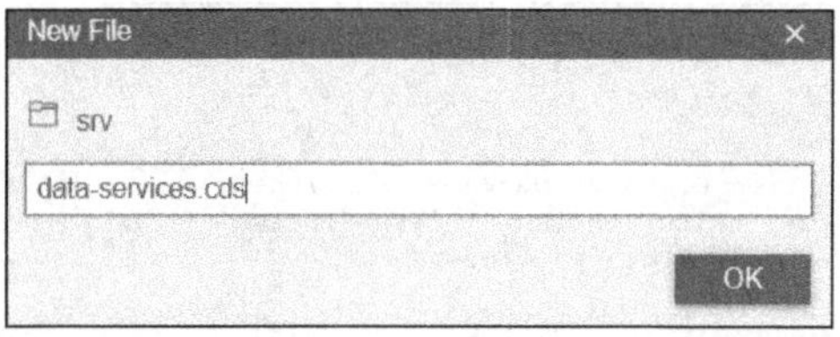

Figure 8.15: Creating a File for the Service Definition

13. Add the following code to the file and save the file by following the menu path **File • Save**:

```
using sample as tbl from '../db/schema';

service dataservices {
@readonly  entity BirthDeathTable as projection on tbl.birth_death;
}
```

14. Run the application by running the command `mvn clean spring-boot:run` and check if the service called **BirthDeathTable** is displayed on the application's page, as shown in Figure 8.16.

Figure 8.16: Service on the Application Page

15. Run the command `cds build` in a new terminal. This step will create the files that will create database artifacts after the application is deployed, as shown in Figure 8.17.

```
Problems    ~/projects/BTP_CAP_JAVA    ~/projects/BTP_CAP_JAVA ×
user: BTP_CAP_JAVA $ cds build

> WARNING: This application uses a very old version 6.1.3 of package '@sap/cds',
> which will no longer be supported with the next major version of '@sap/cds-dk'.
> We strongly recommend an upgrade to latest version 6 of '@sap/cds'.

[cds] - determining build tasks for project [/home/user/projects/BTP_CAP_JAVA].
[cds] - the following build tasks will be executed
[cds] -     {
      "build": {
        "target": ".",
        "tasks": [
          {"for":"hana", "src":"db", "options":{"model":["db","srv"]}},
          {"for":"java", "src":"srv", "options":{"model":["db","srv"]}}
        ]
      }
    }

[cds] - building project [/home/user/projects/BTP_CAP_JAVA], clean [true]
[cds] - cds [6.1.3], compiler [3.1.2], home [/extbin/npm/globals/lib/node_modules/@sap/cds]

[cds] - done > wrote output to:
   db/src/gen/.hdiconfig
   db/src/gen/.hdinamespace
   db/src/gen/CatalogService.Books.hdbview
   db/src/gen/data/my.bookshop-Books.csv
   db/src/gen/data/my.bookshop-Books.hdbtabledata
```

Figure 8.17: Executing the cds build Command

The following files are created after executing the `cds build` command:

- *db/src/gen/CatalogService.Books.hdbview*
- *db/src/gen/dataservices.BirthDeathTable.hdbview*
- *db/src/gen/my.bookshop.Books.hdbtable*
- *db/src/gen/sample.birth_death.hdbtable*
- *srv/src/main/resources/edmx/CatalogService.xml*
- *srv/src/main/resources/edmx/csn.json*
- *srv/src/main/resources/edmx/dataservices.xml*

[»]

Note

For every entity in a domain model, two types of SAP HANA database artifacts are created: `hdbview` and `hdbtable`. These artifacts are deployed in SAP HANA Cloud.

8.3 Creating a Service Instance and Binding It to an Application

In this section, let's bind the resource **BTP_CAP_JAVA-db** to an SAP HANA deployment infrastructure (HDI) container service instance. Then, we'll deploy these database artifacts and view them in the SAP HANA database explorer.

8.3.1 Creating an SAP HANA Deployment Infrastructure Container Service Instance

To create an HDI container service instance, as shown in Figure 8.18, follow these steps:

❶ In the **EXPLORER** pane, expand the **SAP HANA PROJECTS** section.

❷ Click on the bind icon.

❸ Select **Create a new service instance**.

❹ Provide a name for the service instance and press `Enter`.

❺ The message confirming the creation of the service instance in is SAP Business Application Studio appears.

❻ The confirmation message for the service binding in SAP Business Application Studio appears.

❼ You can view the service instance in the SAP BTP cockpit.

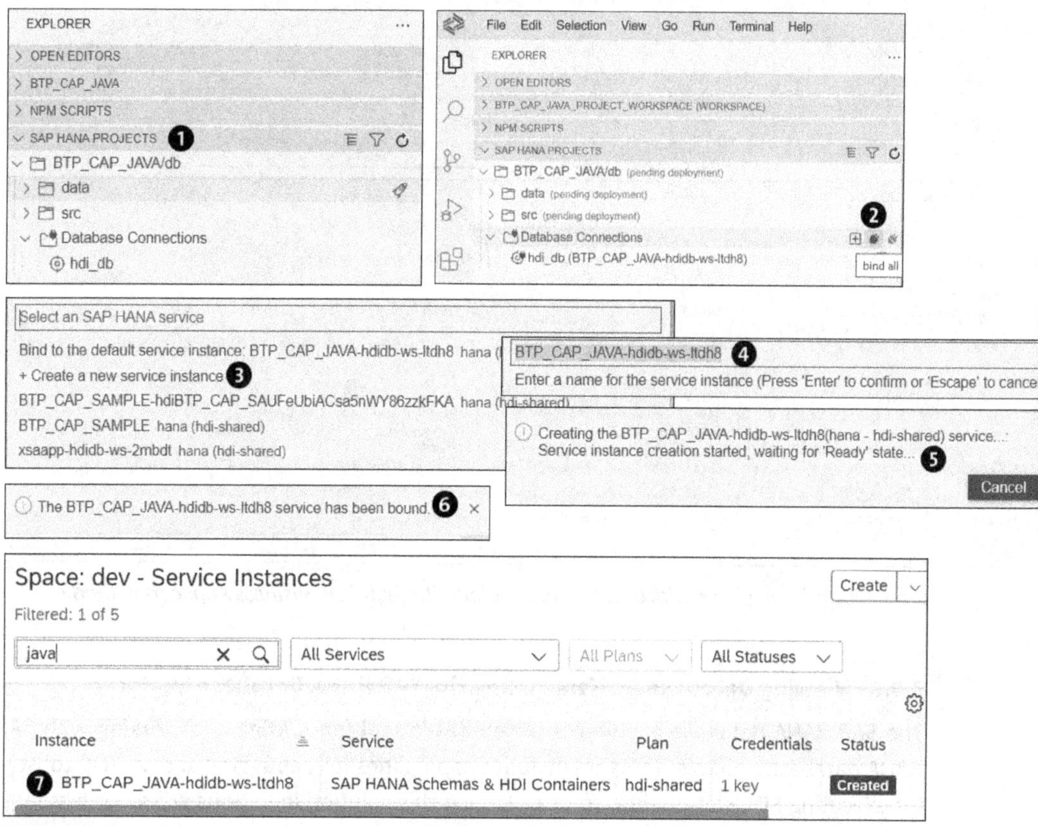

Figure 8.18: Binding a Resource to an SAP HANA Cloud Service Instance

8.3.2 Deploying Artifacts in a Backend Database in SAP Business Application Studio

Now that we've bound the SAP HANA Cloud service instance in the previous section, we're now ready to deploy the necessary objects to the SAP HANA Cloud service through SAP Business Application Studio. As shown in Figure 8.19, to deploy database artifacts to SAP HANA Cloud, follow these steps:

❶ In the **EXPLORER** pane, expand the **SAP HANA PROJECTS** section.

❷ Click on the deploy icon.

❸ You can monitor the status of deployment. The container schema name is generated by `hdi-deploy`. This schema is where all the objects will be deployed.

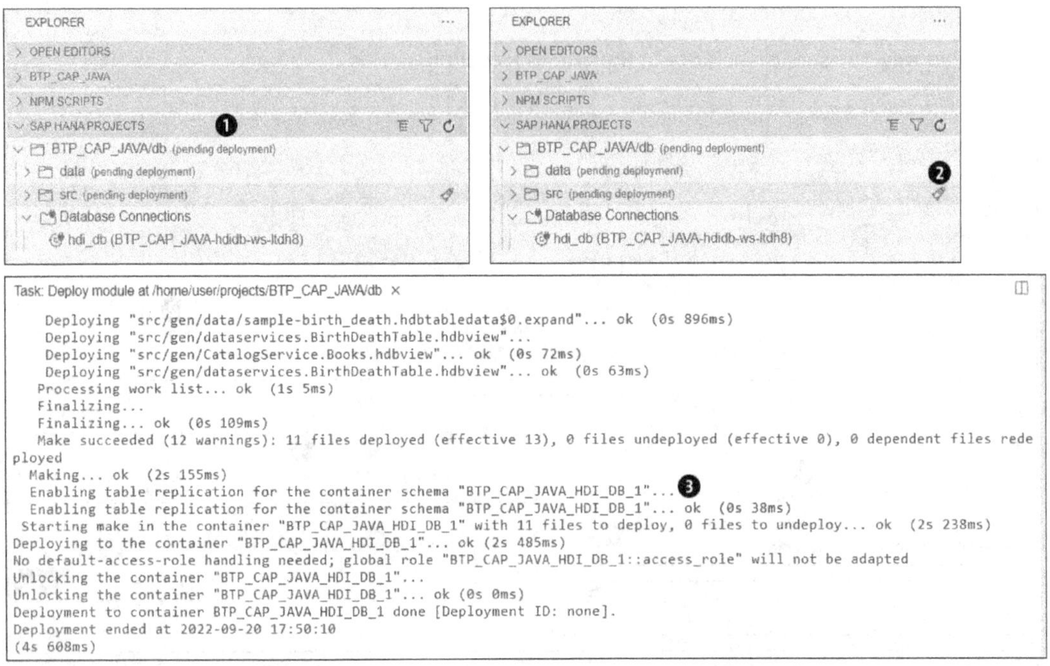

Figure 8.19: Deploying Artifacts to SAP HANA Cloud through SAP Business Application Studio

8.3.3 Viewing Database Artifacts Using the SAP HANA Database Explorer

The SAP HANA database explorer is a web-based tool for viewing and working with SAP HANA database artifacts. This tool includes capabilities for debugging and an SQL console for executing SQL statements. You can launch this solution in various ways, for instance, through the SAP BTP cockpit, but we'll use SAP Business Application Studio to launch the database explorer, as shown in Figure 8.20:

❶ In the **EXPLORER** pane, expand the **SAP HANA PROJECTS** section and click on the **Open HDI container** icon.

❷ The SAP HANA database explorer is launched in a web browser.

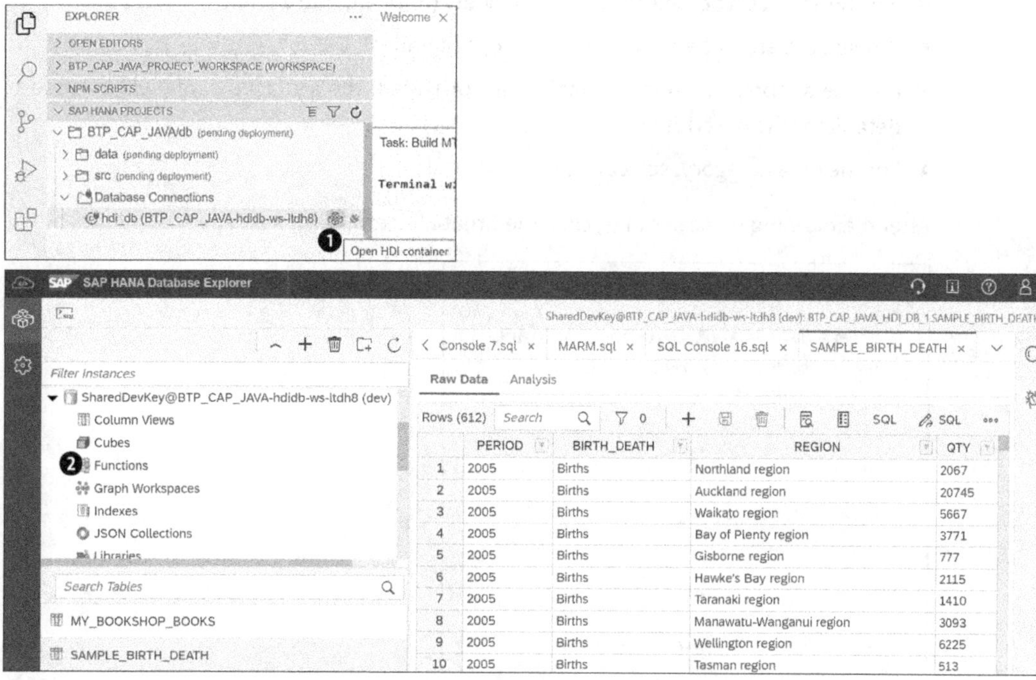

Figure 8.20: Launching the SAP HANA Database Explorer

8.4 Creating Core Data Services Domain Models and Services Based on SAP HANA Artifacts

CDS domain models can be based on persisted database objects. In this section, we'll create an SAP HANA database artifact called a *calculation view*. This calculation view will use the database table we created in Section 8.2. Domain models are created on the calculation view and exposed as services using service definitions.

8.4.1 Creating a Calculation View

A detailed explanation of creating a calculation view is beyond the scope of this book. Our end objective is to expose a calculation view as an OData service endpoint. The source of data for the calculation view is table `SAMPLE_BIRTH_DEATH` in container schema `BTP_CAP_JAVA_HDI_DB_1`. Follow the menu path **View • Find Command** and then enter "database." Select **SAP HANA • Create SAP HANA Database Artifact**. As shown in Figure 8.21, you'll need to do the following:

- Provide the location in the project folder for creating the calculation view file. Typically, the folder is *db/src*.
- Choose **HANA Cloud** as the database version.
- For the artifact type, select **Calculation View (hdbcalculationview)**.
- Provide the artifact a name, for instance, "CV_BIRTH_DATE."
- Provide a label, which is optional. If not provided, the artifact's name will be used by default for the description.
- For the data category, select **CUBE**.

After maintaining these settings, click the **Create** button.

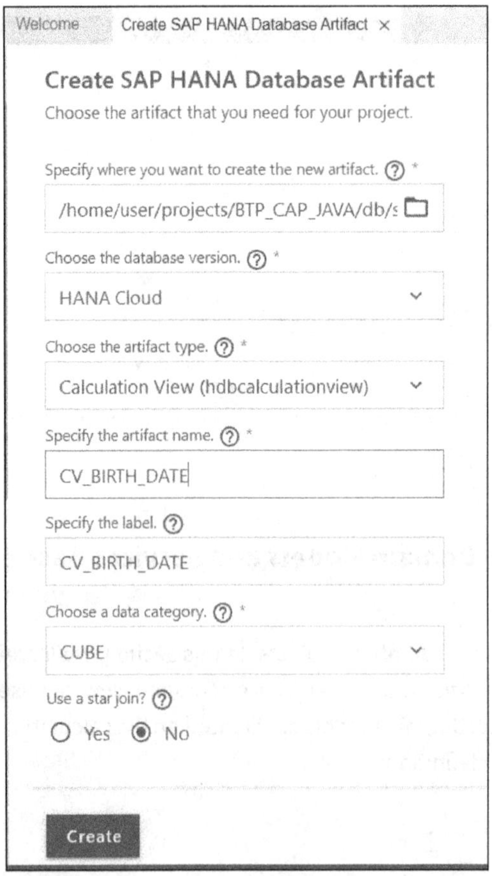

Figure 8.21: Creating an SAP HANA Database Artifact

Calculation views can be deployed to SAP HANA Cloud by clicking on the deploy icon, as shown in Figure 8.22.

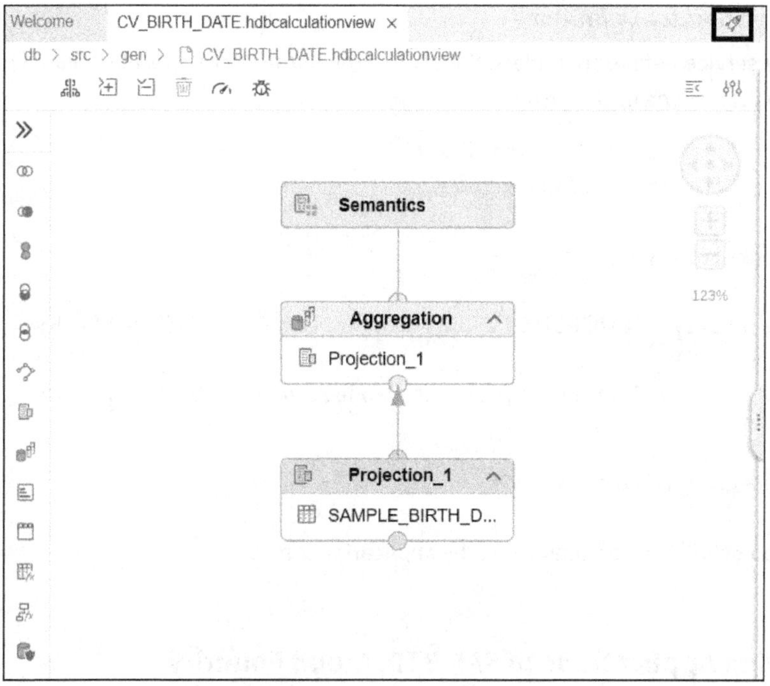

Figure 8.22: Calculation View Details

8.4.2 Creating a Domain Model Based on a Calculation View

To create a domain model based on a calculation view, right-click on the **db** folder under the root folder of your project and select **New File**. For the name of the file, enter "serviceview.cds" and then click **OK**. Enter the code shown in Listing 8.2 into the *serviceview.cds* file.

```
context cv{

    @cds.persistence.exists
    @cds.persistence.calcview
    entity ![BIRTH_DEATH] {
       ![period]: String(4) @title: 'Period' ;
       ![birth_death]: String(30) @title: 'Death Birth' ;
       ![region] : String(300);
       ![qty] : Integer;
    }
}
```

Listing 8.2: Creating a CDS Domain Model for a Calculation View

8.4.3 Creating a Service Definition

Next, to create a service definition, replace the code in the *data-services.cds* file, located in the *srv* folder, with the code in Listing 8.3.

```
using CV as cv from '.../db/serviceview';
using sample as tbl from '../db/schema';

service dataservices {

    @readonly entity BirthDeathCV    as projection on cv.BIRTH_DEATH;

    @readonly  entity BirthDeathTable as projection on tbl.birth_death;
}
```

Listing 8.3: Replace Code in data-services.cds

The OData endpoint will then be seen when the application is run.

8.5 Deploying Applications to SAP BTP, Cloud Foundry

Deploying an SAP Cloud Application Programming Model Java SDK application to SAP BTP, Cloud Foundry environment, can be performed by building an MTA project. Start by right-clicking on the **mta.yaml** file in the root folder of our BTP_CAP_JAVA project and then select **Build MTA Project**, as shown in Figure 8.23.

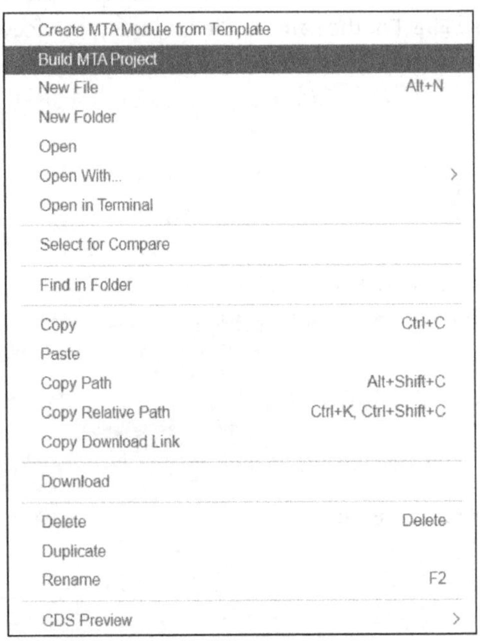

Figure 8.23: Building an MTA for Deployment

A multitarget application archive (MTAR) file with the extension **.mtar** is created in the **mta_archives** folder, as shown in Figure 8.24.

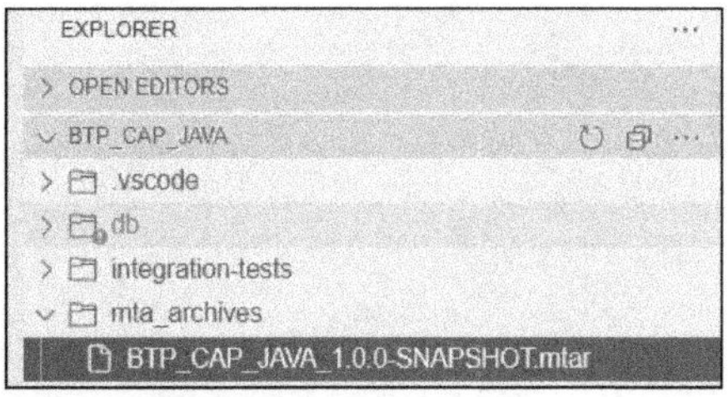

Figure 8.24: MTAR File

Deployment can be completed by right-clicking on the MTAR file and then selecting **Deploy MTA Archive**, as shown in Figure 8.25. The deployment log is shown in Figure 8.26.

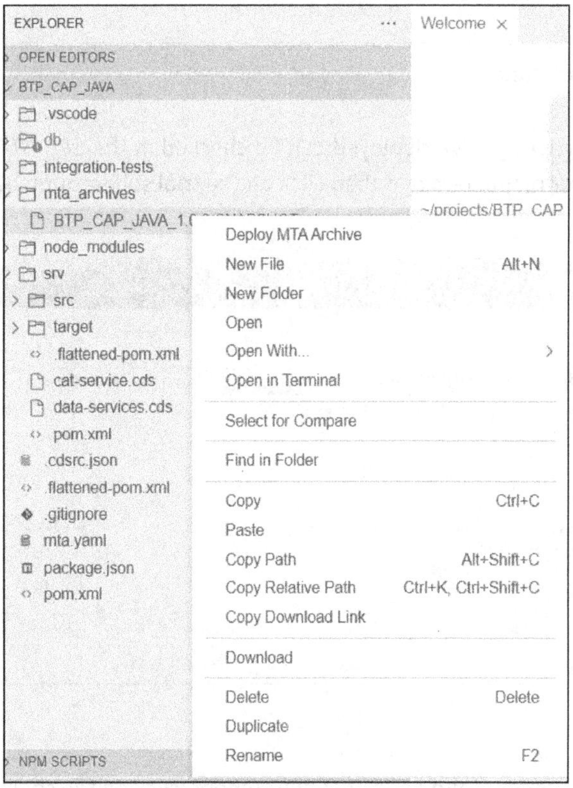

Figure 8.25: Deploying an MTAR File

[+]

Tip

Make sure you're logged on to SAP BTP, Cloud Foundry environment and the correct organization and space are selected.

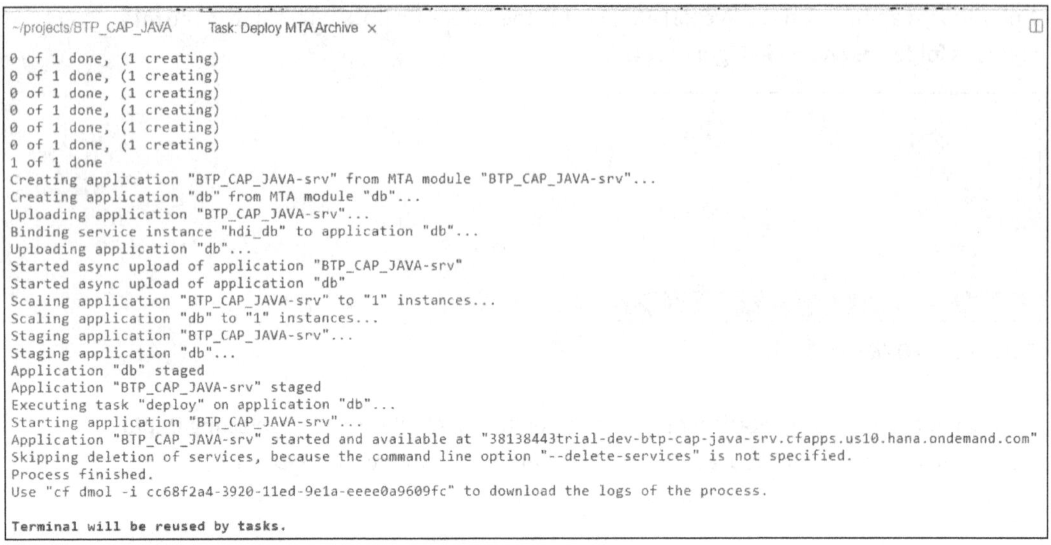

Figure 8.26: Deployment Messages and Status

The statuses of the application services you've deployed can be checked in the SAP BTP cockpit. Select the **Account Explorer** menu item and then click on the **trial** subaccount, as shown in Figure 8.27.

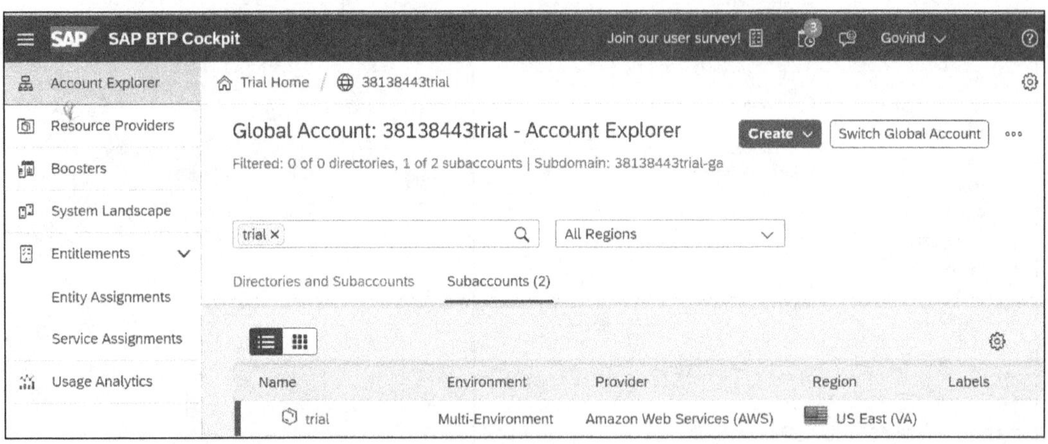

Figure 8.27: Account Page

On the **Subaccount** overview page, under **Cloud Foundry Environment**, click on the space where you want to deploy the application, as shown in Figure 8.28. In our case, we chose the **dev** space.

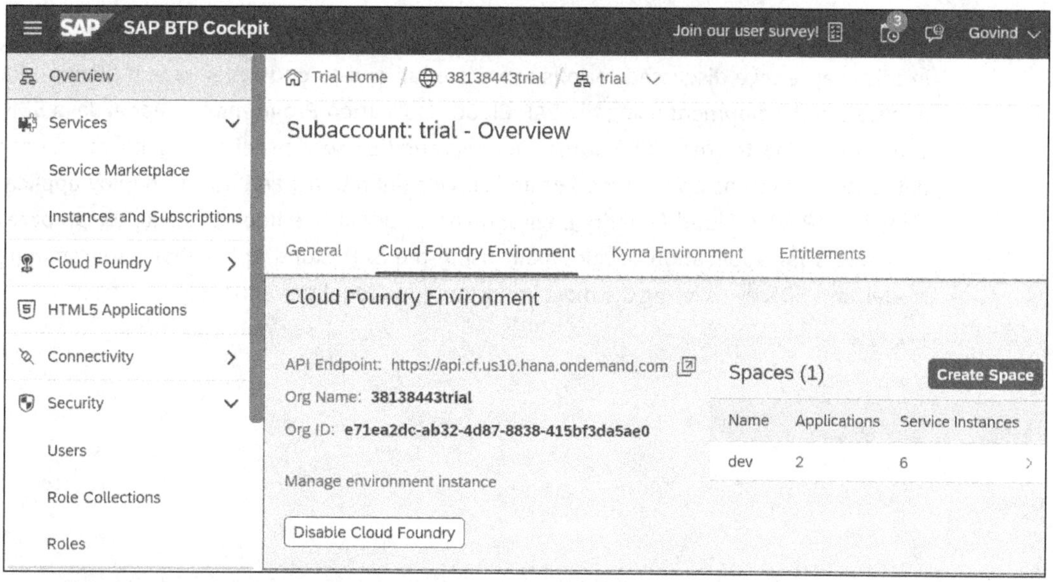

Figure 8.28: Subaccount Overview Page

On the following page, select the **Applications** menu item, as shown in Figure 8.29. This step will list all the applications that have been deployed in the space.

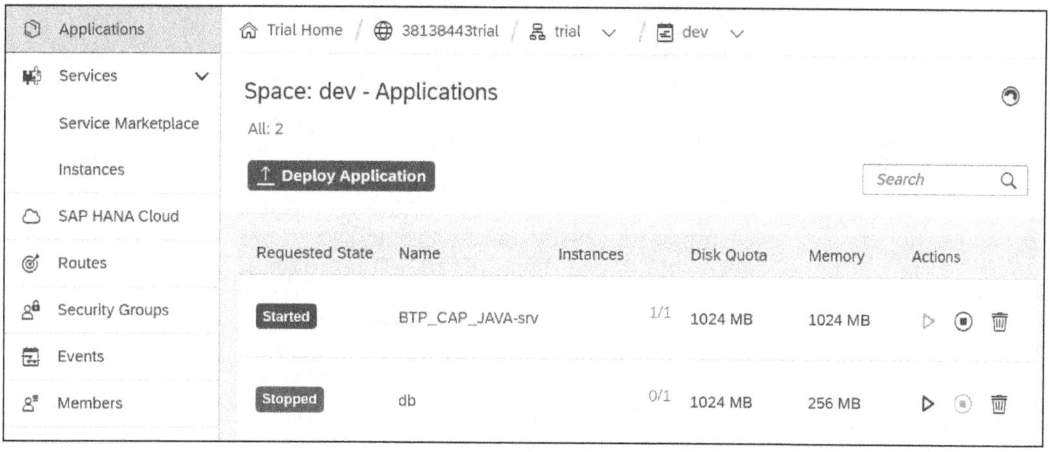

Figure 8.29: List of Applications Deployed in SAP BTP, Cloud Foundry environment

Tip

The **db** service is stopped automatically once its database artifacts are deployed.

8.6 Summary

In this chapter, we discussed various aspects of an end-to-end development lifecycle for application development using the SAP Cloud Application Programming Model Java SDK. You learned how to create CDS domain models and CDS service definitions; how to expose database artifacts as domain models and service definitions; and how to deploy applications in SAP BTP, Cloud Foundry environment, in detail. We hope this chapter prepares developers for application development using the SAP Cloud Application Programming Model Java SDK by covering the most important functionalities and tools.

Chapter 9

Developing ABAP Applications

This chapter will teach you how to develop and deploy ABAP applications in the cloud using SAP Business Technology Platform (SAP BTP) ABAP environment.

ABAP has been *the* essential building block powering all SAP solutions for decades. At TechEd 2017, SAP announced what the ABAP developer community had long been waiting for—the release of SAP BTP ABAP environment. With the introduction of SAP BTP ABAP environment, the large SAP development community can leverage their existing knowledge and experience for modern ABAP development in the cloud. SAP BTP ABAP environment is provided as an additional runtime along with SAP BTP, Cloud Foundry environment, and SAP BTP, Kyma runtime.

SAP BTP ABAP environment can be used in two unique use cases:

- Side-by-side extensions, for building new extensions and migrating existing extensions decoupled from the system core, using application programming interfaces (APIs) for communication while supporting the "clean core" paradigm
- New application development, for developing applications from scratch, integrated with backend systems and leveraging SAP BTP services

In this chapter, we'll begin with an overview of SAP BTP ABAP environment and its architecture. Next, we'll show you how to create a trial instance of SAP BTP ABAP environment and provide an overview of the development tools for Eclipse. Then, we'll dive into the ABAP RESTful application programming model and how it can be used to develop a list report app. We'll cover the differences between a managed app and an unmanaged app and show you how to develop both types. Finally, we'll close the chapter with an overview of developing a web API and will walk you through the steps of transport management.

9.1 SAP BTP ABAP Environment Architecture and Overview

SAP BTP ABAP environment shares the same foundation as SAP S/4HANA Cloud. The programming model is the ABAP RESTful application programming model, and the language is a much-restricted cloud version of ABAP. Restrictions prevent the use of features that are not applicable in cloud environments (file access, system calls, etc.) and unsupported features (Web Dynpro, lists, etc.).

Since its initial release, SAP BTP ABAP environment has been constantly evolving, with many new features introduced with each new version. Even though the release cycle is the same as for SAP's other cloud solutions (i.e., quarterly), SAP is releasing hotfixes in between the releases.

> **Note**
> Steampunk was the internal project name for SAP BTP ABAP environment. However, in 2019, a year after the platform's general availability (GA) release, the name was released to the public and has been commonly used since.

SAP BTP ABAP environment belongs to a group of services falling under the development efficiency pillar of SAP Extension Suite. This environment is available under three service plans:

- Cloud Platform Enterprise Agreement (CPEA): This option is a consumption-based license model.
- Pay-as-you-go: This option is also a consumption-based model without any upfront payment.
- Trial: This option is for platform exploration and evaluation.

Visit SAP Discovery Center, access the ABAP service, and navigate to the **Price** tab to explore more details on the pricing of these service plans. An example price sheet for CPEA, along with persistence memory availability, is shown in Figure 9.1.

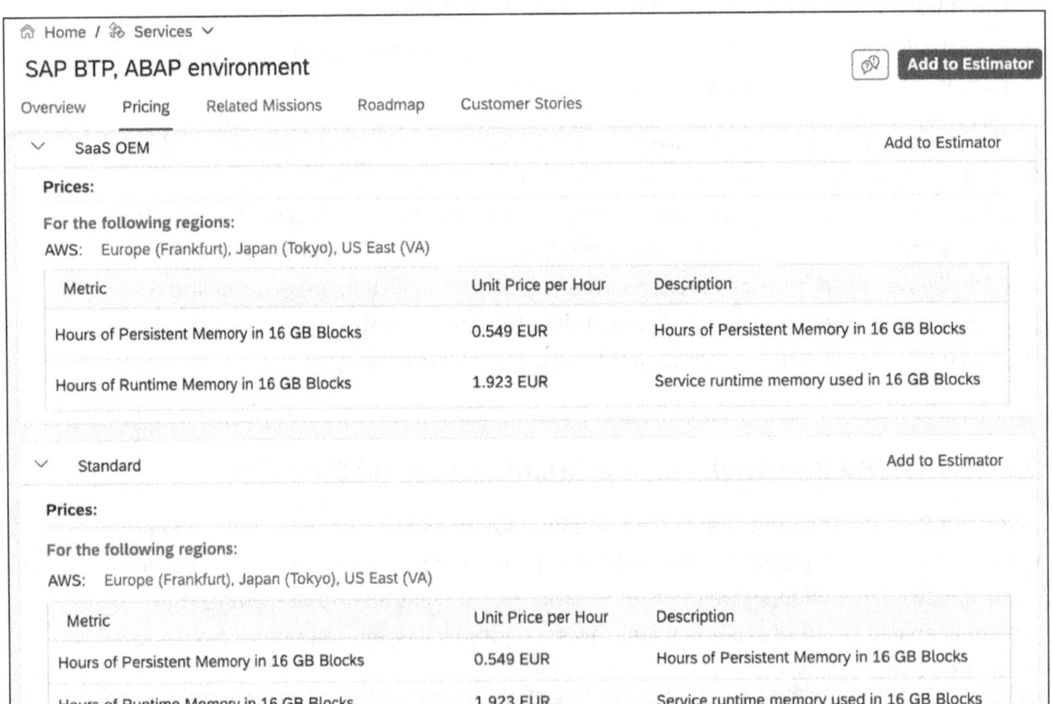

Figure 9.1: CPEA Memory Availability and Pricing

SAP also provides SAP BTP ABAP environment in a trial environment with the following restrictions:

- Shared offering: All users of a trial instance in fact share the same instance. Thus, everyone can see and change everyone else's objects and code.
- For educational purposes only: This account is only for self-education; you cannot use it for other purposes.
- Limited lifetime: The lifetime is the same as for the SAP BTP, Cloud Foundry environment trial account; that is, 90 days.

In the trial account, the following features are limited:

- Limited access to applications on the administrator's SAP Fiori launchpad
  - No apps for identity access management
  - No apps for communication management
  - No app for software component lifecycle
- No support for custom identity providers
- Not possible to use the SAP Destination service or backend connectivity to SAP S/4HANA systems
- Not possible to create and assign business roles

In the next section, we'll show you how to create your own SAP BTP ABAP environment trial account.

9.2 Provisioning an SAP BTP ABAP Environment Trial

To create a trial instance of SAP BTP ABAP environment, you should navigate to your SAP BTP subaccount and select **Service Marketplace** from the menu on the left. From the service marketplace, search for "ABAP" and then click on **Create**, as shown in Figure 9.2.

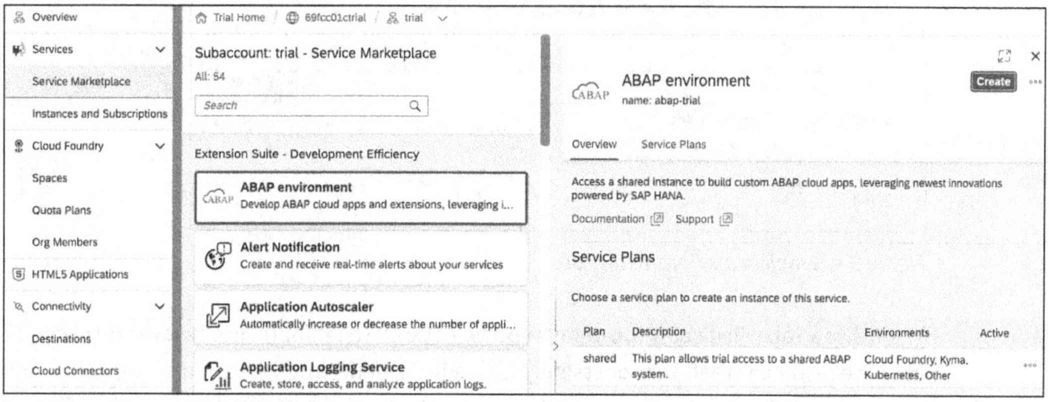

Figure 9.2: Creating an SAP BTP ABAP Environment Trial

On the next screen, shown in Figure 9.3, most values should be prepopulated. Since we're using a trial account, your plan is **shared**, and you should enter an **Instance Name**.

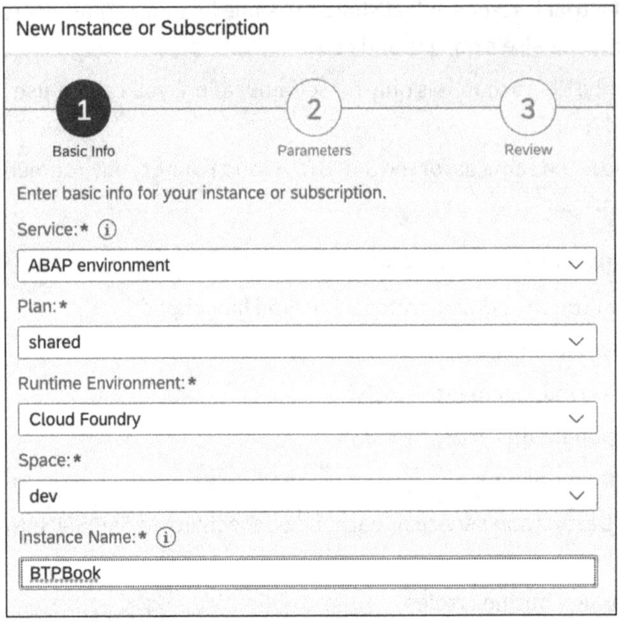

Figure 9.3: Basic Information to Create a Trial Instance

Under the **Parameters** tab, enter the email for your trial account, as shown in Figure 9.4. (We've masked our email address.)

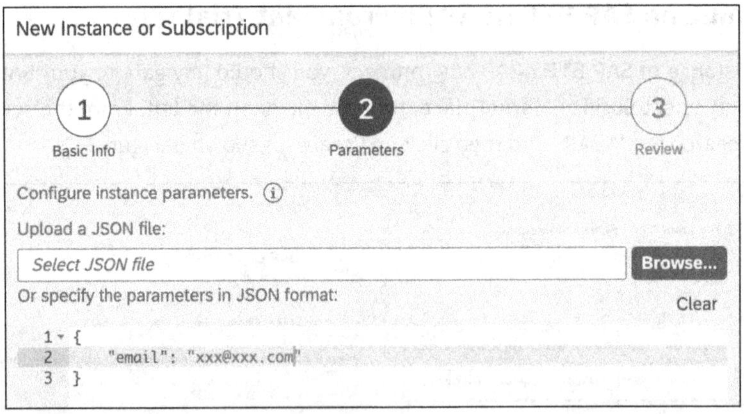

Figure 9.4: Email in the Parameter Tab

The **Review** tab will display the details you've entered so far for review before you click **Create**. The instance creation process may take a little while, and when complete, you should see the created instance from the **Instances and Subscriptions** subaccount menu. The created instance will be displayed, as shown in Figure 9.5.

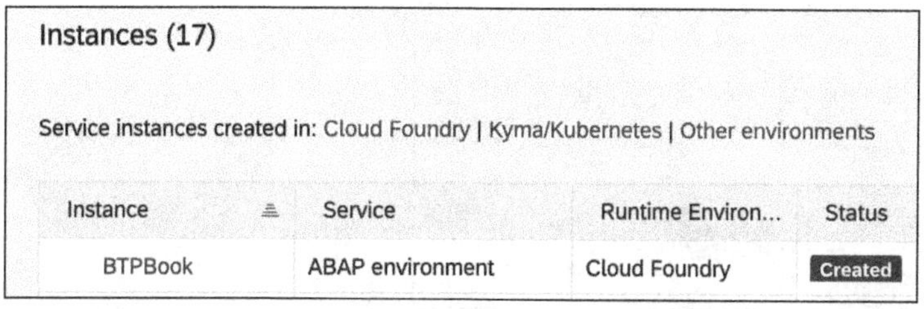

Figure 9.5: Newly Created ABAP Instance

For the next step, you can create the service instance key, which in this case is managed by external applications, for example, Eclipse/ABAP Development Tools (ADT) for development. The process to create a service key is the same as creating any other SAP BTP service: Click on the service instance and then select the **Create** option under **Service Keys**, as shown in Figure 9.6.

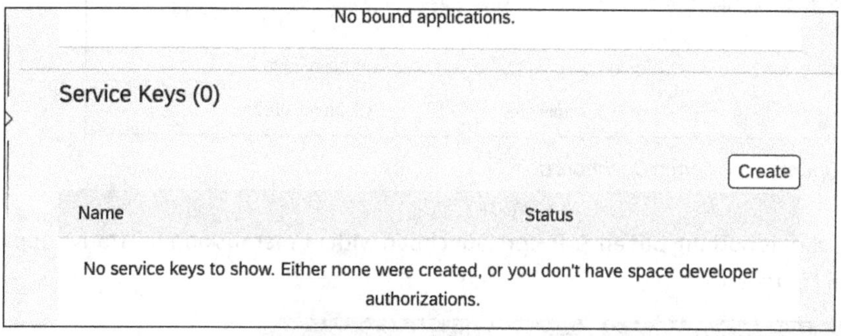

Figure 9.6: Service Key Creation Option

In the service key creation popup window, add a meaningful name, as shown in Figure 9.7, and then click **Create**. The created service key will be displayed under **Service Keys**, as shown in Figure 9.8.

Once the service instance and service key have been created, open the dashboard using the action icon (three dots), as shown in Figure 9.9.

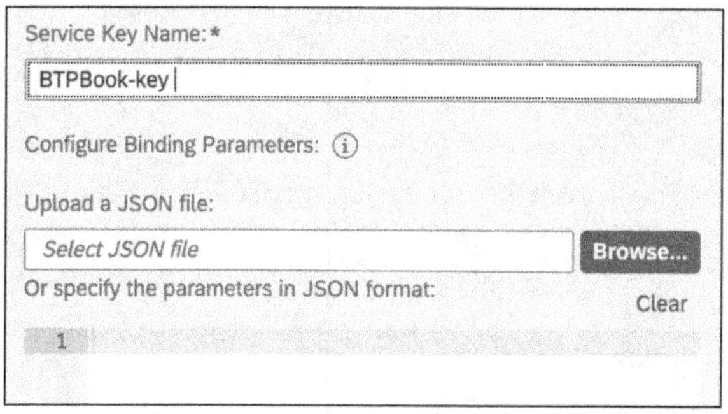

Figure 9.7: Entering a Name for the Service Key

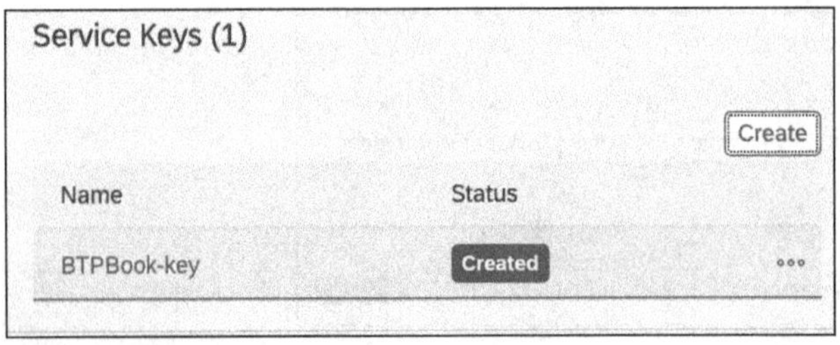

Figure 9.8: Newly Created Service Key

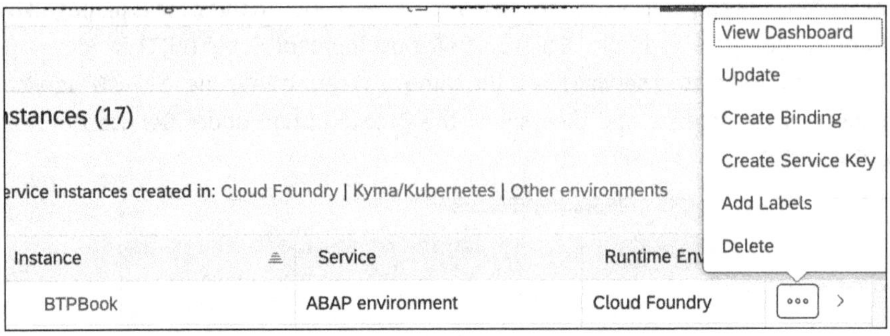

Figure 9.9: Opening the Admin Dashboard

Your dashboard is nothing but an SAP Fiori launchpad with numerous administrative apps, as shown in Figure 9.10.

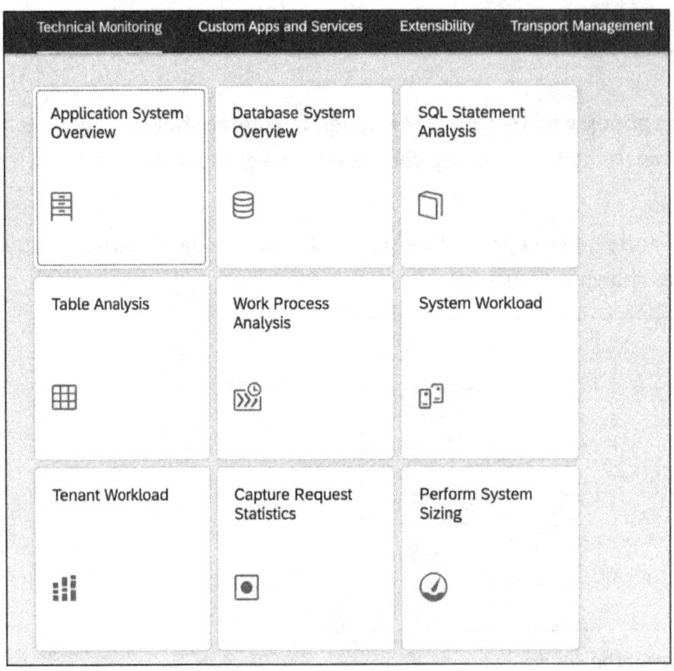

Figure 9.10: Admin Dashboard with Apps

With a trial account, your user will automatically receive administrative access with the proper role. However, for an enterprise account, an administrator must navigate to the subaccount menu of the user to assign the related role collection. You cannot access the admin dashboard without this role collection. If you're logged in to an administrator role for an enterprise account, the creation of the SAP BTP ABAP environment in your enterprise account will be similar, but you'll need to manually add the right role collections to your users.

At this point, we've created and configured the SAP BTP ABAP environment. In the next step, we'll install and set up the Eclipse-based development tools.

9.3 ABAP Development Tools for Eclipse

Eclipse is an open-source IDE built in Java and released under the terms of the Eclipse Public License. Initially developed by IBM as a replacement for IBM VisualAge, in 2001, IBM released it to open source. Since 2004, Eclipse has been governed by the Eclipse Foundation. Over time, Eclipse has become a leading development platform that supports a wide range of languages, including C, C++, COBOL, Java, JavaScript, Perl, Ruby, Python, R, and more. In 2012, ABAP joined the family of languages supported in Eclipse, following the release of the initial version of ADT for Eclipse. Over the next few years, ADT languished in the shadow of SAP's proprietary IDE, ABAP Workbench. However, with the introduction of ABAP core data services (CDS) views, Eclipse became a mandatory development environment. Now, for SAP BTP ABAP environment, Eclipse is the only option for the development of backend components (like classes, tables, CDS views, services, etc.).

In the following sections, we'll provide a deeper exploration of the Eclipse-based ABAP development environment, starting with installation and setup, followed by an overview of SAP Joule for Developers, and concluding with ABAP's AI capabilities.

9.3.1 Installation and Setup

The URL *https://tools.hana.ondemand.com/#abap* is your central access point for installing the development tools for SAP BTP ABAP environment. Instructions for installing Eclipse can be found at *http://s-prs.co/v550406*. Then, you would install ADT as a plugin on top of Eclipse. The installation process is simple and doesn't require significant time, and therefore, we'll now focus on connecting ADT to your SAP BTP ABAP environment system after installing Eclipse.

First, open Eclipse (with ADT installed) and create a cloud-based ABAP project by following the menu path **File • New • ABAP Cloud Project**. This step will open the window shown in Figure 9.11. Here, you can connect to SAP BTP ABAP environment in two ways: using a service key or using SAP BTP, Cloud Foundry environment. Recall that we created a service key (shown earlier in Figure 9.8), and you can retrieve the service key and use it to log on to your

SAP BTP ABAP environment account. However, in this case, we'll use SAP BTP, Cloud Foundry environment to log on. For the next step, shown in Figure 9.12, maintain the **Region**, **Email**, and **Password** fields according to your SAP BTP, Cloud Foundry environment instance.

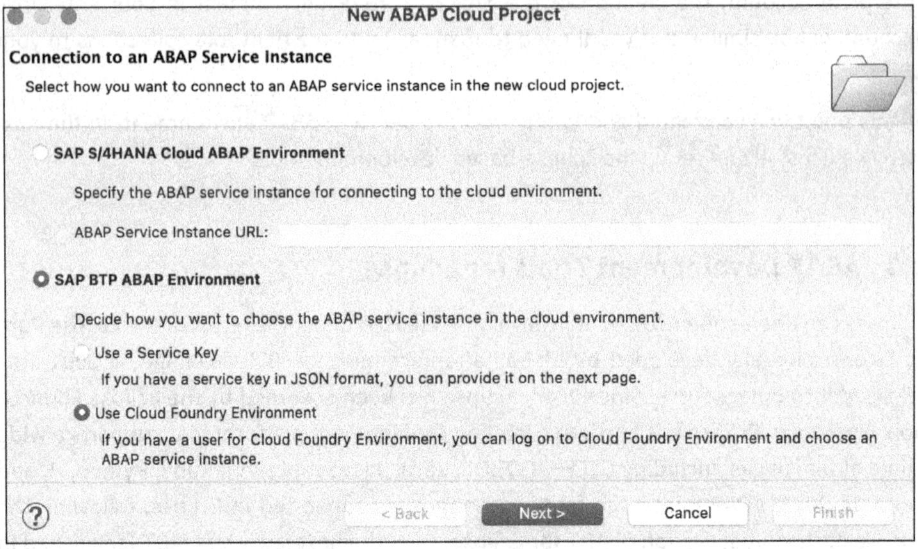

Figure 9.11: SAP BTP ABAP Environment: Project Creation Option

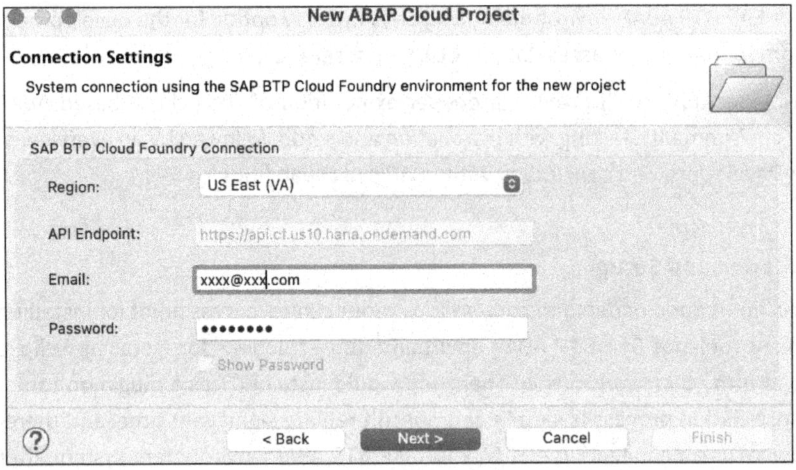

Figure 9.12: Region, Email, and Password for Logon

Next, maintain in the **Organization, Space**, and **Service Instance** fields, as shown in Figure 9.13.

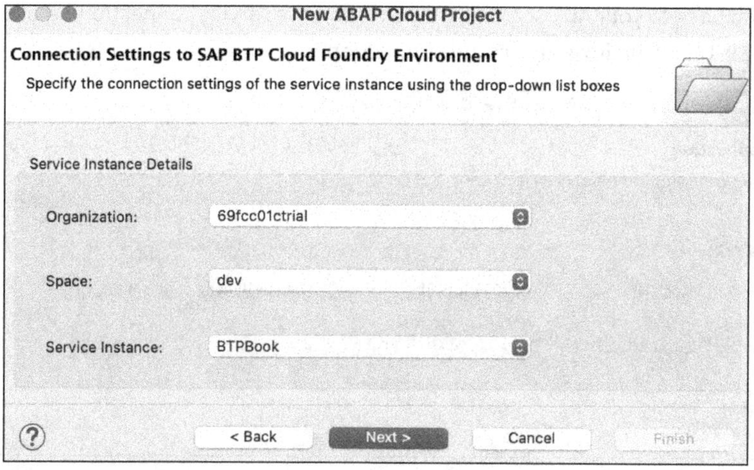

Figure 9.13: Organization, Space, and Service Instance

Next, you'll see the screen shown Figure 9.14 for you to log on to your SAP BTP, Cloud Foundry environment account. You can either copy the URL or click on the **Open Logon Page in Browser** button to log on in a new browser window.

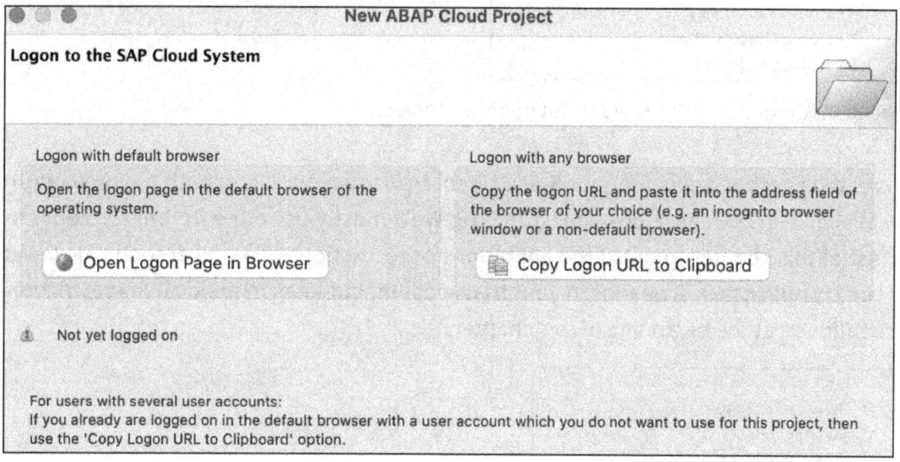

Figure 9.14: Logon URL Option

After successfully entering your SAP BTP, Cloud Foundry environment credentials, you should see a success message in the browser, as shown in Figure 9.15.

Figure 9.15: Logon Successful as Shown in a Browser

Immediately switch back to your ADT screen, and you'll find the service connection details, as shown in Figure 9.16, indicating a successful connection.

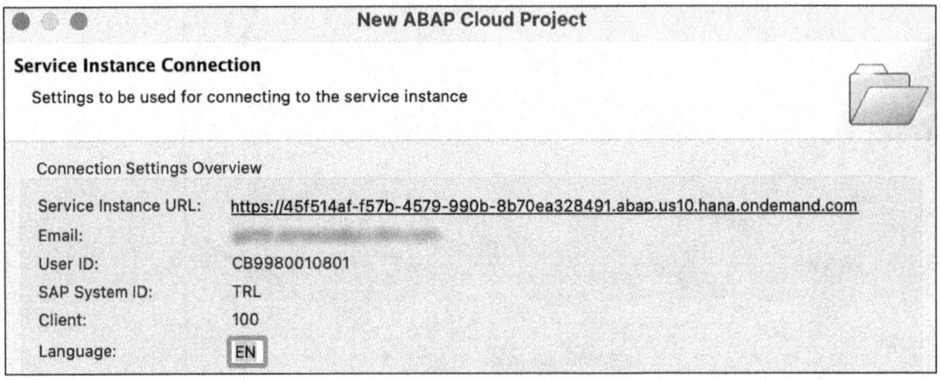

Figure 9.16: Service Connection Details

Click **Next** to complete this connection, and your project will now be displayed in the project explorer, as shown in Figure 9.17.

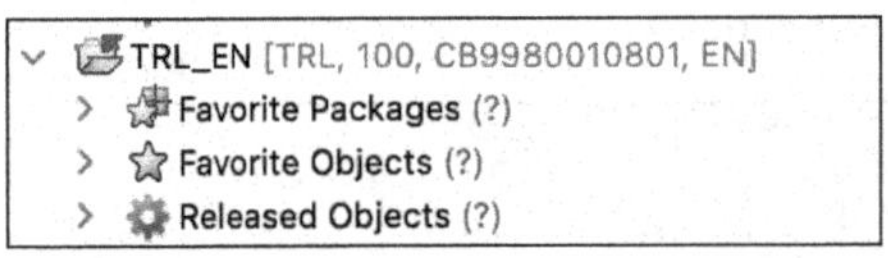

Figure 9.17: Expanded SAP BTP ABAP Environment Project

We created another project for our SAP BTP enterprise account, and both the projects are now available in ADT, as shown in Figure 9.18. We'll mostly use our enterprise account for the rest of this chapter to avoid the limitations or restrictions of a trial account. However, you can try out certain scenarios in your trial account, but keep in mind all the restrictions we mentioned at the beginning of this chapter.

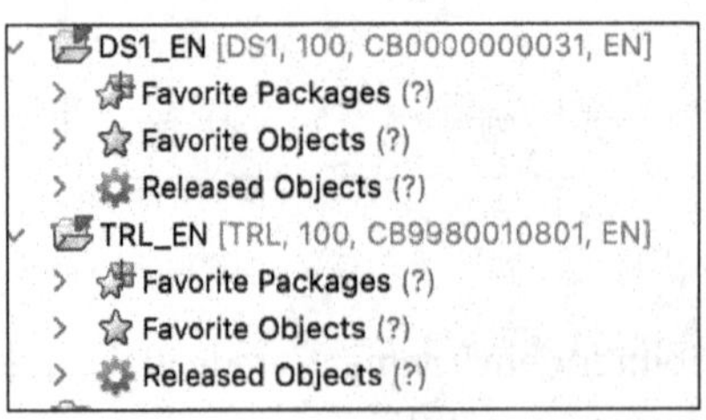

Figure 9.18: One SAP BTP ABAP Environment Project for an Enterprise Account and Another for a Trial Account

Now that the required installation is in place, let us move on to discussing SAP Joule for Developers.

9.3.2 SAP Joule for Developers

As you know already, SAP Joule for Developers is an AI-powered add-on embedded in ADT that augments the daily work of ABAP developers. It's intended to reduce context switching, accelerate learning, and provide just-in-time help for explaining code, generating unit tests, and exploring application artifacts. By combining natural language understanding with ABAP-aware capabilities, Joule closes the gap between consultant intent and technical implementation.

We have observed that in typical projects, developers spend substantial time interpreting unfamiliar objects, tracing dependencies, and seeking guidance from peers and seniors. In such instances, Joule minimizes these interruptions by offering a solution capable of answering questions about ABAP and CDS, clarifying ABAP RESTful application programming model concepts, and even proposing real code based on high-level instructions.

In ADT, Joule appears both as a view and in the context menu. It can explain code, suggest code while you type, and help create tests for ABAP Unit and CDS. The ABAP AI SDK also lets you build custom AI features using SAP AI Core.

Like any other AI tool, Joule's introduction has provided multiple benefits. It reduces onboarding time for new team members and accelerates the work of experienced developers by delivering explanations and examples. It promotes consistent patterns, encourages test-first thinking, and helps teams focus on business value. The result is stronger code quality, shorter cycle times, and more predictable deliveries.

Thus, Joule can help organizations benefit from fewer person-to-person interruptions, better knowledge dissemination, and easier adherence to platform best practices. In combination with the ABAP AI SDK, consultants can also build their own AI features that address domain-specific problems.

To provide some real examples, let's say a developer reading legacy code requests an explanation for database annotations in a table. Joule will return an easy-to-understand summary and link to the exact source context. Another typical example would be a team implementing the ABAP RESTful application programming model that asks Joule for a starter implementation and receives generated artifacts, which are then refined in ADT. Joule also automatically proposes unit tests for the key methods and CDS semantics.

So, organizations and individual teams can measure benefit through reduced time to first commit for new developers and increased unit test coverage, which increases quality and thus necessitates fewer rework cycles during code reviews. Anecdotally, developers get greater confidence when working with unfamiliar modules when they use the on-demand explanations and examples provided by Joule.

Thus, SAP Joule for Developers complements the ABAP toolchain with conversational, context-aware help. Joule also streamlines learning, standardizes patterns, and embeds AI safely into development workflows.

Let's now understand the steps to start SAP Joule for Developers in ADT:

1. Open ABAP Development Tools and locate the search icon in the upper-right corner of the workbench. Type "Joule" in the search field and open the Joule view. On first launch, the assistant greets us and displays the backend system context. We must select a project so that the assistant binds its operations to the correct system.

   The Joule view consists of a chat history, a prompt area for new input, and a header that indicates the active system. All queries are evaluated in context of the selected backend, ensuring that object names and references resolve correctly.

2. Now we can ask questions using complete sentences and natural language. Joule recognizes development concepts and can interpret pasted code when enclosed in backticks or when we use the context menu **Explain** action. For targeted actions, prefix your request with @ to activate a specific skill, such as "@Explain" or ABAP Unit generation.

   The chat maintains conversational context and its history. Any follow-up questions inherit prior history, allowing you to refine the scope or ask for summaries without repeating the original prompt.

[+]

Tip

Before asking for help, ensure the correct project is active. If your organization uses multiple systems, the system indicator in the Joule view helps avoid confusion.

3. If we are planning to share code, include only the relevant snippet to focus the explanation and to reduce unwanted verbiage. If the object is large, we can use the context menu so Joule can attach an object reference rather than pasting the entire source.
4. If we find that Joule is unaware of an object, first we need to verify that we are connected to the right system and that the object exists within the project's visibility. Ambiguous names may prompt Joule to present a selection list; confirm the correct one.
5. If formatting appears odd, ensure our snippet is surrounded with backticks or use the context menu **Explain** entry to provide the code as a recognized source fragment.

9.3.3 ABAP AI Capabilities

The ABAP AI SDK allows custom AI scenarios backed by SAP AI Core. The extensibility AI assistant and documentation features further support developers by helping them find BAdIs and custom fields and by turning SAP Notes into a conversational knowledge base. The ATC **Explain** feature provides plain-language descriptions of ATC findings to accelerate comprehension and remediation.

After Joule explains what an object does, we can request a concrete example of its usage. For instance, once a function module is summarized, we can ask, How do I call this with test data? Joule then provides ready-to-run ABAP code along with an explanation of the parameters and expected outcomes. This property is highly effective for accelerating proofs of concept. It also promotes best practices by demonstrating idiomatic code and proper error handling patterns.

Suppose we need to call a function module to validate a customer ID. We can say, "Show me a minimal example that calls `Z_VALIDATE_CUSTOMER` with a valid ID and checks the Boolean result." Joule then proposes a small program or method snippet that you can paste into your test harness.

We must always review generated code to ensure naming, exception handling, and logging align with development standards. If database access is involved, we need to confirm that the example uses appropriate APIs for the target stack (for example, EML when working with ABAP RESTful application programming model business objects).

We can ask Joule for variations such as negative tests, boundary conditions, and examples with test doubles to round out coverage.

Further examples of AI capabilities can be seen in CDS models, which can become complex through layered views, associations, path expressions, and annotations. With **Explain**, we can quickly verify whether a view meets the requirements of an analytical story, a transactional report, or a service exposure. For calculated fields, we can also ask Joule to detail the formula and provide examples showing how the value changes based on input conditions. When planning changes, we can ask Joule to identify which downstream consumers or business users might be impacted by a field or filter modification. We can then follow up by creating or improving CDS test classes to preserve intended semantics during refactoring. Next, we need to pair the explanation with generated unit tests so that the explanations translate into verifiable behavior.

Predictive code completion helps transform intent expressed in natural language comments into working ABAP code suggestions. By enabling automatic triggering in ADT, Joule can display ghost text as we type. From a class, choose **Joule • Start Chat** for ABAP Unit to begin. Joule inspects the class and indicates whether existing tests are present. We can select methods to cover and request test generation. The assistant returns test class code that can be pasted into the appropriate include or test artifact.

The generated tests typically include setup, method invocation, and assertions for expected outcomes. They provide a foundation that we can expand by adding edge cases, negative scenarios, and coverage for exceptional paths.

Generated CDS test classes include ABAPDoc comments explaining the objective of each test and the assumptions about data. Each method targets a single semantic facet, which keeps failures specific and actionable.

We then set up code that uses the CDS test double framework to inject data into dependent views. Assertions then compare the actual CDS result set with the expected outcome, often expressed through counts and specific field values.

We also use the **OData UI Service from Scratch** wizard to create an ABAP RESTful application programming model business object with all its required repository artifacts. In the

Joule sidebar, we can describe entities, relationships, and prefixes in natural language; Joule then proposes a full model which we can accept and generate.

By opening the extensibility AI assistant and describing our need in plain language, such as "Find a BAdI to check bill of material consistency," the assistant returns candidate BAdIs with documentation links and relevance scores. If a BAdI already has an implementation, the assistant warns us. Otherwise, we navigate directly to the creation screen with the target BAdI preselected and continue with the standard implementation steps.

If we request a new custom field by stating its name, business context, and target APIs, the assistant lists possible business contexts and OData services. After we select the best match, the assistant prepares field metadata which we can accept and finalize on the creation screen. This guided process reduces misconfiguration and ensures that custom fields are correctly enabled for the intended APIs and UIs.

We can add ABAP Unit tests for helper classes and use predictive code completion to quickly draft additional methods. We iterate until the critical paths are thoroughly tested. Throughout, we ask Joule to refactor or clarify code to ensure the implementation remains idiomatic and maintainable. This integrated approach shortens cycle time and increases confidence, making it easier to evolve requirements without destabilizing the application. Therefore, the full lifecycle demonstration highlights how Joule reduces effort, amplifies best practices, and sustains quality from scaffolding to testing and beyond.

ATC findings can be dense and may require cross-referencing multiple resources. With ATC **Explain**, we can right-click a finding to request a plain-language description of the issue, its root cause, and suggested remediation steps. Joule enhances the explanation with links to relevant simplification notes and cookbooks, saving time that would otherwise be spent searching for and correlating guidance.

Overall, Joule enhances development speed, code quality, and maintainability by offering guided automation, consistent test generation, and intelligent assistance across CDS, ABAP, the ABAP RESTful application programming model, and extensibility workflows.

So, our SAP BTP ABAP environment system is up and running, we've installed ADT, connected to our SAP BTP ABAP environment instance, and seen what Joule and AI can offer us in this space. Are we ready for development? Not yet! One unique feature of SAP BTP ABAP environment is the ABAP RESTful application programming model, and you cannot develop applications without understanding the ABAP RESTful application programming model, which we'll cover next, before we explore different development patterns.

9.4 ABAP RESTful Application Programming Model

The ABAP RESTful application programming model is the primary programming model for SAP BTP ABAP environment and enables the efficient end-to-end development of SAP HANA-optimized OData services and SAP Fiori apps. This model supports the development of all types of SAP Fiori applications as well as supports publishing web APIs. The model is based on CDS and follows a layered development approach, as shown in Figure 9.19.

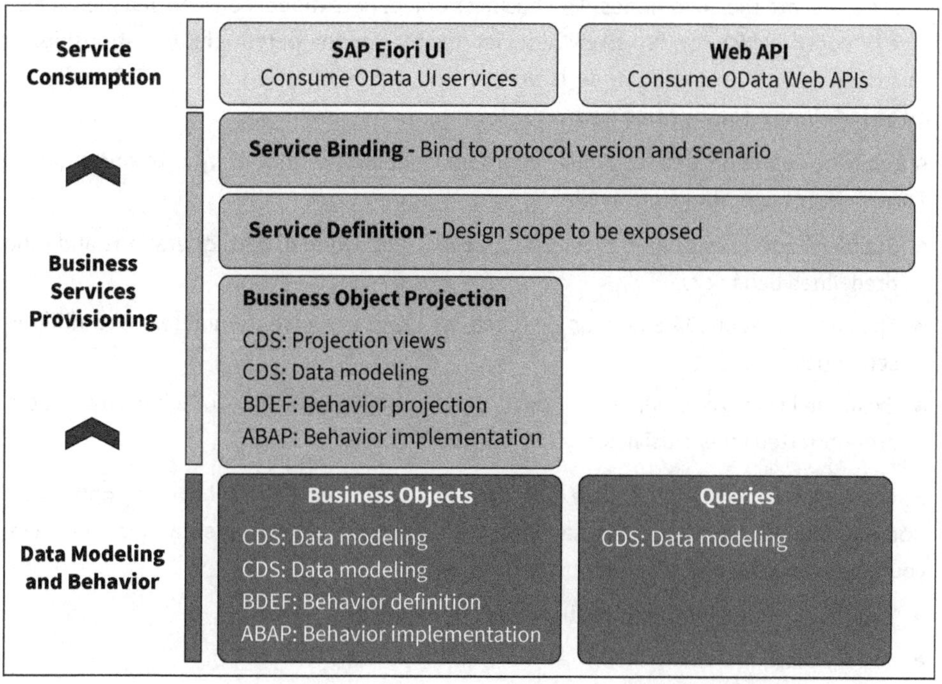

Figure 9.19: Architecture and Components of the ABAP RESTful Programming Model

As shown in Figure 9.19, the ABAP RESTful application programming model is based on three distinct layers: data modeling and behavior, business service provisioning, and service consumption. Let's cover each layer next.

9.4.1 Data Modeling and Behavior

The ABAP RESTful application programming model uses CDS to define data models. CDS provides a framework for defining and consuming semantic data models. Depending on the use case, data models support transactional access or query access to a database. Thus, data models are used as business objects for analytical queries and reports.

Business objects and queries together form a complete data model and application. This combination is a reusable layer consisting of data modeling, database tables, views, and available application logic.

The behavior describes what can be done with the data model, for example, if the data can be updated. In transactional scenarios, the business object's behavior defines which operations and what characteristics belong to a business object. For read-only scenarios, the behavior of the data model is defined by the query capabilities.

Business objects are composed of data model, behavior, and behavior (runtime) implementation. In a business object, `root` is the root node of the business object, while `child1`, `child2`, etc. are the child nodes. The business object data model consists of a hierarchical tree of nodes, which are linked by associations that define parent-child relationships. In this composition tree, each node is an element that is modeled with a CDS entity and arranged along a composition path.

A behavior definition defines which operations should be offered by a business object. Three types of operations are available:

- Standard operations like create, read, update, and delete (CRUD) operations and other predefined behaviors.
- Specific operations like actions or functions that may have parameters and different semantics.
- Exclusive locks are provided for concurrency handling, which is quite important. Etags are supported for exclusive locks.

How behavior should be processed during runtime is defined in the behavior's implementation. A single behavior definition can have multiple behavior implementations. In general, you may encounter one of these possible scenarios:

- You already have complete application code available.
- You are starting from scratch.
- You have some code partially available.

All these scenarios have one thing in common—the runtime business object model. The implementation of a behavior implementation follows three phases:

- Interaction phase
- Transaction buffer
- Save sequence

The *interaction phase* involves three steps: modify, read, and lock. An end user can edit or change data or add data by clicking the **Edit** button. A transaction opens and interacts with backend changes that are stored in the *transaction buffer*, which represents the state of the instance data. A business object always requires a transactional buffer to contain runtime changes. On clicking the **Save** button, the data from the transaction buffer is written to the database. At this point, the interaction phase ends, and the *save sequence* starts, which is used for data persistence.

9.4.2 Business Service Provisioning

In the ABAP RESTful application programming model, business service provisioning projects this data model to certain user interfaces (UIs), which may be role specific or task dependent, or to a web API. Regardless of the provisioning target, the provisioning is based on the same data and functionality contained in the service definition. The provisioning is an artifact, through which the data model and its related behaviors are exposed or projected.

A service definition with CDS entities will be exposed for that service. Your ABAP RESTful application programming model may consist of many entities, but not all of them should be exposed all the time. A service definition helps you provide consumer-specific but protocol-agnostic exposure.

In a service binding, the service must be bound to a protocol and protocol variant, for example, OData v2 or v4, and to a specific use case because the use case defines which capabilities the OData service must fulfill, which is protocol specific. This separation by protocol allows multiple data models and service definitions to be integrated into various communication protocols, without the hassle of reimplementation.

9.4.3 Service Consumption

The service consumption layer defines how your service will be consumed, for example, as an SAP Fiori UI or a web API. Selecting the SAP Fiori UI option tells the framework to consider different additional metadata (e.g., annotations) during service consumption. For a web API, data consumption by an OData client is the main area of focus, and thus SAP Fiori-specific annotations are not considered.

Thus, we've covered all the background you need to start developing in SAP BTP ABAP environment using the ABAP RESTful application programming model. Let's now get to the fun part of this chapter—actual development! We'll start with a basic list report application.

9.5 Developing a List Report App

Let's start creating a simple list report app with some selection criteria that lists values in a tabular format that satisfy the selection criteria. In this app, a user can access the detail page by clicking on a single row in the tabular data. Since SAP BTP ABAP environment may not have any application tables, we'll create a table first and then build various ABAP RESTful application programming model artifacts. Let's start with defining a business requirement as the first step.

9.5.1 Business Requirements

Whenever employees relocate to a new place, other business users (managers) should be able to raise a request where a set of fields (i.e., **Country**, **State**, **City**, **Building**, etc.) can be provided for capturing the details of the new location. The current location's details will also be displayed for easy reference. With this application, we'll display a list report of existing employee move requests. In Section 9.7, while exploring *managed app* scenarios, we'll enhance the same app to offer the option of creating a new move request as well. Additionally, we'll add an option to update the status of one or more employee move requests when the activity is completed. As mentioned earlier, in this section, we'll only focus on displaying data from the table and what the final app will look like. We'll use the preview functionality in ADT to display the data since we don't plan to configure this app in the SAP Cloud Portal service or the SAP Build Work Zone. The final app is shown in Figure 9.20.

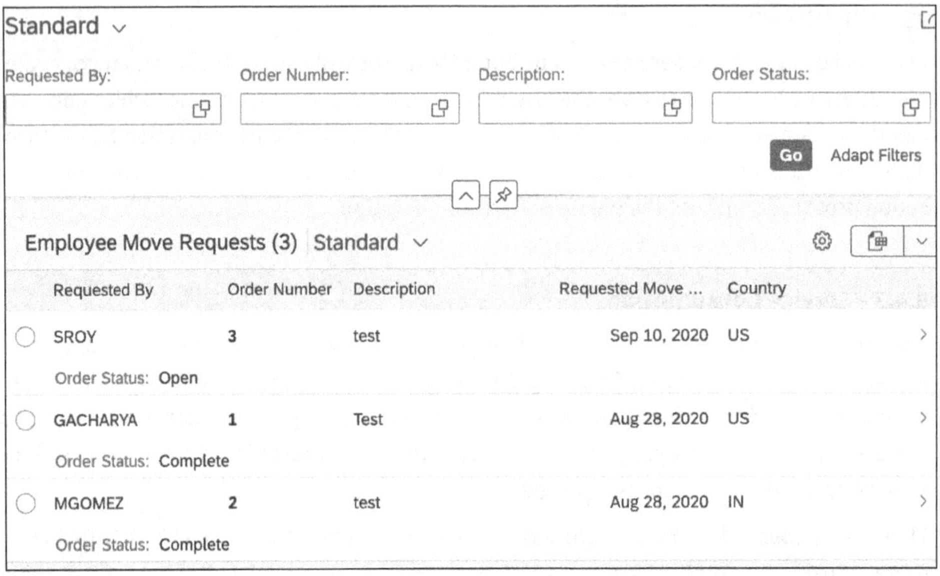

Figure 9.20: Look and Feel of the Final List Report App

Since we're creating our first app, we must create an ABAP package first.

9.5.2 Creating an ABAP Package and Database Table

We won't repeat this step for the other examples in this chapter since these steps are common and can be repeated in a similar way. You should use the same project we created in the previous section, but new packages can be created for different requirements. Right-click on your ABAP project name and select the menu option **New • ABAP Package**. This step will open a new window, as shown in Figure 9.21, where you can maintain the **Name**, **Description**, and **Superpackage** (if any) fields for the package and then select **Development** from the **Package Type** dropdown list.

On the next screen, as shown in Figure 9.22, select **Local** from the **Software Component** dropdown list. Later, in Section 9.11, you'll learn how to create other software components using the Git-enabled Change and Transport System (gCTS).

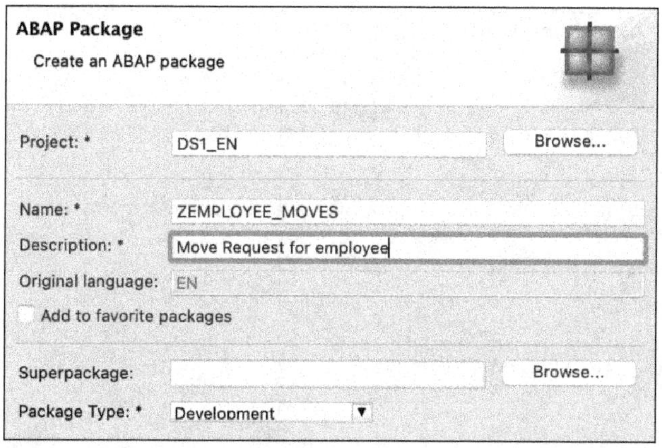

Figure 9.21: New Package Creation: Initial Screen

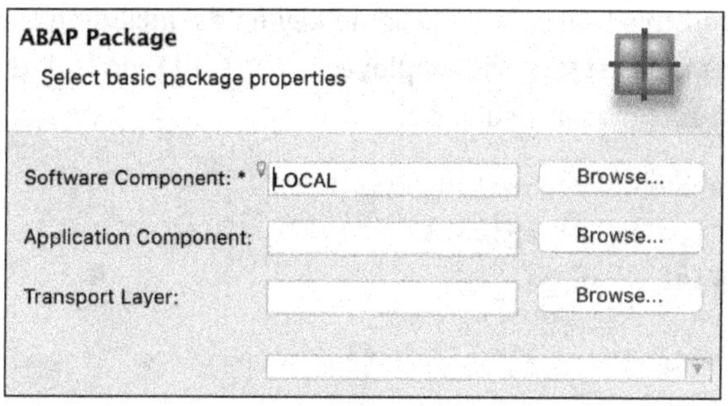

Figure 9.22: Software Component

The next and final screen is for the transport request, as shown in Figure 9.23. You can either create a new transport request or select an existing one and then click **Finish** to create the package.

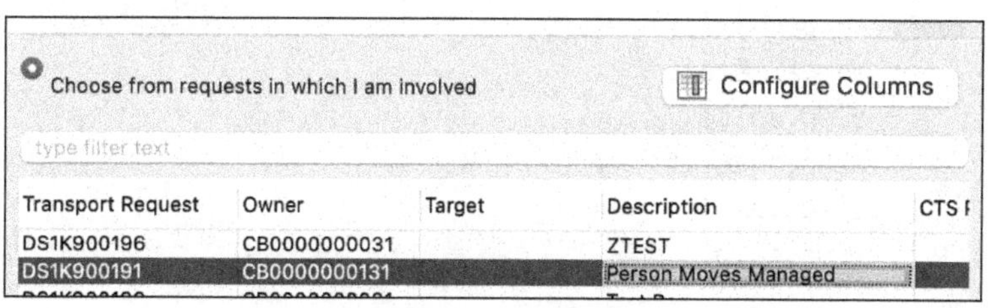

Figure 9.23: Final Step for Creating a Package

We'll also create two tables to store data. Right-click on your package and select **New • Other Repository Objects • Dictionary • Database Table**, as shown in Figure 9.24.

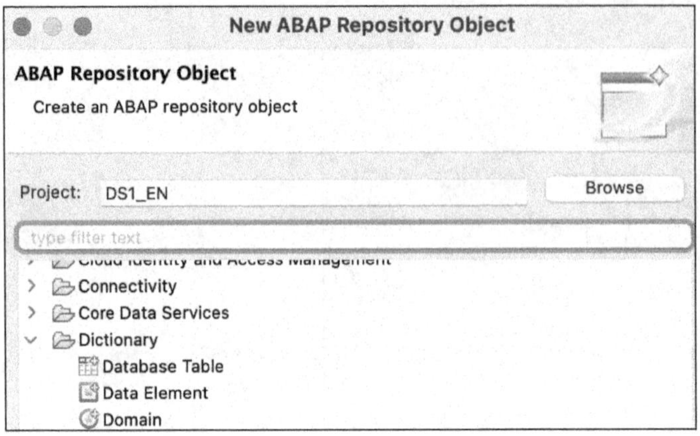

Figure 9.24: Dictionary Object Creation in SAP BTP ABAP Environment

Our first table is the header table, and this table contains some key fields, including the order number for a move request, the move status, the employee's name, the move date, and some other administrative fields, as shown in Listing 9.1.

```
@EndUserText.label : 'Employee Moves Request Header Data'
@AbapCatalog.enhancementCategory : #NOT_EXTENSIBLE
@AbapCatalog.tableCategory : #TRANSPARENT
@AbapCatalog.deliveryClass : #A
@AbapCatalog.dataMaintenance : #RESTRICTED
define table zemp_move_hdr_m {
  key client        : abap.clnt not null;
  key mykey         : sysuuid_x16 not null;
  aufnr             : zaufnr;
  descr             : abap.char(40);
  country           : land1;
  ernam             : abap.char(15);
  status            : zorder_status;
  move_date         : abap.dats;
  created_by        : abap.char(15);
  create_date       : timestampl;
  last_change_date  : timestampl;
  last_changed_by   : syuname;

}
```

Listing 9.1: Header Table

Listing 9.2 shows the item table with the item number, user ID, current location details, and new location information.

```
@EndUserText.label : 'Employee Moves Request Item Data'
@AbapCatalog.enhancementCategory : #NOT_EXTENSIBLE
@AbapCatalog.tableCategory : #TRANSPARENT
@AbapCatalog.deliveryClass : #A
@AbapCatalog.dataMaintenance : #RESTRICTED
define table zemp_move_item_m {
  key client    : abap.clnt not null;
  @AbapCatalog.foreignKey.label : 'GUID'
  @AbapCatalog.foreignKey.screenCheck : false
  key mykey     : sysuuid_x16 not null
    with foreign key [0..*,1] zemp_move_hdr_m
      where mykey = zemp_move_item_m.mykey;
  key posnr     : abap.numc(4) not null;
  aufnr         : zaufnr;
  user_id       : abap.char(15);
  curr_country : land1;
  curr_state    : regio;
  curr_city     : abap.char(40);
  curr_bldg     : abap.char(4);
  curr_floor    : abap.numc(4);
  curr_cube     : abap.char(4);
  next_country : land1;
  next_state    : regio;
  next_city     : abap.char(40);
  next_bldg     : abap.char(4);
  next_floor    : abap.numc(4);
  next_cube     : abap.char(4);

}
```

Listing 9.2: Item Table

We also created a data element and domain for the order number using ADT. Additionally, we created a custom class to create some test data for these two tables. This approach will help view our test data while previewing the corresponding SAP Fiori app in ADT. Once test data is updated, you can preview the data by right-clicking on the table source code and selecting the menu option **Open with • Data Preview**.

So, our tables have been created, and test data has been updated to the table. Let's now create the ABAP RESTful application programming model application to build up this list report.

9.5.3 Building an ABAP RESTful Application Programming Model Application

When we discussed the ABAP RESTful application programming model architecture at the beginning of this chapter, we highlighted CDS as the starting point, and we'll follow the same approach now. Our first CDS is the header-data related CDS, as shown in Listing 9.3.

```
@AbapCatalog.sqlViewName: 'ZEMP_MOVE_H_V'
@AbapCatalog.compiler.compareFilter: true
@AbapCatalog.preserveKey: true
@AccessControl.authorizationCheck: #CHECK
@EndUserText.label: 'CDS for Header table'

@UI: {
headerInfo: {
typeName: 'Employee Move Request',
typeNamePlural: 'Employee Move Requests'
} }

@ObjectModel.semanticKey: ['aufnr']
define  root view ZEMP_MOV_HDR_CDS_V2
  as select from zemp_move_hdr_m as Header
{
@UI.hidden: true
  key mykey                 ,

      @UI.facet: [
      {
      id: 'GeneralData',
      type: #COLLECTION,
      position: 10,
      label: 'Moves Request Header Information'
      },
      {
      type: #FIELDGROUP_REFERENCE,
      position: 10,
      targetQualifier: 'GeneralData1',
      parentId: 'GeneralData',
      isSummary: true,
      isPartOfPreview: true
      },

      {
      type: #FIELDGROUP_REFERENCE,
      position: 20,
      targetQualifier: 'GeneralData2',
```

```
parentId: 'GeneralData',
isSummary: true,
isPartOfPreview: true
}
]

@UI: {
lineItem: [ { position: 10, importance: #HIGH } ],
selectionField: [{position: 10 }],
fieldGroup: [{qualifier: 'GeneralData1',position: 10,importance: #HIGH }]

}
@EndUserText.label: 'Order Number'
Header.aufnr,

@UI: {
lineItem: [ { position: 20, importance: #HIGH } ],
selectionField: [{position: 20 }],
fieldGroup: [{qualifier: 'GeneralData1',position: 30,importance: #HIGH }]
}

@EndUserText.label:'Description'
Header.descr,
@UI: {

lineItem: [ { position: 30, importance: #HIGH } ,
{ type: #FOR_ACTION,
              dataAction: 'statusComplete',
              label: 'Set to Complete' } ],

fieldGroup: [{qualifier: 'GeneralData1',position: 40,importance: #HIGH }]

}
@EndUserText.label: 'Country'
Header.country,
@UI: {
lineItem: [ { position: 10, importance: #HIGH } ],
selectionField: [{position: 10 }],
fieldGroup: [{qualifier: 'GeneralData2',position: 10,importance: #HIGH }]
}

@Consumption.valueHelpDefinition: [{entity: { name: 'zemp_user_VH',
                                            element: 'user_id'  }}]
@EndUserText.label: 'Requested By'
```

```
    Header.ernam,

    @UI: {
    lineItem: [ { position: 30, importance: #HIGH } ],
    fieldGroup: [{qualifier: 'GeneralData2',position: 30,importance: #HIGH }]

    }
    @EndUserText.label: 'Requested Move Date'
    Header.move_date,

    @UI: {
    hidden: true ,

    selectionField: [{position: 20 }]

    }]
    }
    @Semantics.systemDateTime.createdAt: true
    @EndUserText.label: 'Creation Date'
    Header.create_date,
    @UI.hidden: true
    @Semantics.systemDateTime.lastChangedAt: true
    Header.last_change_date,
    @UI.hidden: true
    @Semantics.user.lastChangedBy: true
    last_changed_by,
    @UI.hidden: true
    @Semantics.user.createdBy: true
    created_by,
    @UI: {
    lineItem: [ { position: 40, importance: #HIGH } ,
            { type: #FOR_ACTION,dataAction: 'change_status',label: 'Set
Status to Complete' }],
    selectionField: [{position: 30 }],
    fieldGroup: [{qualifier: 'GeneralData1',position: 20,importance: #HIGH }]

    }
    @EndUserText.label: 'Order Status'
    case  Header.status
    when '1' then 'Open'
    when '2' then 'Partially Processed'
    when '3' then 'Complete'
```

```
        end as status
=
}
```

Listing 9.3: Header CDS

Listing 9.3 may look a little complicated, but let's break it down:

- All the bold lines are CDS fields, which are either taken from the underlying table or are calculated fields (e.g., status).
- The CDS definition starts with the statement `Define root view`, and you must have at least one root view for your business object hierarchy. (You'll learn how to add a child node later when we cover the item-related CDS definition.)
- The rows marked with the @ signs are the annotations, which are primarily responsible for various aspects of the final SAP Fiori application. The `lineItem` annotation represents the tabular data field. The `selectionField` annotation represents the fields available in the filter area, and the `fieldGroup` represents how data will be displayed in the object page (i.e., the page displayed when clicking a particular row), and so on.

In the next step, we'll create another CDS for the item data, as shown in Listing 9.4.

```
@AbapCatalog.sqlViewName: 'ZEMP_MOVE_I_V'
@AbapCatalog.compiler.compareFilter: true
@AbapCatalog.preserveKey: true
@AccessControl.authorizationCheck: #CHECK
@EndUserText.label: 'CDS for Moves Req Item table'
define view ZEMP_MOV_ITEM_CDS_v2
  as select from zemp_move_item_m as Item

  association to parent ZEMP_MOV_HDR_CDS_V2 as _Header on
                                              $projection.mykey = _Header.mykey

{

@UI.hidden: true
  key mykey                 ,
      @UI.facet: [
      {type: #COLLECTION, position: 10, id: 'Item', label: 'Moves Request
Item Data '},
      {type: #FIELDGROUP_REFERENCE, position: 10,
      targetQualifier: 'Item1',parentId: 'Item',
       isSummary: true, isPartOfPreview: true},
      {type: #FIELDGROUP_REFERENCE, position: 20,
      targetQualifier: 'Item2',parentId: 'Item',
```

```
        isSummary: true, isPartOfPreview: true}
        ]
        @UI: {
        lineItem: [ { position: 20, importance: #HIGH } ],
        fieldGroup: [{qualifier: 'Item1',position: 20,importance: #HIGH }]
        }
        @EndUserText.label: 'Item Number'
    key Item.posnr,
        @UI.hidden: true
        @EndUserText.label: 'Order Number'
  Item.aufnr,
        @UI: {
        lineItem: [ { position: 30, importance: #HIGH } ],

        fieldGroup: [{qualifier: 'Item1',position: 30,importance: #HIGH }]

        }
        @EndUserText.label: 'User ID'
        @Consumption.valueHelpDefinition: [{entity: { name: 'zemp_user_VH',
  element: 'user_id'  }}]
        Item.user_id,

        @UI: {
        lineItem: [ { position: 40, importance: #HIGH } ],
        fieldGroup: [{qualifier: 'Item1',position: 40,importance: #HIGH }]
        }
        @EndUserText.label: 'New Country'
        @Consumption.valueHelpDefinition: [{entity: { name: 'Zcountry_CDS_VH',
                                                          element:
 'Country'  }}]
        Item.next_country,
        @UI: {
        lineItem: [ { position: 50, importance: #HIGH } ],
        fieldGroup: [{qualifier: 'Item1',position: 50,importance: #HIGH }]
        }
        @EndUserText.label: 'New State'

        @Consumption.valueHelpDefinition: [{entity: { name: 'Zstate_CDS_VH',
                                                     element: 'State1'  },
                            additionalBinding: [{localElement: 'next_country' ,
                                                 element: 'Country' }]}]
        Item.next_state,
        @UI: {
```

```
lineItem: [ { position: 60, importance: #HIGH } ],
fieldGroup: [{qualifier: 'Item1',position: 60,importance: #HIGH }]
}
@EndUserText.label: 'New City'
@Consumption.valueHelpDefinition: [{entity: { name: 'Zcity_CDS_VH',
                                              element: 'City1'  },
              additionalBinding: [{localElement: 'next_country' ,
                                   element: 'Country' },
                                  {localElement: 'next_state' ,
                                     element: 'State1' }
]
                                                   }]
Item.next_city,
@UI: {

lineItem: [ { position: 70, importance: #HIGH } ],
fieldGroup: [{qualifier: 'Item1',position: 70,importance: #HIGH }]

}
@Consumption.valueHelpDefinition: [{entity: { name: 'Zbldg_CDS_VH',
                                            element: 'Building1'  },
              additionalBinding: [{localElement: 'next_country' ,
                                         element: 'Country' },
                                  {localElement: 'next_state' ,
                                         element: 'State1' },
                                  {localElement: 'next_city' ,
                                         element: 'City1' }
]
                                                 }]
@EndUserText.label: 'New Building'
Item.next_bldg,
@UI: {
lineItem: [ { position: 80, importance: #HIGH } ],
fieldGroup: [{qualifier: 'Item1',position: 80,importance: #HIGH }]
}
@EndUserText.label: 'New Floor'
@Consumption.valueHelpDefinition: [{entity: { name: 'Zfloor_CDS_VH',
                                            element: 'Floor1'  },
              additionalBinding: [{localElement: 'next_country' ,
                                       element: 'Country' },
                                  {localElement: 'next_state' ,
                                       element: 'State1' },
                                  {localElement: 'next_city' ,
```

```
                                                       element: 'City1' },
                                                  {localElement: 'next_bldg' ,
                                                       element: 'Building1' }
]
                                                           }]
Item.next_floor,
@UI: {
lineItem: [ { position: 90, importance: #HIGH } ],
fieldGroup: [{qualifier: 'Item1',position: 90,importance: #HIGH }]
}
@EndUserText.label: 'New Cubicle'
@Consumption.valueHelpDefinition: [{entity: { name: 'Zcube_CDS_VH',
                                        element: 'Cubicle'  },
                  additionalBinding: [{localElement: 'next_country' ,
                                        element: 'Country' },
                                      {localElement: 'next_state' ,
                                        element: 'State1' },
                                      {localElement: 'next_city' ,
                                        element: 'City1' },
                                      {localElement: 'next_bldg' ,
                                        element: 'Building1' },
                                      {localElement: 'next_floor' ,
                                        element: 'Floor1' }
]
                                           }]
Item.next_cube,

@UI: {
lineItem: [ { position: 40, importance: #HIGH } ],
fieldGroup: [{qualifier: 'Item2',position: 40,importance: #HIGH }]
}
@EndUserText.label: 'Current Country'
Item.curr_country,
@UI: {
lineItem: [ { position: 50, importance: #HIGH } ],
fieldGroup: [{qualifier: 'Item2',position: 50,importance: #HIGH }]

}
@EndUserText.label: 'Current State'
Item.curr_state,
@UI: {
lineItem: [ { position: 60, importance: #HIGH } ],
```

```
        fieldGroup: [{qualifier: 'Item2',position: 60,importance: #HIGH }]
        }
        @EndUserText.label: 'Current City'
        Item.curr_city,
        @UI: {
        lineItem: [ { position: 70, importance: #HIGH } ],
        fieldGroup: [{qualifier: 'Item2',position: 70,importance: #HIGH }]

        }
        @EndUserText.label: 'Current Building'
        Item.curr_bldg,
        @UI: {

        lineItem: [ { position: 80, importance: #HIGH } ],
        fieldGroup: [{qualifier: 'Item2',position: 80,importance: #HIGH }]

        }
        @EndUserText.label: 'Current Floor'
        Item.curr_floor,
        @UI: {
        lineItem: [ { position: 90, importance: #HIGH } ],
        fieldGroup: [{qualifier: 'Item2',position: 90,importance: #HIGH }]}
        @EndUserText.label: 'Current Cubicle'
        Item.curr_cube,

    _Header
}
```

Listing 9.4: Item CDS

As shown in Listing 9.4, the item CDS has a similar structure. One additional special annotation is `@Consumption.valueHelpDefinition`, which is responsible for displaying the value help, and we also created the corresponding CDS (not shown). Additionally, using the source code `association to parent…`, we're linking the item entity as the child entity of the header CDS. Also, we added the header association `_Header` at the end of the CDS, as shown in Listing 9.4. You can skip certain annotations or delete them to check their behaviors by previewing the SAP Fiori app.

The item-to-header CDS is linked through association, but now, you must link the header to the items to create a complete business object hierarchy. Switch to the header CDS and add the line `composition [0..*] of ZEMP_MOV_ITEM_CDS_v2 as _Item` after the `select` line and add `_Item` as the last line of the CDS definition. This step will create the complete business object composition to be used in the managed app scenario later.

Next, we'll create a service definition by right-clicking on the package and selecting the **Service Definition** option, as shown in Figure 9.25.

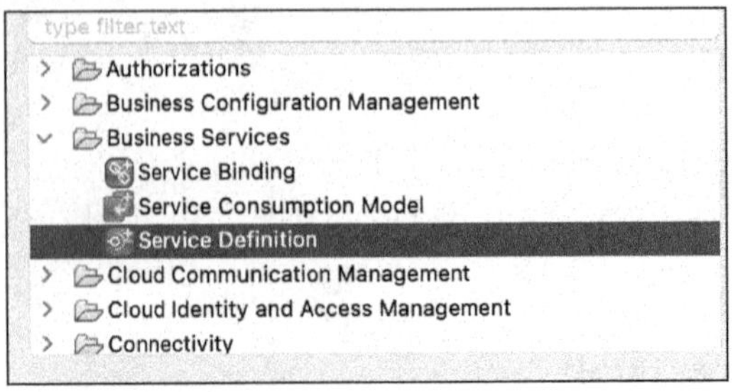

Figure 9.25: ADT: Service Definition Creation Option

Your service definition should look like the code shown in Listing 9.5, where we've exposed both header and item entities. Depending on your scenarios, you can expose other entities.

```
@EndUserText.label: 'Moves service V2'
define service Zemp_moves_service_V2 {
  expose ZEMP_MOV_HDR_CDS_V2;
  expose ZEMP_MOV_ITEM_CDS_v2;
}
```

Listing 9.5: Service Definition Exposing CDS as the Entities

Now, let's create the service binding using ADT (shown earlier in Figure 9.25). When you create a service binding, you must select the service definition and indicate the possible binding types to expose the service, as shown in Figure 9.26.

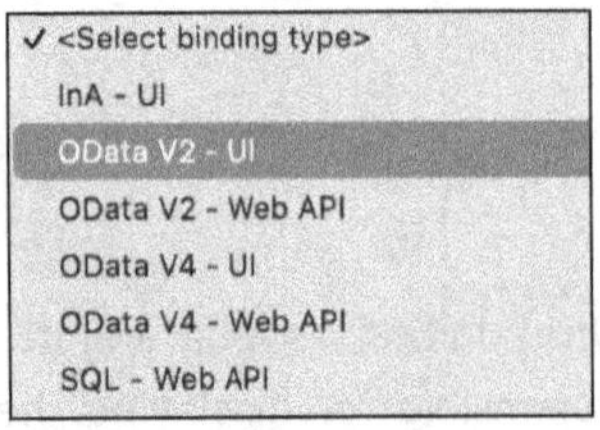

Figure 9.26: Binding Types

As shown in Figure 9.26, you have both OData v2 and v4 options for publishing the service. Additionally, you can also consider the web API option, which will not consider SAP Fiori-related annotations and other frontend-related metadata, instead focusing primarily on exposing the service as an API for external consumption. Once you create your binding, your unpublished binding will resemble the information shown in Figure 9.27.

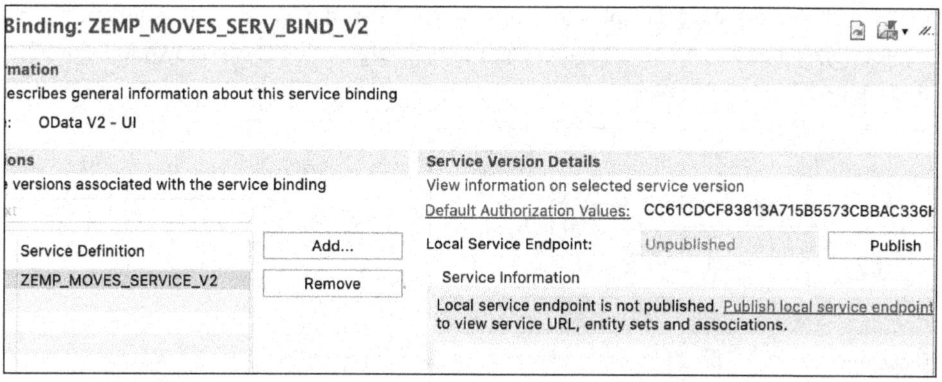

Figure 9.27: Unpublished Service Binding

Clicking on **Publish** will publish the binding as the OData service. You can see the list of entity sets available for the published services along with their associations, as shown in Figure 9.28.

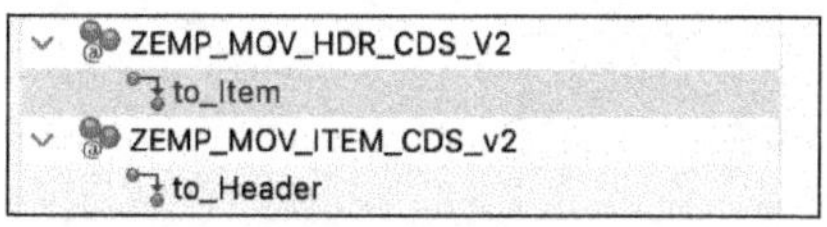

Figure 9.28: Published Entity Sets

Our service is now ready, along with all its UI features, and you can preview the app by using the preview option on the service binding screen. Your previewed app should display the screen shown in Figure 9.29.

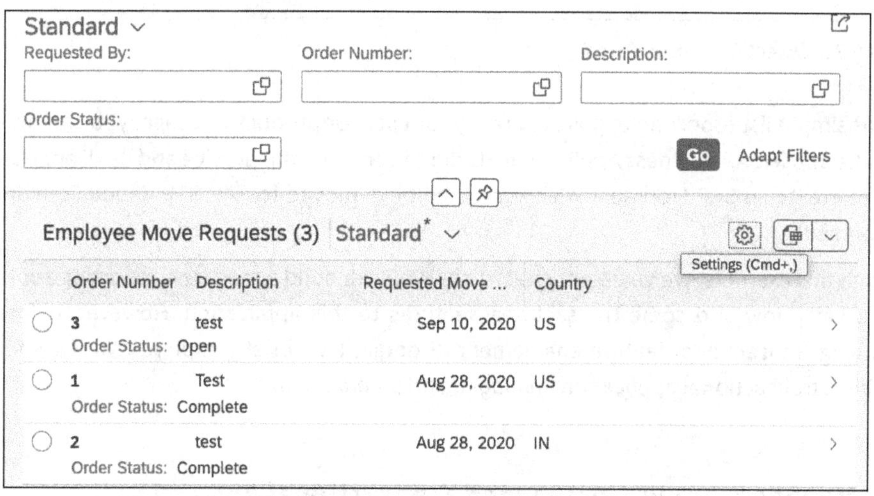

Figure 9.29: List Report Screen of the Previewed App

Clicking on any line should display the object page. Depending on your UI annotations, your object page's header, along with your item details, are available, as shown in Figure 9.30 and Figure 9.31, respectively.

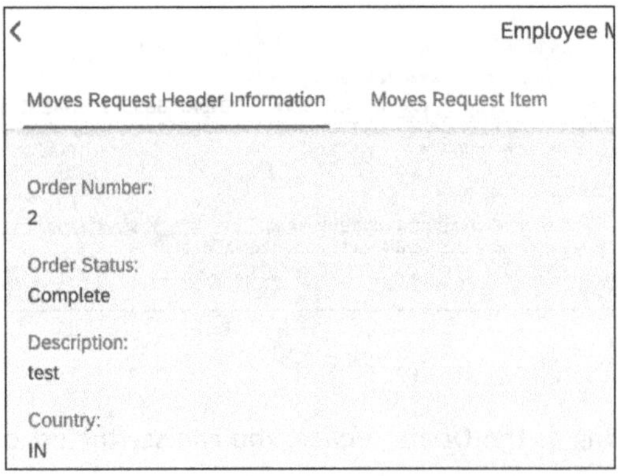

Figure 9.30: Object Page Header with Facets

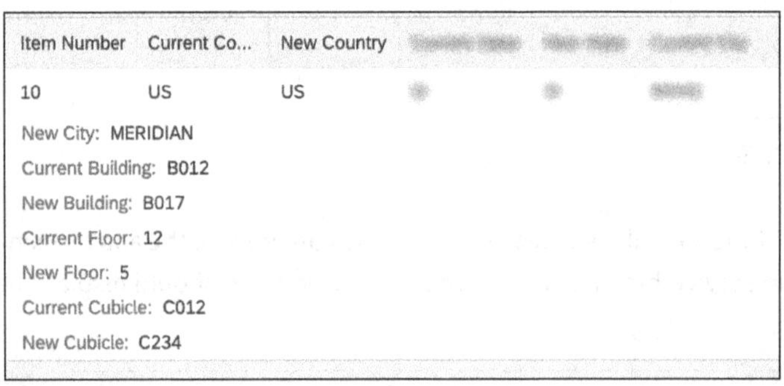

Figure 9.31: Object Page Item Details

So, our simple list reporting app is now ready for use. For productive usage, you can now build the app in SAP Business Application Studio to consume this service and its UI annotations to create an SAP Fiori app, which can then be deployed to SAP BTP, Cloud Foundry environment.

Through this example, we've demonstrated how you can build a read-only reporting application. Let's now add some transactional features to this application. However, before explaining transactional feature enablement in detail, let's briefly explore two ways for building a transactional application: *managed* and *unmanaged*.

9.6 Managed Versus Unmanaged Transactional Apps

When you build any transactional app, you must ensure its data persistency with the utmost care since the incorrect logic may lead to data inconsistencies in the database. Additionally, you must also determine how to validate the data, determine additional data, enable locking, trigger some actions, determine the point of no return, and so on. Naturally, building a transactional application is much more complex than building a read-only application. Via the ABAP RESTful application programming model, SAP provides two primary options for building transactional applications:

- **Managed**
  The term *managed* means the framework and the ABAP RESTful application programming model will manage all the transactional aspects, and thus, as an application developer, you can focus more on the business logic. This scenario is applicable when you need to store your data in custom tables because you're defining entire business objects and functionalities. Even though the framework can take care of many transactional complexities, this approach will allow you to validate the data, calculate additional field values, trigger certain actions, and so on at certain points. If you've previously worked with the SAP Fiori programming model for SAP S/4HANA, these concepts are similar to the Business Object Processing Framework (BOPF) functionality.
- **Unmanaged**
  Unmanaged, as the name suggests, is not managed by the framework, and thus, application developers have much more flexibility for addressing complex requirements. At the same time, all transactional aspects must be handled properly to avoid data inconsistency, deadlocks, etc. An unmanaged application is mostly applicable for situations where you already have some code in your on-premise environment so you can reuse existing code. Another option is a managed application with unmanaged saves; that is, you can start with a managed application, but you then trigger remote function calls (RFCs)/OData when saving the application.

Using a managed application will help you to build the draft functionality, which is a unique feature in SAP S/4HANA that we'll explore later in Section 9.8. However, let's now convert our previously created read-only list report application to develop a more complex managed transactional application.

9.7 Developing a Managed App

The primary elements of an ABAP RESTful application programming model-based transactional app are its behavior definition and its behavior implementation. In the behavior definition, you can specify what transactional functionalities are allowed: For example, you can allow only create and change operations but not delete operations. You can also define certain actions to be executed on the data, for example, setting the status for a business entity and so on. In the behavior definition, you just define these allowed features. Then, you implement them using certain classes, which are called the *behavior implementation*. These classes should implement specific interfaces according to the ABAP RESTful application programming model framework. Again, this concept is quite similar to BOPF-based SAP Fiori application development.

ADT provides the option to create a behavior definition under **Core Data Services**, as shown in Figure 9.32.

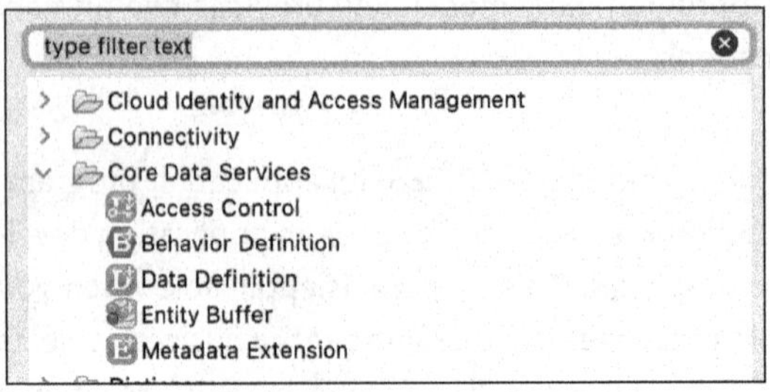

Figure 9.32: ADT: Option to Create Behavior Definition

You must select the root entity while creating a behavior definition, along with deciding between the managed and unmanaged options. For our example, we created one behavior definition for both header and item CDS behaviors, and our entire behavior definition is shown in Listing 9.6.

```
managed;

define behavior for ZEMP_MOV_HDR_CDS_V2 alias Header
implementation in class zbp_emp_mov_hdr_cds_V2 unique
persistent table zemp_move_hdr_m

lock master

{
  // key that will be automatically generated by the framework
  field ( readonly, numbering : managed ) MYKEY;
  mapping for ZEMP_MOVE_HDR corresponding;
  field ( readonly ) aufnr, create_date, status;

  determination calculateFields on modify
  { create; }
  create;
  update;
  delete;
  // instance actions
  action ( features : instance ) statusComplete result [1] $self;
  association _Item { create; }
}

define behavior for ZEMP_MOV_ITEM_CDS_v2 alias _Item
implementation in class zbp_emp_mov_hdr_cds_V2 unique
persistent table zemp_move_item_m
```

```
lock dependent by _Header

{
determination calculateFields1 on modify { create; }
  field ( readonly ) mykey, aufnr, curr_bldg, curr_city, curr_country,
curr_cube,
  curr_floor, curr_state;
  update;
  delete;
}
```

Listing 9.6: Behavior Definition

Let's break down the defined behaviors shown in Listing 9.6: First, we have a managed app in which we've defined two behaviors (one for the header and another for the item) by using two different classes: `ZEMP_MOV_ITEM_CDS_v2` and `zbp_emp_mov_hdr_cds_V2`. ADT provides the option to create a behavior implementation class from a behavior definition via its quick fix functionality. We're allowing create, update, and delete operations for the header but only update and delete operations for the item because item creation will be automatically enabled due to parent-child relationship by using the code `association _Item { create; }`. We also have two determinations (`calculateFields` and `calculateFields1`) to determine the calculated fields on modifying and one instance action (`statusComplete`). The instance action signifies that the action must be triggered for each record, not for the whole entity set. Additionally, we marked some fields as read only and enabled the globally unique identifier (GUID)-based key generation feature for the `mykey` field. Remember, this field will be hidden since the CDS has the annotation `@UI.hidden: true` for this field.

Our behavior implementation class has three primary methods: `STATUSCOMPLETE`, `CALCULATEDFIELDS`, and `CALCULATEDFIELDS1`. These names are automatically determined from the determination or action names. The source code for `CALCULATEDFIELDS` is shown in Listing 9.7, which selects the last order number from the header table and increments this value by 1 to get the new number. This logic is not necessary if you use number ranges, which are supported by SAP BTP ABAP environment.

```
  METHOD calculateFields.

    SELECT MAX( aufnr )
    FROM zemp_mov_hdr_cds_v2
    INTO @DATA(lc_aufnr).

    LOOP AT keys INTO DATA(ls_key).
      lc_aufnr = lc_aUFNR + 1.
      DATA(LC_status) = '1' .
      MODIFY ENTITIES OF zemp_mov_hdr_cds_v2  IN LOCAL MODE
```

```
      ENTITY Header
        UPDATE SET FIELDS WITH VALUE #( ( mykey     = ls_key-mykey
                                          aufnr = lc_aufnr
                                          status = lc_status ) )
        REPORTED DATA(ls_reported).

    ENDLOOP.

  ENDMETHOD.
```

Listing 9.7: Header Calculated Fields

Listing 9.8 displays the item calculated fields. In these fields, we're selecting the employee's current location from another table (zemp_location) and updating the current location's details in the item table.

```
  METHOD calculatefields1.

    TYPES : BEGIN OF ty_location,
              curr_country  TYPE land1,
              curr_state    TYPE regio,
              curr_city     TYPE char40,
              curr_bldg     TYPE char4,
              curr_floor(2) TYPE n,
              curr_cube     TYPE char4,
            END OF ty_location.

    DATA : ls_location TYPE ty_location .

    IF keys IS NOT INITIAL.

      READ ENTITY zemp_mov_hdr_cds_v2\\_Item
      FROM VALUE #(
           FOR <root_key> IN keys ( %key     = <root_key>
                  %control = VALUE #( user_id = if_abap_behv=>mk-on ) ) )
           RESULT DATA(lt_item).

      DATA(lc_user) = lt_item[ 1 ]-user_id .
      TRANSLATE lc_user TO UPPER CASE .

      SELECT SINGLE country,
                    state ,
                    city      ,
                    bldg     ,
                    floor     ,
                    cube
```

```
FROM zemp_location

WHERE user_id = @lc_user
 INTO @ls_location.


    LOOP AT keys INTO DATA(ls_key).

      SELECT SINGLE  aufnr
      FROM zemp_mov_hdr_cds_v2
      WHERE mykey      = @ls_key-mykey
      INTO @DATA(lc_aufnr).

      MODIFY ENTITIES OF zemp_mov_hdr_cds_v2  IN LOCAL MODE
       ENTITY _Item
         UPDATE SET FIELDS WITH VALUE #( ( mykey     = ls_key-mykey
                                           posnr     = ls_key-posnr
                                           aufnr = lc_aufnr
                                 curr_bldg = ls_location-curr_bldg
                                 curr_city = ls_location-curr_city
                                 curr_country = ls_location-curr_country
                                 curr_state = ls_location-curr_state
                                 curr_floor = ls_location-curr_floor
                                 curr_cube = ls_location-curr_cube
                                          ) )
         REPORTED DATA(ls_reported).
    ENDLOOP.
  ENDIF.
ENDMETHOD.
```

Listing 9.8: Item-Level Calculation Logic

Finally, our simple action code, shown in Listing 9.9, sets the status field to a predefined value indicating that an employee location transaction has been completed.

```
DATA(lv_aufnr) = keys[ 1 ]-aufnr.
  UPDATE zemp_move_hdr SET status = '3' WHERE aufnr = @lv_aufnr. " Complete
  IF sy-subrc EQ 0.
    APPEND INITIAL LINE TO result ASSIGNING FIELD-SYMBOL(<ls_result>).
    <ls_result>-aufnr = lv_aufnr.
    <ls_result>-%param-aufnr = lv_aufnr.
    <ls_result>-%param-status = '3'.
  ENDIF.
```

Listing 9.9: Source Code for Action

We also added an annotation, shown in Listing 9.10, to header CDS views that enables the button in the SAP Fiori app. You can add this annotation for any CDS fields. Note that the action name matches the name given in the behavior definition.

```
{ type: #FOR_ACTION,
                    dataAction: 'statusComplete',
                    label: 'Set to Complete' } ],
```

Listing 9.10: Annotation in Header CDS to Enable the Action Button

If you return to your previously created service binding and preview the app, your first screen should now resemble the screen shown in Figure 9.33, where you can see the enabled **Create** button and the disabled **Set to Complete** button.

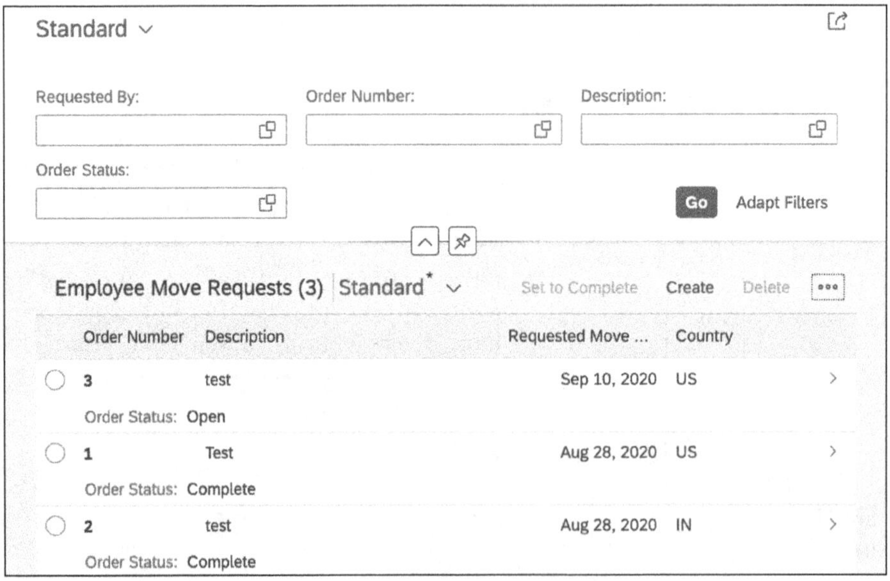

Figure 9.33: First Screen Managed Transactional App

The **Set to Complete** button is disabled since the action is marked as an instance type and we haven't selected a row yet. The moment you select a row, the button should be enabled, along with the **Delete** option, as shown in Figure 9.34.

Clicking on **Create** will display a form, shown in Figure 9.35, where you can enter field values. One limitation is that you must save the header entry first before you create the item entry. You cannot create a header and an item at the same time.

The item creation screen is shown in Figure 9.36. Notice that the current location details are not enabled since they will be retrieved from the location table by using the determination `calculatedfields1`.

So, our managed application is now ready for testing. Before we turn to unmanaged applications, let's briefly look at the draft functionality for managed applications.

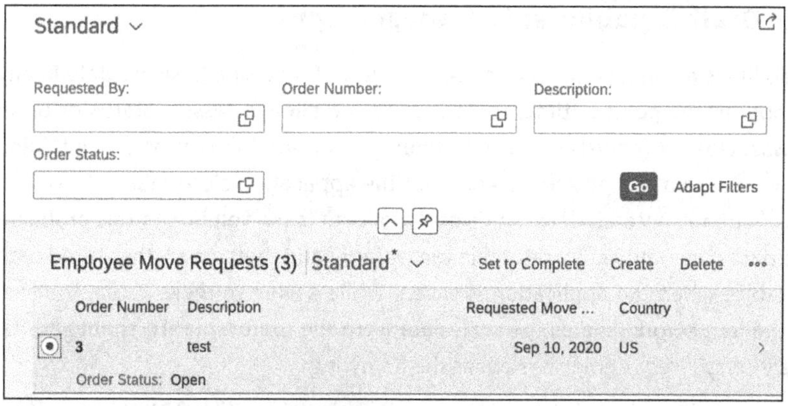

Figure 9.34: Action and Delete Enabled

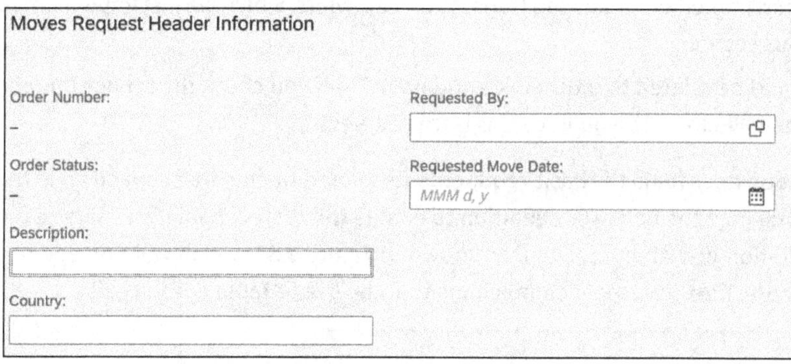

Figure 9.35: Object Screen to Create the Header Entry

Figure 9.36: Item Creation Screen

9.8 Adding Draft Capabilities to Managed Apps

The draft capability is a concept, unique to SAP S/4HANA, for storing business data in any transactional app at any point in time, regardless of the client's session status or of an abnormal browser closure, for instance. Think about a situation where a user is maintaining entries in many fields in a complex application but the application closes abruptly, resulting in the loss of already entered data. SAP's draft concept is the solution to this problem: SAP stores the data constantly in a draft table temporarily and then moves the stored data to the actual tables when the application is saved. While saving to these actual transactional tables, the framework deletes the draft data from the draft table. To enable drafts, you must satisfy certain requirements, such as the following:

- The CDS data model should be based on universally unique identifier (UUID) keys—if you check Listing 9.1 and Listing 9.2, you'll see that our keys are based on UUID only.
- The app should implement managed transaction behaviors, which we just implemented in the previous section.
- This app should be related to a UI service and not API—if you check the service binding shown earlier in Figure 9.27, you'll see this is true as well.

So, our current app does fulfill the basic requirements for adding the draft capability. In the next step, we'll change the behavior definition to enable the draft capability. First, add the `with draft` addition just after `managed`, as shown in Figure 9.37, which will immediately show an error in the next statement complaining that the `draft` table doesn't exist.

```
managed;
with draft;

define behavior for ZEMP_MOV_HDR_CDS_V2 alias Header
implementation in class zbp_emp_mov_hdr_cds_V2 unique
persistent table zemp_move_hdr_m
//etag master last_changed_at
lock master
```

Figure 9.37: Adding with Draft Option and Error Message for Missing Draft Table

So, we added a name, as shown in Figure 9.38, for a table that doesn't exist. Using the quick fix functionality in ADT, accessed by pressing Ctrl/Cmd + Shift + 1, you can create and activate the draft table. We also fixed some additional errors, as shown in Figure 9.38, for Etag functionality.

```
managed;
with draft;

define behavior for ZEMP_MOV_HDR_CDS_V2 alias Header
implementation in class zbp_emp_mov_hdr_cds_V2 unique
persistent table zemp_move_hdr_m
draft table zemp_move_hdr_d
//etag master last_changed_at
```

Figure 9.38: Adding the Draft Table

Our draft table for the header is shown in Listing 9.11, which shows some admin fields automatically added by the framework.

```
@EndUserText.label : 'Draft table for entity ZEMP_MOV_HDR_CDS_V2'
@AbapCatalog.enhancement.category : #EXTENSIBLE_ANY
@AbapCatalog.tableCategory : #TRANSPARENT
@AbapCatalog.deliveryClass : #A
@AbapCatalog.dataMaintenance : #RESTRICTED
define table zemp_move_hdr_d {
  key mandt         : mandt not null;
  key mykey         : sysuuid_x16 not null;
  aufnr             : zaufnr;
  descr             : abap.char(40);
  country           : land1;
  ernam             : abap.char(15);
  move_date         : abap.dats;
  create_date       : timestampl;
  last_change_date  : timestampl;
  last_changed_by   : syuname;
  created_by        : abap.char(15);
  status            : abap.char(19);
  "%admin"          : include sych_bdl_draft_admin_inc;

}
```

Listing 9.11: Header Draft Table

The updated header entity's behavior definition is shown in Figure 9.39.

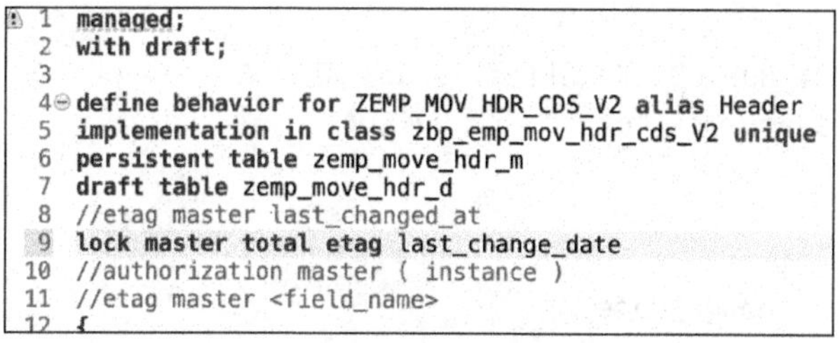

Figure 9.39: Updated Behavior Definition of the Header Entity

Repeat the same process for the item entity as well: Listing 9.12 shows the item draft table, and Listing 9.13 shows the final version of the behavior definition.

```
@EndUserText.label : 'Employee Moves Request Item Data'
@AbapCatalog.enhancement.category : #NOT_EXTENSIBLE
@AbapCatalog.tableCategory : #TRANSPARENT
@AbapCatalog.deliveryClass : #A
@AbapCatalog.dataMaintenance : #RESTRICTED
define table zemp_move_item_m {
  key client   : abap.clnt not null;
  @AbapCatalog.foreignKey.label : 'GUID'
```

```
  @AbapCatalog.foreignKey.screenCheck : false
  key mykey     : sysuuid_x16 not null
    with foreign key [0..*,1] zemp_move_hdr_m
      where mykey = zemp_move_item_m.mykey;
  key posnr     : abap.numc(4) not null;
  aufnr         : zaufnr;
  user_id       : abap.char(15);
  curr_country : land1;
  curr_state    : regio;
  curr_city     : abap.char(40);
  curr_bldg     : abap.char(4);
  curr_floor    : abap.numc(4);
  curr_cube     : abap.char(4);
  next_country : land1;
  next_state    : regio;
  next_city     : abap.char(40);
  next_bldg     : abap.char(4);
  next_floor    : abap.numc(4);
  next_cube     : abap.char(4);

}
```

Listing 9.12: Item Draft Table

```
managed;
with draft;

define behavior for ZEMP_MOV_HDR_CDS_V2 alias Header
implementation in class zbp_emp_mov_hdr_cds_V2 unique
persistent table zemp_move_hdr_m
draft table zemp_move_hdr_d
//etag master last_changed_at
lock master total etag last_change_date
//authorization master ( instance )
//etag master <field_name>
{
  // key that will be automatically generated by the framework
  field ( readonly, numbering : managed ) MYKEY;
  mapping for ZEMP_MOVE_HDR corresponding;
  field ( read only ) aufnr, create_date, status;
  //field (features : instance ) aufnr;
  // determination for calculation of total suppl. price
  // determination calculateFields on save  { field aufnr,create_
date,status ; }
  // determination
  determination calculateFields on modify
  { create; }
```

```
  create;
  update;
  delete;
  // instance actions
  action ( features : instance ) statusComplete result [1] $self;
  association _Item { create; }
}

define behavior for ZEMP_MOV_ITEM_CDS_v2 alias _Item
implementation in class zbp_emp_mov_hdr_cds_V2 unique
persistent table zemp_move_item_m
draft table zemp_move_item_d
lock dependent by _Header
{
determination calculateFields1 on modify { create; }
  field ( readonly ) mykey, aufnr, curr_bldg, curr_city, curr_country,
curr_cube,
  curr_floor, curr_state;
  update;
  delete;
}
```

Listing 9.13: Final Version of the Behavior Definition

[+]

Tip

In our example, we're not using behavior projection, and so no other actions are required. However, if you're using behavior projection, you must adjust the behavior projection to add the draft capability.

All the changes to activate the draft capability have now been implemented. If you preview the SAP Fiori app now using the existing service binding, you'll see an additional field, **Editing Status**, as shown in Figure 9.40. This field will help you filter out the draft data and final data from the list.

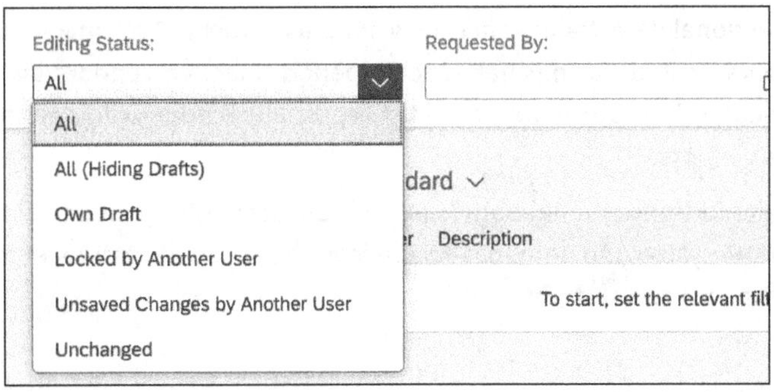

Figure 9.40: Editing Status Field

Now, if you try to create a new entry and enter some data, you'll find that the app is saving the draft data, as indicated by the bottom status bar, as shown in Figure 9.41.

Figure 9.41: Draft Status in the Bottom Status Bar

If you click the back icon in SAP Fiori, you may see a popup window asking you whether you want to save or discard the draft data, as shown in Figure 9.42.

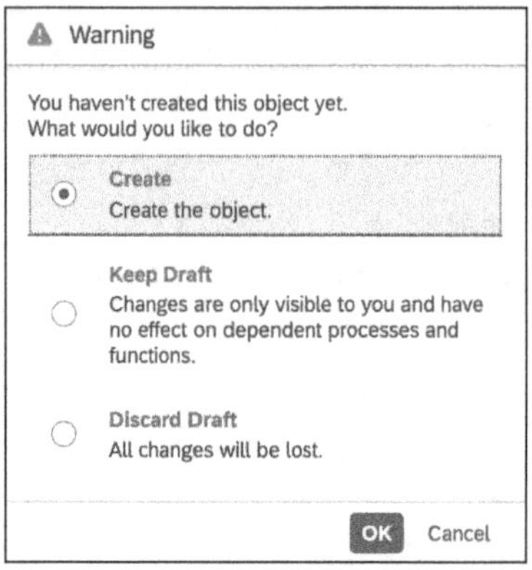

Figure 9.42: Popup Window for Unsaved Draft Changes

Congratulations! You've successfully modified your managed application to create a draft-enabled managed application. As the next step, we'll look at a different example and implement the unmanaged application scenario.

9.9 Developing an Unmanaged App

To learn more about unmanaged applications, let's try to fulfill several requirements through an example.

We want an application that identifies inactive vendors. All vendors are checked for various procurement-related transactional data, for instance, new purchase orders (POs), invoices, credit memos, and payments posted during that specific period. Inactive vendors lack transactions during a defined period. Our unmanaged list report application will identify and centrally block inactive vendors.

Vendor master data and related transactional data (such as POs, accounting documents, etc.) reside in the SAP S/4HANA system. An unmanaged application uses an API call to fetch the following details from the backend system:

- List of vendors
- Last PO creation date for those vendors
- Last accounting document creation date for those vendors

Subsequently, the application logic uses these records to filter out vendors based on the defined period and generate a final list of inactive vendors. To centrally block inactive vendors, the ABAP PaaS logic uses an API to update the vendor master in the SAP S/4HANA system.

Before we describe implementing this scenario in detail, let's go over some key elements for implementing this requirement:

- The backend SAP S/4HANA standard API may not always be sufficient. For example, in this case, we must retrieve PO and accounting document details for a period. The API is available, but when using the standard API, you'll need to pull a high volume of data first and then discard entries that are not relevant. A more efficient approach is to instead create your custom service or extend a standard service to keep data volumes low.
- This app requires a connection to the backend for both querying and updating requirements. We'll use the unmanaged option to address both queries and updates.
- Backend SAP S/4HANA data can be retrieved using OData or RFC. In our example, we'll demonstrate the use of an OData call.

Let's now jump in and implement this unmanaged application for query and update scenarios by following these steps:

1. First, identify all the required OData services available in the backend. If the standard services are insufficient, then you'll need to create one.
2. Once your OData services are identified or created and activated in the backend, download its metadata file from the backend system. For example, as shown in Figure 9.43, you can download the metadata file via Transaction /IWFND/MAINT_SERVICE for an on-premise SAP S/4HANA system.

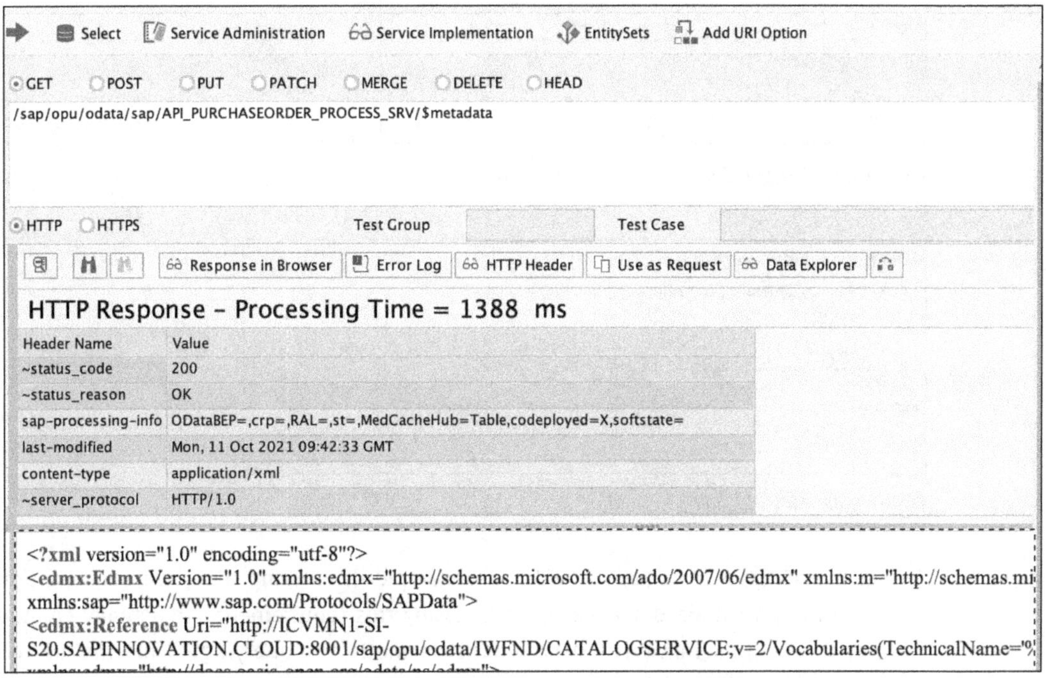

Figure 9.43: Download OData Metadata File from Your Backend, SAP S/4HANA On-Premise Example

3. Now, switch to ADT and create a service consumption model available under the **Service Consumption Model** option, as shown in Figure 9.44.

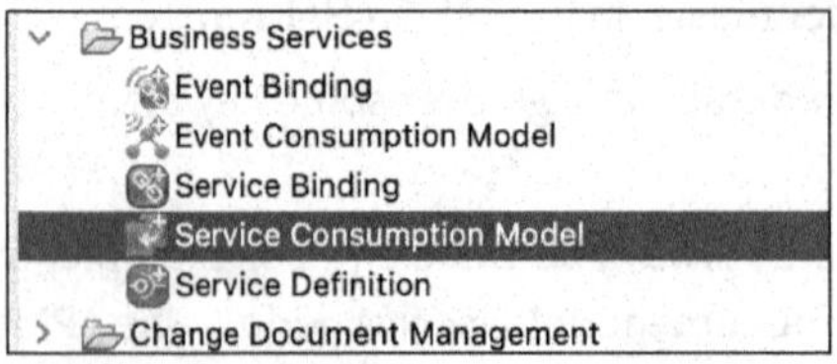

Figure 9.44: ADT: Service Consumption Model Option

4. Figure 9.45 shows the first screen for creating the service consumption model, and you must select whether you're using OData, RFC, or a web service. In our example, we selected the OData option, as shown in Figure 9.45. On the next screen, shown in Figure 9.46, select the metadata file extracted from your backend system and then complete the process of creating the service consumption model.

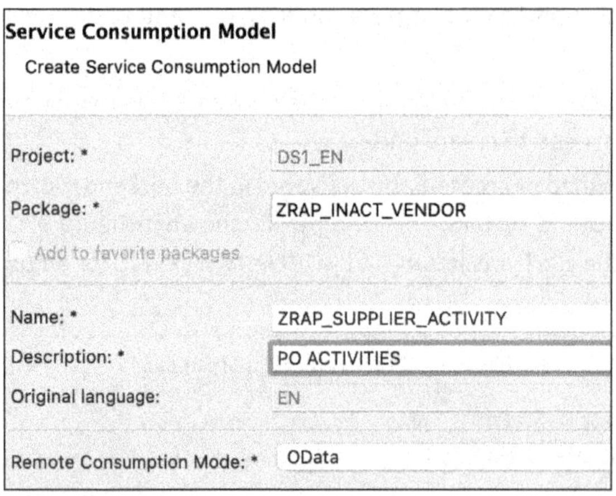

Figure 9.45: Service Consumption Model Creation

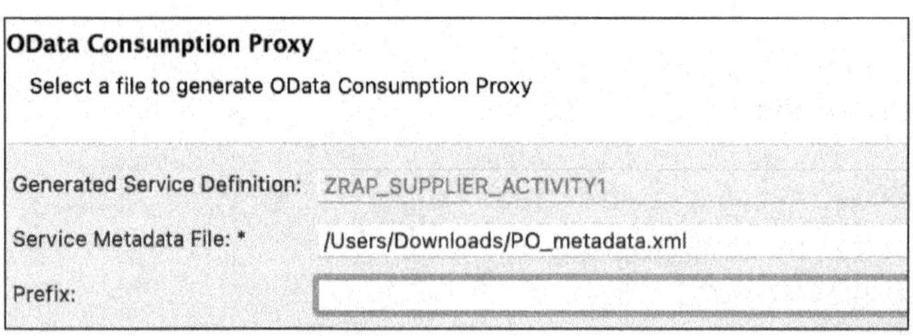

Figure 9.46: Metadata File Selection

5. Figure 9.45, shown earlier, provided an example of creating a service consumption model from a standard service. For addressing this requirement, we actually created three service consumption models: two custom models for querying PO information and

accounting data information and one standard model for blocking the business partner. These three consumption models are shown in Figure 9.47.

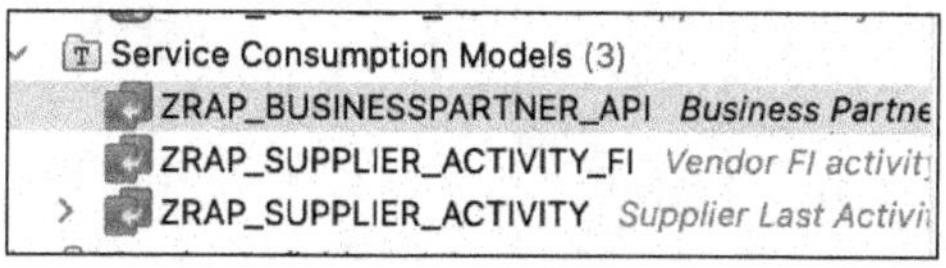

Figure 9.47: List of Consumption Models Created

6. When you open the consumption model in ADT, as shown in Figure 9.48, you'll see some code has been automatically generated that you can copy to the clipboard. We'll use this code later when implementing the query implementation class.

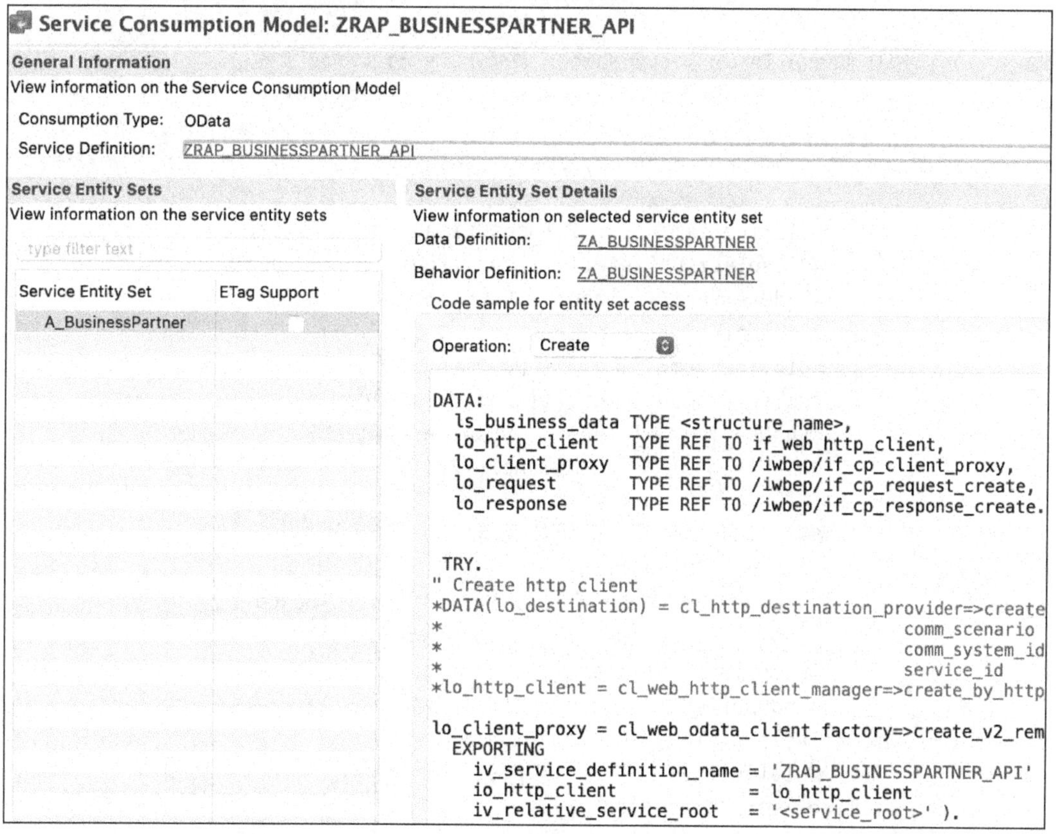

Figure 9.48: Generated Source Code of Service Consumption Model

7. In this next step, we'll create the CDS with a query implementation class since no data source/table exists. This CDS is our data model (template), and the entire query will be executed using the class shown in Figure 9.49. The entire CDS code is shown in Listing 9.14. Note the list of fields, along with annotations for an SAP Fiori application that have been added.

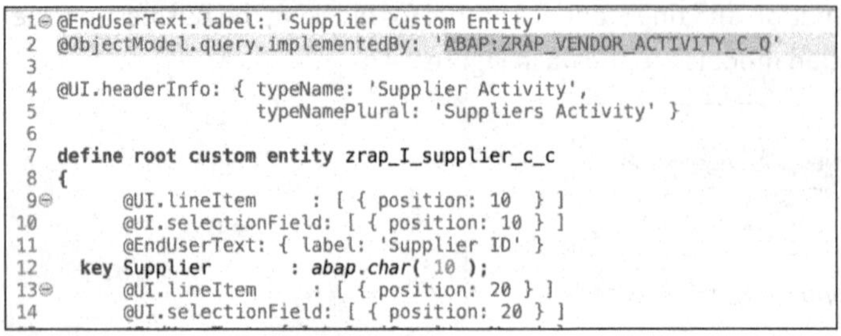

Figure 9.49: Query Implementation Class of the CDS

```
@EndUserText.label: 'Supplier Custom Entity'
@ObjectModel.query.implementedBy: 'ABAP:ZRAP_VENDOR_ACTIVITY_C_Q'

@UI.headerInfo: { typeName: 'Supplier Activity',
                  typeNamePlural: 'Suppliers Activity' }

define root custom entity zrap_I_supplier_c_c
{
      @UI.lineItem     : [ { position: 10  } ]
      @UI.selectionField: [ { position: 10 } ]
      @EndUserText: { label: 'Supplier ID' }
  key Supplier        : abap.char( 10 );
      @UI.lineItem     : [ { position: 20 } ]
      @UI.selectionField: [ { position: 20 } ]
      @EndUserText: { label: 'Supplier Name' }
      name            : abap.char( 220 );

      @UI.lineItem     : [ { position: 30  } ,
      { type: #FOR_ACTION,
                   dataAction: 'set_status_blocked',
                   label: 'Set to Blocked' }]
      @UI.selectionField: [ { position: 30 } ]
      @EndUserText: { label: 'Last Activity' }
      lastactivitydate : abap.dats;
      @UI.lineItem     : [ { position: 40  , value : 'status' ,
criticality: 'stat_icon' }]
      @UI.selectionField: [ { position: 40 } ]
      @EndUserText: { label: 'Is Blocked ?' }
         status : xfeld ;

         stat_icon : abap.int1 ;

}
```

Listing 9.14: CDS with Query Implementation Class

8. Before you can implement the query implementation class, you must create a connection between SAP BTP ABAP environment and your backend SAP S/4HANA system. Assuming that this backend system is on-premise, your administrator must have set up a cloud connector for the connectivity from SAP BTP to the on-premise environment as well as an HTTP destination to connect to this backend system with the proper authentication mechanisms in the SAP BTP subaccount. Our SAP BTP destination is shown in Figure 9.50.

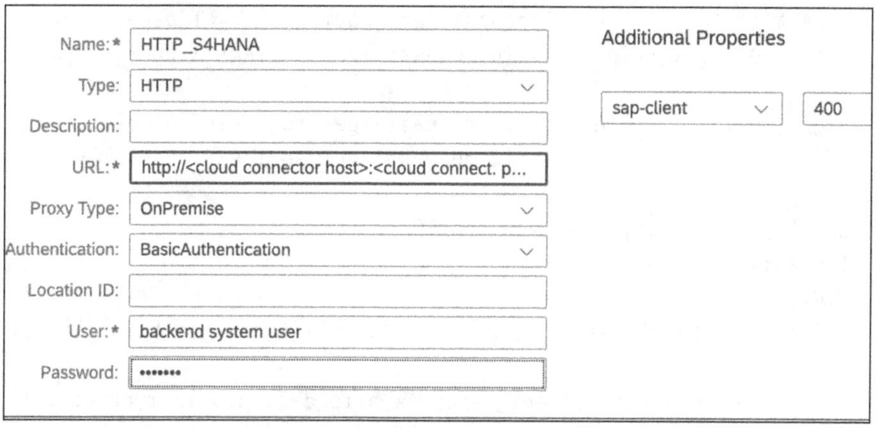

Figure 9.50: SAP BTP Destination to SAP S/4HANA

9. Now, let's create a central class to connect to the backend destination. We created a static method, as shown in Listing 9.15. Also shown is the method implementation where we used the SAP BTP destination created in the previous step and created an HTTP client instance.

```
  CLASS-METHODS:
      get_http_client  RETURNING VALUE(http_client) TYPE REF TO if_
web_http_client
                        RAISING  cx_web_http_client_error
                                 cx_http_dest_provider_error .METHOD
get_http_client.

DATA(http_destination) = cl_http_destination_provider=>create_by_
cloud_destination(
 i_name = 'HTTP_S4HANA'
 i_authn_mode = if_a4c_cp_service=>service_specific
 ).
    http_client = cl_web_http_client_manager=>create_by_http_
destination( http_destination ).

  ENDMETHOD.
```

Listing 9.15: Method Declaration

10. Listing 9.16 shows part of the query provider class implementation, where we are first creating an HTTP client instance by calling the static method created in the last step. The rest of the logic we copy/pasted from the code for our two custom service consumption models. Additionally, we added the code for OData top, skip, fields, and sort logic, which are mandatory to include.

```
CLASS ZRAP_VENDOR_ACTIVITY_C_Q IMPLEMENTATION.
  METHOD if_rap_query_provider~select.
     DATA(lv_top)      = io_request->get_paging( )->get_page_size( ).
          DATA(lv_skip)     = io_request->get_paging( )->get_offset(
).
          DATA(lt_fields)  = io_request->get_requested_elements( ).
          DATA(lt_sort)     = io_request->get_sort_elements( ).

data : lv_filter_property type char20.
"""Instantiate Client Proxy
    TRY.
       DATA(http_client) = zrap_cl_vendor_aux_c_cp=>get_http_client(
).
      CATCH cx_web_http_client_error cx_http_dest_provider_error INTO
DATA(error_creating_http_client).

        EXIT.
    ENDTRY.

    DATA(lo_client_proxy)  = cl_web_odata_client_factory=>create_v2_
remote_proxy(
    EXPORTING
      iv_service_definition_name = 'ZRAP_SUPPLIER_ACTIVITY'
      io_http_client             = http_client
      iv_relative_service_root   = '/sap/opu/odata/sap/ZRAP_LASTPO_
DATE_BY_LIFNR_CDS/' ).
```

Listing 9.16: Initial Portion of Query Provider Implementation

11. After adding the standard code copied from our two service consumption models, we also added some custom code, shown in Listing 9.17, to merge the query and create the table to be displayed on the final SAP Fiori screen.

```
SORT lt_vendor_fi BY supplier .

    DATA : lt_final TYPE STANDARD TABLE OF zrap_I_supplier_c_c .

    LOOP AT lt_vendor_po ASSIGNING FIELD-SYMBOL(<lfs_po>) .
      APPEND INITIAL LINE TO lt_final ASSIGNING FIELD-SYMBOL(<lfs_
final>) .
```

```
  <lfs_final>-supplier =  <lfs_po>-supplier .
  <lfs_final>-name =  <lfs_po>-name .
  <lfs_final>-lastactivitydate = <lfs_po>-lastpodate .
  <lfs_final>-status = <lfs_po>-status .

  IF <lfs_final>-status = ' ' .
    <lfs_final>-stat_icon = 3.
  ELSE.
    <lfs_final>-stat_icon = 1.
  ENDIF.

  READ TABLE lt_vendor_fi ASSIGNING FIELD-SYMBOL(<lfs_fi>)
             WITH KEY supplier = <lfs_po>-supplier BINARY SEARCH
.
  IF sy-subrc  = 0.

    IF <lfs_fi>-lastdate > <lfs_po>-lastpodate .
      <lfs_final>-lastactivitydate = <lfs_fi>-lastdate .
    ENDIF.
  ENDIF.

ENDLOOP.


LOOP AT lt_vendor_fi ASSIGNING <lfs_fi> .
  READ TABLE lt_final TRANSPORTING NO FIELDS WITH KEY
        supplier = <lfs_fi>-supplier .

  IF sy-subrc <> 0.
    APPEND INITIAL LINE TO lt_final ASSIGNING <lfs_final>.
    <lfs_final>-supplier = <lfs_fi>-supplier .
    <lfs_final>-name = <lfs_fi>-name .
    <lfs_final>-lastactivitydate = <lfs_fi>-lastdate .
    <lfs_final>-status = <lfs_fi>-status .
    IF <lfs_final>-status = ' ' .
      <lfs_final>-stat_icon = 3.
    ELSE.
      <lfs_final>-stat_icon = 1.
    ENDIF.

  ENDIF.
ENDLOOP.
```

```
IF rt_date[] IS NOT INITIAL .
  DELETE lt_final WHERE lastactivitydate  NOT IN rt_date .
ENDIF.
io_response->set_data( lt_final ) .

DATA : li_lines TYPE int8 .
li_lines = lines( lt_final ) .

IF io_request->is_total_numb_of_rec_requested( ).
  io_response->set_total_number_of_records( li_lines ).
ENDIF.
```

Listing 9.17: Code to Merge Data from 2 OData Call

12. As the next step, we created an unmanaged behavior definition. As shown in Listing 9.18, this behavior definition includes an action to block the vendor via its business partner data in the backend SAP S/4HANA system. This behavior is the unmanaged update we need to implement.

```
unmanaged implementation in class zbp_rap_i_supplier_c_c unique;
define behavior for zrap_I_supplier_c_c alias Vendor_CE
{
field ( read only ) lastactivitydate ,name ,supplier , status ;

  action ( features : instance ) set_status_blocked result [1] $self;
}
```

Listing 9.18: Behavior Definition for Blocking

13. Our behavior implementation class will call the business partner service using the code generated by the service consumption model and some additional processing logic for connecting to the backend to block the supplier business partner, as shown in Listing 9.19.

```
DATA: ls_entity_key                 TYPE za_businesspartner,
      ls_business_data              TYPE za_businesspartner,
      lo_http_client                TYPE REF TO if_web_http_client,
      lo_client_proxy            TYPE REF TO /iwbep/if_cp_client_proxy,
      lo_resource                TYPE REF TO /iwbep/if_cp_resource_entity,
      lo_request_resp            TYPE REF TO /iwbep/if_cp_request_update,
      lo_response_resp           TYPE REF TO /iwbep/if_cp_response_update,
      lo_request_read            TYPE REF TO /iwbep/if_cp_request_read,
      lo_response_read           TYPE REF TO /iwbep/if_cp_response_read,
      ls_vendor_set_status_blocked LIKE LINE OF et_vendor_set_status_
blocked.
```

```
    TRY.
        DATA(http_client) = zrap_cl_vendor_aux_c_cp=>get_http_client( ).
      CATCH cx_web_http_client_error cx_http_dest_provider_error
INTODATA(error_creating_http_client).

        EXIT.
    ENDTRY.

    LOOP AT it_vendor_set_status_blocked ASSIGNING FIELD-SYMBOL(<lfs_
blocked>) .
      TRY.
          lo_client_proxy = cl_web_odata_client_factory=>create_v2_remote_
proxy(
                  EXPORTING
                  iv_service_definition_name = 'ZRAP_BUSINESSPARTNER_API'
                  io_http_client              = http_client
                  iv_relative_service_root    = '/sap/opu/odata/sap/API_
BUSINESS_PARTNER/' ).

          DATA(lc_supplier) = |{ <lfs_blocked>-supplier ALPHA = IN  }| .
          " Set entity key
          ls_entity_key = VALUE #(
                              businesspartner = lc_supplier
                          ).
          " Navigate to the resource
          lo_resource = lo_client_proxy->create_resource_for_entity_set(
'A_BUSINESSPARTNER' )->navigate_with_key( ls_entity_key ).

          " Execute the request and retrieve the business data
          lo_response_read = lo_resource->create_request_for_read( )-
>execute( ).
          lo_response_read->get_business_data( IMPORTING es_business_data
= ls_business_data ).

        CATCH /iwbep/cx_cp_remote INTO DATA(lx_remote).
        " Handle remote Exception
        " It contains details about the problems of your http(s) connection

        CATCH /iwbep/cx_gateway INTO DATA(lx_gateway).
          " Handle Exception

      ENDTRY.
      ls_business_data-businesspartnerisblocked = abap_true .
```

```
      TRY.
        " Navigate to the resource and create a request for the update
operation
          lo_resource = lo_client_proxy->create_resource_for_entity_set(
'A_BUSINESSPARTNER' )->navigate_with_key( ls_entity_key ).
          lo_request_resp = lo_resource->create_request_for_update(
/iwbep/if_cp_request_update=>gcs_update_semantic-put ).

          lo_request_resp->set_business_data( ls_business_data ).

          " Execute the request and retrieve the business data
          lo_response_resp = lo_request_resp->execute( ).

          " Get updated entity
          CLEAR ls_business_data.
          lo_response_read->get_business_data( IMPORTING es_business_data
= ls_business_data ).

          ls_vendor_set_status_blocked-supplier = ls_business_data-
businesspartner .
          ls_vendor_set_status_blocked-%param-name = ls_business_data-
businesspartnerfullname.
          ls_vendor_set_status_blocked-%param-supplier = ls_business_data-
businesspartner  .
          APPEND ls_vendor_set_status_blocked TO et_vendor_set_status_
blocked.

        CATCH /iwbep/cx_cp_remote INTO lx_remote.
          " Handle remote Exception
          " It contains details about the problems of your http(s)
connection

        CATCH /iwbep/cx_gateway INTO lx_gateway.
          " Handle Exception
      ENDTRY.
```

Listing 9.19: Logic to Connect to the Backend to Block the Business Partner in SAP S/4HANA

14. We'll now create a service binding using the context menu of the package and navigating to **New • Other ABAP Repository Objects • Business Services • Service Binding**. Publish the service binding. The final published service binding is shown in Figure 9.51.

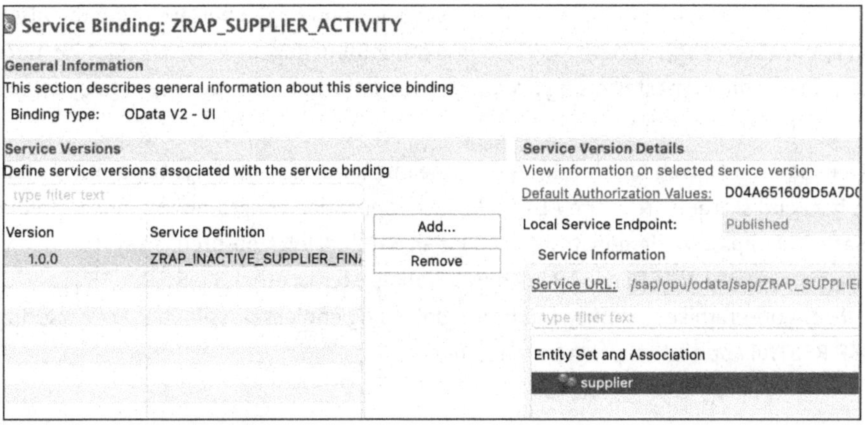

Figure 9.51: Service Binding

15. If you now preview your app from the service binding, your data should appear from the backend, as shown in Figure 9.52.

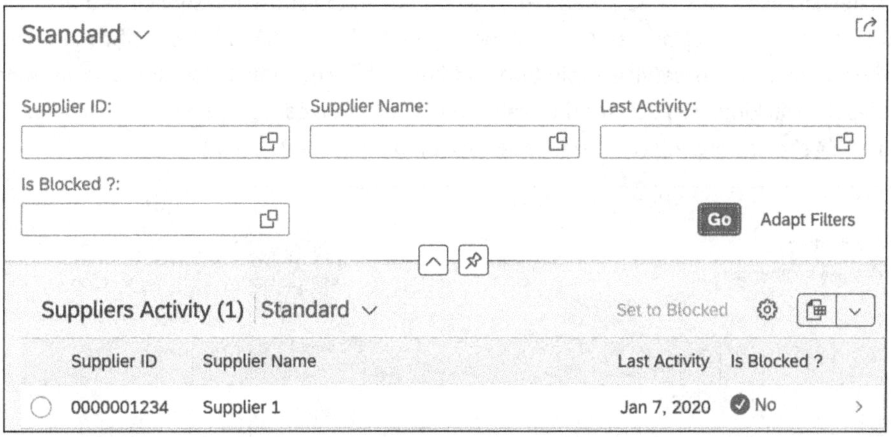

Figure 9.52: SAP Fiori App Preview, Data from Backend SAP S/4HANA

16. If you now select any row, the **Set to Blocked** button should be enabled, as shown in Figure 9.53. Clicking on this button will trigger the action implemented in the behavior implementation class and will ultimately call the business partner OData API to block the business partner in the backend SAP S/4HANA system.

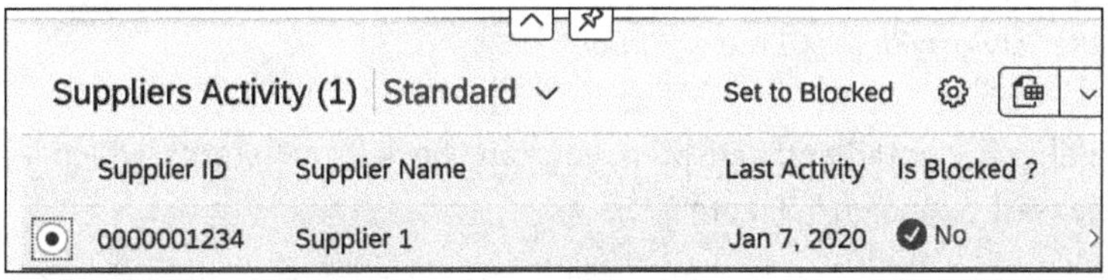

Figure 9.53: Blocking Business Partner/Supplier in the Backend System

In this section, we focused on implementing a specific unmanaged pattern using the ABAP RESTful application programming model. Some other unmanaged patterns might be more complex, where you might start using a managed scenario and then save the data by calling an API in the backend system.

Before we move on to web API development, one last thing to mention is the ability to call various backend systems from the ABAP RESTful application programming model. In our unmanaged example, we demonstrated calling an SAP S/4HANA on-premise system. However, you can also call APIs from SAP S/4HANA Cloud systems, other SaaS APIs, RFC calls to SAP S/4HANA on-premise systems, and more, unlocking endless possibilities for exploring the ABAP RESTful application programming model.

9.10 Developing a Web API

So far, we've shown you how to create service bindings in multiple places, starting with a read-only list reporting application. In each case, we created the binding as a UI. However, if you need to use a service as a web API, you don't need to expose the service as a UI at all. Publishing your services as web APIs will mean you don't need SAP Fiori-specific metadata or annotations in the service definition. Additionally, exposing these services as APIs enables the tracking of API states (i.e., released for external consumption or not released). Figure 9.54 shows how you can create a service binding using OData v2.

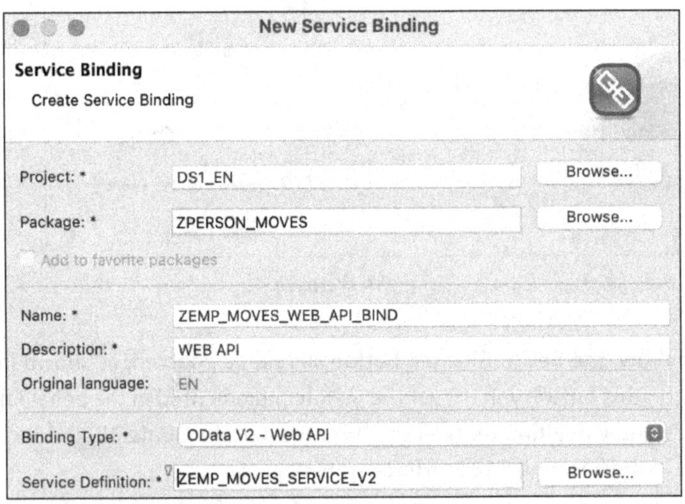

Figure 9.54: Binding the Service with the OData v2 Option

Once the service binding is created and activated, you can check its API status, which initially will be **Not Released**, as shown in Figure 9.55.

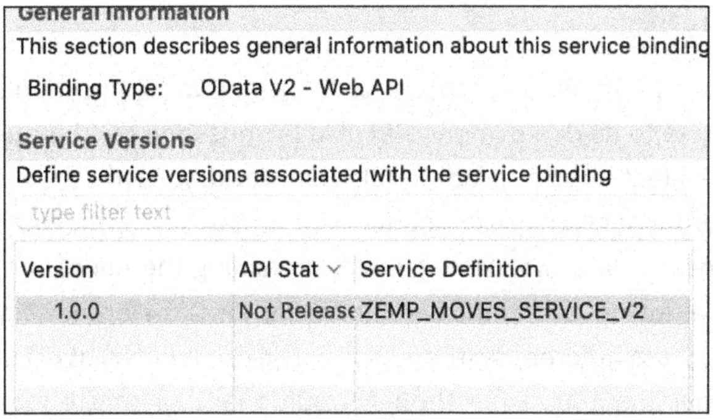

Figure 9.55: API Not Released

You can release the API via the context menu for the service, as shown in Figure 9.56. Then, the API status will be changed to **Released**, as shown in Figure 9.57.

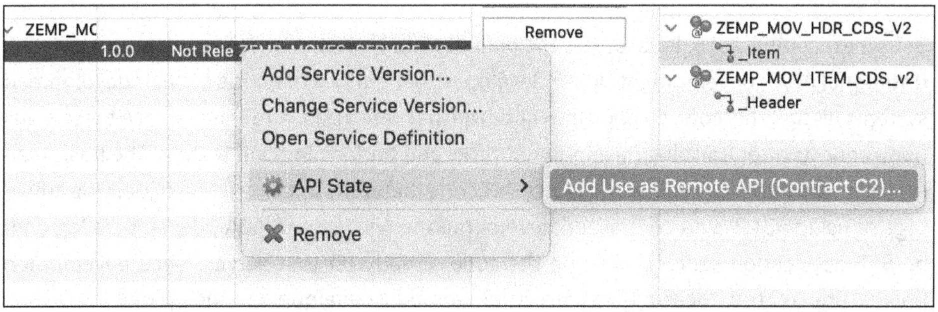

Figure 9.56: Releasing an API

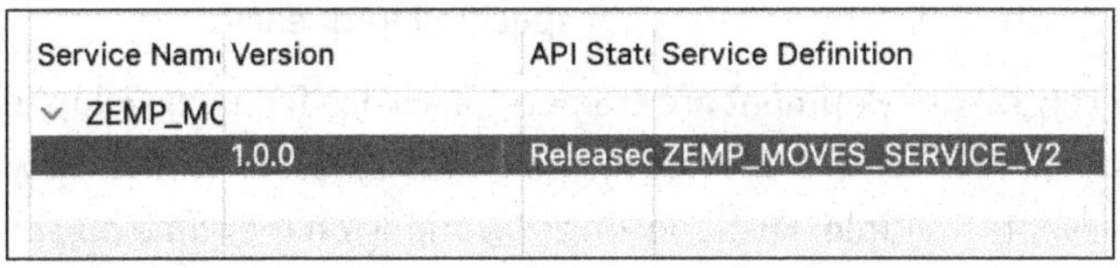

Figure 9.57: API Status Changed to Released

Except for UI features, the service, including all its implemented features, is now exposed as an API. Thus, its implemented behaviors (i.e., create/change/delete/action, etc.) are also exposed in this API.

We've so far explored many development patterns with the ABAP RESTful application programming model in SAP BTP ABAP environment. As a developer, you'll also need to understand transport management and how to move the objects from one environment to another—our next topic of discussion.

9.11 Transport Management

Application lifecycle management is primarily the process of managing an application's lifecycle from the development phase to its deployment and also includes the continuous management of its changes in an efficient way. In a traditional ABAP world, you would use a transport management solution for managing application lifecycles. In this section, you'll learn how to reuse traditional concepts in a modern way while retaining the benefits of using abapGit and gCTS. abapGit was introduced to the SAP Community in 2014 for the easy exchange and installation of code across different systems that may exist in different organizational cloud or on-premise environments. Even though traditional copy/paste may work for some scenarios, ABAP community members saw the need for a consistent way of exchanging ABAP objects to avoid repetitive installation efforts and manual errors. The result was abapGit, which has become quite popular since. Even though abapGit is an open-source project started by SAP Community members, SAP has been contributing to abapGit since 2018.

Before explaining gCTS, let's look at the benefits and limitations of lifecycle management in traditional ABAP development. ABAP developers generally write code to their development systems, assign this code to packages and transport requests, and then release these transport requests after finishing development tasks and performing unit testing. The Basis team monitors the import queue and imports this code manually or using a batch job. One advantage of using ABAP is its centralized development environment. ABAP developers log on to an SAP application server and write the code directly on that server. Once development objects are created, these objects are visible to other developers in the system and can be reused after activation. This concept is completely different compared to other traditional, non-SAP development environments, where individual developers might write code in their local environments and then deploy this code to servers at specified frequencies.

The advantages provided by central development are tremendous since fellow developers can quickly check what's happening in related developments and raise issues earlier. However, central development prevents multiple developers from working on the same objects simultaneously. By introducing gCTS, SAP is aiming for the best of both worlds, that is, repository-based development where concurrent work on the same object is possible, while retaining all the benefits of central development.

gCTS was formally introduced as a part of SAP S/4HANA 1909. However, gCTS is also used in SAP BTP ABAP environment, under the hood. The high-level process flow for gCTS is shown in Figure 9.58.

Using the gCTS process flow, shown in Figure 9.58, any ABAP object can be enabled in gCTS and can be stored in a Git repository. Objects can be attached to transport requests/tasks, just as in traditional development. Changes attached to a task will be committed to a local repository and then finally be pushed to a remote repository. Finally, objects in the remote repository can be transferred to the target system's local repository and can then be imported by using traditional tools.

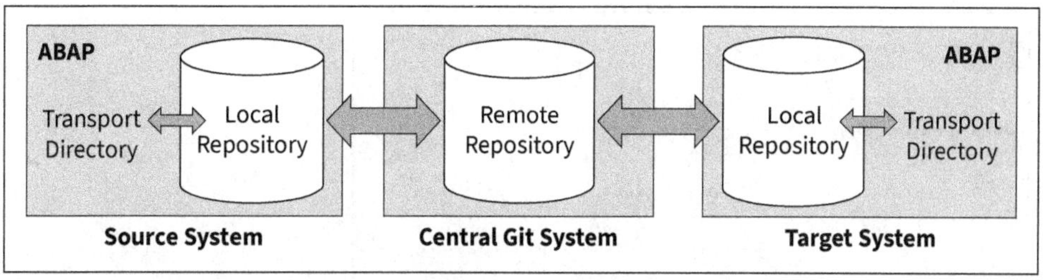

Figure 9.58: gCTS Process Flow

The Manage Software Components app is an SAP Fiori app available for administrators to configure, execute, and monitor the lifecycle management of your ABAP objects in SAP BTP ABAP environment. This app can be displayed in the administrator dashboard of the service instance. For example, Figure 9.59 shows two ABAP services running in our SAP BTP, Cloud Foundry environment development space. In this section, you'll also learn how to move objects from one instance to another.

Name	Service	Plan	Referencing Applications
1	abap	standard	None
2	abap	standard	None

Figure 9.59: 2 SAP BTP ABAP Environment Instance

Clicking on any instance name will display the **Open Dashboard** option for that instance. This dashboard is simply an SAP Fiori launchpad primarily used by administrators. The Manage Software Components app is available under the **Software Component Lifecycle Management** group.

This app is the bread and butter of lifecycle management in SAP BTP ABAP environment and uses some gCTS components under the hood. If you're working with gCTS in SAP S/4HANA, then some configuration is required, along with the creation of the GitHub repository. These configuration and repository creation steps are not required for SAP BTP ABAP environment, which takes care of all this complexity behind the scenes. Let's say you have two ABAP instances in your SAP BTP, Cloud Foundry environment organization/subaccount. To transport objects from one instance to another, follow these steps:

1. **Create a software component**
   Open the Manage Software Components app and click on **Create.** Enter a name and description for the software component and select an option from the **Type** dropdown list (**Development** or **Configuration**), as shown in Figure 9.60. Then, click **Save**.

   The creation of a software component automatically creates a GitHub repository behind the scenes. Thus, you should **Clone** and **Pull** the repository, as shown in Figure 9.61, before starting your development.

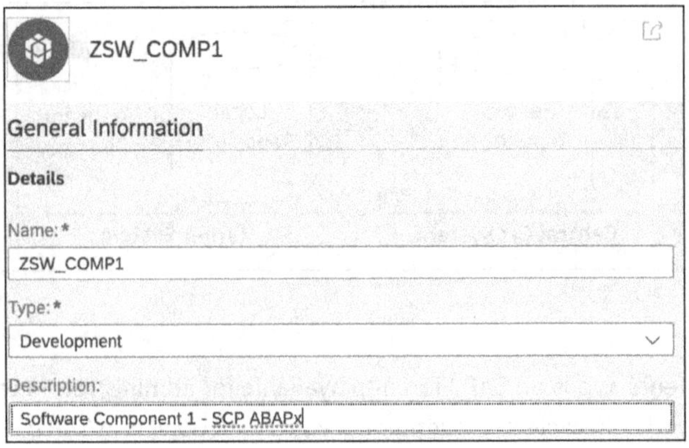

Figure 9.60: Creating a Software Component

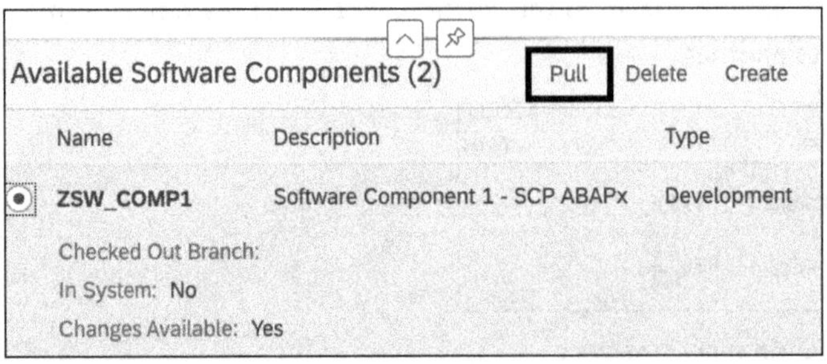

Figure 9.61: Cloning and Pulling from an Automatically Created Repository

After saving the software component, you should see its details under its **History** and **Branching** tabs, as shown in Figure 9.62. You won't be able to create any branches until a development object is assigned to this software component, a process we'll describe later in this section.

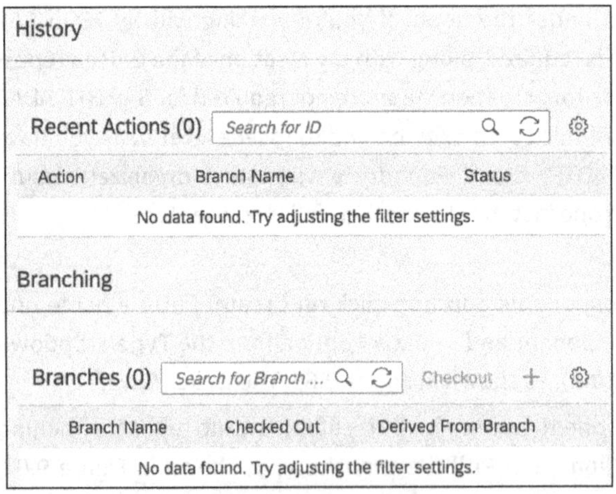

Figure 9.62: Newly Created Software Component's History and Branching

2. **Create an ABAP package**
   In this next step, you'll create a package under the software component we just created. Open ADT and select **Favorite Package • Add Package** under your SAP BTP ABAP environment system. Then, search for your software component, as shown in Figure 9.63.

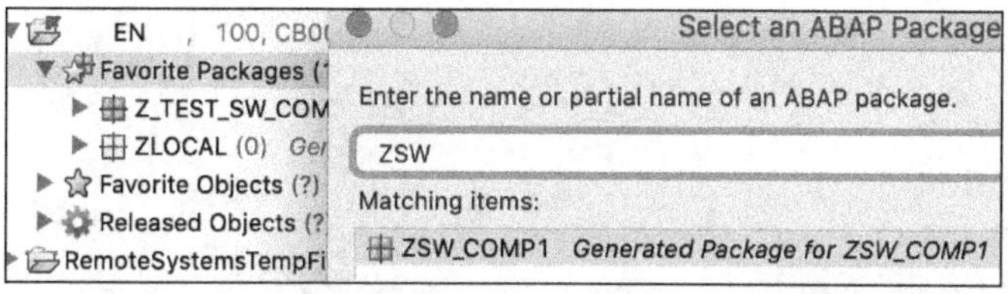

Figure 9.63: Software Component (Structure Package) in the Package Hierarchy

You can now create a new package under this structure package/software component, as shown in Figure 9.64. Create a new transport request or use an existing transport request while creating this package. This new package is the package where you'll assign your ABAP objects.

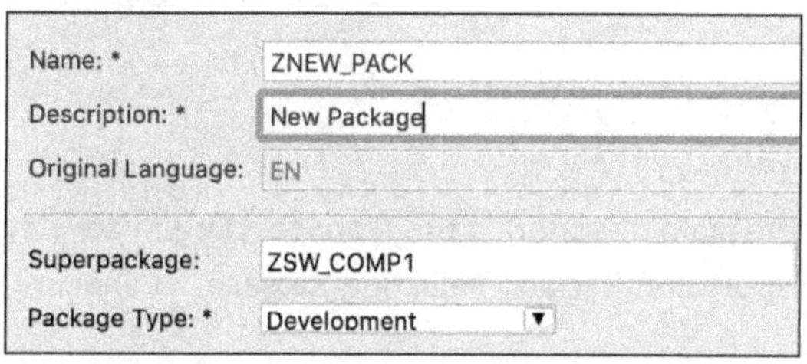

Figure 9.64: New Package Creation

The newly created package should now be visible in the package hierarchy, as shown in Figure 9.65.

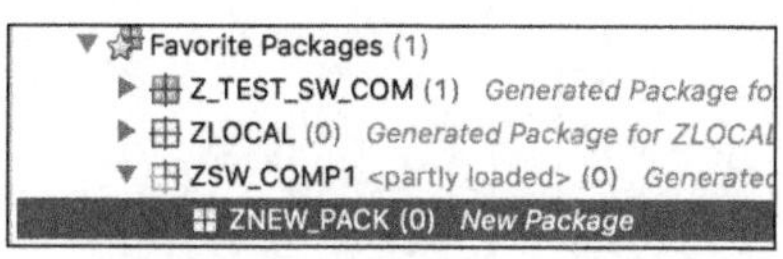

Figure 9.65: New Package Available under the Software Component

3. **Create a new ABAP class**
   Now, let's create a new class under our newly created package **ZNEW_PACK**, as shown in Figure 9.66. Our class is for demonstration purposes; however, in practice, you can use any of the objects created previously (e.g., a CDS definition, a behavior definition, a behavior implementation, a query implementation class, etc.).

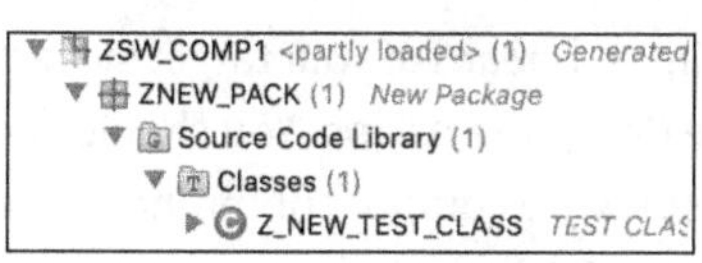

Figure 9.66: New ABAP Class

For illustration purposes, we've provided some example logic for a simple ABAP class in Listing 9.20.

```
CLASS z_new_test_class DEFINITION
  PUBLIC
  FINAL
  CREATE PUBLIC .
  PUBLIC SECTION.
    METHODS test_meth.
  PROTECTED SECTION.
  PRIVATE SECTION.
    data v_var1 type c.
ENDCLASS.
CLASS z_new_test_class IMPLEMENTATION.
  METHOD test_meth.
     v_var1 = 'A'.
  ENDMETHOD.
ENDCLASS.
```

Listing 9.20: Test ABAP Class Logic

4. **Release the transport request**
   You can now release your task(s) and the main transport in the **Transport Organizer** view in ADT by right-clicking on your transport number and selecting **Release**, as shown in Figure 9.67.

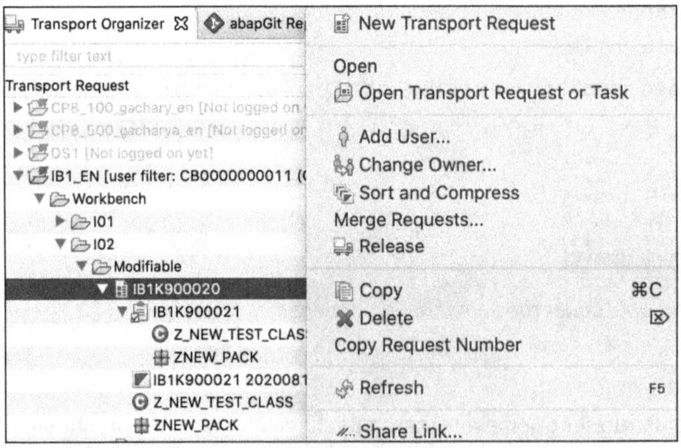

Figure 9.67: Releasing Tasks and the Main Transport

5. **Pull from target system**
   Once the transport is released, you can open the admin dashboard of your target system (e.g., your test system) and open the Manage Software Components app. You should see your software component with an indicator that the software component has not been created in this system (**In System: No**), as shown in Figure 9.68.

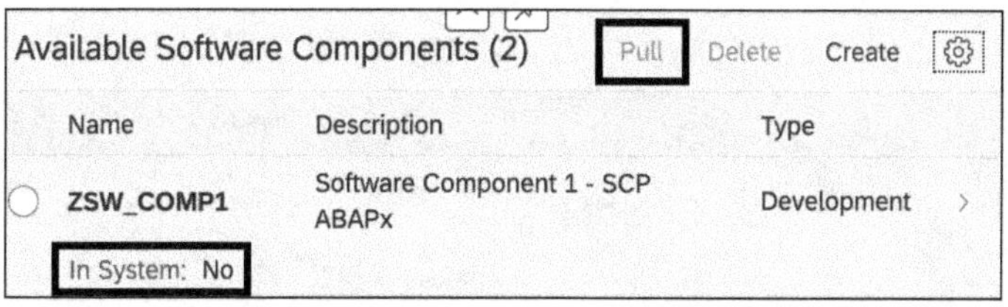

Figure 9.68: Software Component in the Target System

You can now pull objects from the repository by selecting your software component and clicking the **Pull** button, as shown in Figure 9.68.

6. **Check your object(s) in the target system**
   After pulling the repository, your transported ABAP object(s) should be visible under the target system in ADT, as shown in Figure 9.69.

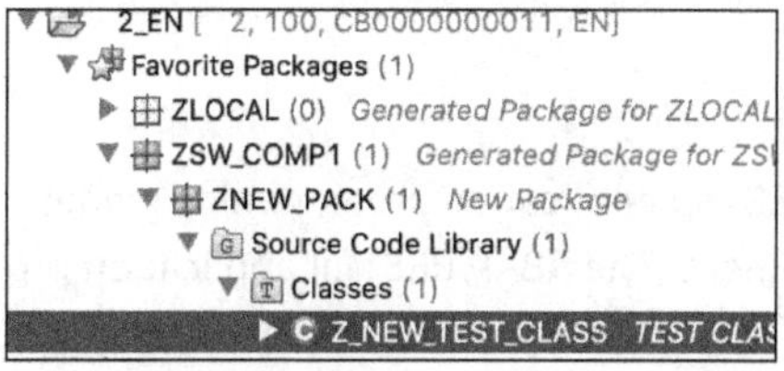

Figure 9.69: ABAP Class Available in the Target System

At this point, you've successfully transported an object from a source system to a target system using the Manage Software Components app, which internally uses gCTS.

Even though you now have sufficient information for most transporting scenarios, you probably have some questions about branching and merging possibilities in gCTS. Branching and merging are supported in the latest version of gCTS in SAP S/4HANA. We'll now briefly explain the branching capability in SAP BTP ABAP environment.

To work with multiple versions of the code simultaneously, you'll need to create multiple branches. When you create a software component, its master branch is created automatically. To create other branches, go to the software component and pull it once. After the pull request is successful, you can create further branches by clicking the **Create** option. Figure 9.70 shows two branches: the **master** branch and the **Maintenance** branch.

You can now check out individual branches by clicking the **Checkout** button. Once you check out any branch, that system will have the version available in that branch. Let's test that. Make sure your master branch code is released and transported and then create a maintenance branch. You can check out the maintenance branch and add some additional code using ADT. Since you've checked out the maintenance branch, these additional changes will be kept in that branch only. You can now release the new transport request created for the maintenance branch. After releasing the transport, you can now selectively check out either the master branch or the maintenance branch and then try to display the source code in ADT. You'll now see the version of code corresponding to the branches.

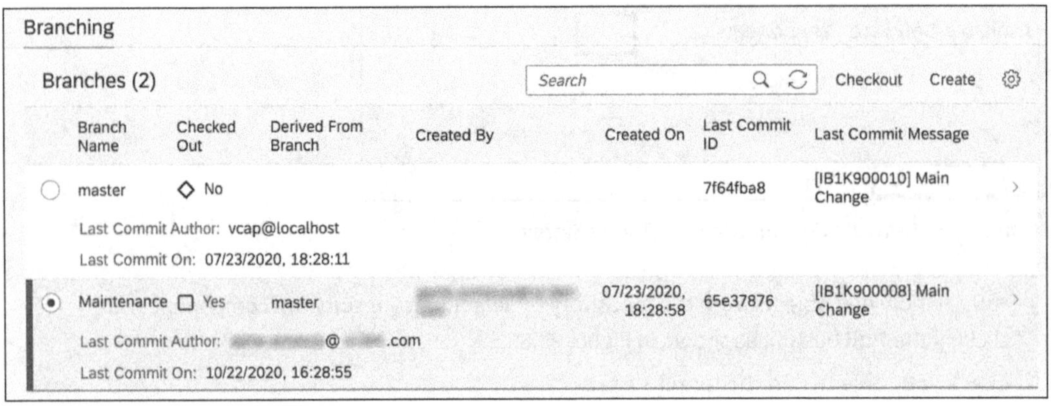

Figure 9.70: Master Branch and Maintenance Branch

9.12 Summary

SAP BTP ABAP environment provides the ABAP RESTful application programming model as the programming model for application development. The ABAP RESTful application programming model supports read-only reporting applications, managed and unmanaged transactional applications, draft-based managed applications, web APIs, and more. The Manage Software Components app is responsible for moving content from one SAP BTP ABAP environment system to another. This concept is different than transporting SAP Cloud Application Programming Model applications, where you would use multitarget application archives (MTARs) and SAP Cloud Transport Management for transport.

The ABAP RESTful application programming model is not only available for SAP BTP ABAP environment but also for SAP S/4HANA (from 1909) and the latest version of SAP S/4HANA Cloud (through developer extensibility). We hope to see significant new functionalities in SAP BTP ABAP environment and in the ABAP RESTful application programming model due to their popularity among ABAP developers. Now that we've wrapped up our ABAP chapter, let's move on to a more niche topic in the next chapter—Python applications.

Chapter 10

Developing Python Application Using SAP Business Application Studio

In the previous chapters, we have experienced building Node.js, Java, and ABAP applications using SAP Business Application Studio and Eclipse on SAP Business Technology Platform (SAP BTP). We will now experience the process of building a Python application using Jupyter Notebook integrated with SAP Business Application Studio. As Python is a key programming language for building AI-driven solutions, it's important to have a basic understanding not only of the language but also of how to effectively leverage it for designing and deploying enterprise-grade applications.

Since the paradigm shift in the AI world, not only are AI models evolving, but we are also experiencing a simultaneous shift in the skills and technologies that support them. Programming languages like Node.js and Python are becoming increasingly critical because they enable developers to build, integrate, and scale AI-driven solutions more efficiently on platforms such as SAP BTP and other hyperscalers.

As one of the most diverse and popular programming languages, Python has already established itself as a "go-to" language, especially for AI applications regardless of platform or tools. After a brief introduction to Python and its relevant libraries, this chapter will introduce Jupyter Notebook, followed by a step-by-step process for creating a Python-based application on SAP BTP.

10.1 Python Overview and Libraries

As a programming language, Python is simple and easy to learn, especially for beginners. Over the years, Python has evolved beyond a programming language into an ecosystem powered by the widest possible range of libraries and frameworks. Whether for a simple web application, complex data analysis, or developing a machine learning model, it meets the needs of every user—including beginners, pro-code developers, and data scientists—and helps them turn their ideas into production-ready solutions. Additionally, strong community support and continuous innovation keep the language relevant, resilient to real-world challenges, and future-ready. Some of the most popular Python libraries that can help empower your enterprise applications include:

- **TensorFlow**
  TensorFlow is one of the most widely used Python libraries, particularly for building and training machine learning models. It supports almost every aspect of machine learning models, ranging from basic models to large-scale deep learning models.
- **PyTorch**
  PyTorch is next in line and well known for its simplicity and flexibility, especially for beginners. It feels more "Pythonic" and easier to debug compared to others.
- **Python Data Analysis Library**
  Python Data Analysis Library (popularly known as pandas) is an open-source Python package that falls under the data cleansing and transformation category. It provides high-performance, easy-to-use data structures and data analysis tools for labeled data. Using the Python programming language's, persuasive syntax and rich functionality enable you to deal with missing data.
- **Matplotlib**
  Matplotlib is one of the most popular plotting libraries for visualization in Python. This library helps generate data visualizations, including two-dimensional diagrams and graphs such as histograms, scatterplots, and non-Cartesian coordinate graphs. With a bit of effort, Matplotlib can create nearly any visualization, including line plots, scatterplots, area plots, bar charts, histograms, pie charts, stem plots, and more.
- **scikit-learn**
  scikit-learn is one of the fastest-growing Python libraries. Developed by Google, this open-source symbolic math library for numerical computation uses data flow graphs.
- **SciPy**
  Scientific Python, also known as SciPy, is another free and open-source Python library used extensively in data science for high-level computations. This library encapsulates modules for efficient mathematical routines like linear algebra, interpolation, optimization, integration, and statistics.
- **NumPy**
  NumPy, the backbone for numerical computing in Python, handles arrays, matrices, and mathematical operations efficiently. Most AI libraries depend on NumPy under the hood.
- **Flask**
  Flask is a lightweight web framework in Python generally used to quickly create prototype web applications and APIs without much effort. It empowers developers by providing core tools like routing and request handling, allowing flexibility to add only what's needed. Later in the chapter, we will leverage Flask to create a Python-based prototype application with minimal effort.

- **LangGraph**
  LangGraph provides a set of tools and libraries to create, run, and optimize generative AI models in a scalable and efficient manner. At its core, LangGraph uses graph-based architectures to model and manage the intricate relationships between various components of an AI agent workflow.
- **SAP Cloud SDK for AI (Python)**
  SAP Cloud SDK is one of the most powerful toolkits especially for building SAP AI Core-based machine learning applications. It connects seamlessly with the generative AI hub and SAP AI Core, providing developers access to leading large language models from providers such as OpenAI, Amazon, and Google.

Beyond these, several other Python-based libraries are readily available in the market for various purposes within the specified categories, and you can choose any of them for your Python-based experiments. For all libraries, you need to install the necessary modules inside a notebook using the `pip` command, which is a package manager for Python packages or modules.

10.2 Jupyter Notebook

Jupyter, a next-generation notebook, is one of the most popular among data scientists and is integrated into many data science platforms, including SAP Data Intelligence. For your machine learning data—whether from a raw comma-separated values (CSV) file, SAP HANA Cloud, or other supported formats—you can directly inject it into the notebook to run various machine learning experiments. Jupyter Notebook is an open-source web application for creating and sharing documents that contain live code, equations, visualizations, and text. Jupyter Notebook ships with the IPython kernel, which allows you to write programs in Python, but currently, more than a hundred other kernels are also available.

Once created, your notebook becomes your playground for all experiments, as shown in Figure 10.1. Since Jupyter Notebook is language agnostic, it supports over 40 programming languages—including Python, R, Julia, Scala, and PySpark—and has become one of the preferred environments for data scientists and engineers to explore and analyze data.

A notebook's cells default to using code when first created, and each cell uses the kernel you selected when starting the notebook. In our case, we began with Python 3 as the kernel, so we'll write Python code in the notebook's cells.

Since your initial notebook has only one empty cell, it can't really do anything. Thus, to verify that everything is working, let's add some Python code to the cell and try running it. Add the following code to the first cell, as shown in Figure 10.1:

```
print('Hello Jupyter!').
```

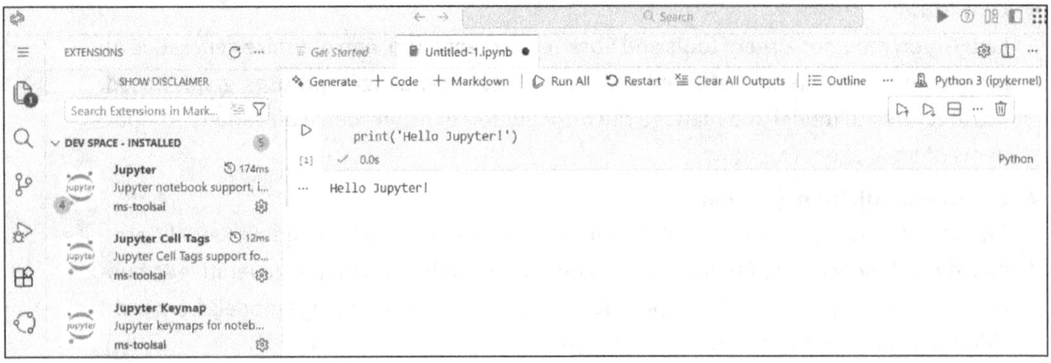

Figure 10.1: Hello Jupyter

In this context, “running it” means “executing the cell’s contents.” To execute a cell, select it and click the run icon in the row of buttons at the top. Alternatively, you can press Shift+Enter. If your notebook contains multiple cells, running them in order shares variables and imported data across the cells. As a result, you can easily separate your code into logical chunks without having to reimport libraries or recreate variables or functions in each cell.

As shown in Figure 10.2, when you run a cell, notice that some square brackets appear next to the word to the left of the cell. The square brackets will autofill with a number indicating the order in which you ran the cells. For example, if you open a fresh notebook and run the first cell at the top, the square brackets will display the number 1. But wouldn’t it be interesting to know what happens behind the scenes when you work with the notebook?

Figure 10.2: Running a Cell

In addition to running code, Jupyter Notebook stores code, output, and markdown notes together in an editable document called a notebook. When you save the document, your browser sends this file to the notebook server, which saves it on disk as a JavaScript Object Notation (JSON) file with an *.ipynb* extension. The notebook server, not the kernel, handles saving and loading notebooks, allowing you to edit them even without the kernel for that language—you simply won't be able to run code. Additionally, the kernel has no knowledge of the notebook's contents: Code cells are sent for execution only when the user runs them.

Now that you understand how Jupyter Notebook works, let's focus on the dedicated Python SDK for building scenarios using Jupyter Notebook within the machine learning scenario manager.

10.3 Creating a Python Application

To build the Python application we are going to use Jupyter Notebook inside SAP Business Application Studio. We'll walk you through initial setup, creating your project, installing the necessary Python extension, installing the Flask library, and deploying your Python application.

10.3.1 Setup

To begin, launch SAP Business Application Studio by selecting the relevant dev space, as shown in Figure 10.3. In our case, we have created a dev space of type **SAP Fiori**.

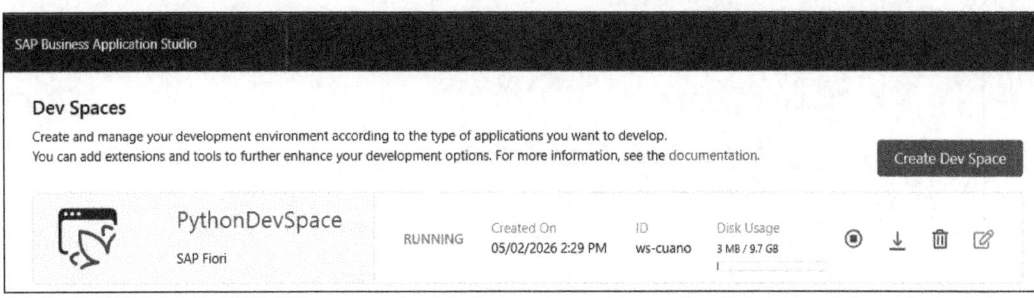

Figure 10.3: Python Dev Space

Once inside SAP Business Application Studio, the first thing that we are going to check is what version of Python is installed. Open a new terminal using **Terminal • New Terminal**, as shown in Figure 10.4, and execute the following `--version` command (see Figure 10.5) to check the version:

```
user: ~ $ python --version
```

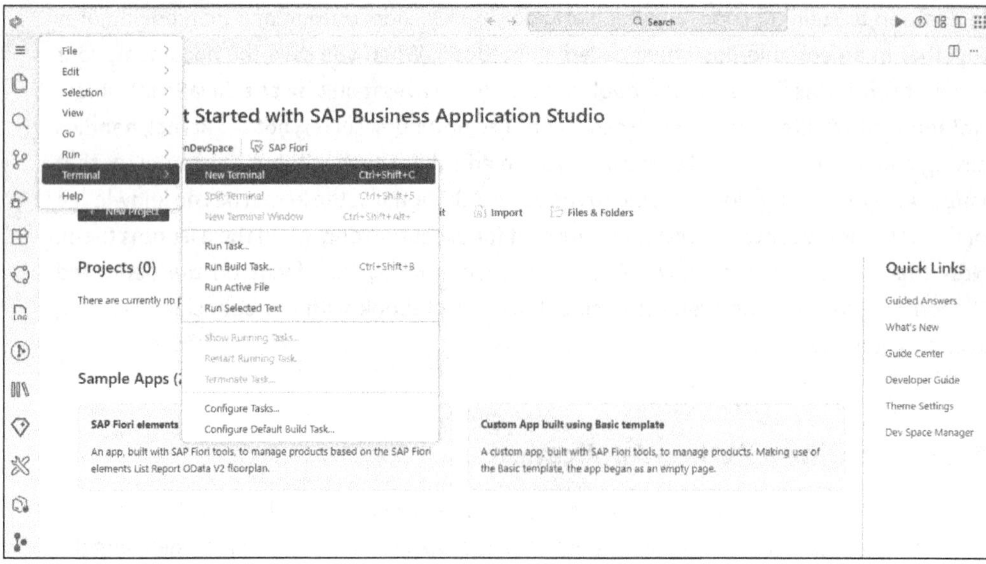

Figure 10.4: Launch New Terminal

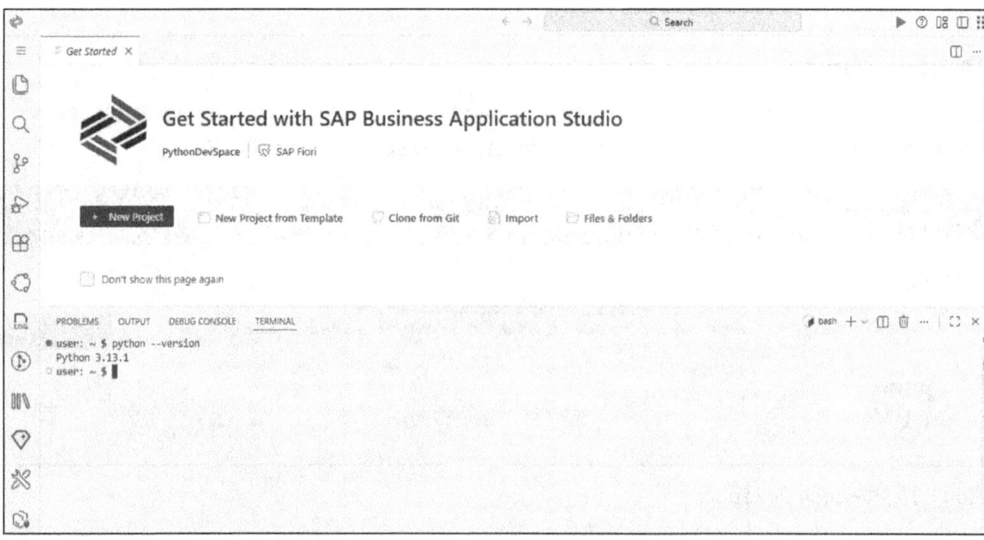

Figure 10.5: Validate Python Version

After ensuring that you have the latest version, the next step is to execute the `pip` command, as shown in Figure 10.6. This command downloads the official `get-pip.py` script from the Python Packaging Authority and uses it to install `pip` for Python. This ensures that installed packages can be executed globally, as follows:

```
curl https://bootstrap.pypa.io/get-pip.py > get-pip.py && python3 get-
pip.py &&  echo "export PATH=/home/user/.local/bin:$PATH" >> .bashrc &&
source ~/.bashrc
```

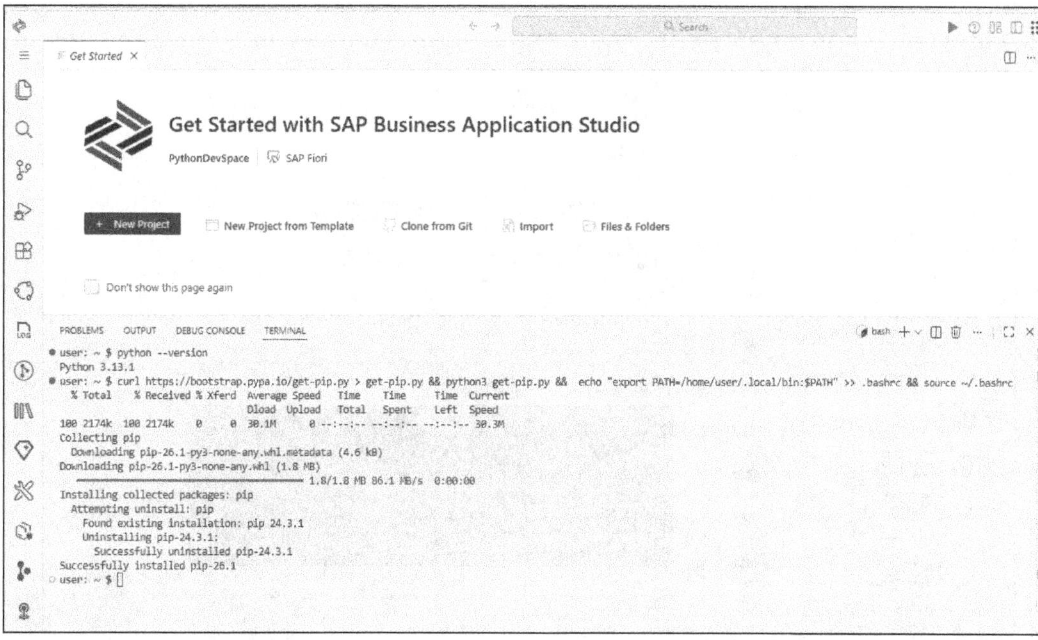

Figure 10.6: Install pip

As we are planning to leverage Jupyter Notebook as part of our Python application development, we need to install the associated extension inside SAP Business Application Studio. Click the **Extension** icon from the left side panel to launch the **Extension: Marketplace** and search for "Jupyter", as shown in Figure 10.7.

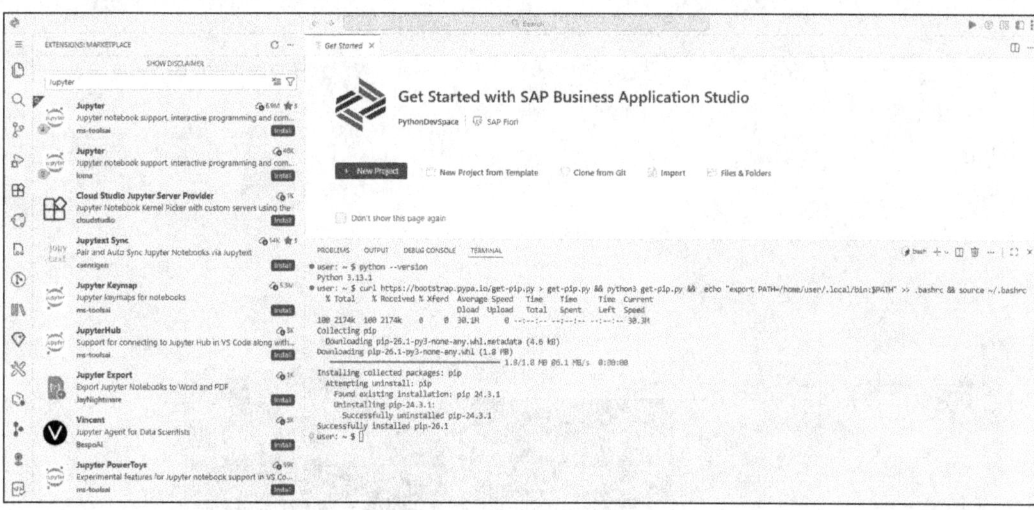

Figure 10.7: Search for the Jupyter Notebook Extension

Select the **Jupyter** extension and click **Install** to initiate the installation, as shown in Figure 10.8.

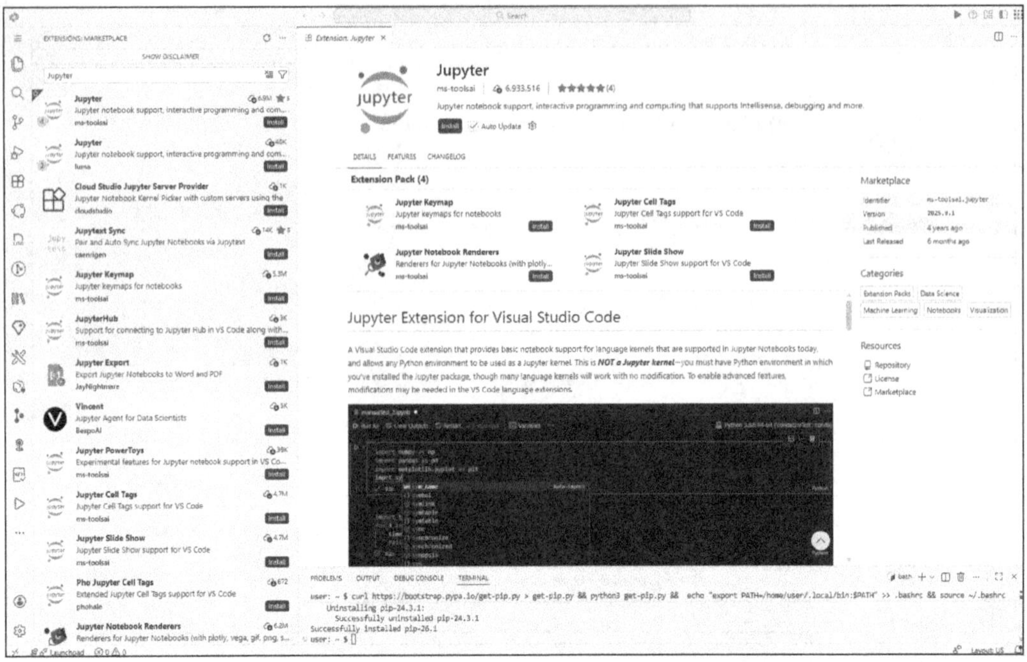

Figure 10.8: Install the Jupyter Notebook Extension

The installation triggers a trust-publisher acknowledgement message, as shown in Figure 10.9. Click on **Trust Publisher & Install**.

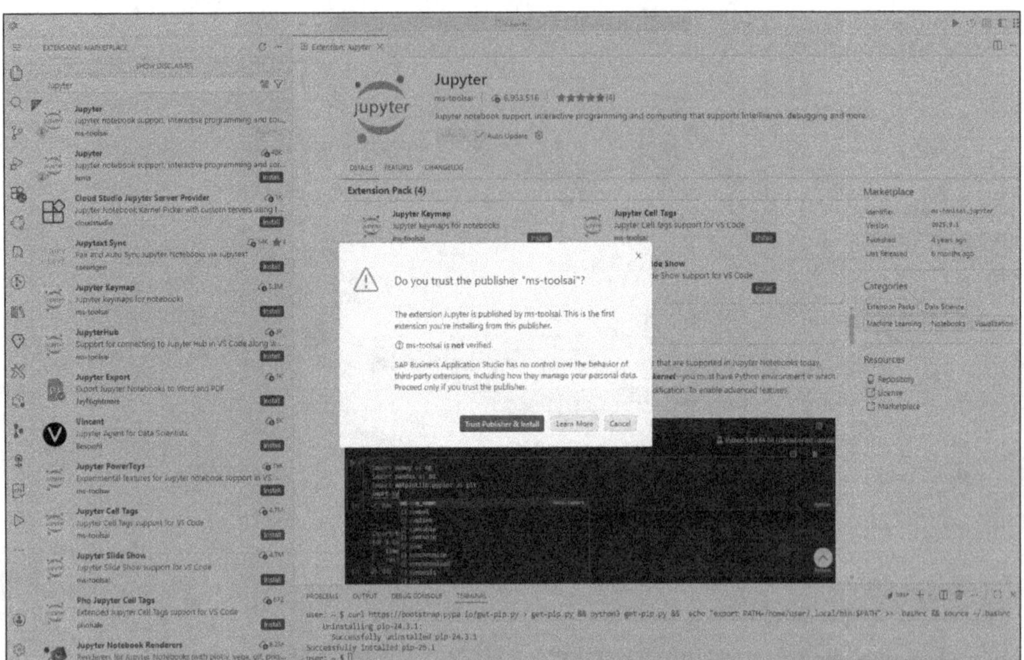

Figure 10.9: Installation Continued

As soon as the installation is completed, you can launch a notebook through the command palette. To open the command palette you need to press `Ctrl` + `Shift` + `P` and search for and select the **Create New Jupyter Notebook** command, as shown in Figure 10.10.

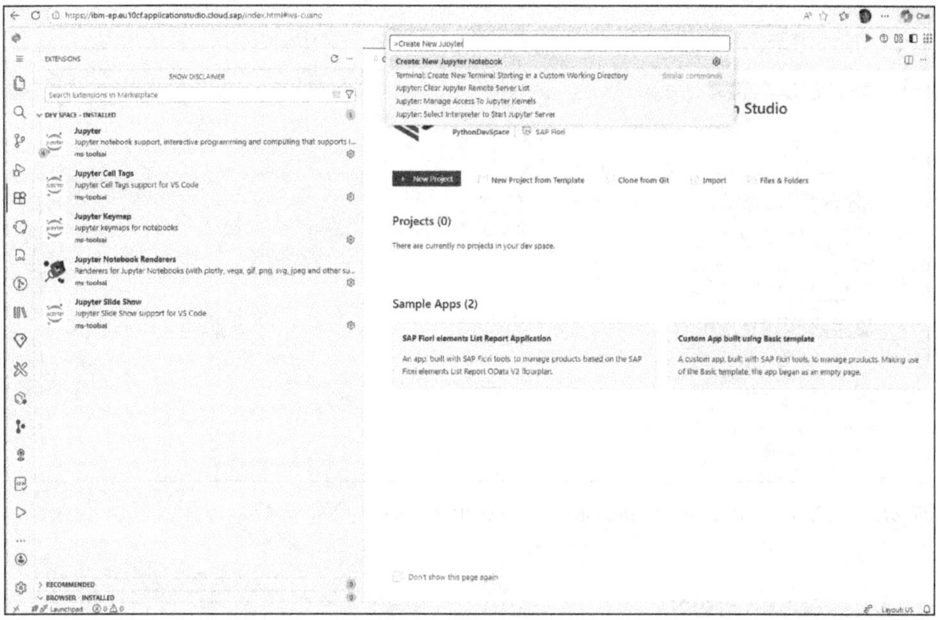

Figure 10.10: Launch a Notebook

The *.ipynb* file extension, as shown in Figure 10.11, is used inside Jupyter Notebook to execute code, text, visuals, and outputs in one place through an interactive, cell-based interface. This kind of cell-based execution is ideal for exploration, data analysis, and conducting experiments.

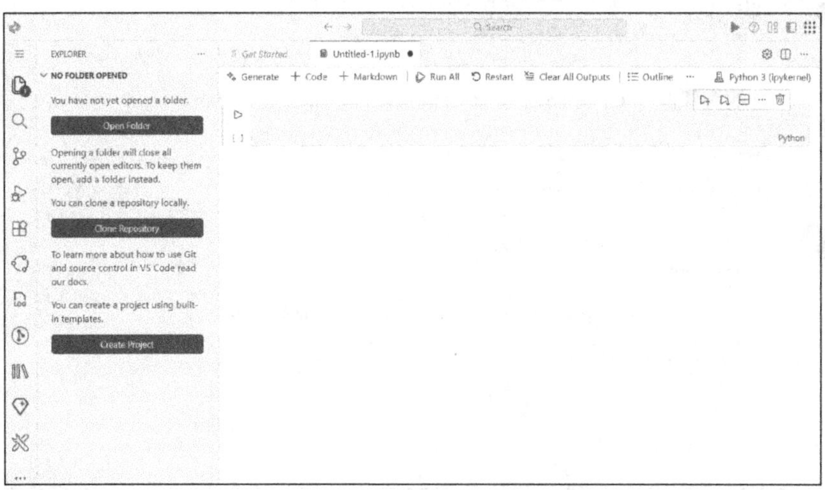

Figure 10.11: An Empty Notebook for Experiments

Jupyter Notebook is not limited to Python; it also supports additional programming languages such as HTML, JavaScript, Perl, XML, and more, as shown in Figure 10.12.

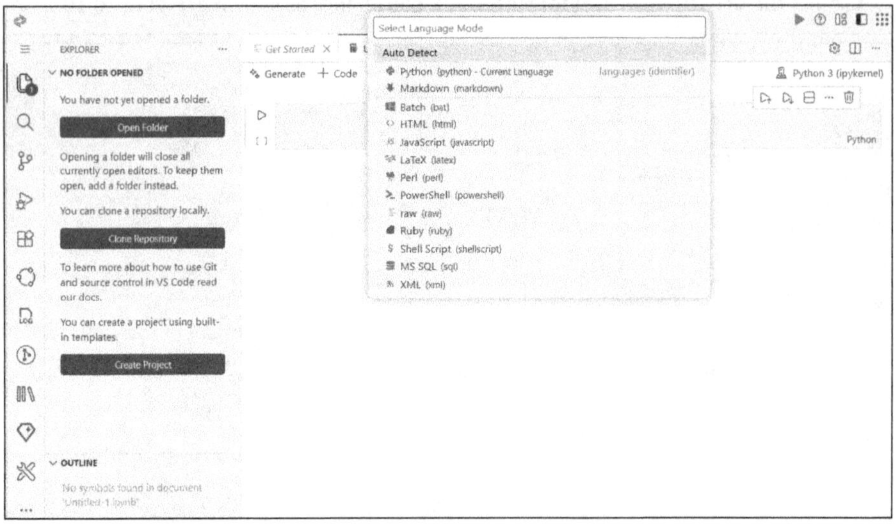

Figure 10.12: Various Languages Inside Jupyter Notebook

10.3.2 Create a Project

Now that we've completed our setup, let's create an empty project using the command palette by typing in "Create empty project", as shown in Figure 10.13.

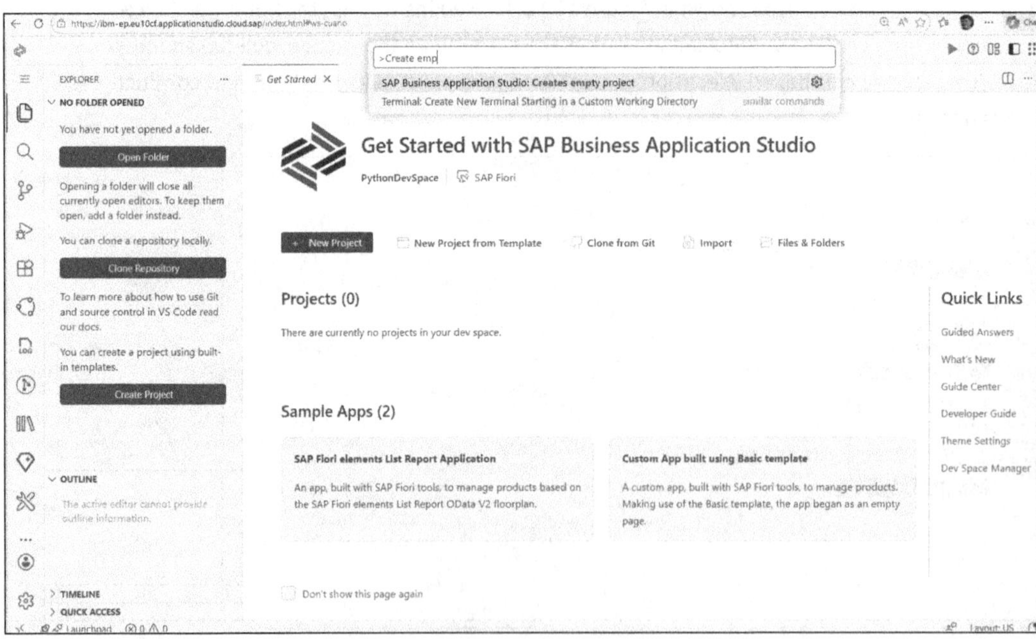

Figure 10.13: Create an Empty Project

Enter a project name as shown in Figure 10.14—in our case, “pythonsampleapp”—and then press [Enter].

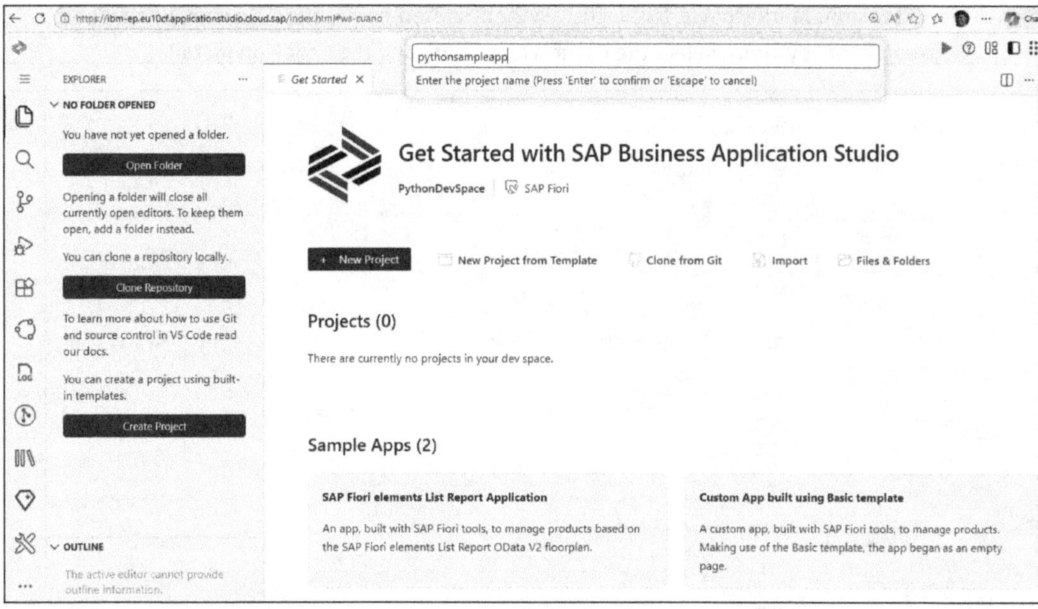

Figure 10.14: Enter Project Name

Next, select **Runtime** inside the command palette and choose **Set Default**, as shown in Figure 10.15.

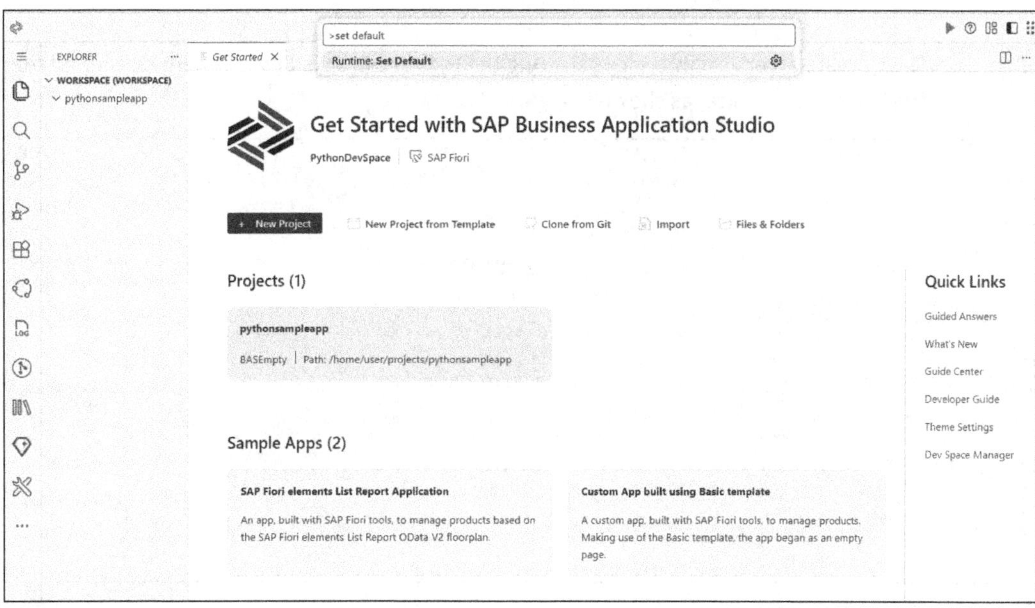

Figure 10.15: Set Runtime

10.3.3 Install Python Extension

Now, before moving to the next steps, we need to install one more extension from the **Extension: Marketplace** to execute Python-related commands. As shown in Figure 10.16, search for "python", select the **Python** extension, and then click on **Install**.

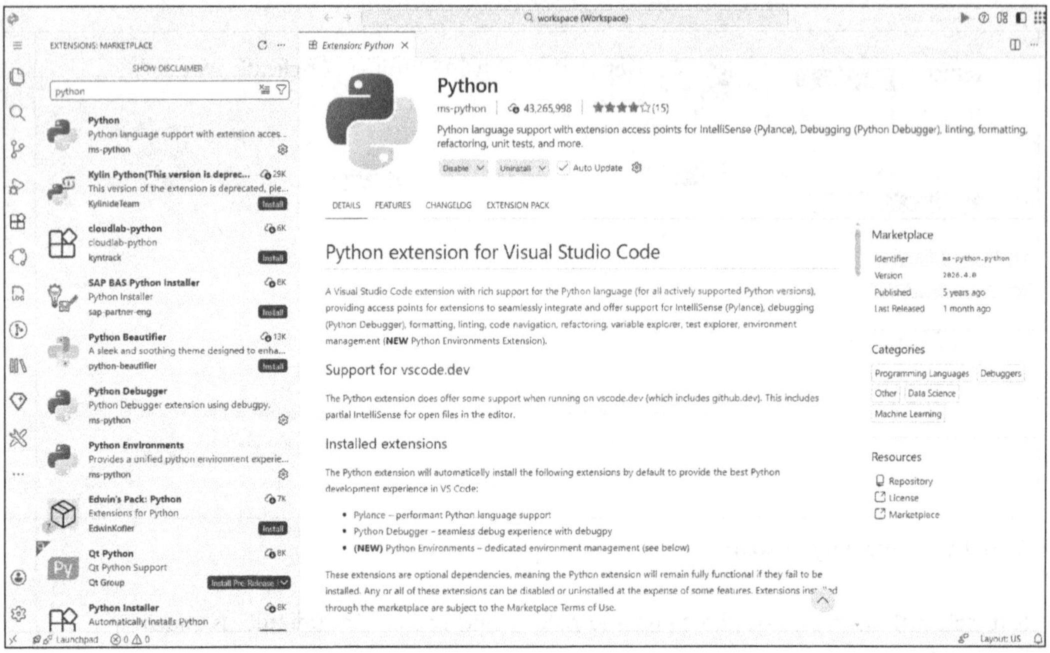

Figure 10.16: Install Python Extension

Next, create an environment by selecting the **Python: Create Environment** command from the command palette, as shown in Figure 10.17.

Figure 10.17: Create Environment

Choose **Venv Creates a '.venv' virtual environment in the current workspace** from the command palette, as shown in Figure 10.18. Within Jupyter Notebook, this helps you to maintain a consistent and reproducible development environment for executing notebooks and managing the required libraries.

Figure 10.18: Select Preferred Environment

As shown in Figure 10.19, you'll now choose your interpreter path, which in our case is **Python 3.13.1 64-bit ~/.asdf-inst/shims/python3.13.**

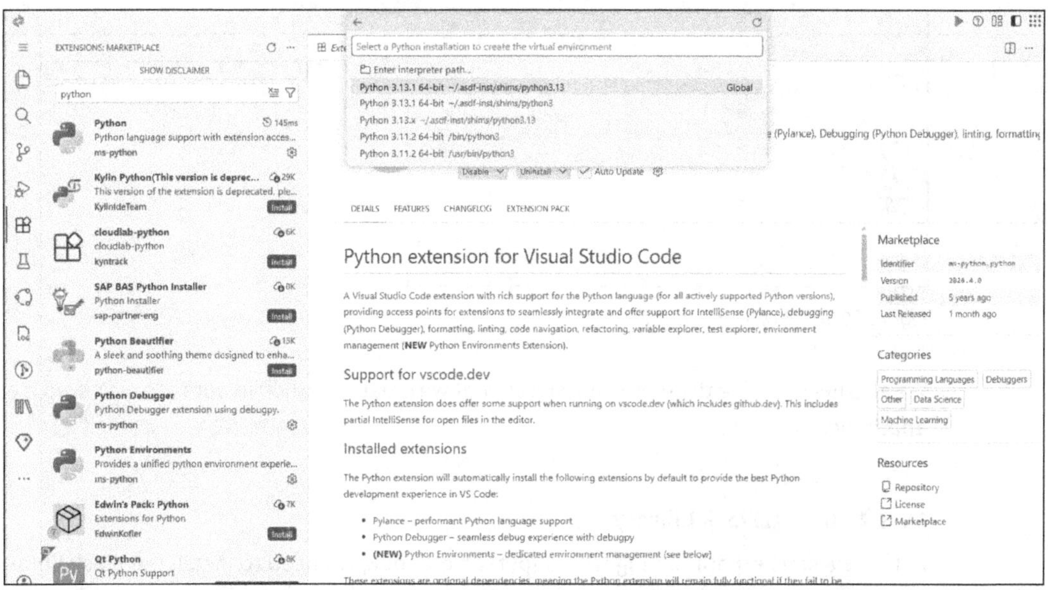

Figure 10.19: Select Python Version

As shown in Figure 10.20, inside the project structure we need to create the following files as part of the root folder of the project:

- *app.py*
- *manifest.yml*
- *requirements.txt*
- *runtime.txt*
- *Procfile*

To create these files, all you need to do is right-click on the root project folder and select the **New File** option.

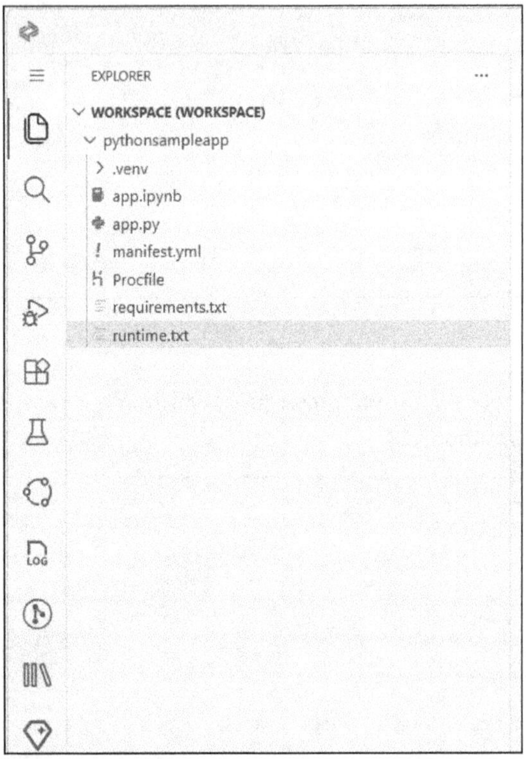

Figure 10.20: Project Structure

We also need to save the empty notebook that we created earlier; in our case we saved it as **app.ipnyb**.

10.3.4 Install Flask Library

Before we start experimenting with Jupyter Notebook, we need to install the Flask library using the `pip install` command, as shown in Figure 10.21. You can simply paste the command into one of the cells and click the play ▷ button which will appear next to the cell.

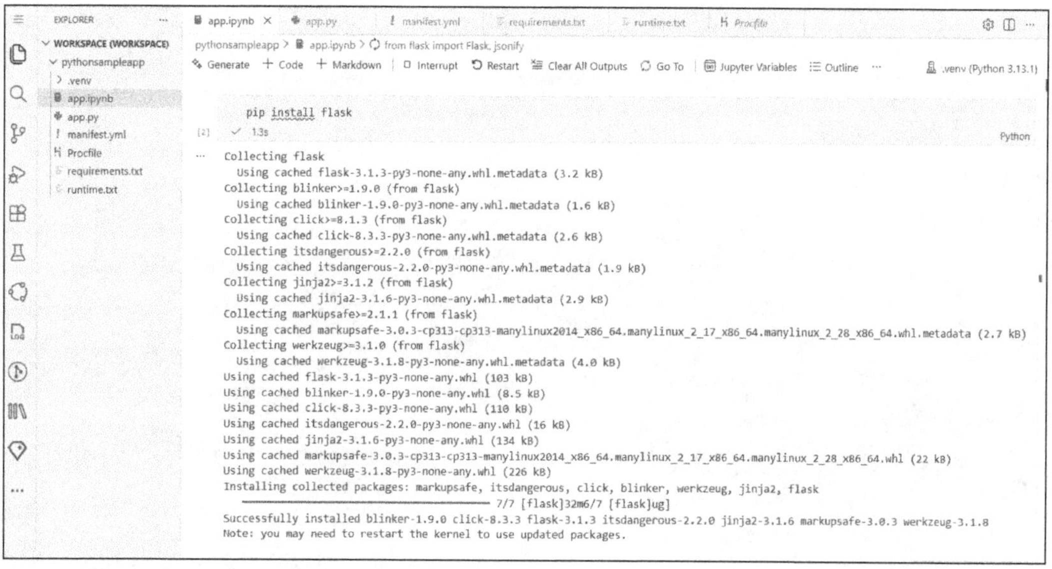

Figure 10.21: Install Flask Library Inside Notebook

Once the library is installed, insert a new cell into the notebook and paste in the code from Listing 10.1, as shown in Figure 10.22.

```
from flask import Flask, jsonify
app = Flask(__name__)
@app.route('/')
def home():
    return "Hello from Flask on SAP BTP Cloud Foundry!"
@app.route('/health')
def health():
    return jsonify(status="UP")
if __name__ == '__main__':
    app.run(host='0.0.0.0', port=8080)
```

Listing 10.1: Web Application Python Script

This Python script uses the Flask framework to create a simple web application that can run on the SAP BTP, Cloud Foundry environment. It initializes a Flask app and defines two routes: the `root` (`/`) returns a basic text response indicating that the app is running, while the `/health` endpoint returns a JSON response used for health checks. The application is configured to run on all network interfaces (`0.0.0.0`) and listens on port `8080`, making it accessible externally.

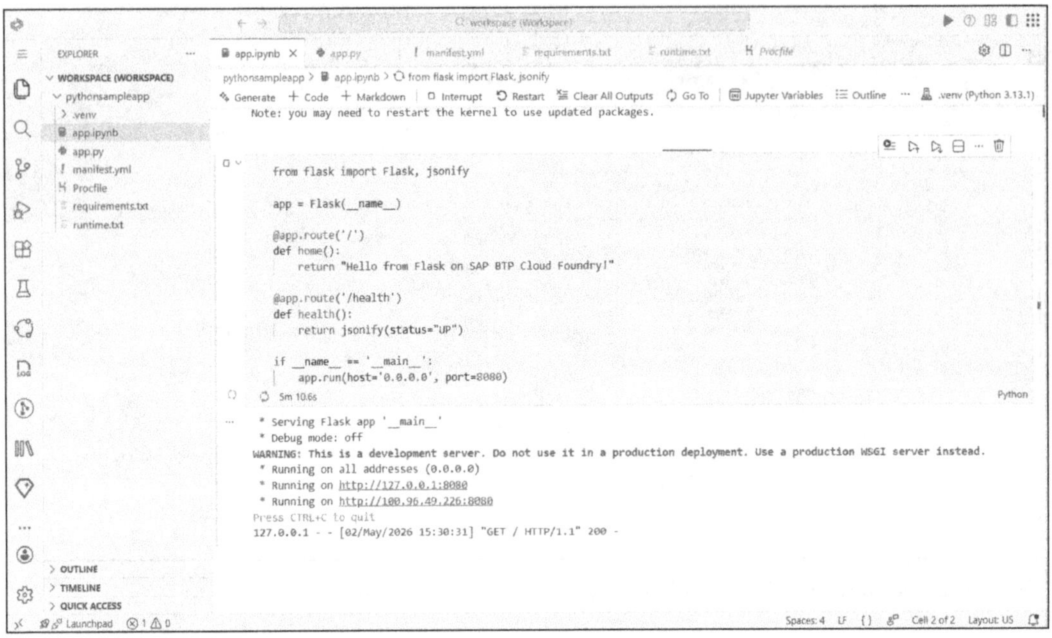

Figure 10.22: Experiment with Code

After pasting in the code snippet shown in Listing 10.1, you can test the code snippet by clicking the ▷ button next to the notebook cell. You should see the output shown in Figure 10.23.

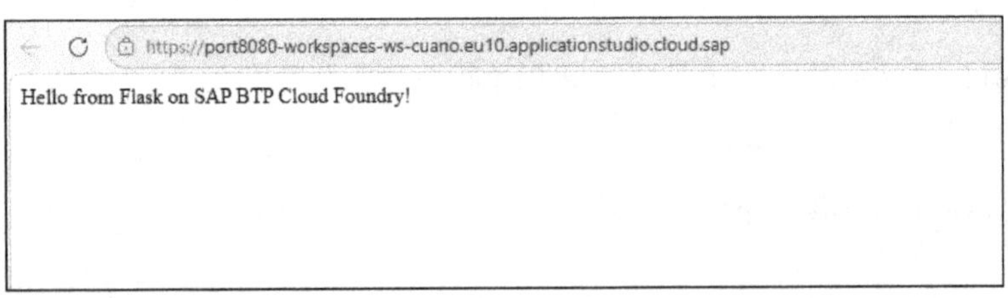

Figure 10.23: Notebook Cell Output

Now that our notebook-based experiment is complete, we will reopen the Python terminal via the command palette by selecting **Python: Create Terminal**, as shown in Figure 10.24.

Next, we will leverage the same code from our experiment (Listing 10.1) and save it inside *app.py* as a Python script ready for deployment. Then, inside the terminal, type the `pip install Flask` command to install the Flask library, as shown in Figure 10.25. This will ensure availability of the Flask library during both design and runtime.

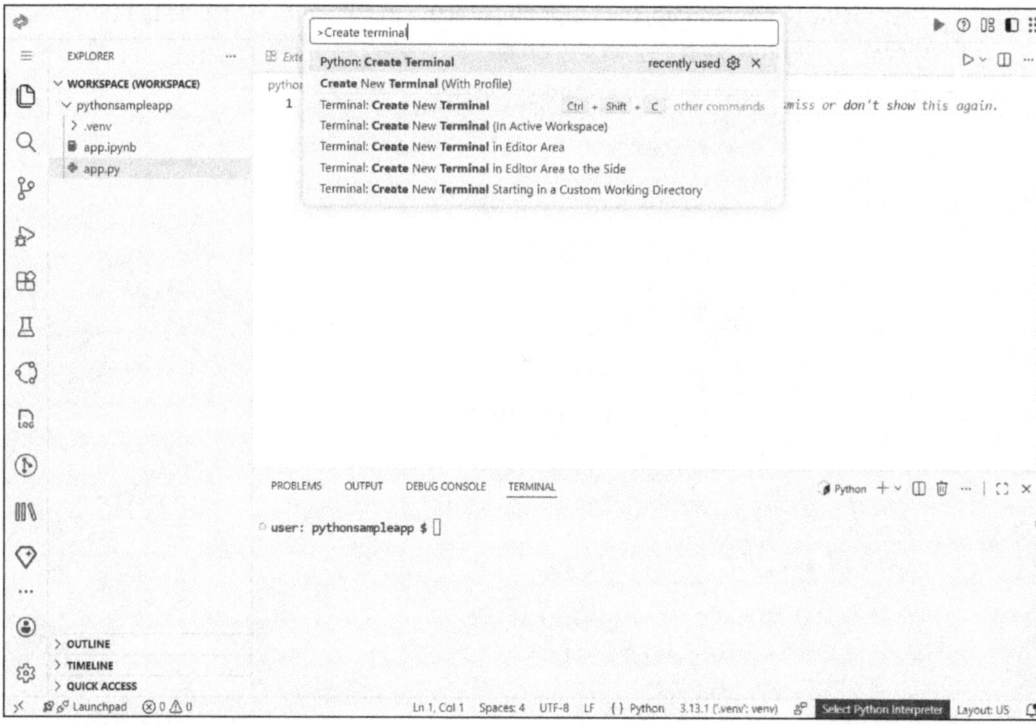

Figure 10.24: Create Python Terminal

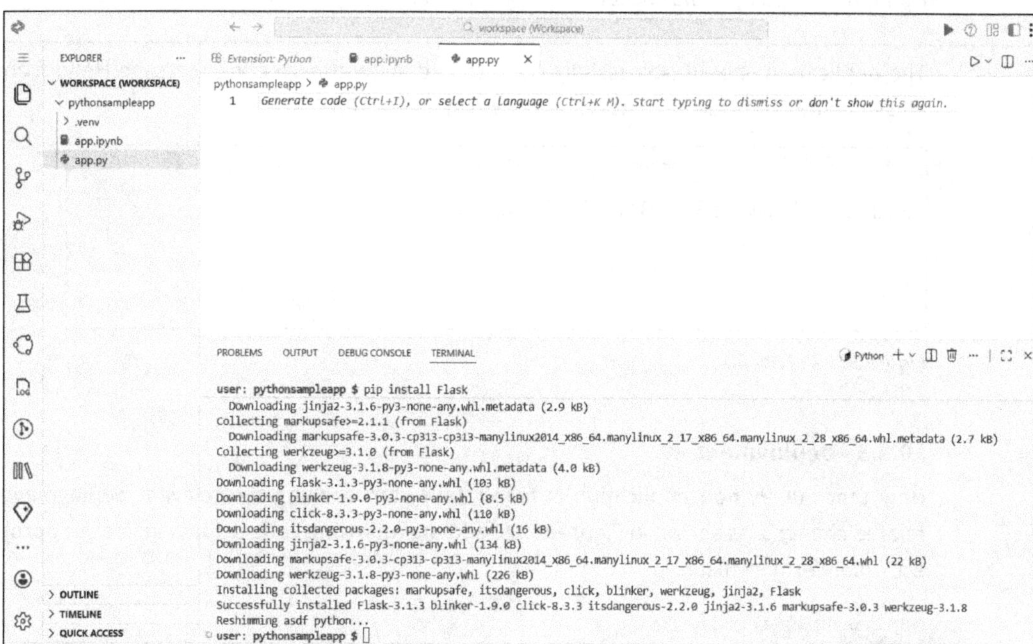

Figure 10.25: Install Flask

After saving the Python *app.py* script file, execute `python app.py` inside the terminal, as shown in Figure 10.26.

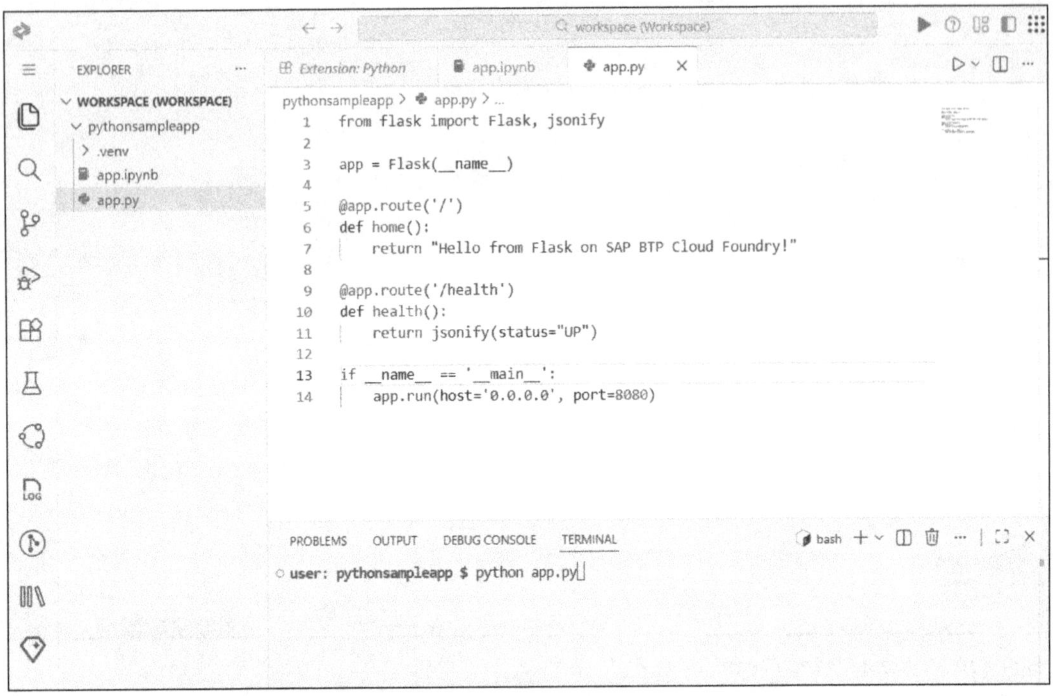

Figure 10.26: app.py Python Script

The application should successfully execute and show the greeting message **Hello from Flask on SAP BTP Cloud Foundry!**, as shown in Figure 10.27.

Figure 10.27: Trial Run

10.3.5 Deployment

Now that our Python application is ready, let's start making the relevant deployment-related changes. As shown in Figure 10.28, add the following code snippet inside your project's *manifest.yml* file:

```
applications:
- memory: 128MB
  disk_quota: 512MB
  random-route: true
```

The configuration shown in Figure 10.28 allocates the required memory and disk space to the application for execution.

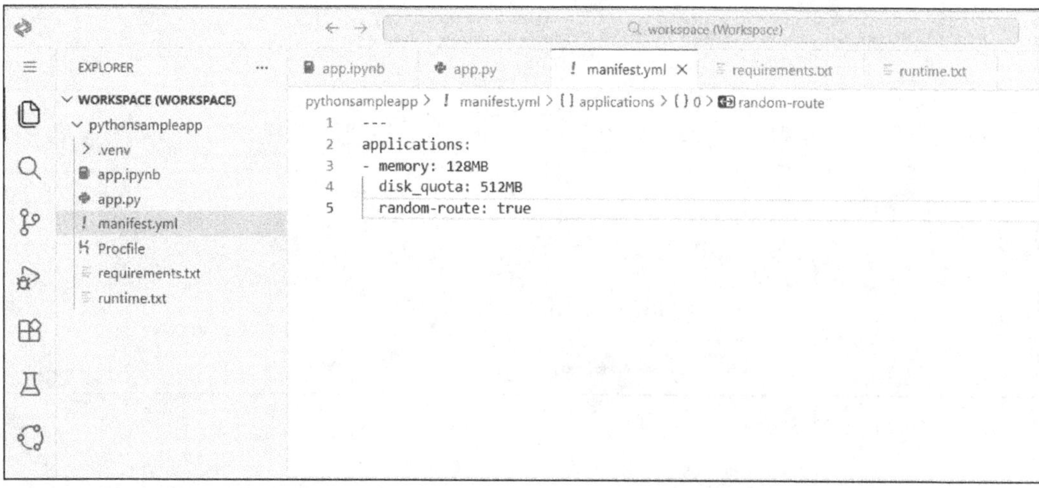

Figure 10.28: manifest.yml File

As shown in Figure 10.29, the *runtime.txt* file is used to specify the exact runtime environment (e.g., the language version) that the application should use during deployment, which in our case is `python-3.13.11`.

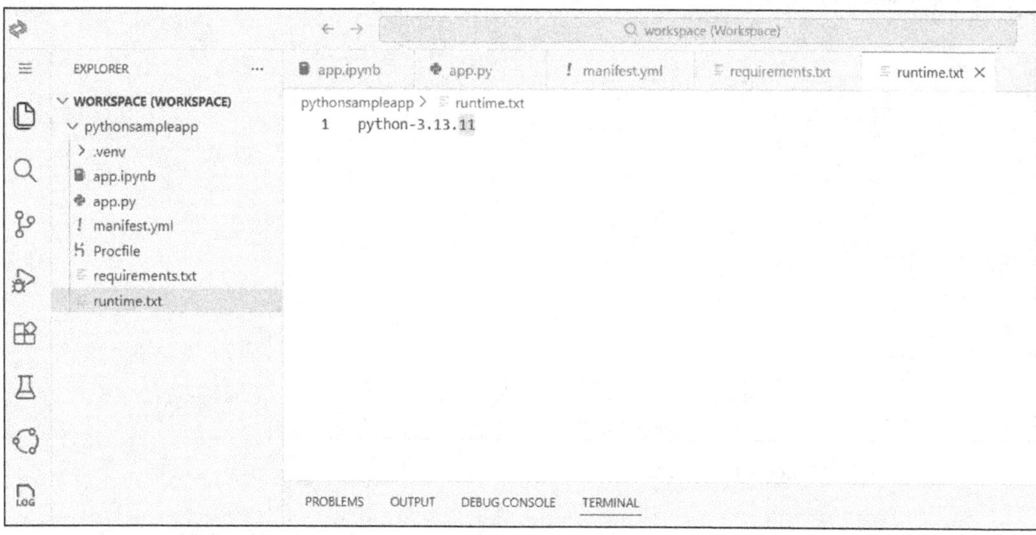

Figure 10.29: runtime.txt

The *Procfile*, as shown in Figure 10.30, tells the platform how to start your application, as the Cloud Foundry runtime does not automatically know which command should run the app in production.

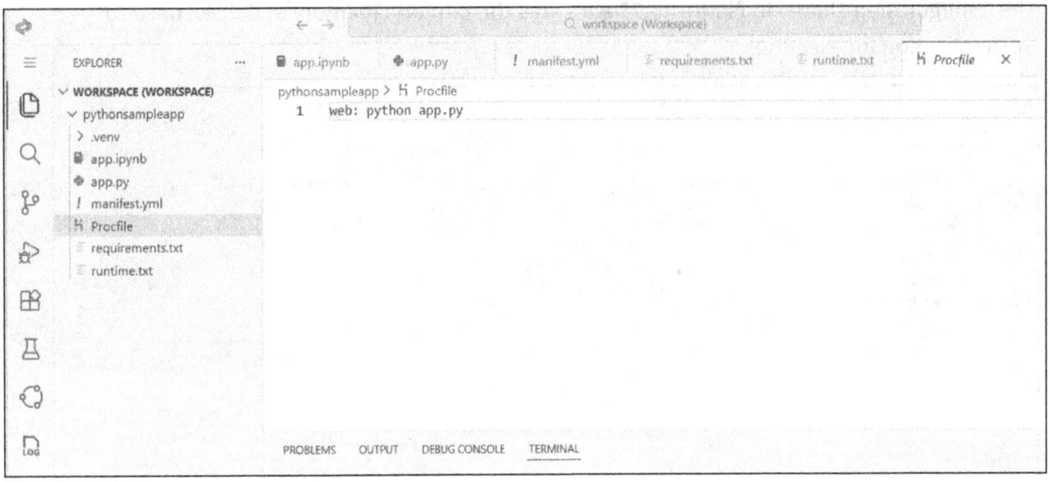

Figure 10.30: Procfile

Now that we're done with the files, we need to use **CF: Login to Cloud Foundry** from the command palette to log in, as shown in Figure 10.31.

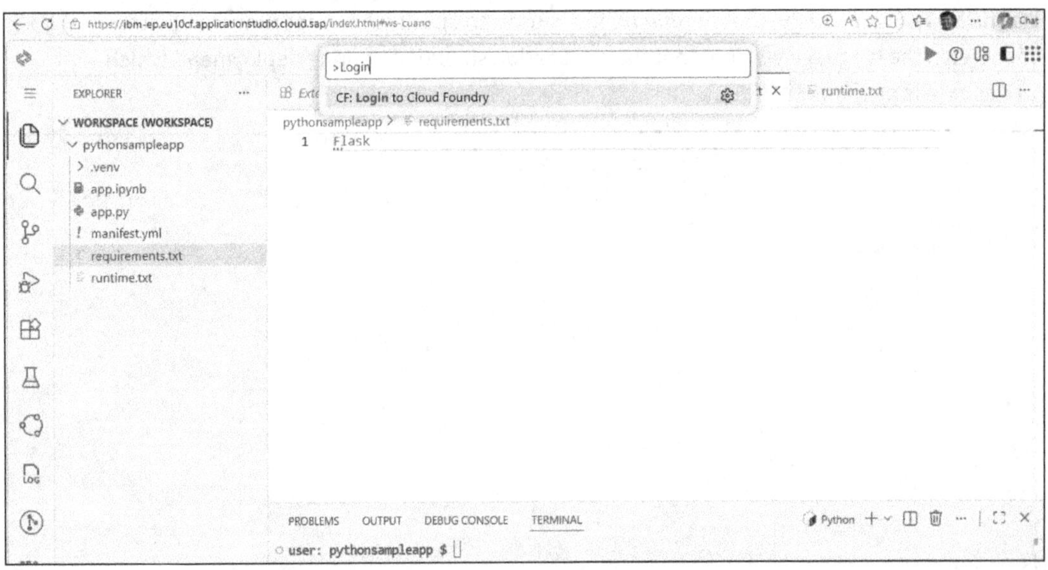

Figure 10.31: Login Command

As shown in Figure 10.32, enter your credentials to log in to Cloud Foundry for further deployment tasks.

Once login is successful, fill in the **Select Cloud Foundry Organization** and **Select Cloud Foundry Space** fields, as shown in Figure 10.33, and click **Apply**.

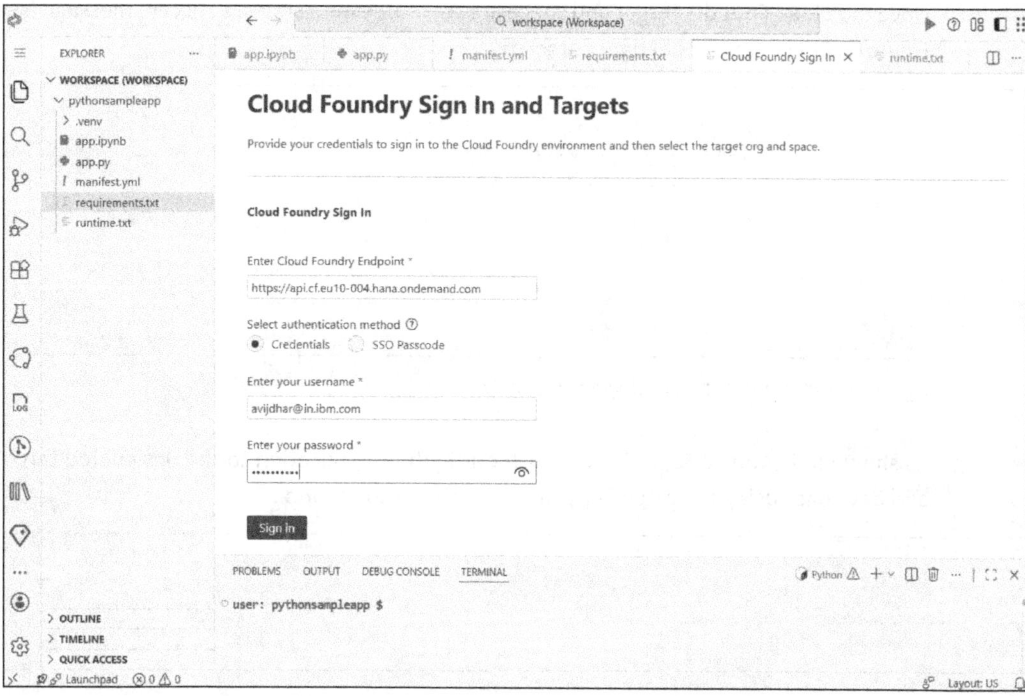

Figure 10.32: Login to Cloud Foundry

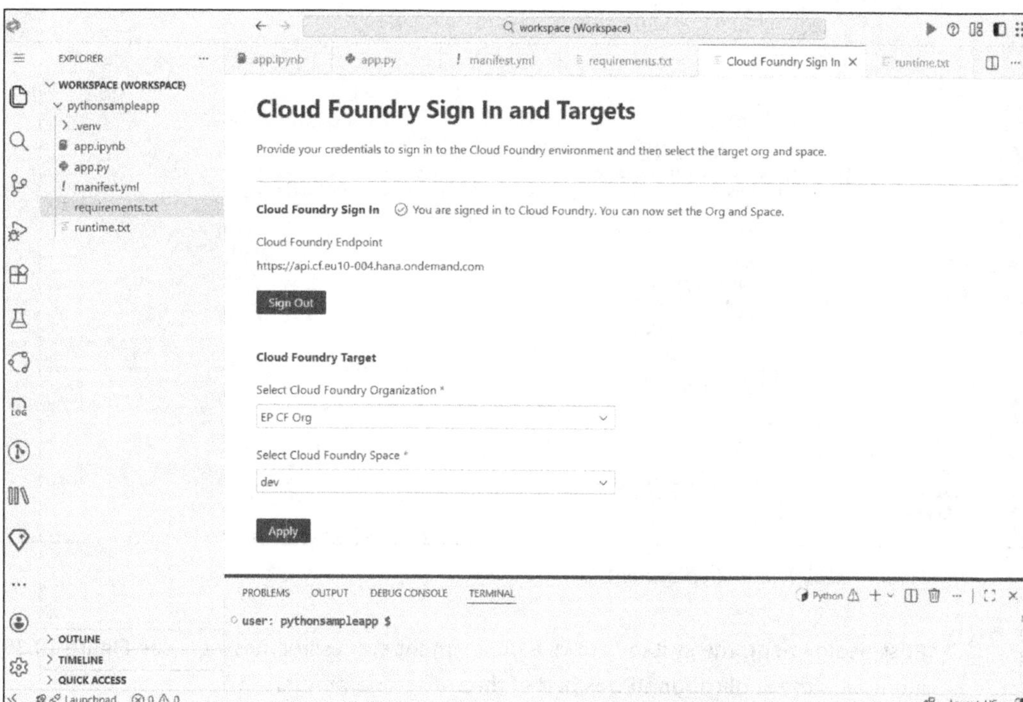

Figure 10.33: Select Target and Space

Once the Cloud Foundry target and space are set, the system shows a success message (see Figure 10.34) in the bottom right corner.

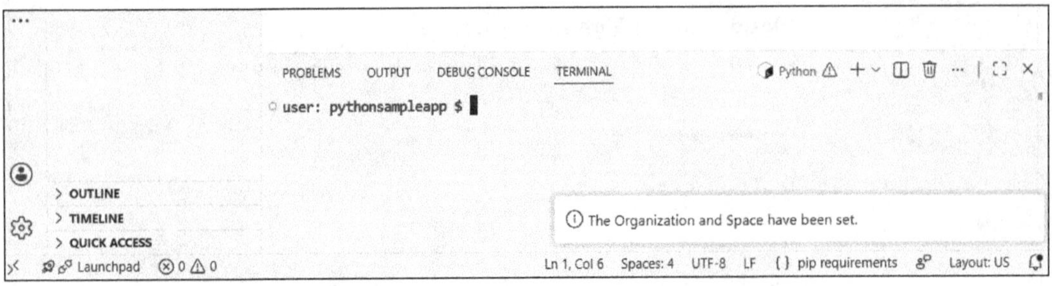

Figure 10.34: Organization and Space Are All Set

As shown in Figure 10.35, let's now push our Python application to the designated target and dev space using the `cf push pythonsampleapp` command.

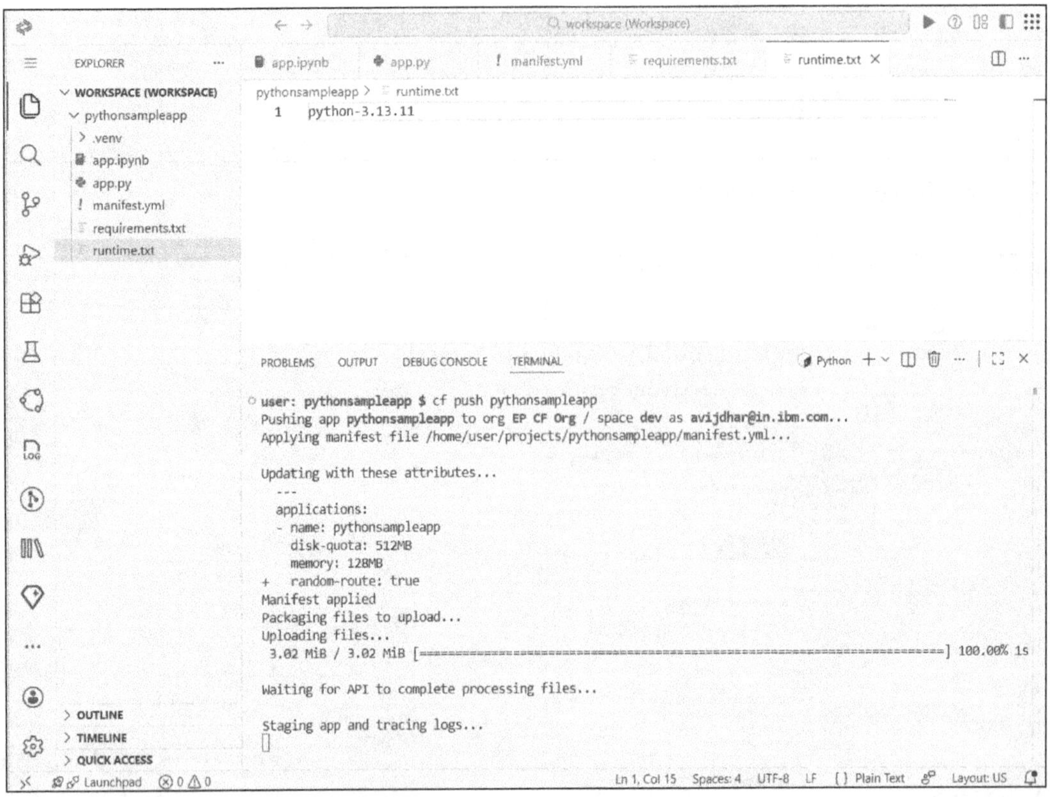

Figure 10.35: Trigger Deployment

Post-deployment, the system shows a deployment successful message (see Figure 10.36) along with the application URL as part of the `routes` parameter.

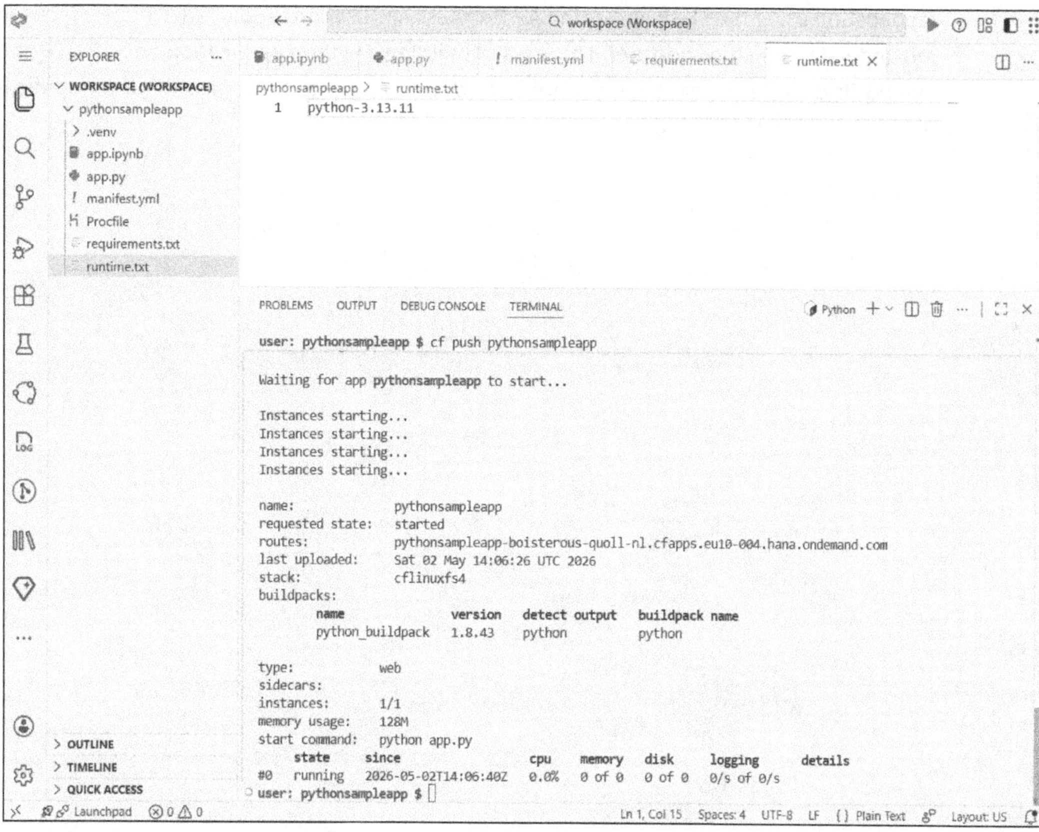

Figure 10.36: Deployment Successful

Use the application URL obtained in the previous step to access the application and you'll again see the greeting message **Hello from Flask on SAP BTP Cloud Foundry!**, as shown in Figure 10.37.

Figure 10.37: Final Output

10.4 Summary

As we conclude the chapter, let's quickly recap what we have learned so far. We began with a brief introduction to Python and Jupyter Notebook, then proceeded step by step to set up the required environment for building a Python application, followed by creating a Python application using the Flask library in SAP Business Application Studio. The chapter focused on understanding the end-to-end process of building a Python application on SAP BTP and aimed to provide the necessary context for developing enterprise-grade Python applications.

Chapter 11
Developing Applications in SAP Business Technology Platform, Kyma Runtime

In this chapter, we'll introduce you to SAP Business Technology Platform (SAP BTP), Kyma runtime, starting with an introduction to its containerization concept and its components and the potential use cases you can address using SAP BTP, Kyma runtime. You'll also experience the step-by-step process for deploying and exposing an SAP Commerce Cloud-based mock application programming interface (API) using SAP BTP, Kyma runtime.

Containers nowadays are front and center in many modern IT landscapes and present an opportunity to decompose applications into smaller, more focused units to enable developers and IT professionals to deploy applications across environments and cloud platforms with little or no modification. The abstraction layer provided by an individual container image enables you to fundamentally rethink how distributed applications are built. With the growth in adoption rate, however, many organizations soon realized that containers started surfacing everywhere, and monitoring, scaling, and orchestrating certain functions across many containers became difficult. Here is where Kubernetes, as a container-orchestration system, comes to rescue for the deployment, scaling, and management of containers. According to its official definition, Kubernetes is a portable, extensible, open-source platform for managing containerized workloads and services that facilitates both declarative configuration and automation. Kubernetes basically facilitates the modularization of containers using *pods*, which are nothing but co-located containers that can share resources like file systems, kernel namespaces, and IP addresses. Cloud-native technologies, and especially Kubernetes, enable the required automation and visibility necessary to manage applications at scale and with high velocity.

Now that you have a fair understanding of Kubernetes, let's face a new set of challenges. With the growing demand in cloud-native technologies, developers need not just a development platform but also resources for storage, backup, monitoring, and logging as well as a service mesh to enforce policies on data in transit. Each of these individual systems must be properly configured and deployed, as well as logged, monitored, and backed up on its own. Here is where Kyma shines! But what is Kyma?

Kyma is SAP's open-source project on Kubernetes and is comprised of a collection of projects united to simplify the extension and integration of monolithic software while providing seamless connectivity to enterprise systems. It provides a platform for extending applications with serverless functions following a microservice-based architecture by stitching event consumption, authentication, logging, tracing, monitoring, and alerting all together in one place. Kyma is specially designed to provide the tools you need to write end-to-end, production-grade cloud-native applications. Its service catalog allows for easy consumption of cloud services made available from the major hyperscalers using open service brokers and access to services you have connected to Kyma using the application connector.

After a detailed introduction to Kyma, in subsequent sections, we'll dive more deeply and talk about SAP BTP, Kyma runtime, including potential use cases. Finally, we'll demonstrate the basic steps for setting up SAP BTP, Kyma runtime, and for building an SAP Commerce Cloud application extension.

11.1 SAP BTP, Kyma Runtime

SAP BTP, Kyma runtime, is a fully supported SAP BTP service with a consumption-based billing model that uses the Kyma project (open source) as its platform. Kyma is integrated into SAP BTP and runs alongside other runtimes such as Neo or Cloud Foundry. SAP BTP provides a managed control plane to securely connect SAP portfolio solutions and other business services. Behind the scenes, the control plane uses Gardener (another SAP open-source project) to provision and manage k8s clusters.

As shown in Figure 11.1, the critical components of SAP BTP, Kyma runtime, include the following:

- **Development framework**
  SAP BTP, Kyma runtime now fully supports SAP Cloud Application Programming Model applications in a cloud-native, Kubernetes-based environment, which was previously limited to the Cloud Foundry environment. This comprehensive deployment capability provides the opportunity for scaling applications across diverse platforms and taking advantage of containerized microservices while maintaining seamless integration with a wide range of SAP BTP services.
- **Built-in serverless engine**
  Kyma runtime leverages an open-source serverless computing framework called *Kubeless* running on top of Kubernetes, which is suitable for building lightweight use cases including support for auto-scaling, routing, real-time logging, monitoring, and troubleshooting.
- **Built-in service mesh**
  Kyma's service mesh provides all the required components to support secure service-to-service communication using mutual Transport Layer Security (mTLS) along with proxying, service discovery, traceability, faster development, and testing and deployment cycles for cloud applications—all in a simple, reliable, and visible way.

- **Built-in API microgateway**
  This component provides value-extracting governance, discovery, observability, and stability features in a reusable agent. Added value is provided to operators by exposing microservices securely including support for various injectable authentication mechanisms and configurable identity providers.
- **Service catalog**
  The service catalog groups reusable, integrated services from service brokers registered in Kyma. Its purpose is to provide an easy way to access services that service brokers manage so you can use them in your applications. For example, you can easily consume services provided by different hyperscalers like Microsoft Azure, Amazon Web Services (AWS), or Google Cloud Platform (GCP).
- **Eventing**
  The Kyma event bus allows you to integrate various external solutions with Kyma. To achieve successful integration, the event bus uses the publish-subscribe messaging pattern, which allows Kyma to receive business events from various solutions, enrich this data, and trigger business flows using lambda functions or services defined in Kyma.
- **Logging**
  This tracing solution provides a holistic view of the request flow so you can understand the system and make informed decisions regarding troubleshooting and performance optimization. Tracing in Kyma uses Jaeger as a backend, which serves as the query mechanism for displaying information about traces.
- **Monitoring**
  Kyma comes bundled with tools that provide the most accurate and up-to-date monitoring data. The Prometheus open-source monitoring and alerting toolkit provides this data, which is consumed by different add-ons, including Grafana for analytics and monitoring and Alertmanager for handling alerts
- **SAP system connectivity**
  The runtime now offers seamless integration with SAP solutions including SAP S/4HANA Cloud, SAP Commerce Cloud, SAP Field Service Management, and SAP Cloud for Customer through built-in application connectors. This allows for side-by-side extensibility without touching the core SAP systems. Because SAP Cloud SDK is now fully supported on the SAP BTP, Kyma runtime, enterprise applications can take advantage of features like multitenancy, resilience, caching, and more, along with SAP S/4HANA connectivity. The SDK seamlessly integrates with Kyma's Kubernetes-based architecture and, in parallel, maintains compatibility with Cloud Foundry-specific modules and classes, which ultimately help developers deploy existing applications without code modification.
- **Enhanced multicloud support**
  Over the last few years, SAP BTP, Kyma runtime has been enhanced to support the three major hyperscalers: AWS, Microsoft Azure, and Google Cloud Platform. Hyperscaler selection is possible at the subaccount level and region availability has been extended to 22 regions globally.

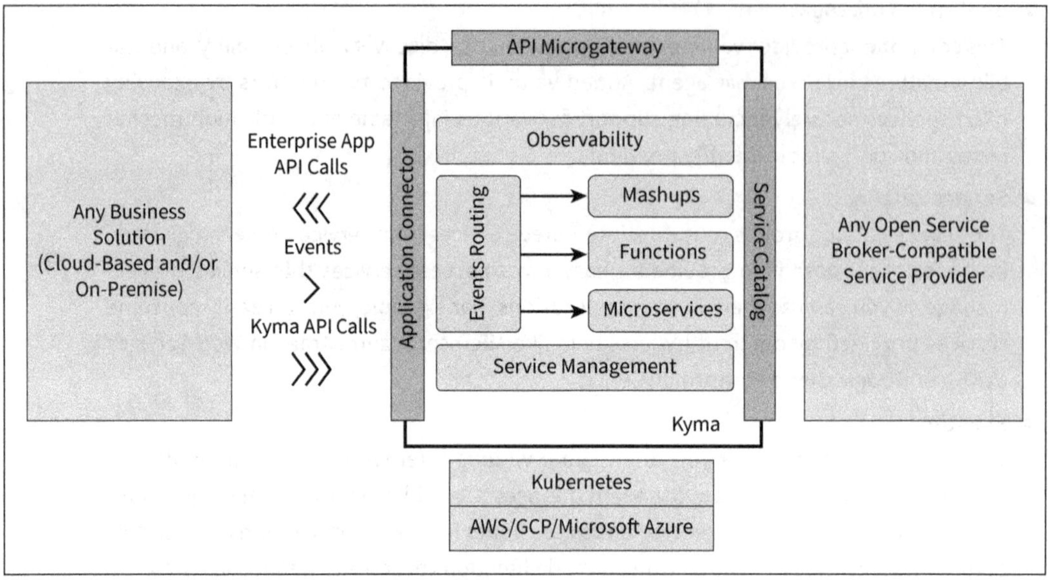

Figure 11.1: SAP BTP, Kyma Runtime

Now that you have a fair understanding of containers and Kubernetes and of SAP BTP, Kyma runtime, and its architectural components, let's now focus on use cases and scenarios where using SAP BTP, Kyma runtime, can be suitable and bring value to the table.

11.2 Use Cases

SAP BTP, Kyma runtime, is the recommended solution for business cases requiring you add business logic as extensions. It supports asynchronous extensions using event-driven architectures as well as synchronous extensions with APIs and microservices. Let's explore some potential use cases that you can address using Kyma runtime, such as the following:

- Out-of-the-box connectivity with SAP systems using an application connector and management plane
- Ability to consume services from hyperscalers as well as from SAP BTP via the service catalog
- Deployment of microservices and lambda functions to implement extensions for customizing business logic
- Achieving side-by-side extensibility with SAP BTP, Kyma runtime, either through business events or cloud-native extensions
- Building truly cloud-native extensions where operational and security concerns such as monitoring, tracing, alerting, auto-scaling, and secure interservice communications are already resolved out of the box

Now, imagine you have an SAP business solution such as SAP Commerce Cloud. You can connect this solution with SAP BTP, Kyma runtime, through a simple registration mechanism that creates a secure connection where both asynchronous business events can be subscribed to using microservices or functions running in SAP BTP, Kyma runtime. Alternatively, all exposed commerce APIs can be called from these services or functions. This flexibility has the benefit of offloading non-core commerce processing to the Kyma cluster, which improves performance and "keeps the core clean" so that no custom extensions impede updates of the core commerce system. For example, let's say we have a special pricing function that looks up customer-specific prices and product availability in real time to display on the product detail page. This feature can be a complex and resource-intensive process to create. SAP BTP, Kyma runtime, allows you to build out a service that can take care of this requirement efficiently, using scalable microservices or even functions maintained as source code (in JavaScript or Python). In this way, you can leverage backing services from SAP BTP or from hyperscalers, including high-performance datastores like SAP HANA or Redis to deliver complex results quickly.

In the next section, we'll extend some SAP Commerce Cloud functionality using SAP BTP, Kyma runtime, and gain some hands-on experience

11.3 Building with Kyma

To demonstrate the basic steps of setting up SAP BTP, Kyma runtime, and extending an application, we'll take advantage of the mock application, which is basically a simulated version of SAP Commerce Cloud. In this context, note that SAP Commerce Cloud is a cloud-native, omnichannel commerce solution for business-to-business (B2B), business-to-consumer (B2C), and even business-to-business-to-consumer (B2B2C) companies delivering personalized experiences from acquisition through to retention. The solution unifies all touchpoints, whether online, mobile, point of sale (POS), call center, social, or print, on a global scale.

11.3.1 Prerequisite Configuration

Before you start, you must set up the required prerequisites, including service instance creation, obtaining the code setup from the Git repository, and more. The source code we are leveraging to build this application extension can be downloaded from *http://s-prs.co/v550404*. Save this project to your local system for further use in the following steps. Once you have code, the next step is to enable the Kyma runtime service from your SAP BTP, Cloud Foundry environment, subaccount as shown in Figure 11.2.

Also, to demonstrate these steps, we are using an SAP BTP, Cloud Foundry environment, trial account, which you can easily obtain from *http://s-prs.co/v550405*.

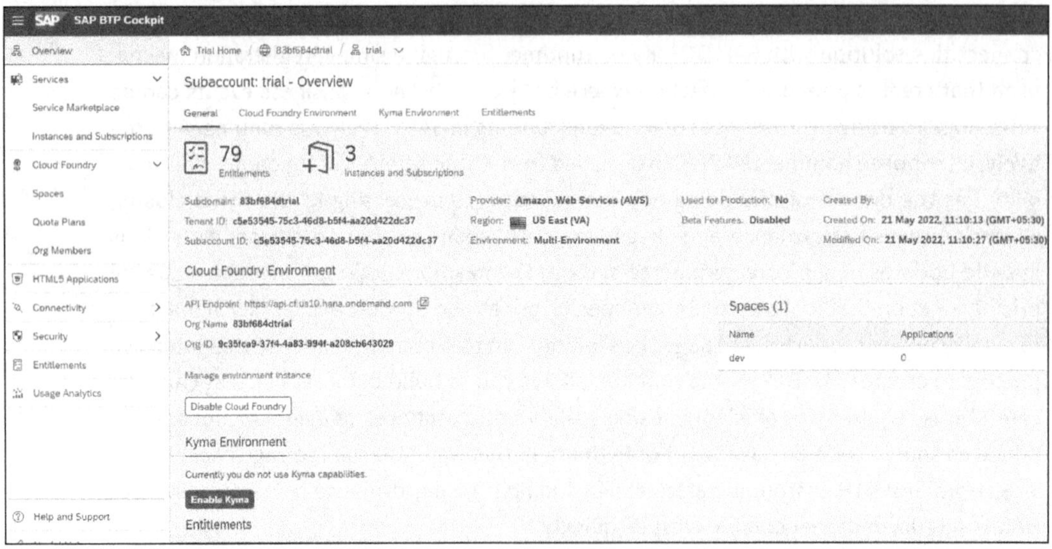

Figure 11.2: Enabling Kyma on an SAP BTP Subaccount

To enable SAP BTP, Kyma runtime, first navigate to your SAP BTP, Cloud Foundry environment, subaccount and click **Enable Kyma** inside the **Kyma Environment** section, as shown in Figure 11.3.

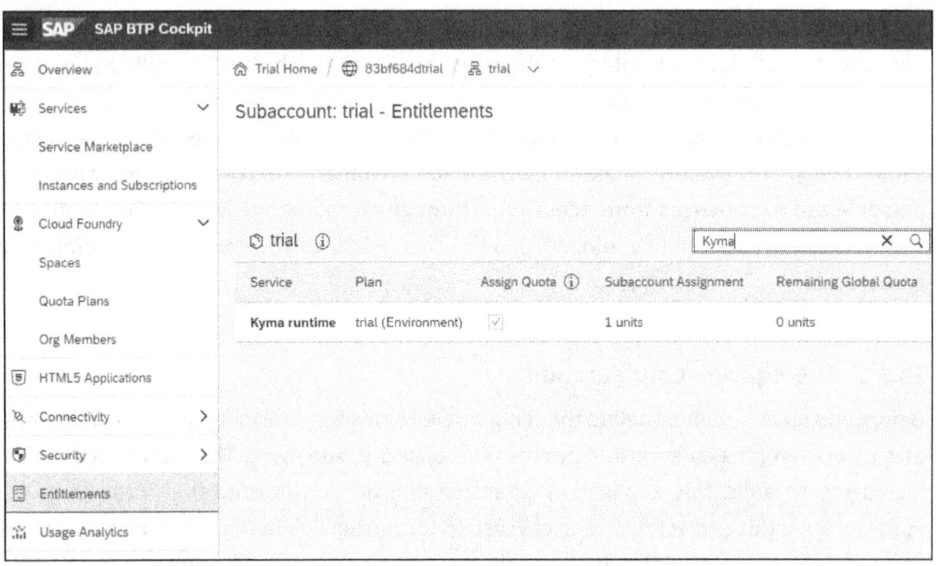

Figure 11.3: Entitlements

If you already have SAP BTP, Kyma runtime, enabled on your SAP BTP, Cloud Foundry environment, account, you'll see it listed in the **Entitlements** section, as shown in Figure 11.4.

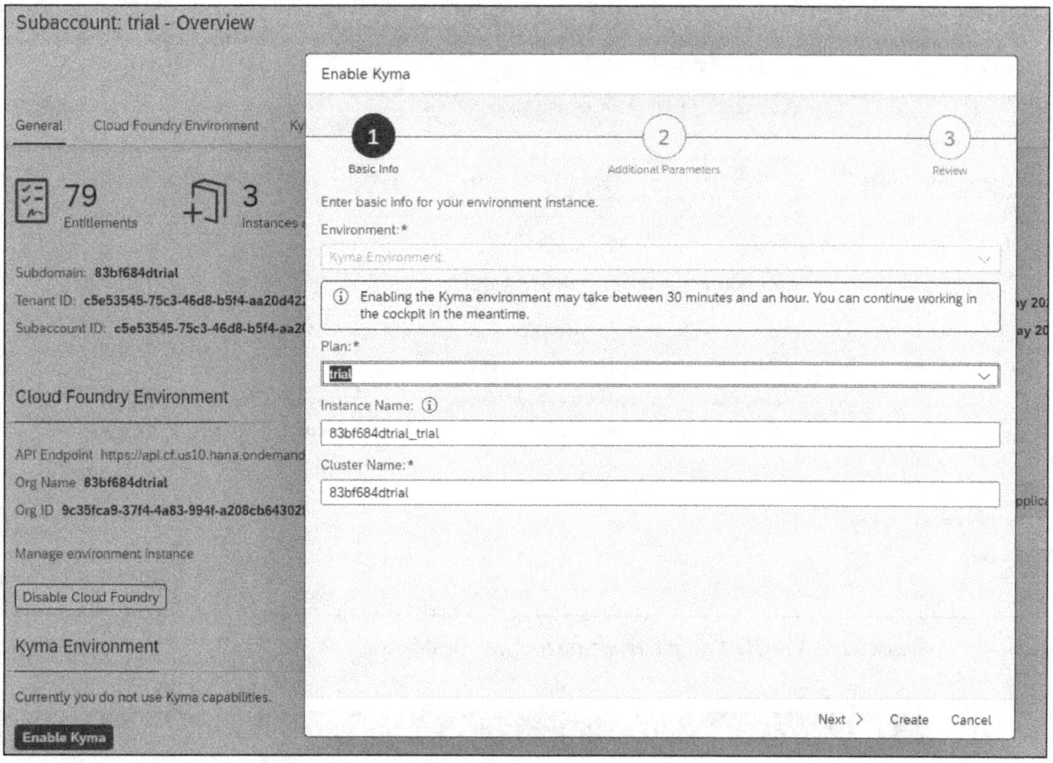

Figure 11.4: Enabling SAP BTP, Kyma Runtime

Clicking the **Enable Kyma** inside the **Kyma Environment** section, shown in Figure 11.3, launches the **Enable Kyma** popup window, where you'll need to enter basic info in the **Plan**, **Instance Name**, and **Cluster Name** fields, as shown in Figure 11.4, and then click **Create**.

Once the creation process is finished, SAP BTP, Kyma runtime, will be listed under the **Environment** tab, along with your SAP BTP, Cloud Foundry environment, instance.

11.3.2 Configuring SAP BTP, Kyma Runtime, and Service Deployment

Once you've completed the initial setup for instance creation, you can now navigate to the Kyma dashboard by choosing the option **Go to Dashboard**, as shown in Figure 11.5. The Kyma dashboard provides instant access to work with multiple clusters; you can switch between them, add clusters, and get a better view of the apps running and the resources involved.

In the Kyma dashboard, click the **Namespaces** menu item to show a list of existing namespaces. To create a new namespace, click **Create Namespace +**, as shown in Figure 11.6.

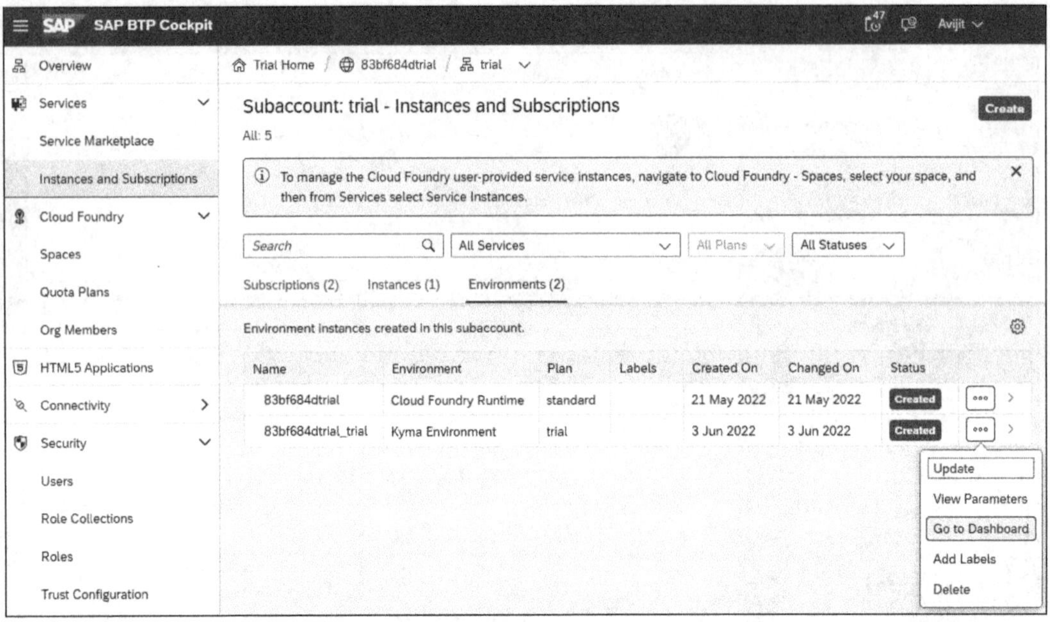

Figure 11.5: SAP BTP Cockpit: Navigate to Kyma Dashboard

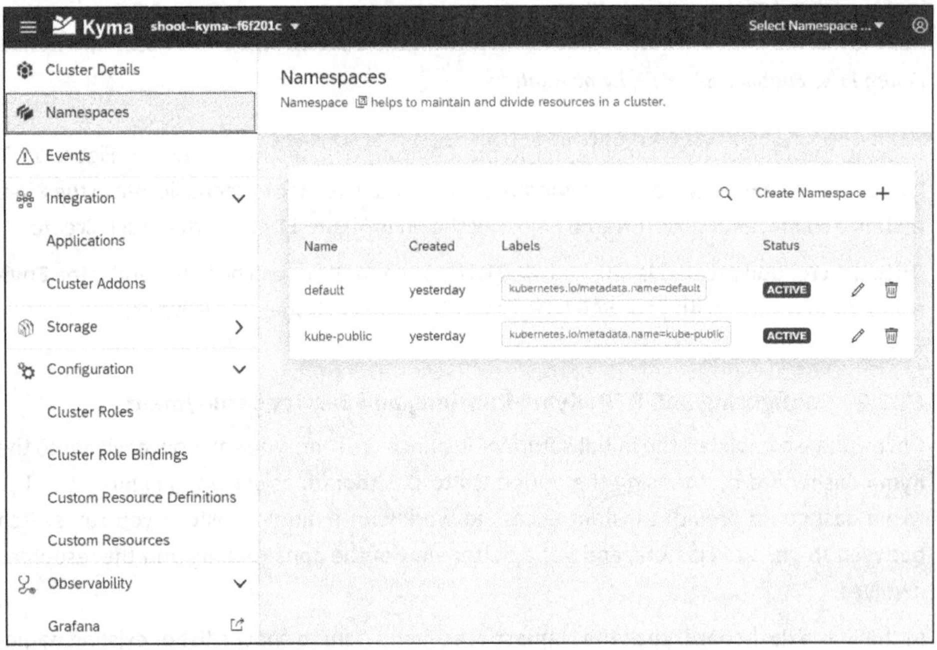

Figure 11.6: SAP BTP, Kyma Runtime: Namespaces

Inside the **Create Namespace** popup window, shown in Figure 11.7, enter a name for your namespace. The key purpose of namespace creation is to organize objects in a cluster and provide a way to share clustered resources among several users. By having access to resources within a namespace, these users have permission, which also ensures security.

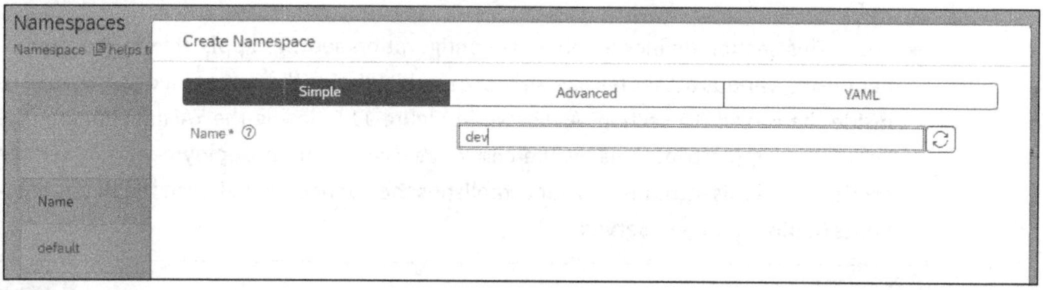

Figure 11.7: Creating a Namespace

Since we are utilizing the ready-to-use SAP Commerce Cloud mock application, immediately after creating the namespace, we can further navigate inside the namespace by choosing the name. Start the process of creating a deployment by clicking **Deploy new workload +** and choosing the **Create Deployment** option, as shown in Figure 11.8.

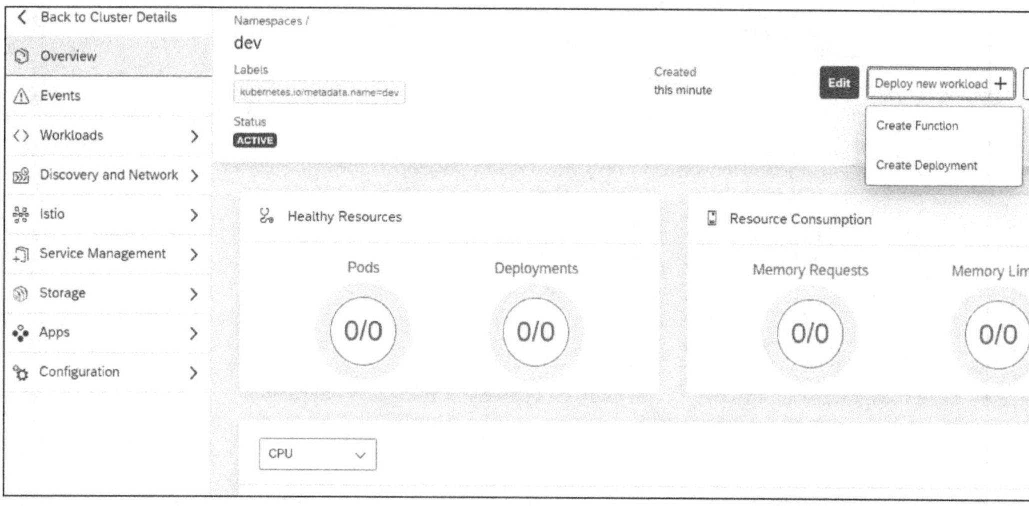

Figure 11.8: Creating a Deployment

Clicking the **Create Deployment** option launches **Create Deployment** window where you should choose the **YAML Ain't Markup Language (YAML)** option. Before we prepare the required YAML file, let's first try to understand the underlying YAML file specification, which primarily includes four sections:

- `apiVersion`: This section holds the version information (in our case, `apps/v1`).
- `kind`: This section mentions the applicable Kubernetes components (deployment, service, volume, etc.). In our case, we have Deployment chosen since we are defining a deployment YAML.

- `metadata`: This section holds the metadata specification for components. The connection between the components of a Kubernetes app is established by `labels` and `selectors`.
- `spec`: This section defines all possible configuration settings applicable to the component using various attributes like `selector`, which holds the same value as the `labels` inside the `metadata` section. As shown in Figure 11.9, inside the YAML definition, the `selectors` app attribute has value `commerce-mock` in both **Deployment** and **Service** sections, and this attribute is what establishes the connection between the two components **Deployment** and **Service**.

Figure 11.9: Deployment: YAML Definition

Copy the contents of the *k8s-deployment.yaml* and *k8s-service.yaml* files from the local folder where you have downloaded the *xf-application-mocks* application from GitHub and paste this code into their respective panes in the **Create Deployment** window, as shown in Figure 11.9. Then, click the **Create** button to create the specified service.

Now that we've created the service, the next step is to create a `PersistentVolumeClaim` (PVC) for the deployed service. But what is a PVC? According to Kubernetes' official definition, a `PersistentVolume` (PV) is a piece of storage in the cluster that has been provisioned by an administrator or dynamically provisioned using storage classes. This is resource in the cluster just like a node is a cluster resource, and a PVC is a request for specific storage by a user, which is similar to a pod, but pods consume node resources while PVCs consume PV resources. Pods can request specific levels of resources (CPU and memory). PVCs can request specific sizes and access modes (e.g., they can be mounted `ReadWriteOnce`, `ReadOnlyMany`, `ReadWriteMany`, etc.). As shown in Figure 11.10, in our case, we are using the access mode `ReadWriteOnce`, which means the volume can be mounted as read-write by a single node. The `ReadWriteOnce` access mode allows multiple pods to access the volume when the pods are running on the same node.

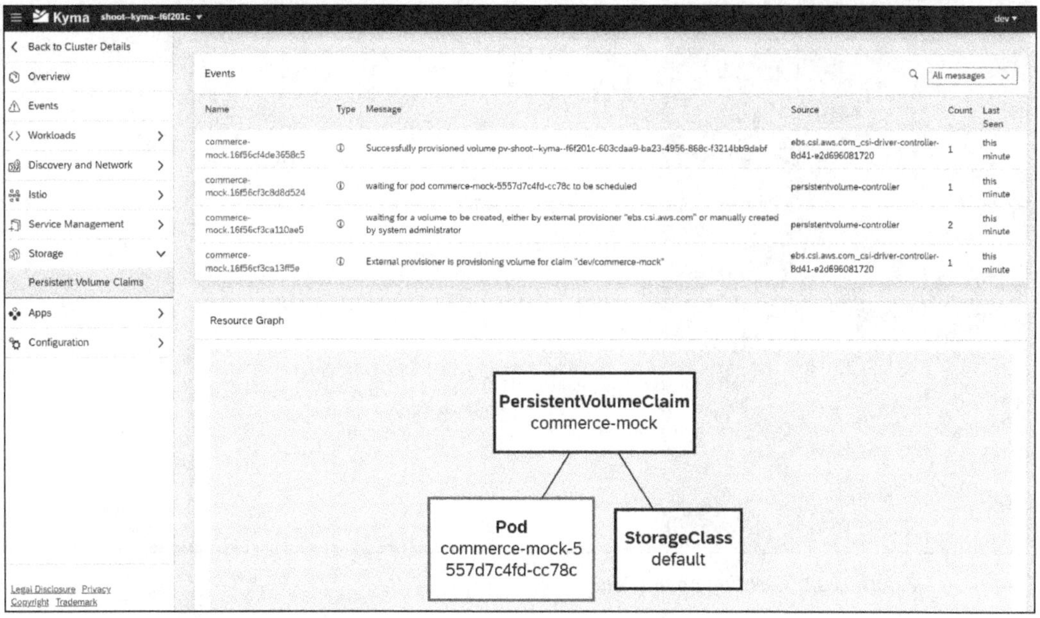

Figure 11.10: Persistent Volume Claim

Copy the contents of the file */xf-application-mocks/commerce-mock/deployment/k8s-pvc.yaml* to the **Create Persistent Volume Claim** section on the screen shown in Figure 11.10 and then click **Create**.

In this next step, we'll now create an API rule, but first, let's try to understand the key purpose of API rules. API rules in Kyma are a component responsible for exposing services and supports several security strategies, for instance, JSON Web Tokens (JWTs) and OAuth2 to secure exposed services using tokens issued by Kyma Dex for JWTs or issued by the Kyma OAuth2 server for OAuth2 tokens. Based on this feature, you can secure an entire service or secure selected endpoints. Alternatively, you can even leave the service unsecured if you so choose. Follow the menu path **Discovery and Network • API Rules • Create API Rule** to initiate the API rule creation screen. Inside the **Create API Rule** window, enter "commerce-mock" in the **Name** field, choose **commerce-mock** in the **Service** dropdown list, and enter "commerce" in the **Subdomain** field. Finally, don't forget to enable each of the desired operations in the **Methods** section and then click **Create**, as shown in Figure 11.11.

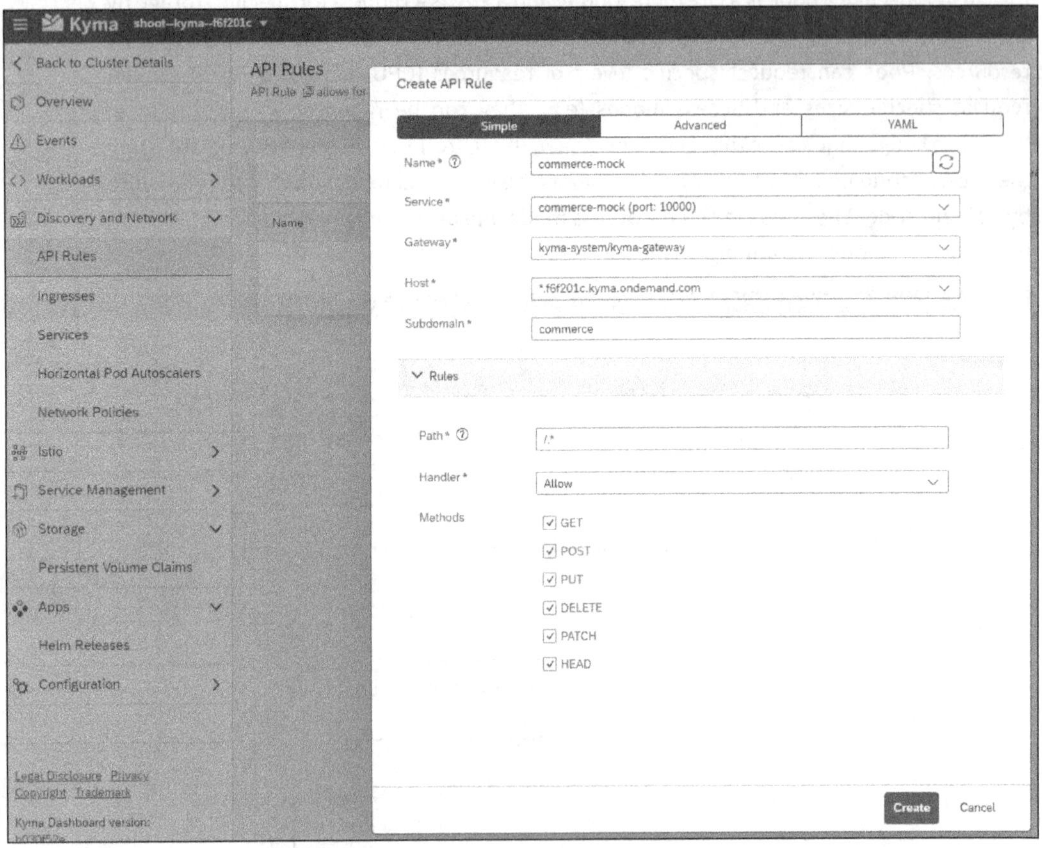

Figure 11.11: Creating an API Rule

Once created, the new API rule appears in the table shown in Figure 11.12, with the **Host** column showing the URL and **Status** columns describing the status. The URL will further be used to launch the mock application.

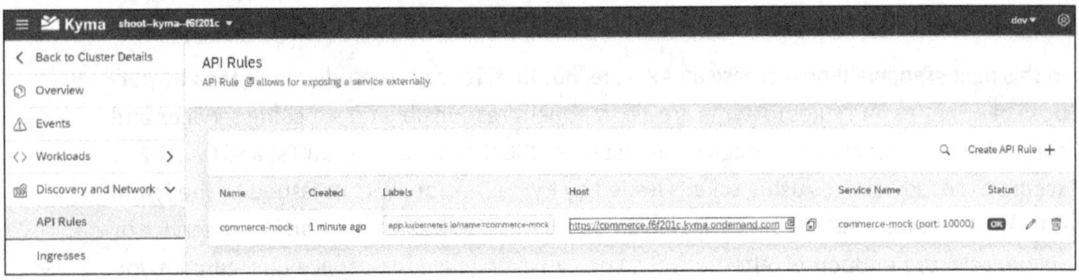

Figure 11.12: API Rules

Click the URL shown in the form of *https://commerce.*******.kyma.ondemand.com* and open the SAP Commerce Cloud mock application, as shown in Figure 11.13.

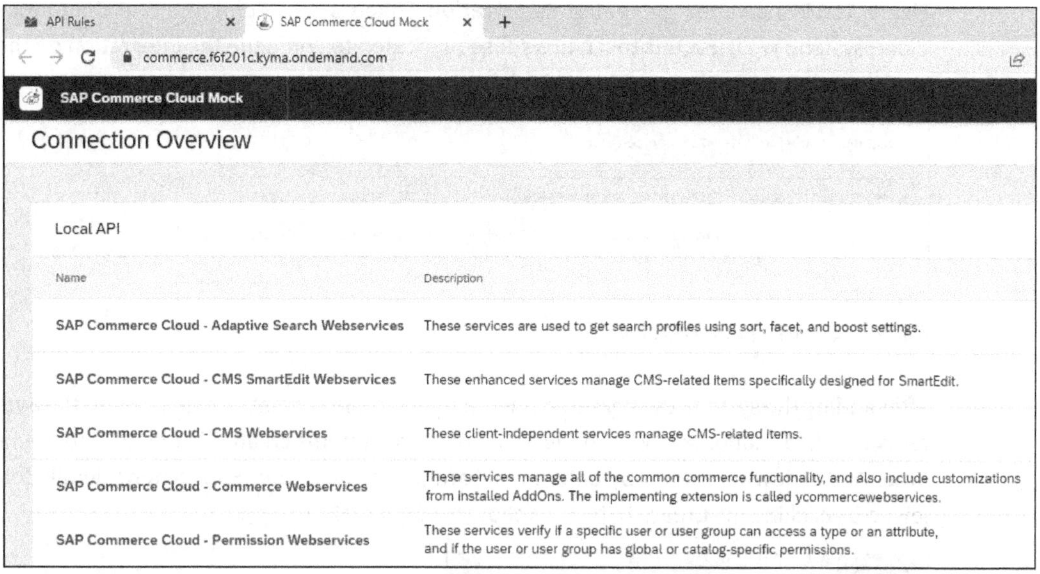

Figure 11.13: Commerce Mock APIs

Now, we need to create a system by navigating to the **System Landscape** section of the SAP BTP cockpit. Then, click **Add System**, and select SAP Commerce Cloud from the **Type** drop-down list, as shown in Figure 11.14. When adding a new system, make sure that you're a global account administrator or that you're a system landscape administrator of the global account where you want to register your SAP system.

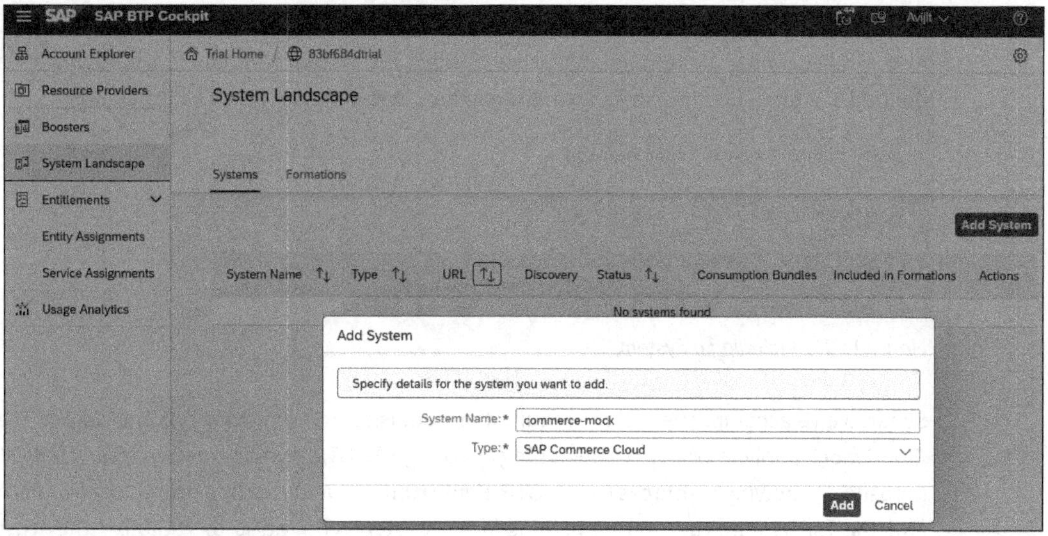

Figure 11.14: Adding a System

The process of adding a new system is based on an integration token that is used for pairing the system with the corresponding global account.

Figure 11.15 shows the generated **Registration Token** after adding the SAP Commerce Cloud system. We'll use this token in a subsequent step to configure the integration on the corresponding SAP solution side, so don't lose it.

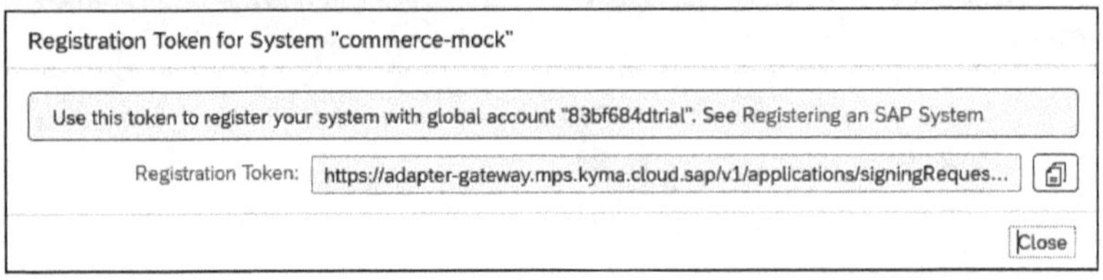

Figure 11.15: Generating a Token

After adding the system, the next step is to navigate to the **Formations** tab, next to the **Systems** tab. A formation lets you connect one or more systems created in SAP BTP to a runtime. Add a new formation by entering a **Formation Name** and **Subaccount** details followed by clicking the **Create** button, as shown in Figure 11.16.

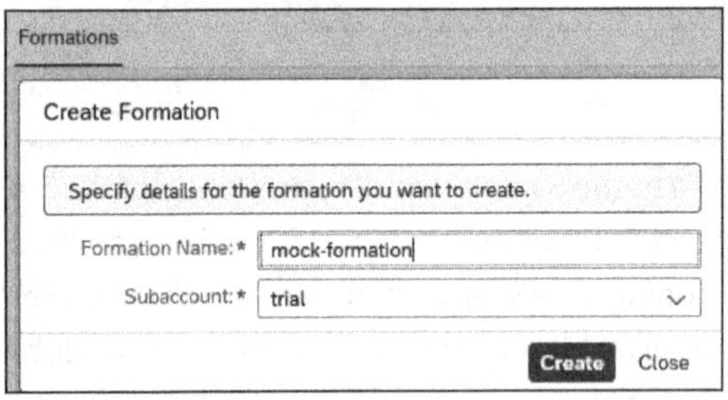

Figure 11.16: Creating a Formation

Also, while defining the formation, you must select the system **commerce-mock**, which we defined in a previous step, as an included system, as shown in Figure 11.17.

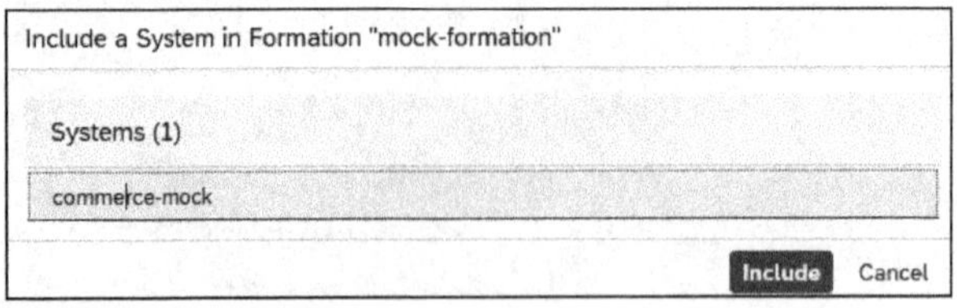

Figure 11.17: Including a System

So far, we've accomplished all the required prerequisites for deploying our SAP Commerce Cloud mock application service, including enabling SAP BTP, Kyma runtime; cloning the Git repository; applying resources to SAP BTP, Kyma runtime; and creating the required system and the formation. The only remaining process step required is to establish the trust between the SAP Commerce Cloud mock application and SAP BTP, Kyma runtime.

11.3.3 Creating Pairing and Application Programming Interface Rules

Through the pairing process, we are basically connecting and registering all the commerce mock application APIs that we exposed when defining the API rule in the previous step on SAP BTP, Kyma runtime. In this process, we'll leverage the token (**Registration Token**) that we obtained in a previous step. To perform the pairing process, navigate to the host URL generated earlier in your preferred browser and then click **Connect**, as shown in Figure 11.18.

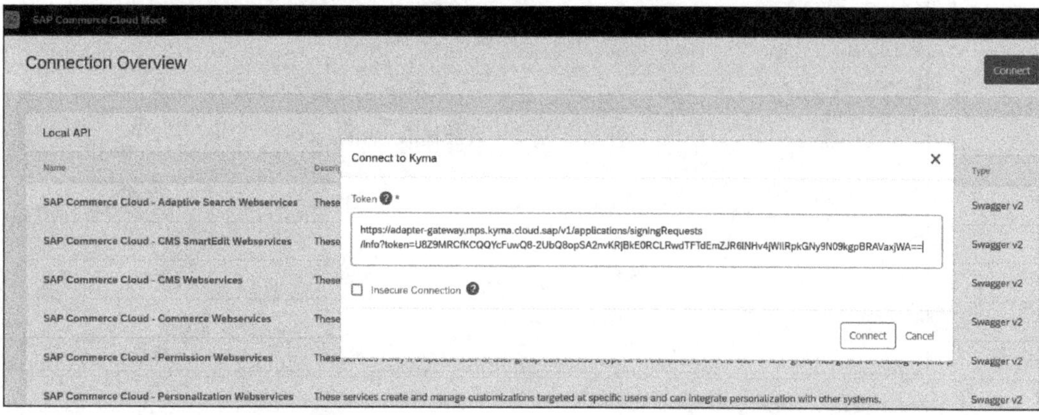

Figure 11.18: Connecting to Kyma

Clicking this button launches the **Connect to Kyma** popup window, where you'll need to enter the registration token in the **Token** field and then click **Connect**. Next, choose **Register All** to register all the APIs and events from the mock application, as shown in Figure 11.19.

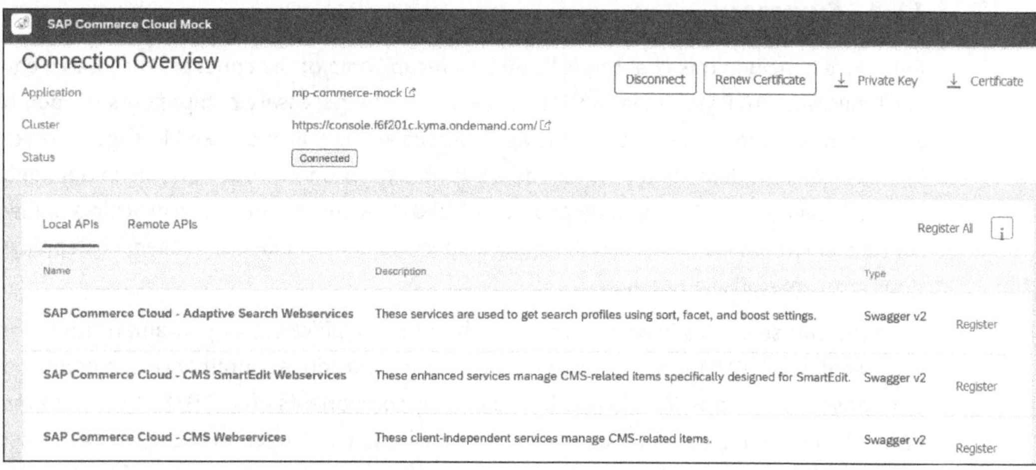

Figure 11.19: Kyma Runtime: Registering APIs

After completing the pairing process, you must now verify if the APIs have been successfully deployed. For this task, navigate back to SAP BTP, Kyma runtime, workspace, as shown in Figure 11.20.

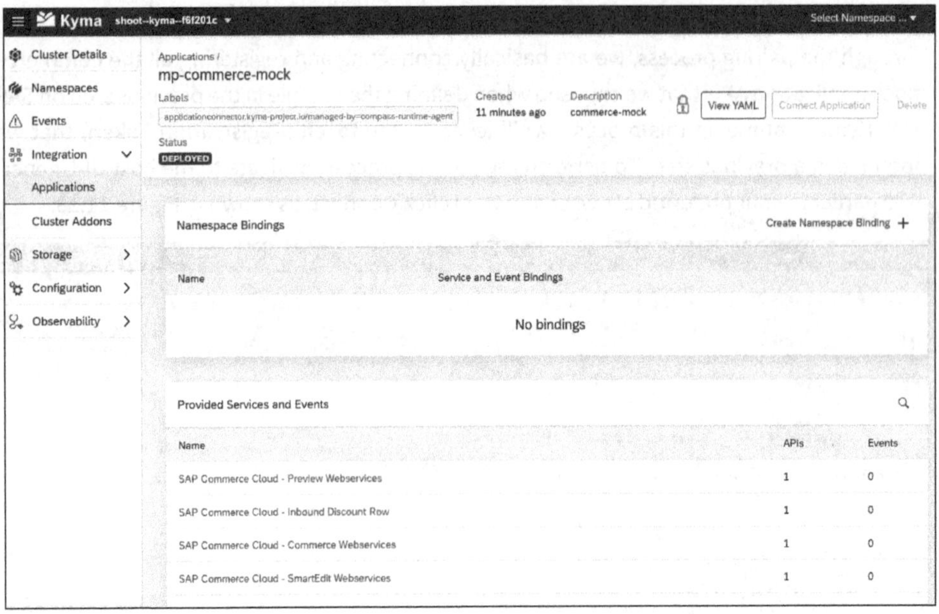

Figure 11.20: Deployed Application

Finally, navigate to **Integration • Applications** to show the deployed APIs and their statuses (ideally, **Deployed**). This status indicates that you've successfully completed the process of deploying the SAP Commerce Cloud mock application on SAP BTP, Kyma runtime.

11.4 Summary

Before we conclude this chapter, let's quickly recap some of the concepts that we've covered thus far. SAP BTP, Kyma runtime, provides all the necessary components needed to easily connect and extend monolithic applications and can also be used to trigger serverless functions or microservices to extend a business function without worrying about application dependencies. The primary goal with SAP BTP, Kyma runtime, is to provide a unified runtime environment with everything developers need to manage microservices applications in an enterprise environment.

Over the course of this chapter, we highlighted the significance of containerization and illustrated how SAP BTP, Kyma runtime, can help you introduce containers in the context of enterprise applications. We also explained the key components of SAP BTP, Kyma runtime, and explored some potential use cases and business challenges that can be addressed using SAP BTP, Kyma runtime. Finally, we covered the step-by-step process of exposing an SAP Commerce Cloud-based mock API using SAP BTP, Kyma runtime. In the next chapter, you'll get more familiar with developing SAP Fiori applications, including a brief overview, architectural review, and finally a step-by-step look at developing and extending different types of SAP Fiori applications.

Chapter 12

Developing SAP Fiori Applications

In this chapter, we'll provide an overview and explore the architecture of SAP Fiori applications. You'll learn all about developing and extending both freestyle applications and SAP Fiori elements applications using SAP Business Application Studio.

Now that you understand the various backend programming options using the SAP Cloud Application Programming Model (Node.js or Java) and the ABAP RESTful application programming model from previous chapters, we'll show you how SAP Fiori applications can be generated from annotations, without a single line of frontend coding. After all this technical backend focus, you must be eager to learn more about SAP Fiori development! So, let's explore SAP Fiori's design concepts, architecture, development, and extensions with the help of the various SAP Business Technology Platform (SAP BTP) components and services described in this chapter.

User experience (UX) concepts have evolved considerably over the last decade and continue to be a growing field in software design and development. With the new simplified look and feel of SAP Fiori apps, the UX in the SAP world has been redefined. Many businesses realized that SAP Fiori could help users engage more in their daily activities. The concept behind SAP Fiori also benefited developers. With the help of a set of new tools and templates, SAP Fiori has empowered developers to create and enhance apps for any platform much more quickly—all while providing a consistent, unified, and innovative experience to meet market demands.

SAP Fiori remains the main standard for user experience in SAP, so whenever we talk about modern SAP applications, we still rely on SAP Fiori design and UX principles. That part hasn't changed. What has changed significantly is how we build and deliver SAP Fiori apps.

Earlier, we mainly used tools like SAP Business Application Studio. Now, SAP has brought everything under a new umbrella called SAP Build, which is part of SAP BTP. This is more of a complete ecosystem rather than just a development tool. So, in simple terms, if you are a developer, you will mainly use SAP Build Code. SAP Build Code gives you everything in one place—coding, lifecycle management, and team collaboration—to build complex SAP Fiori applications.

For accessing applications, instead of the traditional SAP Fiori launchpad, SAP now recommends using SAP Build Work Zone. This acts as a central entry point where users can access all their apps, and it even adds collaboration features like sharing content or working together. On top of all this, SAP has introduced Joule, an AI solution. This is quite powerful because you can simply describe what you want, and it can help generate things like data models, services, or logic. So, it really speeds up development and reduces repetitive work.

In short, SAP has evolved from merely providing UI tools to offering a complete, AI-enabled platform for building and running applications.

In this chapter, we'll start by exploring SAP Fiori concepts and its architecture in general before we jump into SAP Fiori development using SAP BTP.

12.1 Overview and Architecture

SAP Fiori design principles are meant for developing enterprise-grade frontend applications with a better and richer UX. SAP Fiori simplifies business processes by making the applications role based, simple, tailored to users' tasks, and available in all leading technologies and mobile devices.

This core goal is reflected in its five design principles: *role based*, *adaptive*, *coherent*, *simple*, and *delightful*. As SAP Fiori is a set of design principles, you might wonder about its underlying technology. The SAP standard uses SAPUI5 as the core technology behind developing SAP Fiori applications. However, the SAP Fiori concept has been expanded over time, and multiple other technologies have been wrapped into SAP Fiori-specific themes. These wrapper tools are also called SAP Fiori applications and include, for example, SAP Fiori-themed SAP GUI applications, embedded analytics applications, and more. In this book, we'll only use SAPUI5 as the technology for our SAP Fiori developments to keep things simple. To run any SAP Fiori app, you must have the following components in your landscape:

- **Frontend server**
  SAP Fiori applications are primarily arranged in tiles, and these tiles can be configured in the frontend server, which is also responsible for activating the OData service for the consumption of SAP Fiori apps. The frontend server can be an ABAP application server with specific SAPUI5-specific software components installed, or a frontend server can be hosted in SAP BTP as a service. Sometimes, you might have both, for example, with SAP Build Work Zone, standard edition.
- **Backend system**
  The backend system is the business system that provides the relevant OData services.
- **SAP Web Dispatcher**
  SAP Web Dispatcher is the first entry point for end users accessing an SAP Fiori architecture. This solution handles all HTTP/HTTPS requests from end users coming from mobile devices or desktop devices, accepts or rejects these requests, acts like a load balancer to route requests to different application servers, and adds an additional layer of security.
- **SAP Fiori launchpad**
  The SAP Fiori launchpad is the entry point to all SAP Fiori apps on mobile and desktop devices. The launchpad starts with a homepage with multiple tiles. Each tile corresponds

to an application that the user is authorized to execute. Since SAP Fiori follows role-based design principles, the homepage displays only tiles that the user has the authorization to execute.

- **Cloud connectors**
  A cloud connector helps an SAP Business Application Studio developer connect to an on-premise system to explore and access its backend services.
- **SAP Destination service**
  This service works with a cloud connector for on-premise connectivity and also helps define various cloud-based and on-premise destinations for service consumption.

Overall, SAP's approach to development has changed. Earlier, SAP Business Application Studio was solely a development tool, but now SAP is moving toward providing a complete developer experience. Think of it this way: SAP Business Application Studio is your core development tool (IDE), and SAP Build Code is a comprehensive package built on top of it that makes development easier and faster.

If you use SAP Business Application Studio alone, you get a flexible environment for building SAP Fiori or SAP Cloud Application Programming Model applications. With SAP Build Code, SAP provides everything prearranged—AI, services, and lifecycle tools—so you don't have to spend time setting things up. A significant new addition is Joule. With Joule, you can describe what you want, and it helps generate code, creates unit tests automatically, and even explains or improves your existing code. Additionally, SAP Build Code improves productivity by offering many preconfigured services (reducing manual work) and providing a shared space (SAP Build lobby) where developers and business users can collaborate. It provides visual tools to design apps before you begin coding.

To explore the components involved in running SAP Fiori applications, let's examine the SAP Fiori tools available in SAP Business Application Studio.

12.2 SAP Fiori Tools in SAP Business Application Studio

SAP Fiori tools is a set of tools available as a part of SAP Business Application Studio that you can use to simplify your SAPUI5-based SAP Fiori developments. SAP Fiori tools are particularly helpful for using SAP Fiori elements applications, as you'll learn in this section. However, you can also use this tool for freestyle SAP Fiori development. SAP Fiori tools act as extensions to SAP Business Application Studio so you can create applications, visualize navigation, automatically generate code, and more. Used in combination with SAP Fiori elements, these extensions can increase development productivity, ensure UX consistency between applications, and help you build scalable experiences. SAP Fiori tools include the following components:

- **Application wizard**
  Enables an interactive approach, shown in Figure 12.1, to quickly generate basic SAP Fiori elements and SAPUI5 freestyle applications based on defined floorplans.

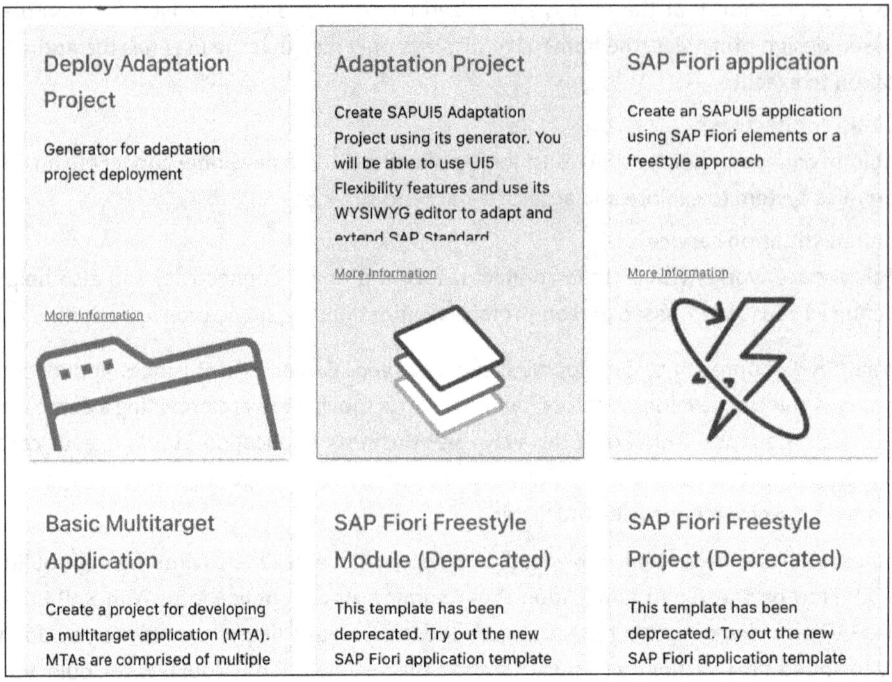

Figure 12.1: Application Wizard

- **Application modeler**
  This tool is used to visualize pages, navigation details, and service entities and can be helpful for configuring those properties based on your needs. The application modeler is shown in Figure 12.2.

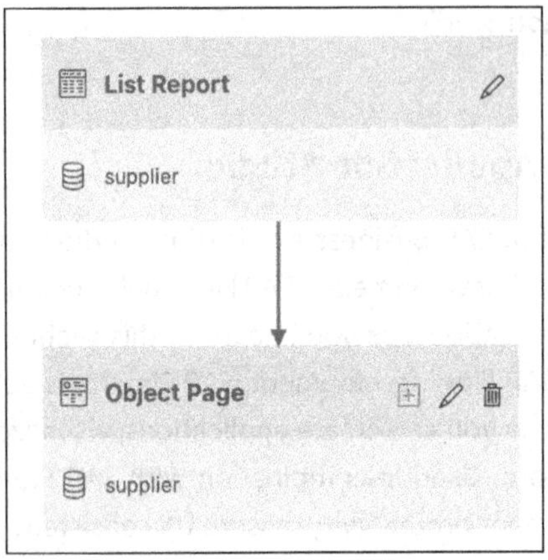

Figure 12.2: Application Modeler

- **Guided development**
  This is our favorite part, and you'll see why soon. As the name suggests, you can follow the steps required to implement a feature and then use the guided development

approach to make the required changes in your project without writing any code. Some guided development options are shown in Figure 12.3.

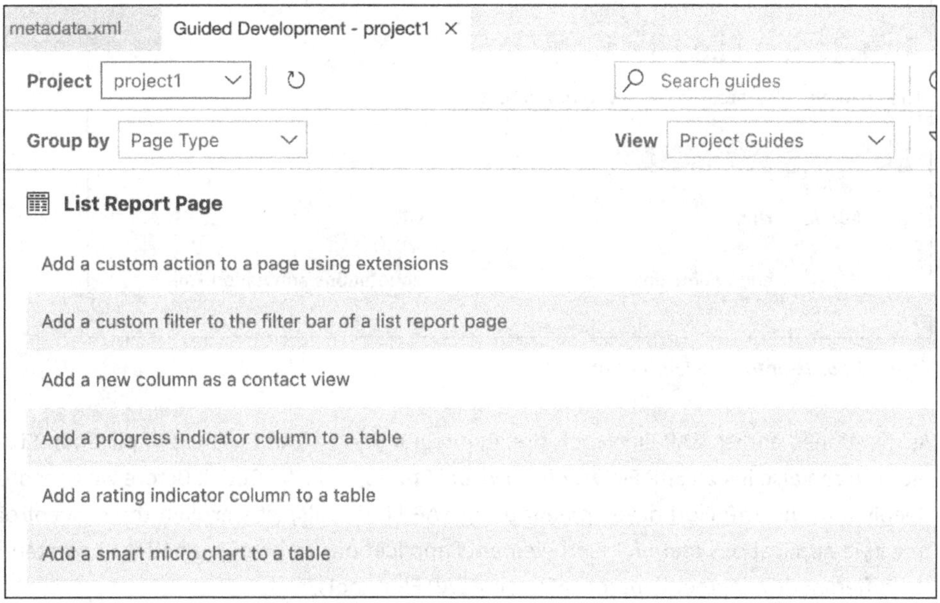

Figure 12.3: Guided Development

- **Service modeler**
  For the visualization of the OData service's metadata files, as shown in Figure 12.4.

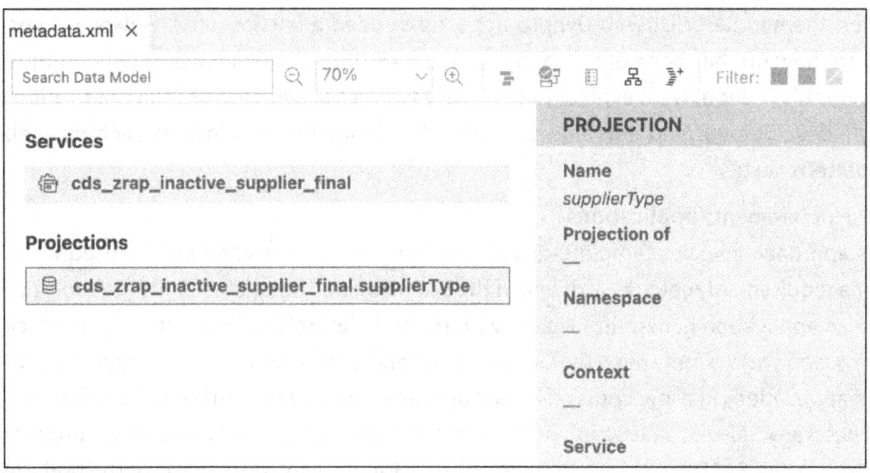

Figure 12.4: Service Metadata File

- **Annotation file manager**
  Access to resources that help to define annotations in the code editor, thus improving application development by reducing effort and maintaining code consistency. The annotation maintenance screen is shown in Figure 12.5.

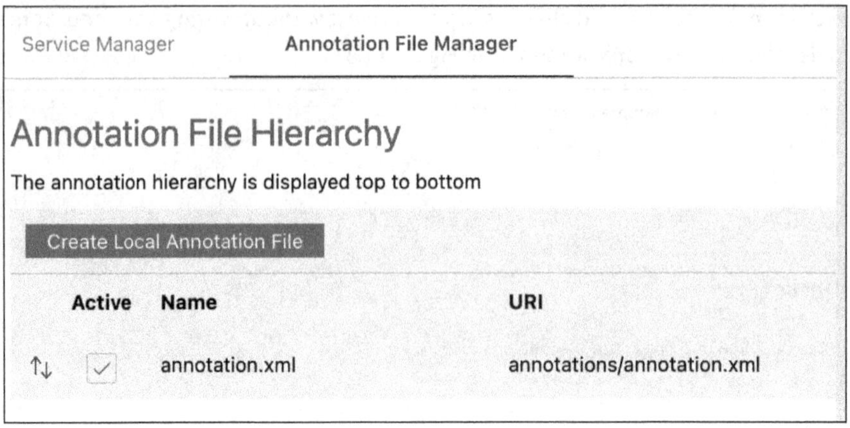

Figure 12.5: Maintaining Annotations

As mentioned earlier, SAP Fiori tools are an integral part of SAP Business Application Studio. You can also install SAP Fiori tools on Visual Studio Code (VS Code). Before we go more deeply into the SAP Fiori development using SAP Fiori tools, let's explain the concept of freestyle applications and SAP Fiori elements applications to help you better understand the different development approaches using SAP Fiori tools.

12.3 Freestyle SAP Fiori versus SAP Fiori Elements

As mentioned earlier, beyond SAPUI5, you can achieve an SAP Fiori UX in other ways (e.g., SAP Fiori-themed SAP GUI/Web Dynpro apps, embedded analytics, etc.). However, these topics are beyond the scope of this book. (Other SAP PRESS books are available on these topics.) In this section, we'll explain two primary ways of developing SAPUI5 applications: freestyle development and SAP Fiori elements development. Let's look at each development pattern next:

- **SAP Fiori elements applications**
  This approach involves template-driven SAP Fiori applications and can be used whenever a requirement matches with one of the SAP-provided templates. SAP Fiori tools provide an application generation wizard with multiple floorplans for commonly used scenarios, and you'll simply need the OData service and a set of annotations to generate SAP Fiori applications in a low-code/no-code approach. Yes, you read that correctly! You may not need any SAPUI5, JavaScript, or CSS coding at all when your environment is properly designed. The SAPUI5 runtime interprets the metadata and annotations of the underlying OData service and generates the UI and navigation for the SAP Fiori app at startup. Since the frontend is generated by the framework, it strictly follows SAP Fiori design guidelines to ensure consistency and high quality. This option does not need deep SAPUI5 knowledge.

- **Freestyle applications**
  The freestyle approach, as the name suggests, provides more freedom to developers that may or may not use templates. SAP Fiori tools include options for generating SAP Fiori apps using the SAPUI5 freestyle approach. You can select the type of template required, along with the relevant OData service endpoint and entities to create the data binding to start the initial application development process. Then, you would add your frontend coding in SAPUI5 according to your requirements. The initial template selection is optional, and you can select an empty template as well, which provides flexibility. This option requires more background knowledge and expertise in the SAPUI5 programming language.

Table 12.1 compares the two approaches. In general, our recommendation is to use SAP Fiori elements unless your screen layout must follow some strict requirements. Using SAP Fiori elements can significantly save you time and money. For additional flexibility, you can always enhance SAP Fiori elements with additional screen settings and logic, instead of writing a full-blown freestyle application from scratch.

Requirements	SAP Fiori Elements Approach	Freestyle Approach
Requirements demand more customization and flexibility in the UI.		X
Requirements match one or more existing SAP Fiori elements templates.	X	
Requirements need to be addressed quickly, and cost is the primary driver.	X	X
You must develop a native iOS or Android app.		X
The app will run mainly as a web app on desktops, tablets, and mobile devices.	X	
Offline capabilities are needed.		X
The development team has limited SAPUI5 experience.	X	

Table 12.1: Compare SAP Fiori Elements Versus Freestyle

Now that you understand the differences between freestyle applications and SAP Fiori elements applications, let's cover the steps for building SAP Fiori applications in each development approach. We'll start with freestyle development first and then explain SAP Fiori elements, so that you can understand the true advantages of using SAP Fiori elements.

12.4 Developing Complex Freestyle SAP Fiori Applications

The SAPUI5 freestyle approach was the development approach initially provided by SAP. This approach gives you complete flexibility regarding an application's look and feel. Even if you have complete flexibility, however, you can also select among some predefined templates to generate an initial SAP Fiori app with some default code and then add additional code to meet your specific requirements. You can select the type of template required based on your needs, along with the relevant OData service endpoint, to start the application development process.

With freestyle SAPUI5 application development, the following floorplans are available to jump start your development:

- **SAP Fiori worklist applications**
  As the name suggests, a worklist application displays a collection of items available to users to take some action. Users can review the details and can take action to complete the task.
- **SAP Fiori list-detail applications**
  The list-detail template implements a flexible column layout, which offers different layouts with up to three columns. In the template, two columns can be used, which are list and detail. The split-screen layout is optimized for displaying and processing a list of items. On the left side of the screen, users can quickly search and navigate through the list. On the right, they can see the details about the selected item and trigger related actions or edit the data. Select this template if users need to review and process a collection of data quickly with minimal navigation.
- **SAPUI5 applications**
  The basic template is intended for scenarios that do not fit into the previous two scenarios, and thus, developers need to create an SAPUI5 app from scratch. With this basic template, a blank canvas is available to start coding right away. The basic file structure is set up according to best practices.

It used to be possible to treat SAP Fiori and SAPUI5 as the same, but that's no longer the case. Now, SAP Fiori focuses on design and user experience, while SAPUI5 is just one way to build it. A major change is the introduction of UI5 web components. These lightweight UI elements function like simple HTML tags but already adhere to the SAP Fiori design guidelines. The advantage is that they don't require the full SAPUI5 framework.

As a result, developers are no longer limited to SAPUI5. They can use modern frameworks such as React, Angular, or Vue and still build applications that look and function like SAP Fiori apps. Essentially, SAP Fiori is no longer limited to a single technology or programming model. It has become a flexible design standard applicable across various development approaches.

Since SAP Fiori list-detail applications are the most common use of the freestyle approach, in the next section, we'll walk you through the steps for building a list-detail application based on the Northwind OData service.

12.4.1 Developing List-Detail Applications

As mentioned, the SAP Fiori list-detail application template implements a typical split-screen layout, one of the design patterns specified by the SAP Fiori design guidelines. The split-screen layout is optimized for displaying and processing a list of items. For our example, we'll use the Northwind OData service to build an application to track sales orders.

Assuming that your SAP Business Application Studio is up and running and you're already inside the SAP Fiori *dev space*, you can create an SAP Fiori freestyle project from SAP Business Application Studio by following the menu path **File • New Project from Template**. Then, select the **SAP Fiori application** tile, as shown in Figure 12.6, and click **Start.**

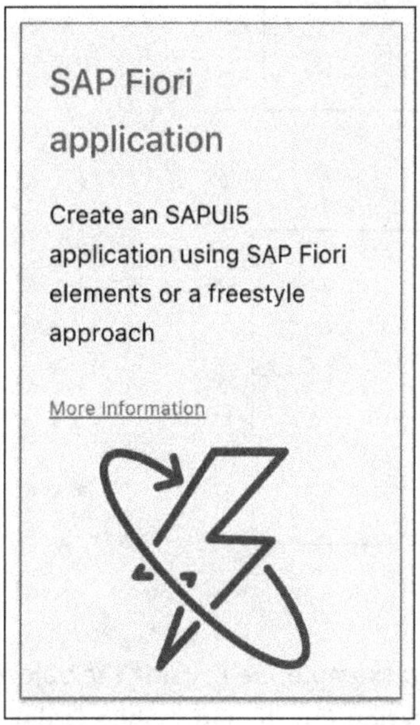

Figure 12.6: Selecting the SAP Fiori Application Option in SAP Business Application Studio

In the next step, select the **SAPUI5 freestyle** option from the **Application Type** dropdown list and then click on **SAP Fiori List-Detail Application**, as shown in Figure 12.7. Then, click **Next.**

In the **Data source** dropdown list, choose **Connect to an OData Service** and enter the Northwind OData service URL *https://services.odata.org/v2/northwind/northwind.svc*, as shown in Figure 12.8.

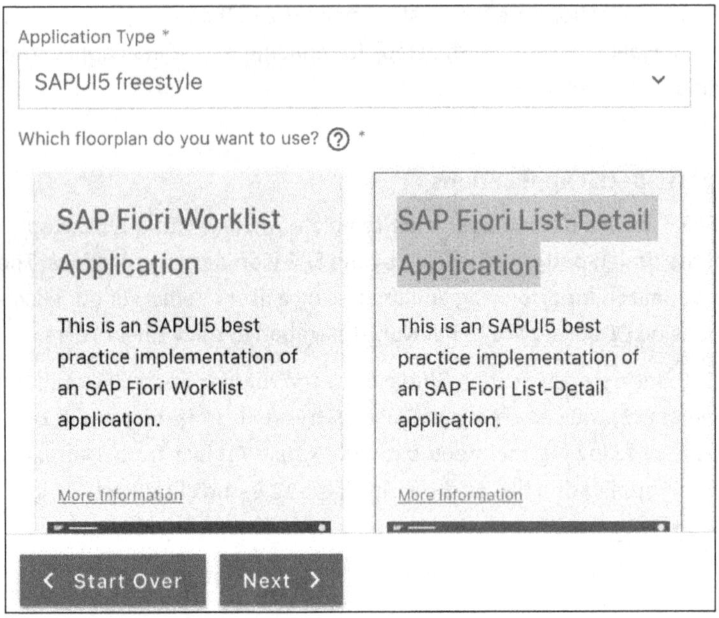

Figure 12.7: SAPUI5 Application and List-Detail

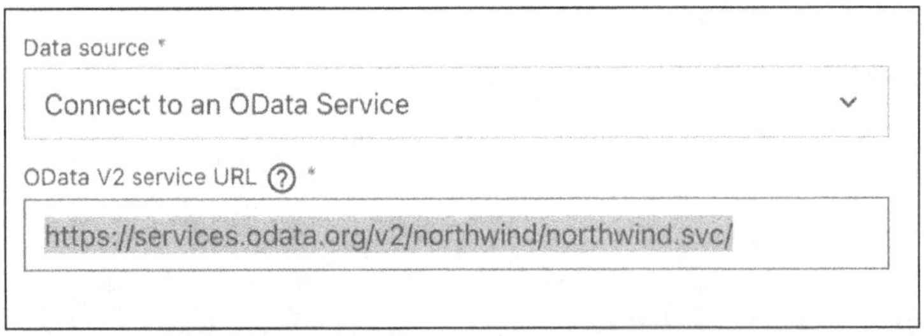

Figure 12.8: Selecting a Data Source

Maintain the entity's details, as shown in Figure 12.9. In our example, we're using the object collection option **Orders**, which will appear in the list section (on the left), while **Order_Details** is the line-item collection that will appear in the detail section (on the right) of the SAP Fiori screen.

After entering a module name, application title, application description, and optionally adding a deployment descriptor, as shown in Figure 12.10, you can finish this project. If you select the **Yes** radio button when asked to add the deployment configuration to a multitarget application (MTA), you'll need to select a deployment target, as shown in Figure 12.11. Select **Cloud Foundry** as the deployment target and select **None** for the destination since we're not using any backend system as the destination in this case (remember, we're using the Northwind OData service URL without creating a destination). We'll describe the deployment process and MTAs in more detail in Section 12.6.

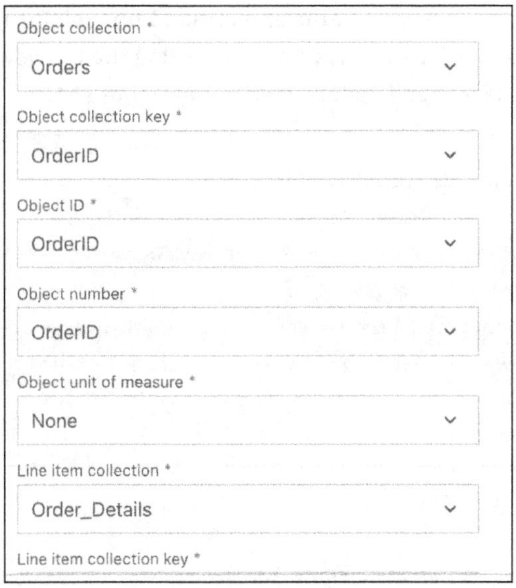

Figure 12.9: Entity Details

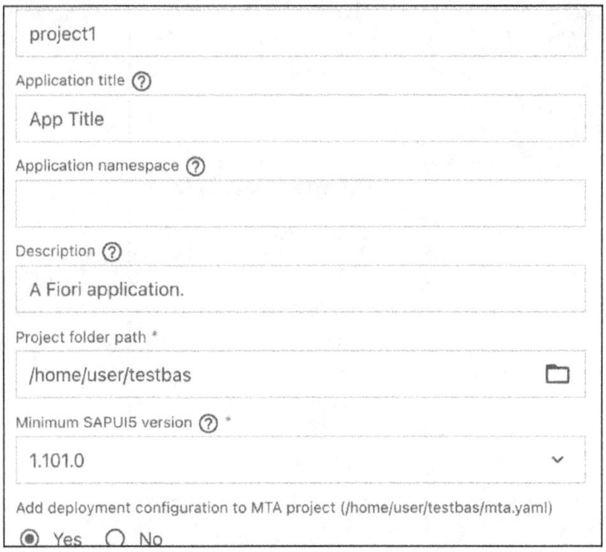

Figure 12.10: Other Details, Including MTA, for Creating the Project

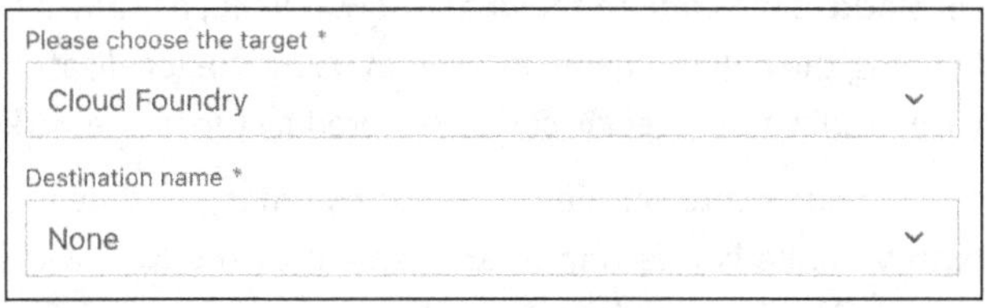

Figure 12.11: Cloud Foundry as the Deployment Target

The default application is generated, and you can preview the application from SAP Business Application Studio via the context menu of the application by selecting the **Preview** option and selecting the sandbox SAP Fiori launchpad option, as shown in Figure 12.12.

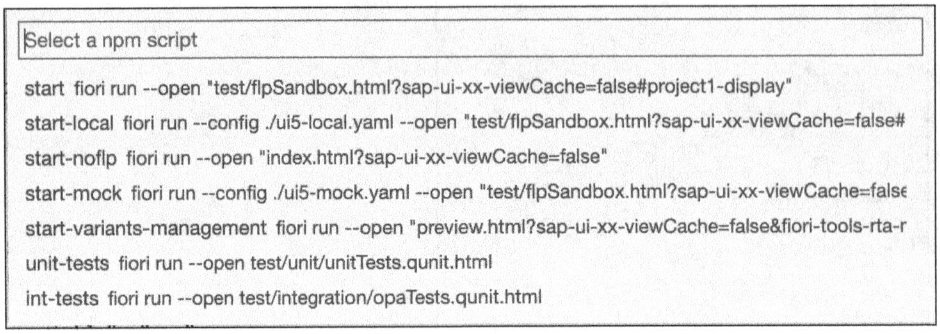

Figure 12.12: Selecting the Sandbox SAP Fiori Launchpad

The generated application will be displayed, as shown in Figure 12.13.

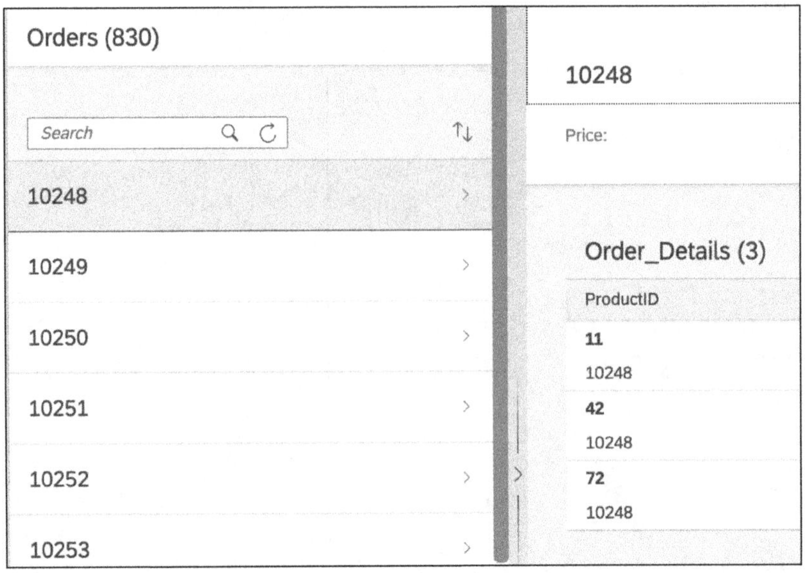

Figure 12.13: Previewing an Application

You can now enhance this application in different ways. We'll add a few columns to the detail page of the split screen. Let's add some code to the *detail.view.xml* file, as shown in Figure 12.14. We're simply hard coding the column name instead of using the localization file *i 18n*. However, you can also use localization-specific code as offered by the framework.

Now, let's add the code to populate values into our newly added columns, as shown in Figure 12.15. In our case, the path variable is referring to an element of the Northwind OData service, which you can easily explore in a web browser.

```
            </Toolbar>
        </headerToolbar>
        <columns>
            <Column>
                <Text text="{i18n>detailLineItemTableIDColumn}"/>
            </Column>
            <Column
                minScreenWidth="Tablet"
                demandPopin="true"
                hAlign="End">
                <Text text="{i18n>detailLineItemTableUnitNumberColumn}"/
            </Column>
            <Column>
                <Text text="Price"/>
            </Column>
            <Column>
                <Text text="Quantity"/>
            </Column>
            <Column>
                <Text text="Discount"/>
            </Column>
        </columns>
```

Figure 12.14: Adding Columns to the Detailed View

```
        <items>
            <ColumnListItem>
                <cells>
                    <ObjectIdentifier
                        title="{ProductID}"
                        text="{OrderID}"/>
                    <ObjectNumber id ="id1"
                    number="{
                        path: 'UnitPrice',
                        formatter: '.formatter.currencyValue'
                    }" unit="USD" />
                    <ObjectNumber id ="id2"
                    number="{
                        path: 'Quantity',
                        formatter: '.formatter.number'
                    }" />
                    <ObjectNumber id ="id3"
                    number="{
                        path: 'Discount',
                        formatter: '.formatter.number'
                    }" />
                </cells>
            </ColumnListItem>
        </items>
```

Figure 12.15: Populating Column Values in the Detail View of the List-Detail App

Your *detail.view.xml* content should resemble the code shown in Listing 12.1 with added column names and column content mapped from the OData service. Additionally, the formatter helps you format content in different ways (e.g., `.formatter.number` is displaying content in bold, etc.).

```
<mvc:View
    controllerName="project1.controller.Detail"
    xmlns="sap.m"
    xmlns:semantic="sap.f.semantic"
    xmlns:mvc="sap.ui.core.mvc">

    <semantic:SemanticPage
        id="detailPage"
        busy="{detailView>/busy}"
        busyIndicatorDelay="{detailView>/delay}">
        <semantic:titleHeading>
            <Title
                text="{OrderID}"
                level="H2"/>
        </semantic:titleHeading>
        <semantic:headerContent>
            <ObjectAttribute title="{i18n>priceTitle}"/>
            <ObjectNumber />
        </semantic:headerContent>
        <semantic:content>

            <Table
                id="lineItemsList"
                width="auto"
                items="{Order_Details}"
                updateFinished=".onListUpdateFinished"
                noDataText="{i18n>detailLineItemTableNoDataText}"
                busyIndicatorDelay="{detailView>/lineItemTableDelay}">
                <headerToolbar>
                    <Toolbar>
                        <Title
                            id="lineItemsTitle"
                            text="{detailView>/lineItemListTitle}"
                            titleStyle="H3"
                            level="H3"/>
                    </Toolbar>
                </headerToolbar>
                <columns>
                    <Column>
                        <Text text=
"{i18n>detailLineItemTableIDColumn}"/>
                    </Column>
                    <Column
                       minScreenWidth="Tablet"
```

```
                        demandPopin="true"
                        hAlign="End">
                        <Text text=
"{i18n>detailLineItemTableUnitNumberColumn}"/>
                    </Column>
                    <Column>
                        <Text text="Price"/>
                    </Column>
                    <Column>
                        <Text text="Quantity"/>
                    </Column>
                    <Column>
                        <Text text="Discount"/>
                    </Column>
                </columns>
                <items>
                    <ColumnListItem>
                        <cells>
                            <ObjectIdentifier
                                title="{ProductID}"
                                text="{OrderID}"/>
                            <ObjectNumber id ="id1"
                            number="{
                                path: 'UnitPrice',
                                formatter: '.formatter.currencyValue'
                            }" unit="USD" />
                            <ObjectNumber id ="id2"
                            number="{
                                path: 'Quantity',
                                formatter: '.formatter.number'
                            }" />
                            <ObjectNumber id ="id3"
                            number="{
                                path: 'Discount',
                                formatter: '.formatter.number'
                            }" />
                        </cells>
                    </ColumnListItem>
                </items>
            </Table>
        </semantic:content>
        <!--Semantic ShareMenu Buttons-->
        <semantic:sendEmailAction>
            <semantic:SendEmailAction
```

```
                id="shareEmail"
                press=".onSendEmailPress"/>
        </semantic:sendEmailAction>

        <semantic:closeAction>
            <semantic:CloseAction
                    id="closeColumn"
                    press=".onCloseDetailPress"/>
        </semantic:closeAction>

        <semantic:fullScreenAction>
            <semantic:FullScreenAction
                    id="enterFullScreen"
                    visible="{= !${device>/system/phone} &amp;&amp;
!${appView>/actionButtonsInfo/midColumn/fullScreen}}"
                    press=".toggleFullScreen"/>
        </semantic:fullScreenAction>
        <semantic:exitFullScreenAction>
            <semantic:ExitFullScreenAction
                    id="exitFullScreen"
                    visible="{= !${device>/system/phone} &amp;&amp;
${appView>/actionButtonsInfo/midColumn/fullScreen}}"
                    press=".toggleFullScreen"/>
        </semantic:exitFullScreenAction>

    </semantic:SemanticPage>

</mvc:View>
```

Listing 12.1: Detail.view.xml Content

You can now preview your application, and the new columns should be visible in the detail section. As shown in Figure 12.16, the **Order_Details** section has been populated with additional columns and their values. So, you should now have a fair understanding of creating a list-detail application following a freestyle approach. You can develop a worklist application or an SAPUI5 application in a similar way by selecting the corresponding template. When creating an SAPUI5 application, an empty application will be generated with very minimal coding upon which you can design your own full-blown application according to your requirements. Before we move on to discussing SAP Fiori elements applications, let's quickly review some key files generated for all SAPUI5 applications.

Figure 12.16: Previewing an Application to Display Additional Fields

12.4.2 Basic Files in SAPUI5 Applications

When you leverage templates during SAPUI5 freestyle development to generate an application, three important files are created automatically, namely, the following:

- *manifest.json*: This file is the deployment descriptor and contains the data model for the OData service, libraries and components, the root view of your application, and the routing configuration. Routing helps you navigate from one view to another.
- *component.js*: This file acts as an index to encapsulate details about the application.
- *app.view.xml*: This file defines the root view of the app and contains an `App` control or a `SplitApp` control as a root control.

Now that you understand how to build a freestyle application and some key files that are generated for SAPUI5 applications, let's shift our focus to building an SAP Fiori elements application.

12.5 Developing Complex SAP Fiori Elements Applications

As mentioned earlier, development using SAP Fiori elements may not need any front-end/SAPUI5 coding at all since this development process follows a low-code/no-code approach. These applications receive backend data from OData services, and the UI screens are automatically generated by the framework based on annotations added in the backend. You can also add local annotations in SAP Business Application Studio based on your requirements. The SAPUI5 runtime interprets the metadata and annotations of the underlying OData service and uses corresponding views in the SAP Fiori app to display this information to the user, as shown in Figure 12.17.

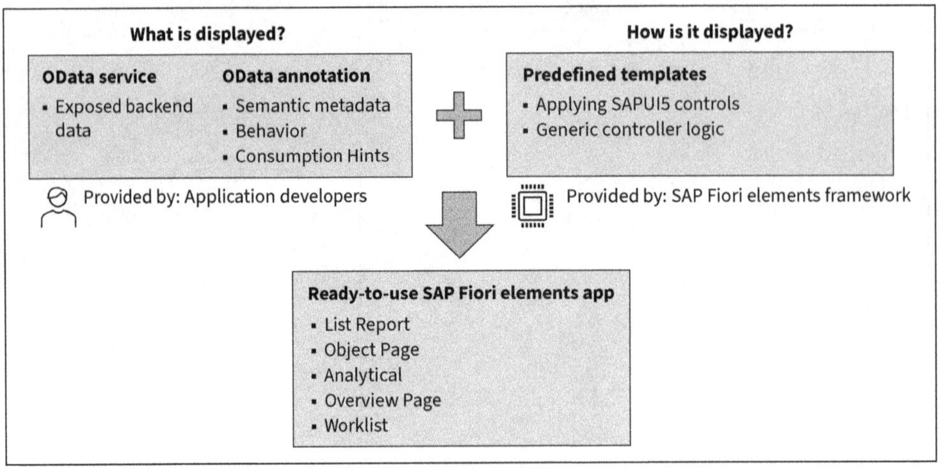

Figure 12.17: How SAP Fiori Elements Applications Are Generated

SAP Fiori elements provide the following advantages:

- Less development and maintenance effort since little to no UI coding is involved.
- Since you're using predefined templates, the look and feel of similar types of applications are consistent. Also, the latest changes in SAP Fiori design guidelines are automatically taken care of by the framework.
- The metadata-driven development model uses different annotations and significantly reduces the amount of code required for UI creation, which helps backend developers concentrate more on business logic.

In the following sections, we'll explore the available floorplans/templates included for SAP Fiori elements application development.

12.5.1 Determining Floorplans

At the time of writing, SAP Fiori elements provide five floorplans out of the box:

- **List report and object page**
  Use the list report floorplan when you need to display a list of items in a tabular format. It also enables the use of filters to narrow down the list.
  An object page is used to view the details of a particular object, edit an existing object, or create new objects. For example, you can use a list report to display a list of sales orders and then view the details of a specific order on the object page. The list report is typically used in conjunction with an object page.
- **Analytical list page**
  This floorplan is used to build analytical applications. You can view data from different

perspectives and drill down to multiple levels, which helps you find the root cause of an issue. The purpose of the analytical list page is to identify problem areas within datasets using data visualization and business intelligence. From this page of the application, you can view the data as charts as well as lists. An added advantage is graphical filtering, which is not available in a list report. Visualization helps users to recognize situations and reduces the number of interaction steps required to gain insights or to identify significant issues. Chart visualization increases the productivity of users and enables users to spot relevant data more quickly.

- **Worklist**
  This floorplan is helpful when you want to display a collection of work items assigned to a user and the user needs to take some action on them. You generally don't need sophisticated filtering for this type of application.
- **Overview page**
  With this dashboard-type floorplan, you can display different aspects of data in multiple cards. In these cards, you might display data in charts or lists, or you can create your own custom cards as well. This user-friendly experience makes viewing, filtering, and acting on data quick and easy. You also can navigate to a different application for further analysis.
- **Form entry object page**
  This new addition to the SAP Fiori elements family allows you to enter data into the object page for data entry.

When you create an SAP Fiori elements application via SAP Business Application Studio, you should see numerous floorplan options, as shown in Figure 12.18.

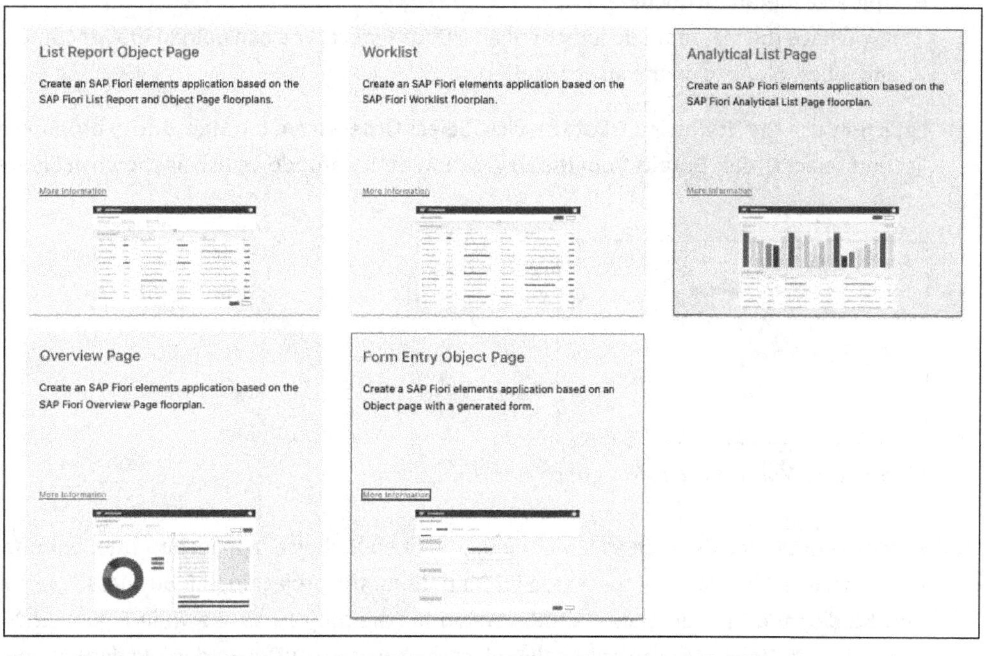

Figure 12.18: Available Floorplans for Creating an SAP Fiori Elements Application

For our example, we selected the **List Report Object Page** option because we want to build a list report and an object page. Once you select a floorplan, your next step is to select the necessary data sources.

12.5.2 Determining Data Sources and Other Details

Different types of data sources can be used when building SAP Fiori applications. As we did earlier when building a freestyle application, we'll use the Northwind OData service again. However, let's first explain the other types of data options available:

- **Connect to a system**
  To use a service from a backend on-premise or cloud system, this option will help you to select the related destination maintained through SAP BTP.
- **Connect to an OData Service**
  Select this choice if you want to directly connect to a service that takes care of authentication. We'll use this option for our Northwind example since we'll focus on the authentication.
- **Connect to SAP API Hub**
  Select this option to test any application programming interface (API) from SAP Business Application Studio. This option should not be used for productive scenarios.
- **Use a local cap project**
  Select this option to build an SAP Fiori application based on an SAP Cloud Application Programming Model project available in your SAP Business Application Studio.
- **Upload a metadata document**
  If you have the metadata document of an OData service, you can upload that document and follow along to create an application.

Let's now use the Northwind OData service. Select **Orders** from the **Main entity** dropdown list and select **Order_Details** from the **Navigation entity** dropdown list, as shown in Figure 12.19.

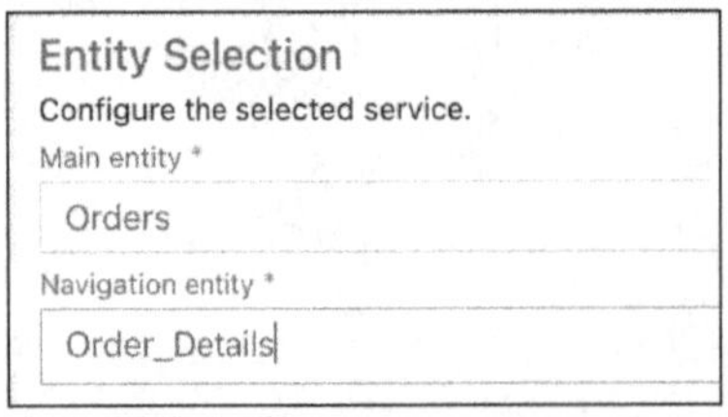

Figure 12.19: Main Entity and Navigation Entity

Enter the other details, such as a module name of your choice, application title, optional namespace, a description of the app, a folder path for the project in SAP Business Application Studio, and the minimum SAPUI5 version for the project, as shown in Figure 12.20. Then, click on **Finish**. You can optionally select the **Yes** radio button under **Add deployment**

configuration if you want to add the deployment configuration automatically while creating the project. However, for our example, we'll add the deployment configuration separately and thus select the **No** radio button. This choice will create a project that you can further enhance without writing any code via the guided development tool.

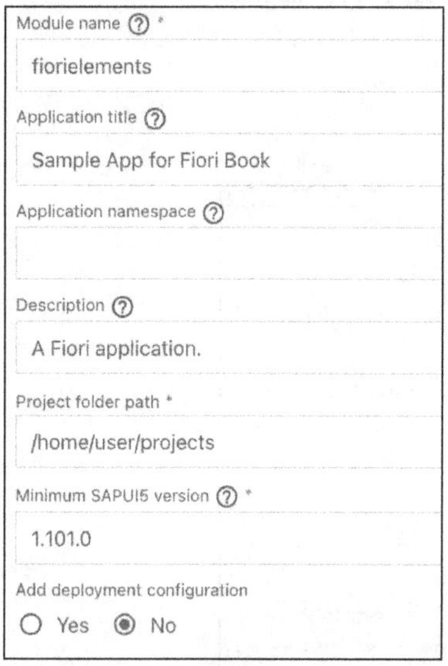

Figure 12.20: Other Details for Creating the Project

12.5.3 Guided Development

After the project is created, you'll need to work on its UI elements. For this task, we'll use the guided development tool, which can be launched by following the menu path **View • Find Command • Fiori: Open Guided Development Tool**. This tool provides multiple options for enhancing your applications using a low-code/no-code approach. Some options for list reports are shown in Figure 12.21.

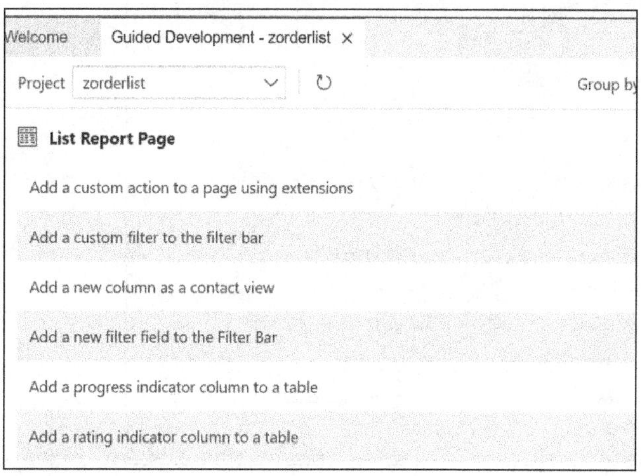

Figure 12.21: Guided Development Options

Let's add some fields to display in the report. First, select the **Add and edit table columns** option and then select the correct entity type from the dropdown list, as shown in Figure 12.22. Then, add the columns and click the **Insert Code Snippet** button, which will generate the annotation lines and add those lines to the annotation file.

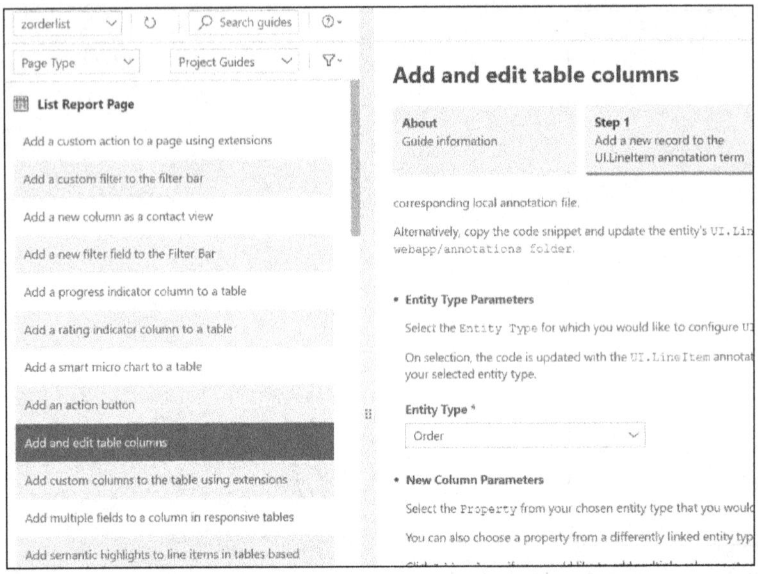

Figure 12.22: Adding Table Columns

You can also select how the column sequence should be ordered in your output table and the importance of the fields, as shown in Figure 12.23. Depending on the importance of the field, it may show up differently depending on the device.

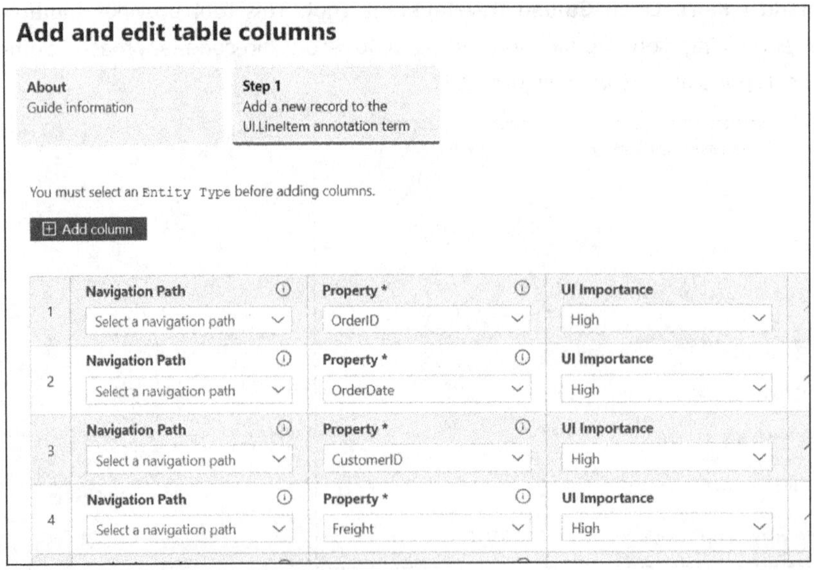

Figure 12.23: Column Sequence and Importance

Let's now add selection fields to use as the filter criteria, and for this task, we'll select the option **Add a new filter field to the Filter Bar**, as shown in Figure 12.24. Follow the steps accordingly to create the selection options using filter fields.

Figure 12.24: Adding the Fields for the Search Criteria

At this time, your screen output should resemble the screen shown in Figure 12.25 when you preview your application.

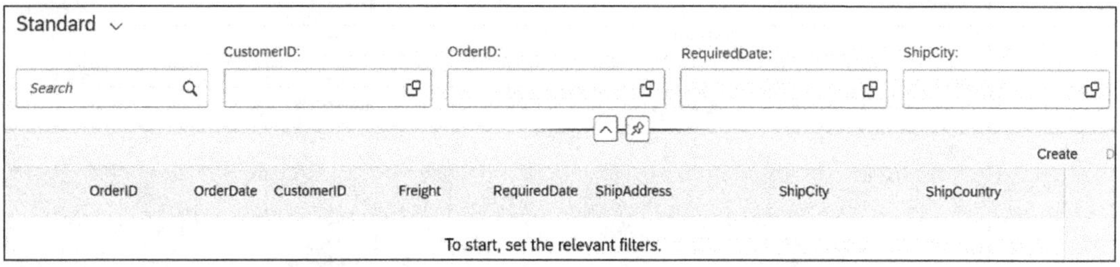

Figure 12.25: Initial Preview of the Application

As shown in Figure 12.25, the output is an empty list. If you click the **Go** button, it will display the output. To change this behavior so that the output is displayed on the initial screen instead, you'll need to adjust the annotation file, as shown in Figure 12.26.

Now, when you preview the application, you'll see the data is loaded automatically without needing to click the **Go** button, as shown in Figure 12.27.

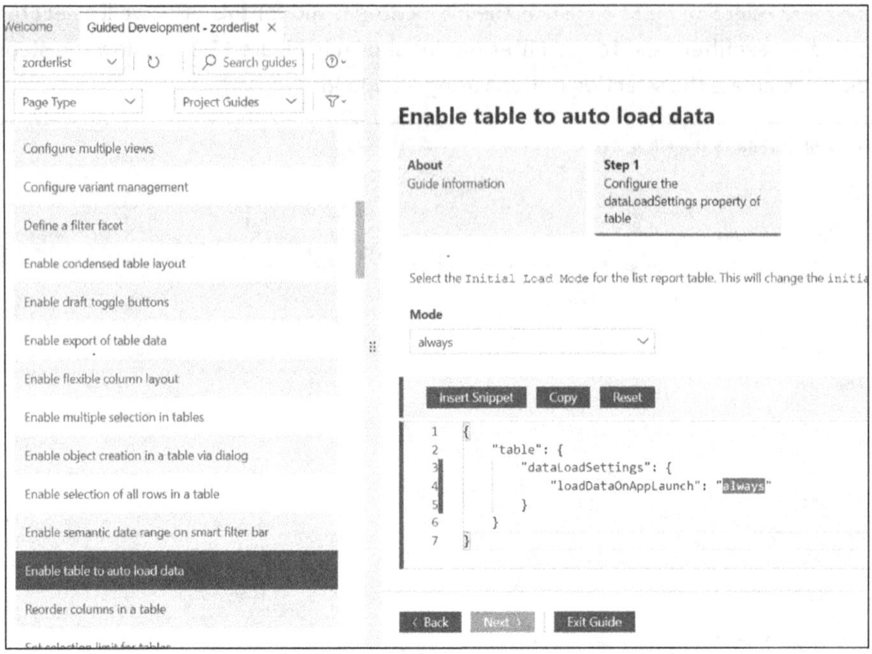

Figure 12.26: Initial Data Load

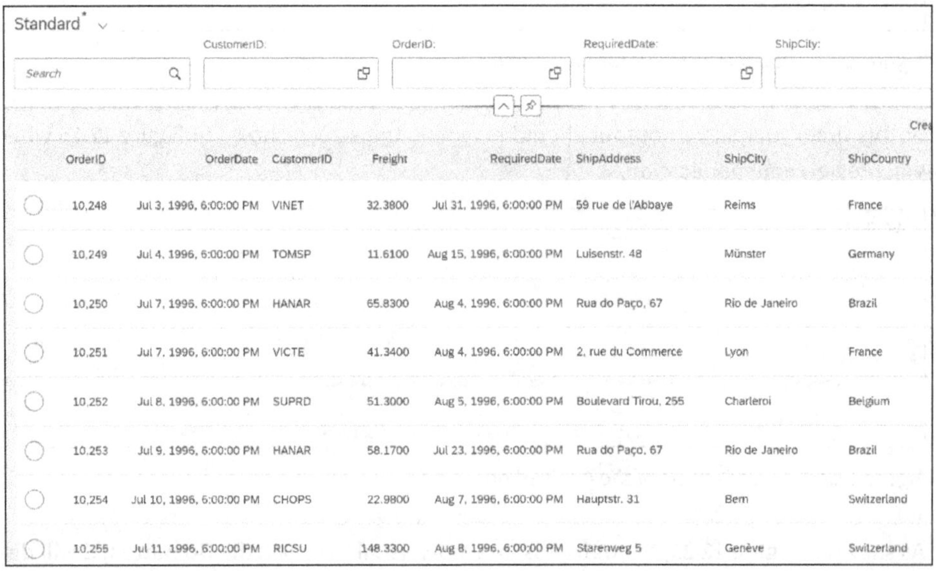

Figure 12.27: Automatic Data Load Without Using the Go Button

You can further enhance this functionality by selecting the **Enable multiple selection in tables** option, as shown in Figure 12.28.

So far, we've described only a few guided development options, and many other options are available to explore and apply to meet your requirements. We'll now shift our focus to object page development, which is closely related to list report development, but you can navigate to an object page by clicking on a row. We'll enhance our application with different options available through guided development.

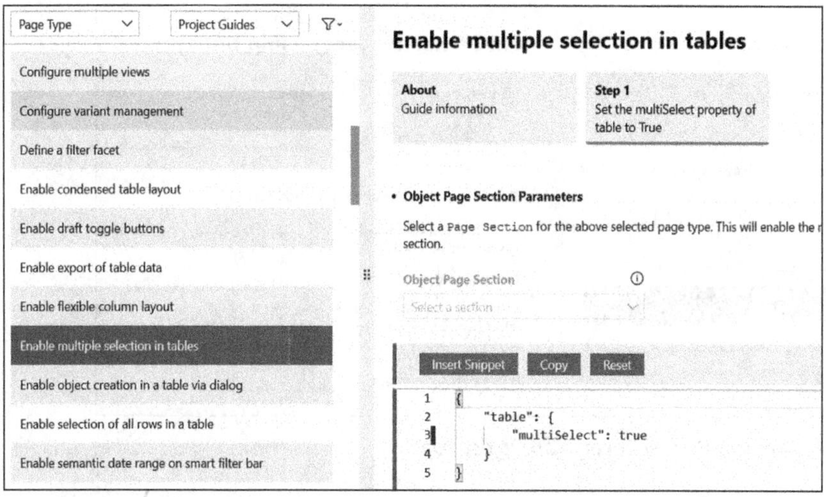

Figure 12.28: Enabling Multiple Selection

Let's start with building the header of the object page. In the guided development tool, select **Configure object page header**, populate the required values, and click the **Insert Code Snippet** button, as shown in Figure 12.29.

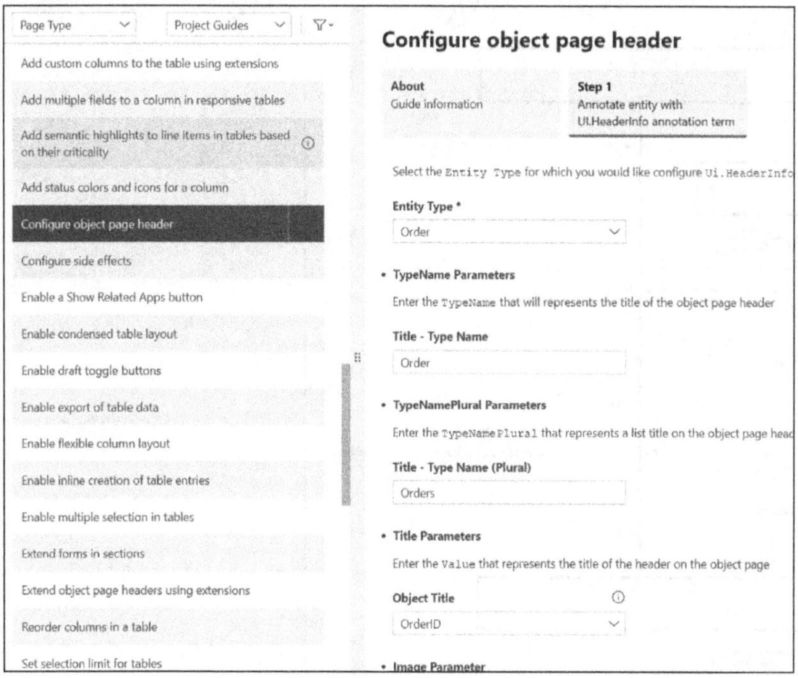

Figure 12.29: Configuring an Object Page

Let's now enhance this header section further by adding a few more fields in the header area. For this task, select the **Add a header facet using data points** option in the guided development tool, as shown in Figure 12.30.

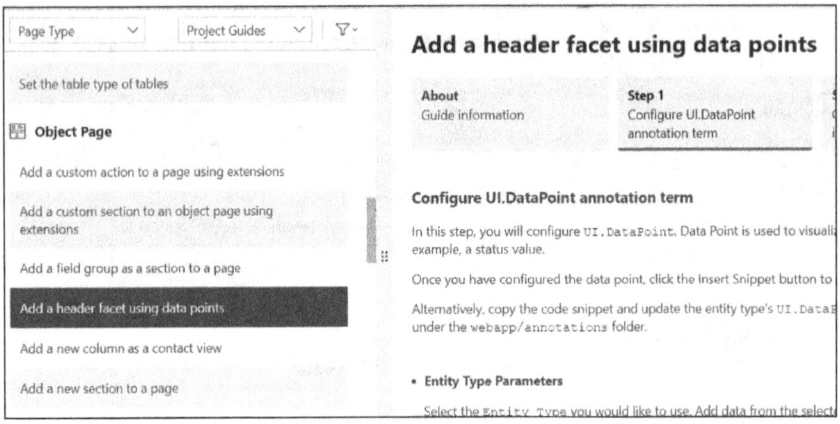

Figure 12.30: Adding a Header Facet in an Object Page

Additional configurations for the data point, such as its unique qualifier, must be maintained for the entity type, as shown in Figure 12.31, where we used HDR001 as the data point qualifier. Additionally, Figure 12.31 shows that the data point qualifier HDR001 is also referenced while defining the facet parameter for the **Order** entity type.

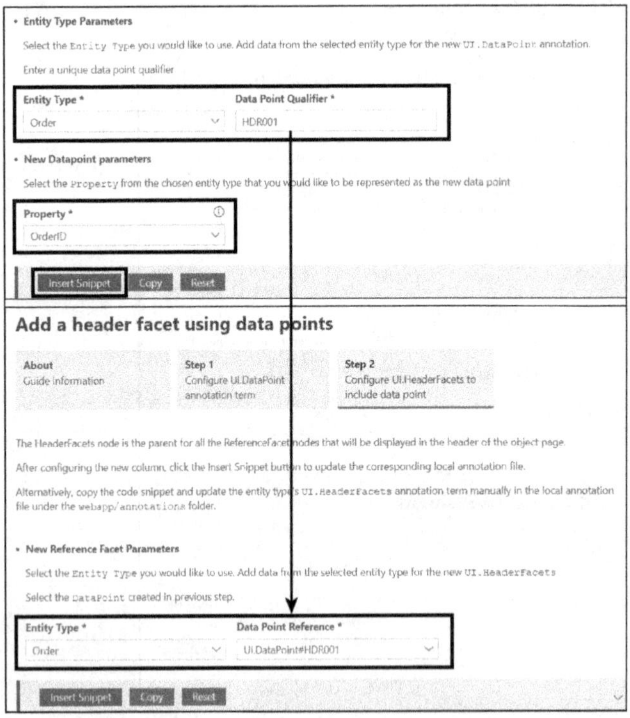

Figure 12.31: Additional Configuration for the Header Facet

Now, we've added two additional fields in the annotation file to be displayed on the object page, as shown in Figure 12.32. Note that the property `Title` was added manually using the code completion feature to populate the screen labels.

```
<Annotation Term="UI.DataPoint" Qualifier="HDR001">
    <Record Type="UI.DataPointType">
        <PropertyValue Property="Value" Path="OrderDate"/>
        <PropertyValue Property="Title" String="{@i18n>orderPlacedOn}"/>

    </Record>
</Annotation>
<Annotation Term="UI.DataPoint" Qualifier="HDR002">
    <Record Type="UI.DataPointType">
        <PropertyValue Property="Value" Path="RequiredDate"/>
        <PropertyValue Property="Title" String="{@i18n>orderRequiredBy}"/>
    </Record>
```

Figure 12.32: Annotation File with Additional Fields

If you preview the app now, you should see the changes reflected on the object page, as shown in Figure 12.33.

Figure 12.33: Output Preview with Additional Fields in the Header Facet

You can add other features to the object page and list report by leveraging other options from the guided development tool, and each time, the annotation file will be updated with corresponding details. Have fun exploring the numerous options available.

So far, we've explained in detail how a list report and object page can be built with annotations and thus without writing any frontend code. Don't forget that numerous other types of floorplans are available, and you can build applications with any of these floorplans and follow a similar low-code/no-code approach.

Earlier, when we used SAP Fiori elements, we had a limitation—you either stayed with the standard templates or switched completely to freestyle SAPUI5 if you needed customization. Now, that problem is solved with the flexible programming model. You start with standard SAP Fiori elements templates like list report or object page, which provide most of the application, and then add custom parts only where your business needs them.

For customization, you can create custom sections within your pages and incorporate building blocks such as tables or forms. These ready-made components already follow SAP Fiori design guidelines and work seamlessly with OData v4, so you don't have to build everything from scratch. Additionally, development has become simpler with macros. Instead of writing extensive UI code, you simply describe what you want, and the framework generates the UI using metadata and annotations. As such, SAP Fiori elements is no longer just a no-code tool for simple apps. It's more of a low-code tool with an extension model—you get most of the app automatically and then enhance it where needed.

Now that you understand various development aspects of building SAP Fiori applications, let's briefly cover the deployment process for custom SAP Fiori applications.

12.6 Deploying an Application

Once you've built your custom SAP Fiori application in SAP Business Application Studio, you can deploy your application either in a backend ABAP system or in SAP BTP. In this book, we'll only cover deployment to SAP BTP, but other books available from SAP PRESS show you how to deploy an SAP Fiori application to an ABAP system.

For deployment to SAP BTP, however, you would follow these steps:

1. The prerequisite for deployment in SAP BTP is to install the MTA tool for your application. This tool is for exploring and validating the MTA descriptor file (*mta.yaml*). You can install this tool globally by running the command `npm i -g mta`.
2. After the MTA tool is successfully installed, log on to the SAP BTP, Cloud Foundry environment, instance where you want to deploy the application. In SAP Business Application Studio, follow the menu path **View • Find Command** and select the **CF: Login to Cloud Foundry** option. SAP Business Application Studio also provides a menu on the left to connect to a Cloud Foundry target. Regardless of your option, you must select your SAP BTP, Cloud Foundry environment URL, email, password, subaccount, and space.
3. Add the deployment configuration by going to your project's root folder in the SAP Business Application Studio terminal and running the command `npx fiori add deploy-config`. You'll need to enter your deployment target (i.e., Cloud Foundry, not ABAP) and select any destination you're using.
4. Now, in the project explorer of SAP Business Application Studio, you'll see a new file named *mta.yaml*. Right-click on this file and build an MTA project by selecting the **Build MTA Project** option in the context menu.
5. After the build process is successful, a new *.mtar* file will be created in a new folder **mta_archives**. Now, right-click on the *.mtar* file and choose the **Deploy MTA archive** option, which will kick off the deployment process.
6. After the deployment process is finished, we should see the application has been deployed in the Cloud Foundry space we selected earlier.

Changes have also occurred in deployment; it has become a much more important part of the overall architecture—it's no longer just an afterthought. Today, on SAP BTP, we use a multi-runtime approach, allowing us to choose where to run our application based on our needs. Cloud Foundry remains the standard and simplest option. It's primarily used for standard SAP Fiori applications where SAP manages the infrastructure.

Kyma (Kubernetes) is the newer and more advanced option. We use it when we need more flexibility—for example, building complex applications, developing microservices, or using different programming languages like Python. It also performs well in event-driven scenarios.

Another important change is that deployment has become more flexible. SAP Fiori apps can be packaged as Docker containers and run in Kyma or use serverless functions for smaller logic components. Additionally, the way we deploy has changed. Instead of manually deploying via tools such as SAP Business Application Studio, we now use CI/CD pipelines. Tools like GitHub Actions or SAP's own SAP Continuous Integration and Delivery automate the process by handling tasks such as code quality checks, testing, and deployment to various environments (dev, test, prod). So, deployment is no longer manual, but an automated part of a DevOps approach.

So far, we've covered the development of new applications and their deployment to the SAP BTP, Cloud Foundry environment. In the next two sections, we'll explore another common scenario—the extension of the SAP-delivered SAP Fiori applications.

12.7 Extending Freestyle Applications

An extension is nothing more than an enhancement of a standard SAP-provided app without copying or cloning the app itself. Building extensions properly can save a lot of effort when compared to building a new custom SAP Fiori app from scratch. SAP provides several options, tools, and techniques to extend standard SAP Fiori apps in a structured way. SAP Business Application Studio and SAP Fiori tools are integral parts of this extension concept. As a prerequisite, make sure you have a destination created in your SAP BTP subaccount to access the standard app for your system. Figure 12.34 shows an example of a destination connected to an on-premise system via a cloud connector. Note that this kind of extension is not supported for SAP S/4HANA Cloud Public Edition.

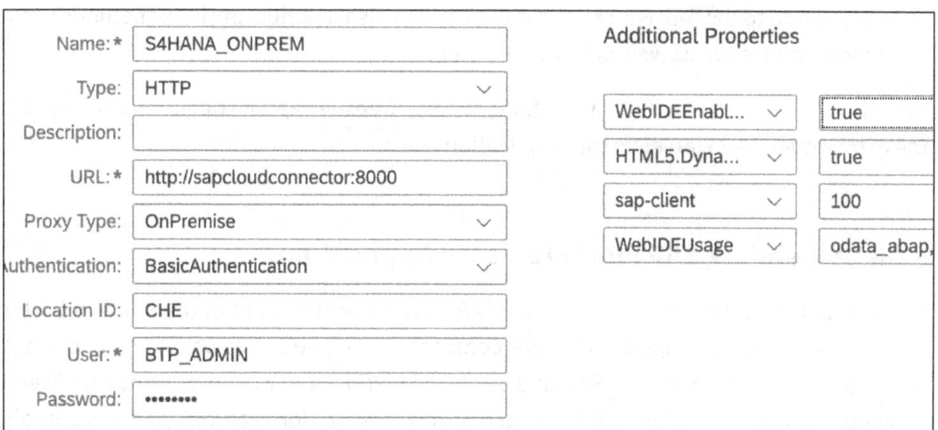

Figure 12.34: Destination in an SAP BTP Subaccount

Once you create the destination, you can then create the SAP Fiori adaptation project via SAP Business Application Studio, as shown in Figure 12.35.

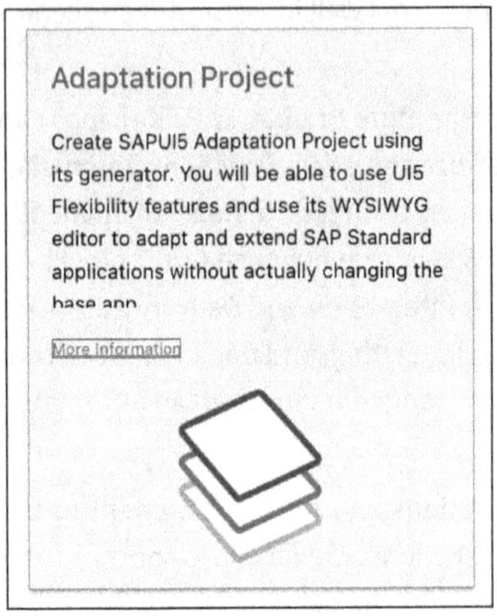

Figure 12.35: Adaptation Project in SAP Business Application Studio

An SAP Fiori freestyle application offers several different types of extension options, such as the following:

- **View Extension**: If a view of the standard app provides extension options to extend the view using a fragment specification, you can add additional fields to the view by adding fragments using an adaptation project.
- **View Modification**: Select this option hide any SAPUI5 control in the custom application. By the way, the same functionality can be achieved using a UI adaptation as well.
- **View Replacement**: Select this option if you want to replace the existing view with a completely new view.
- **Controller Extension**: Controller extensions allow you to add functionality to a fragment that you add to the SAP Fiori application. You can also override any base methods of the original application as well as lifecycle methods.

Now that you understand the options for extending freestyle applications, let's also turn to the extension of SAP Fiori elements applications next.

12.8 Extending SAP Fiori Elements Applications

SAP has delivered thousands of standard SAP Fiori applications out of the box as a part of SAP S/4HANA, and new applications are constantly being added as part of new releases. However, using a standard SAP Fiori app out of the box is not always feasible due to diverse business requirements. During a fit-to-standard workshop, for example, perhaps you discovered that a standard application could fulfill the majority of a requirement. In this case, you can try to tailor that application with minor enhancements to meet your specific business needs by applying the strategies described in Table 12.2.

Scenario	Examples	Extension Options
Simple screen enhancements in the applications specific to an end user	■ Adding fields in filter criteria ■ Adding/removing columns in the output	End users personalizing SAP standard apps
Less complex changes that do not require any logic change and are applicable for a group of users	■ Changing screen labels/titles ■ Changing the order of fields/group of fields ■ Embedding external content in the application	Key users adapting SAP standard apps
Complex changes that need coding knowledge and are applicable for a wide range of users	■ Adding new logic ■ Adding new business objects	Developers enhancing standard SAP apps

Table 12.2: Different Strategies for Extending SAP Fiori Elements Apps

The approach for extending an SAP Fiori elements application is the same: You'll create the destination where the app is available and then create an adaptation project in SAP Business Application Studio to use the various options available for SAP Fiori elements. Let's describe some frequently used enhancement options for SAP Fiori elements applications, such as the following:

- Use extension points in the object page. The object page provides the extension points before and after the header facet, for instance, for replacing the header facet or for adding a new section to the header facet or before or after subsections of the object page.
- Add custom logic while the table in the list report or object page is loaded.
- Add columns to the list report and object page tables.
- Add a custom page (only applicable for OData v4).
- Add custom views for the list report.
- Add custom actions.
- Add custom fields to the filter bar.
- Adapt logic before saving.
- Add custom logic before standard operations.

So, when extending an SAP Fiori element app, the possibilities are endless, and you can learn more about this topic at *http://s-prs.co/v550430.*

[»]

Note

You can deploy your adaptation project only on the system where the original app is available. If your original app is available in SAP S/4HANA, you must deploy the adaptation project in that system. You cannot deploy that extension app in SAP BTP, for example.

For modern SAP Fiori development, the direction is very clear—OData V4 combined with the ABAP RESTful application programming model is now the standard you should use. If you are building new applications, you should use the ABAP RESTful application programming model because it keeps all business logic in the backend and drives the UI through metadata instead of manual coding. You should use OData V4, because it is more efficient, faster, and supports advanced features such as improved filtering and optimized data transfer. Together, they enable the use of SAP Fiori elements with the flexible programming model, allowing you to use standard building blocks while still extending them when needed.

Regarding OData V2, it is still widely used and fully supported, especially in existing applications. However, it is considered a legacy option, meaning SAP is no longer focusing on innovation within it. New features—such as AI-based capabilities, draft handling, and advanced UX improvements—are coming exclusively in OData V4.

SAP is moving toward a more backend-driven, efficient, and modern architecture, and OData V4 with ABAP RESTful application programming model at the center of it.

12.9 Summary

SAP provides SAP Business Application Studio as an SAP BTP service, and SAP Business Application Studio includes SAP Fiori tools to easily build and extend your SAP Fiori applications. Complex developments and frontend extensions can be built through metadata-driven guided development and thus using low-code/no-code approaches. Additionally, SAP Build Work Zone can be used as your central launchpad for all your backend applications, and the content can be federated. When we talk about SAP BTP cloud-native development, we mostly mean backend service development using ABAP, Java, or Node.js. However, frontend SAP Fiori development is an integral part of full-stack application development, and we hope that you now have a fair understanding of this topic. With this chapter, we conclude our review of the various extension aspects of SAP BTP that fall under the SAP Extension Suite umbrella.

PART III
Operating Applications

Chapter 13
DevOps Implementation

In this chapter, we'll cover the concepts behind DevOps including continuous integration (CI)/continuous delivery (CD) and pipeline engineering. We'll also describe the different tooling options that can be leveraged for applying DevOps on SAP cloud-native applications, including tools and services available on SAP Business Technology Platform (SAP BTP). We'll also share some architectural examples.

DevOps, as the name suggests, is a combined cycle of development and operations using specific methodologies, tools, and processes while leveraging multi-skill teams. In any landscape, whether a custom application or an enterprise solution like SAP, delays and bottlenecks can occur across the development cycle. Lack of adequate testing (often due to time constraints), lack of skilled resources, or low participation from business users can lead to poor quality, lower customer satisfaction, and negative impacts on the business. Lack of visibility into application and business process performance can also compound overall quality degradation.

A lack of CD can impact an entire business. With new types of requirements coming from systems of engagement that are fast evolving and that must integrate with and access existing system of record applications, the complexity of these solutions dramatically increases. These challenges are further aggravated with changing business needs, such as the following:

- The need to experiment in the market to judge the success of products and services
- The inability for lines of business to iterate quickly in response to the market because of IT timelines
- The need to maintain system stability
- The inability for operations to keep pace with changes pushed into the system
- The need to quickly iterate functionality
- The inability to access production-like environments to validate application changes during development

To address these challenges, the concept of DevOps was developed nearly a decade ago and has since been adopted in the SAP world, particularly for building innovative solutions on SAP BTP. For cloud-native development, including microservices, DevOps is the expected way of working; otherwise, the core tenets of being cloud native—such as responsiveness to business requirements and regular, easy updates with less maintenance—cannot be truly achieved.

DevOps offers several advantages over traditional development and maintenance methods. Let's explore how DevOps has changed working methods and how it supports application development and maintenance.

13.1 DevOps Fundamentals

In this section, we'll explain what we mean by *DevOps* and how you can use a framework to adopt a structured approach to derive the most out of this concept. The framework, used for enterprise applications, is useful for SAP applications, especially cloud-native applications. Note that many variations exist in the market today using different terms like DevSecOps, DevBizOps, and so on.

SAP Continuous Integration and Delivery enables us to embed security directly into the development lifecycle by adopting a "shift left" approach, where security checks occur as early as possible. By integrating these controls into automated pipelines, every code change is validated for quality and security immediately upon commitment. This process includes running ABAP test cockpit (ATC) checks for functional and performance quality, using SAP Code Vulnerability Analyzer to detect issues such as SQL injection and XSS, and executing automated compliance stages like malware scanning and static application security testing (SAST). In addition, SAP Continuous Integration and Delivery enables us to integrate third-party tools such as Checkmarx, SonarQube, and WhiteSource (Mend), creating a comprehensive DevSecOps workflow that doesn't treat security as an afterthought.

For our purposes, of course, security is an important consideration, but we'll stick to the term DevOps only. In this section, we'll also explain the principles of DevOps: automation, CI, and the breakdown of team silos. Then, we'll explain how the tools help in each phase of a DevOps cycle and what it means to move from one phase to the other. We'll touch on the shift-left concept of testing and the benefits of DevOps. Finally, we'll explore the tools available in SAP BTP and the other complementary tools for implementing DevOps.

13.1.1 Core Tenets of DevOps

DevOps is a way of working with a team performing both development and operations in a continuous cycle, often using the Agile methodology for development and as much automation as possible.

The concept behind DevOps is breaking down silos among the various IT teams to promote efficiency and enhance responsiveness to business changes. Many companies have developed their own DevOps framework, and for product companies and hyperscalers, their products align to those frameworks, or their services are aligned to the main concepts of DevOps. For example, IBM uses a DevOps framework, which, as shown in Figure 13.1, has six dimensions with twenty-three characteristics for enterprise applications, referred to as the 6C Framework. Other DevOps frameworks are similar to this model, which we will use as our reference for defining the different tools and processes that enable DevOps.

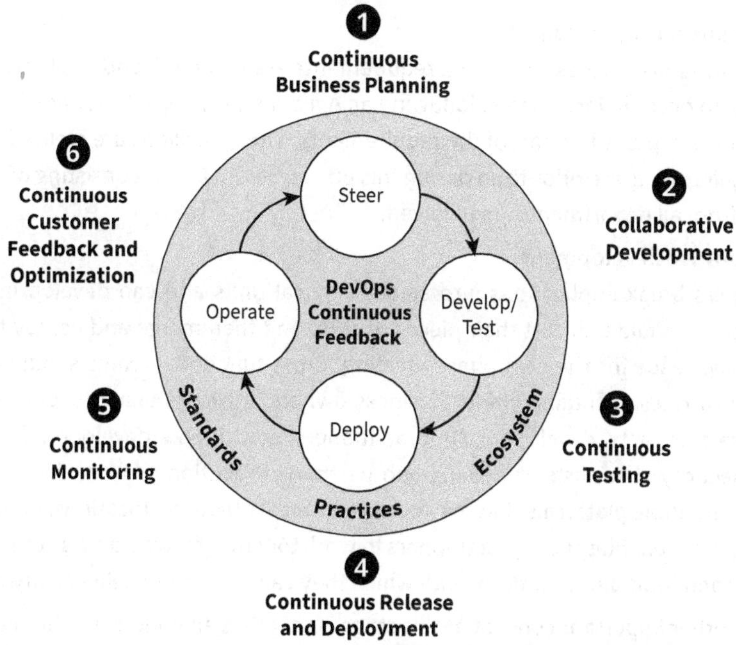

1
1. Capture, Analyze, and Prioritize Business Requirements
2. Project Planning
3. Measure to Project Metrics
4. Traceability
5. Dashboard Portfolio Measures

2
6. Configuration Management
7. Collaborative Development
8. Build Management
9. Release Planning
10. Change Management
11. Dashboard and Traceability

3
12. Test Management and Execution
13. Test Automation
14. Traceability
15. Test Data Management

4
16. Release Management
17. Environment Management (Provisioning Automation)
18. Deployment Automation (Configuration, Code, Patches, and Upgrades)

5
19. Monitor Capacity and Optimize
20. Monitor Performance and Optimize
21. Monitor User Experience and Optimize
22. Event and Incident Management

6
23. User Feedback and Optimization

Figure 13.1: IBM DevOps Framework for Enterprise Applications

Before we further discuss the 6C Framework shown in Figure 13.1, let's review three core principles you must adhere to for each phase:

- Automate various steps as much as possible.
- Every phase should occur as a continuous cycle in an iterative way; agility is key.
- Working effectively requires breaking down team barriers, reducing handoffs, and including business representatives on the team.

Now, let's discuss the six dimensions shown in Figure 13.1:

❶ **Continuous business planning**

With continuous business planning, requirements are gathered, and implementation is planned to occur in increments, following an Agile methodology. Feedback from earlier iterations is captured as part of the requirements. Thus, backlogs are created/updated, and requirements are prioritized during this phase. Scrum teams consisting of representatives from all departments are involved.

❷ **Collaborative development**

Developers break applications into smaller logical units and can develop in parallel. Developers should unit test their pieces of code and then merge and deploy the entire code, now ready for the next phase—testing. Cross-functional teams should include a number of roles, including line of business owners, business analysts, enterprise and software architects, developers, QA practitioners, consultants/developers from operations, security specialists, suppliers, and partners. Practitioners from these areas can work on multiple platforms that may be spread across multiple locations. Collaborative development enables these practitioners to work together by providing a common set of process and tools and a platform with which they can create and deliver software.

Also, another important concept at this stage is that CI is an underlying thread connecting many development activities across different phases: development, testing, quality gates, deployment, defect logging, and more. We'll cover these activities in more detail later in this chapter. Collaboration between developers and other roles in the development team is important, as teams are more virtual and have an unprecedented geographical spread nowadays. This aspect must be considered when selecting tools.

❸ **Continuous testing**

Continuous testing is typically an area where DevOps adoption has been rather prevalent, through the adoption of automated testing. However, some additional considerations are required; for example, the shift-left concept of testing has gained enormous importance. This concept means that testing should be conducted as early in the software development cycle as possible to detect errors and problems in the code and to get feedback on expected outcomes. We all know that bugs, if detected later in the development cycle, for instance, in the preproduction environment, are more costly to fix and retest because the code and configuration must pass through the different environments again before the solution can be retested in the environment where the error was detected. Thus, you'll need to identify scenarios early so that these scenarios can be tested in the development environment itself. Of course, doing so can result in enormous challenges where new ways of working, and the necessary tooling, must be adopted. For developers to unit test most scenarios, these scenarios must be identified, test data supporting the scenarios must be made available, and automation tools must be ready for testing code. As you can imagine, you cannot possibly simulate all scenarios in a dev environment at all, and some scenarios can only be tested in later phases like integration testing scenarios, user role testing scenarios, etc. In these cases again, the shift-left concept for security design has evolved. So now, even user authorizations are expected to be designed and developed during application development, which will require user role testing in the dev environment.

Test-driven development (TDD) is a related concept in which code development revolves around testing scenarios. The shift-left method can also be adopted through automated testing, ensuring that every line of code is executed. This testing is therefore not scenario based but is performed through embedded test scripts that simulate inputs and expected outcomes for each code block. Also, developers must perform test automation using the necessary tools, which requires additional effort. Therefore, IT managers may need to change their mindset to account for the additional time and effort required for test automation and bug fixing. Investing at the right time can yield significant benefits by reducing defects in later phases, shortening deployment time, lowering costs without sacrificing quality, shortening overall timelines, and preventing deployment delays relative to production dates. Thus, continuous testing that begins early in the implementation cycle is key.

❹ **Continuous release and deployment**

To understand the concept behind continuous release and deployment, let's first reexamine the concept of CI in some detail. CI can be implemented at the time of development, as mentioned earlier. A large team of developers can work on cross-technology components in multiple locations to deliver software in an Agile manner by ensuring that each team's work is continuously integrated with that of other development teams, with quality ensured through checks in the CI toolset. This concept, when implemented properly, reduces risk and identifies issues earlier in the software development lifecycle. The various activities from development to deployment are all interconnected and automated using tooling, and these activities will include steps like development, (automated) unit testing, (automated) quality checks, the creation of the deployment unit, and preparation for deployment. In Section 13.2.7, we'll show you how to set up the tooling for CI.

Continuous release and deployment involve automating, as much as possible, the release of code from a development environment to a test or quality system to production, given some conditions (like no critical open errors, etc.) and automating the deployment of the application. How much automation is possible depends on which tools are available, the integration among the various tools, and, of course, how much manual control is required. The goal of continuous release and deployment is to release new features to customers and users as soon as possible, without any disruption to existing business.

CI/CD typically go hand in hand for releasing error-free, high-quality code quickly into the next environment. In fact, this CI/CD process can be designed to include steps to create containers, deploy to containers, deploy servers for serverless coding, and more.

❺ **Continuous monitoring**
Continuous monitoring entails monitoring applications in different stages of deployment. The purpose of this step is to be vigilant about all your applications in all stages of the deployment cycle so that any plan adjustments can be made immediately, any enhancements to performance or usability can be taken into account, and any adjustments to the system parameters can be made easily.

❻ **Continuous customer feedback and optimization**
Continuous customer feedback and optimization entails getting end user feedback and categorizing this feedback so that any new requirements or changes are added to the overall requirements collected in continuous business planning. Feedback is obtained directly from users through various means, including incident recording, social media, or user forums. Alternately, feedback can also be derived by monitoring your applications. This feedback is utilized to improve and enhance the functionalities of your applications.

13.1.2 DevOps Inner Circles

Next, let's look at the four inner circles that define a DevOps architecture:

- Steer: This inner circle is about planning and prioritizing requirements and also conducting meetings with stakeholders to provide direction and feedback to the team.
- Develop/test: This inner circle is about developing and testing applications iteratively and continuously adding features based on new or refined requirements or on feedback.
- Deploy: This inner circle is about deploying features for productive use to business users.
- Operate: This inner circle is about performing operations like providing support, bug fixing, monitoring, and gathering feedback.

The DevOps concept has been applied to cloud-native application development, initially in the custom (non-enterprise) application space. Gradually, the benefits of adopting DevOps principles have become evident even for enterprise applications. These advantages are enhanced by the cloud platforms now available to support the enterprise application space, such as SAP BTP.

Thus, "continuity across the entire lifecycle" is the main mantra of DevOps. Continuity is enhanced by reducing barriers, minimizing handoffs, automating activities and integration, and continuously incorporating feedback to support quick software releases by facilitating the identification of needs (plan), development (dev/test) of features supporting those needs, releasing (release/deploy) features to production as soon as possible, gathering feedback through monitoring, and optimizing or iterating to achieve the desired effects.

Team composition is now crucial because DevOps requires many different types of skills. However, most importantly, the mindset change required for activities like automated unit testing, incremental development, and breaking down barriers among teams is often easier said than done.

These concepts can be fully realized only with ample tool support, which we'll discuss in the next section.

> **Note**
> Most major product companies and cloud companies have adopted DevOps; some have defined their own frameworks.

13.1.3 Continuous Integration/Continuous Delivery

The term CI/CD is often used synonymously with DevOps. While the two terms are not the same, CI/CD is a key concept in the implementation of the DevOps concept in the IT world.

CI is a software development practice where members of a team integrate their work frequently; usually, each person integrates at least daily, which thus leads to multiple integrations per day. Each integration is verified by an automated build (including testing) to detect integration errors as quickly as possible.

By implementing CI practices in your software delivery process, you can merge code changes from multiple developers often, usually several times a day, in an automated and risk-free way.

SAP Continuous Integration and Delivery is recommended when you're looking for a simple, ready-to-use solution that doesn't require much DevOps expertise and involves no infrastructure management. It's particularly useful for teams that want to follow SAP Continuous Integration and Delivery best practices without having to design and maintain complex pipelines from scratch; following are some examples:

- If the team has limited CI/CD experience: This is a good option when the team isn't very familiar with tools like Jenkins or GitLab CI.
- When you need a quick setup: This works well when there's a need to rapidly configure and run predefined pipelines for common SAP development scenarios, such as SAP Cloud Application Programming Model or SAP Fiori.
- If you want managed infrastructure: It works well when you want to avoid the effort of installing, patching, and maintaining build servers and CI/CD tools.
- For tight SAP ecosystem integration: This is especially effective when seamless, out-of-the-box integration with other SAP BTP services is important, such as SAP Cloud Transport Management and SAP Discovery Center.
- To follow SAP best practices by default: Using SAP Continuous Integration and Delivery helps ensure that the build, test, and deployment processes automatically align with SAP-recommended workflows.

Let's look at the basic workflow for CI/CD next:

1. A developer writes code.
2. A developer pushes code changes into a central source code management system.

3. The central source code management system triggers the CI server.
4. The CI server runs automated build and tests.
5. The CI server sends feedback to the developer.

CD involves the validation of every change and involves preparing changes after automated technical and functional testing so that a revised solution can be shippable. The term "shippable" refers to being ready for deployment to production. Gauging this readiness for deployment will depend on your company's deployment strategy.

Let's consider some common CI/CD practices used for custom development, although some entries in the following list are more relevant for technologies other than ABAP:

- **Keeping everything under version control**
  You should always use multiple feature branches to minimize disruption to your main branch.
- **Automating the build**
  Automate the build in a centralized system, not in any of the local machines.
- **Running unit tests in the build**
  Identify and set up your unit test scenario early using the TDD approach.
- **Committing early and often**
  Avoid any bulk merging for your master branch.
- **Building each change**
  Implement feature branches to keep your master branch clean.
- **Fixing build errors immediately**
  Avoiding delays helps keep the development cycle going.
- **Keeping the build fast**
  You may need to use automation to achieve this goal.
- **Testing in a clone of the production environment**
  Set up your test environment in a reproducible way by using an "infrastructure as code" approach.
- **Making it easy to get the latest build results**
  Your CI server should be capable of easily displaying the latest build result.
- **Ensuring that the build process is transparent to everyone**
  Transparency is the key to trust, which is essential for DevOps to work.
- **Automating deployment**
  Since automation is the underlying principle of DevOps implementations, automating deployment is an important step.

In the next section, we'll discuss what it entails to implement DevOps, especially for cloud-native SAP scenarios.

13.2 Implementing Tooling for DevOps

To implement DevOps, the following three aspects must be considered:

- **People**
  This dimension is about having the right skills/training in your team, with the right mindset for a way of working that will be completely different. This change in mindset may also require a change in team structures and collaborating closely with people across multiple teams, including business and IT.
- **Process**
  Set up the process to be followed, perhaps starting with an Agile development methodology and close cross-team collaboration, and then expanding to automation and the necessary controls and beyond.
- **Tools**
  Set up the tools that support automation and continuous planning; collaborative development, deployment, and monitoring; and CI.

In this section, we'll focus on the tooling required for DevOps. Implementing the tooling for DevOps could take 6 to 8 months, depending on which phases you want to implement, which tools are being set up, whether a DevOps tooling platform is available for the technology you want to leverage, and other factors. For example, SAP BTP, as a platform as a service (PaaS), provides DevOps services. However, in practice, you may already have some licenses or may have already set up some of the tools for test automation, process modeling, or solution monitoring. As a result, leveraging these existing tools makes sense, and you can use platform services for other tasks. The advantage of using the tools and services on the platform is to reduce capital expenditures and avoid the additional headache of maintaining this set of tools. An additional point to consider is that all these tools must be integrated and automated, in terms of their invocation and use, as much as possible.

Diving more deeply, let's follow a day in the life of a team adopting DevOps and review some DevOps tools available in the market.

13.2.1 Continuous Everything

The bottom line is that you should enable "continuous everything" in every phase and automate as much as possible. Let's look at how SAP's portfolio is ready to help you realize the vision of continuous everything in application development. The flow shown in Figure 13.2 shows how activities can occur continuously and iteratively in a loop.

Although the auto-generation of process models and implementation artifacts from requirements is not currently available, Focused Build for SAP Solution Manager offers capabilities to associate business requirements with business process hierarchies and to report on traceability.

Further, process discovery tools are available that can either scan through your existing systems or monitor your users to generate process models. However, the auto-generation of configuration settings from process models or requirements is not available today.

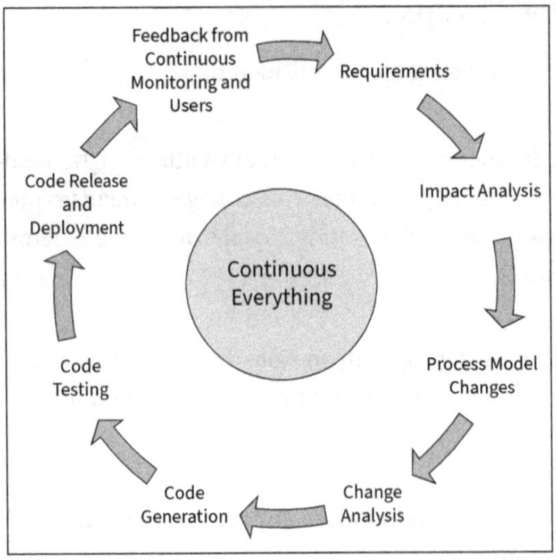

Figure 13.2: Continuous Everything

The auto-generation of code from models is not common in ABAP but is the standard in the development workbenches/IDEs for SAP HANA and some newer technologies. In fact, some low-code/no-code tools have this code generation capability in their solution backends, although that code may not be always visible to the developers/power users. SAP offers documentation generators for certain kinds of code.

The identification and auto-provisioning of environments is possible using tools like SAP Landscape Management. In this scenario, an infrastructure as code model is followed in which operators provision and manage systems through code, rather than manually. SAP Landscape Management is available as a solution for hybrid environments including both on-premise systems and cloud-based solutions. A subset of this solution is available, called SAP Landscape Management Cloud, which is a solution as a service, a combination of product and services, and focuses on managing SAP system operations on cloud infrastructures. Other automation products like Ansible have advanced capabilities in this regard.

The identification of what needs to be tested based on changes in code/configuration is available through SAP Solution Manager's Business Process Change Analyzer (BPCA), custom code analyzers, and more. This approach currently works for the on-premise systems like SAP S/4HANA. Also, some prerequisites must be fulfilled for this analysis to happen, for instance, technical bills of materials must be configured in SAP Solution Manager, before this change analysis can work. Test data management—identifying the data required to test a scenario and extracting that data from a different system—can be managed through various tools, but the maturity of these tools varies. However, automatically changing/developing new requirements based on feedback from business process managers is not currently available.

The following section delves deeper into each of the major steps of an application lifecycle management.

13.2.2 Requirements Management

Most projects now use some form of Agile methodology since waiting months or even years to meet business requirements is not acceptable. Rather, business users want to start using some minimum viable product (MVP) immediately, which also enables easier change management and faster adoption. Of course, the MVP at go-live must fulfill the basic business need. Sometimes, the Agile methodology is applied only for the development stage of the project, while requirements are being gathered, maybe through some design thinking activities, but generally using discovery workshops at the beginning of the project. These requirements can be captured and planned, as we'll discuss next.

Agile Methodology Application

Generally, some best practices of the Agile methodology include the following:

- Create a business process hierarchy in an imported branch while designing. You can use SAP Best Practices content, which includes some end-to-end processes, to jump start your development.
- Release the application to the development branch after finalization and as soon as the design is fixed.
- Release the application to the production branch during project go-live.
- Use the maintenance branch for run and support.

Figure 13.3 shows an example business process hierarchy and Figure 13.4 shows the result of this hierarchy in SAP Solution Manager.

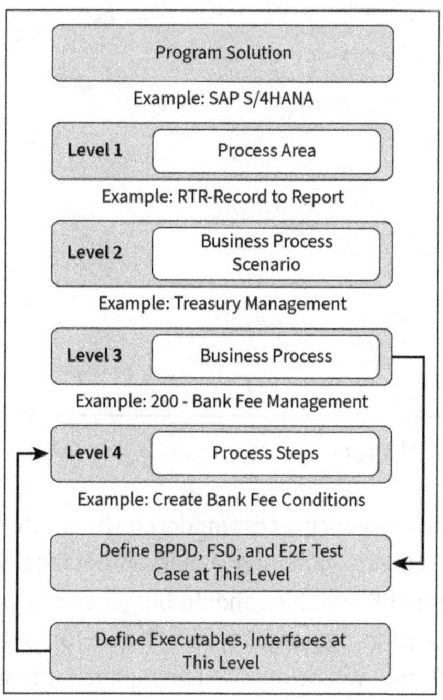

Figure 13.3: Business Process Hierarchy

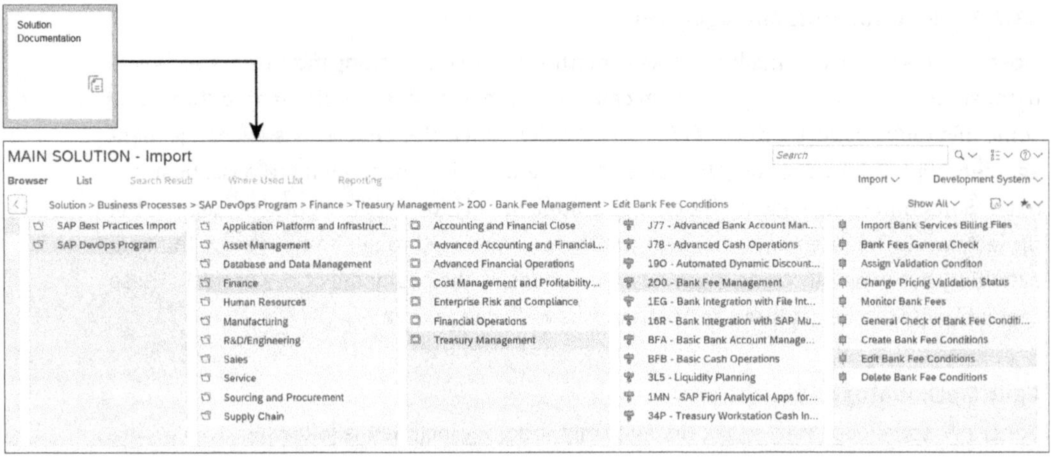

Figure 13.4: Result in SAP Solution Manager

Figure 13.5 shows a business process hierarchy created in SAP Solution Manager while Figure 13.6 shows its corresponding Agile plan captured in Focused Build for SAP Solution Manager.

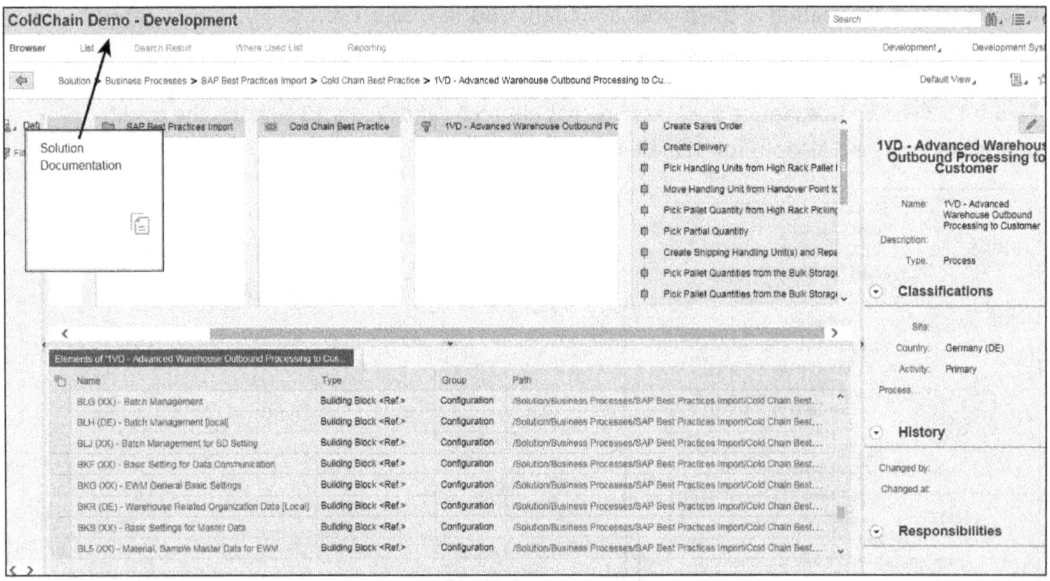

Figure 13.5: Business Process Hierarchy in SAP Solution Manager

Now that your requirements are defined and prioritized, your process model can be created or designed. In an ideal state of continuity, requirements gathering should automatically generate the relevant process model. Process model hierarchies should be generated in SAP Solution Manager, while configuration and code in the target system (SAP ERP, SAP S/4HANA, or cloud-native applications on SAP BTP, etc.) will be required to implement that process model in SAP. As a result, no gap will exist between what is required and what is implemented, and nothing is lost in translation, while the speed of implementation has been accelerated because of the available automatic capabilities. The process model should also automatically generate the documentation required to maintain the system and to meet legal, regulatory, and auditing requirements. The process model should also identify what environments are required to implement the process and automatically provision these environments as required. Not only will implementation be expedited, but also the relevant environments will be ready in time for each phase (development, testing/quality, preproduction, production, etc.), which brings the Dev (development) and the Ops (operations) together.

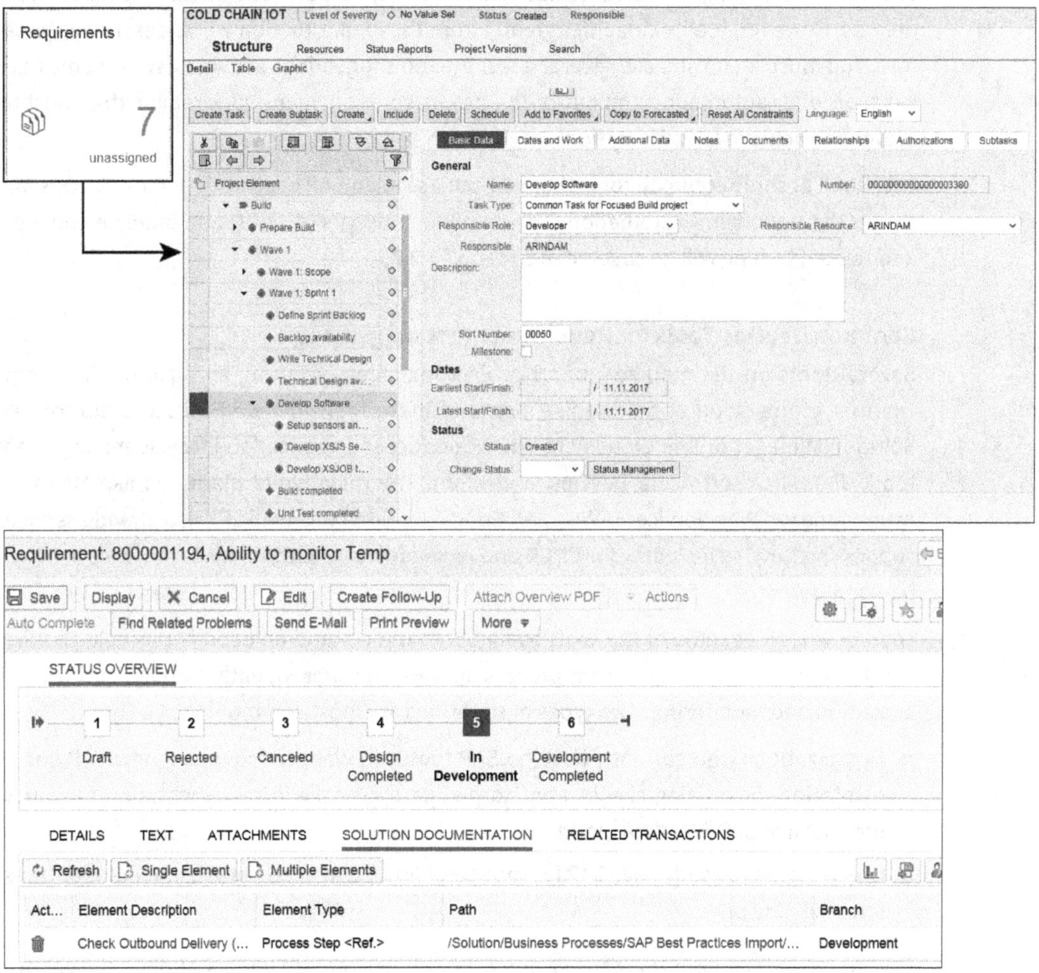

Figure 13.6: Project Plan in Focused Build

The process model and the resulting configuration/code should determine automatically what needs to be tested and what test data is required while also ensuring the readiness of the test environment by auto-provisioning the environment (hardware, software, configuration, code, and data). As tests occur automatically, once ready, code and configuration are released to production.

So far, we've covered the build and deploy phases of development, but during runtime, the process model should also drive how the business processes are managed and monitored to ensure that processes are executed in a way that meets requirements and targets.

By identifying new ways to optimize and improve, feedback can prompt changes in requirements. Thus, your vision should be to automate the entire lifecycle across all activities that may occur in different tools during various phases of the implementation cycle, as discussed earlier. Then, the release becomes extremely fast and seamless, helping to avoid defects. However, reality can differ greatly from the ideal situation we've described; therefore, you must set up and use several disparate tools for various activities. While some market tools offer automation and out-of-the-box integration, many gaps remain that must be bridged to achieve full automation.

Let's look at some common tools that you can use in an SAP context to pursue DevOps and set up the underlying automation. The detailed concepts of CI/CD and pipeline concepts and related tooling will be covered in Section 13.2.7.

Common DevOps Tools for Requirements Management

Several tools on the market support project management using an Agile methodology, whether in the cloud or on-premise. Some of these tools include Focused Build for SAP Solution Manager or SAP Cloud ALM (the cloud focused application lifecycle management tool), Jira, Microsoft Azure DevOps, and more. The main cloud platforms like Microsoft Azure, Amazon Web Services (AWS), and Google Cloud Platform (GCP) also provide services supporting DevOps including for CI/CD and Agile project management techniques like Kanban boards.

Let's now look at some of the tools available to understand and codify business requirements, convert them into business process models, and subsequently manage the business execution and monitoring. Two types of situations are most common:

- Organizations who currently have no SAP footprint want to move to a new SAP implementation, or existing clients want to pursue a greenfield implementation for a new product like SAP S/4HANA Cloud.
- Organizations already have SAP footprint and now need to add new capabilities or transform existing ones.

For the second scenario, you can use process discovery tools to understand the current process models in your existing SAP systems. For the first scenario, your requirements, as collected during the discovery phase of the project, form the main input.

In this way, for both scenarios, a new blueprint and new business process model can be designed and used for business process execution and subsequent monitoring.

SAP Solution Manager

SAP Solution Manager provides features for performing process modeling and requirements management (with help from Focused Build). These tools can capture requirements or fit-gaps and link them to the business process hierarchy in SAP Solution Manager.

For clients with an existing SAP landscape, SAP Solution Manager's documentation and library generation cockpit can help you during business process discovery.

SAP Solution Manager 7.2 has a built-in process modeling tool where graphical process modeling can be performed. The output of process discovery from certain partner companies of SAP can be uploaded into the SAP Solution Manager 7.2 process modeler. The process model can drive the business process execution and monitoring through SAP Focused Run.

However, at this time, no prebuilt process model is available in SAP Solution Manager 7.2. The alternative solution is to use SAP Signavio Process Insights tools, which are part of the SAP Signavio Process Transformation Suite. This solution has multiple uses, but what is relevant in this context is that it includes several process models and business process KPIs out of the box. You can then adopt or create new process models from these resources.

SAP Cloud ALM

As mentioned earlier, SAP's strategy, aligned to today's business needs, is for all products to be available in the cloud. Along those lines, SAP has come up with SAP Cloud ALM, a software as a service (SaaS) product enabling application lifecycle management for cloud-based products. While SAP Solution Manager focuses on on-premise and private cloud landscapes, SAP Cloud ALM is continuously building its capability for change and process management solutions for hybrid landscapes.

An example use case is described later in Section 13.2.7, where we discuss a typical customer landscape that requires hybrid DevOps and change management solutions based on SAP Solution Manager and SAP Cloud ALM.

SAP Solution Manager and SAP Cloud ALM tools have functions during other cycles of DevOps as well, which we'll touch on in each phase. Please note that SAP Solution Manager remains relevant for on-premise and hybrid landscapes. SAP Cloud ALM is the strategic direction for cloud-first and SaaS-heavy landscapes.

13.2.3 Collaborative Development Tools

For developing pieces of business logic in the SAP world as well, several tools can generate logic using drag-and-drop interfaces where you move different pieces (i.e., workflow components or business logic behind a UI), and these activities generate code in the backend. We've touched on several such tools throughout this book. However, for complex requirements, you must create code using SAP Business Application Studio for SAP BTP, Cloud Foundry environment, for example. Such IDEs are integrated with code versioning tools like Git, which we'll discuss next.

Git and Its Terminology

Git is a free and open-source distributed version control system designed to handle everything from small to very large projects with speed and efficiency.

In essence, Git is a version control system that stores your development artifacts in an automated version-controlled way. Git allows multiple people to work on the same software and can integrate the work of many different developers. If you're an ABAP developer, you may not be too familiar with Git since usually you're developing code in a central ABAP editor (e.g., Transaction SE80). Using Git is not required while working in ABAP Development Tools (ADT) with the exception of SAP BTP ABAP environment.

Git is lightweight, and you can easily install Git for a local project. For corporate projects, various Git hosting services are available. GitHub is one web-based hosting service based on Git that is open to all and is quite popular among full-stack developers. Other popular hosted Git services are Gerrit, GitLab, and Bitbucket (Atlassian).

Let's quickly look at some common terminology for Git and GitHub to help you better understand CI/CD and the example we'll demonstrate in this chapter:

- **Repository**
  A repository is equivalent to a project's folder and includes many files in their various versions.
- **Commit**
  A commit is a revision or change to a file. Developers actually save files to a local Git repository. Each save after a change can be considered a commit.
- **Master**
  The master is the main branch of the codebase currently running for the user.
- **Branch**
  A branch is a parallel version of the repository, which can help you work freely without impacting the master or primary branch.
- **Merge**
  This process refers to combining the changes of one branch with another (e.g., merging changes from a branch with the master).
- **Fetch**
  A fetch refers to retrieving the most recent changes without merging them into the codebase.
- **Pull**
  A pull refers to fetching and merging with the current branch you're requesting from.
- **Push**
  A push refers to sending your locally committed changes to a centralized or remote repository (e.g., GitHub).

- **Pull request**
  A pull request consists of the proposed changes to a repository as requested by a developer. A collaborator can accept or reject these changes.
- **Clone**
  Cloning is the act of copying a remote repository to your local computer in a newly created directory.

Next, let's consider abapGit.

abapGit

abapGit is an open-source Git client meant for storing and importing ABAP code. For ABAP code for SAP BTP, written using the Eclipse IDE, you can link your ABAP system to an abapGit repository. Once the package is ready, you can use the **Stage and Push** option to sync it to the abapGit repository. In the **Staging** option, you can see which ABAP objects have been changed, deleted, or created and which are ready to be transferred to the abapGit repository. More ABAP objects can be added to the **Staging** area. Once the objects are ready for moving to the repository, you can use the **Commit and Push** option, which will start the actual code synchronization with the abapGit repository.

abapGit is a good option for the various scenarios described in detail in SAP Help Portal documentation. For scenarios using SAP BTP ABAP environment, abapGit is a good option for keeping ABAP objects from the development environment (or your SAP BTP subaccount) in sync with your test or production subaccounts.

13.2.4 Deployment Options in the Cloud

When we talk about deployment, a few things must be considered. One consideration is how to package your developments. The second is what tool to deploy your applications with. These aspects are native to the technology stack and to the specific SAP product. We'll dive into these questions in this section. Next, you must consider how overall traceability can be achieved. The options for an SAP-centric landscape are to use SAP Solution Manager or the SAP Cloud ALM tool for more cloud-centric application landscapes. These tools can be augmented by the tools available from hyperscalers (e.g., Microsoft Azure DevOps tools). The other aspect is how to automatically deploy the packages through CI tools, which we'll discuss further in Section 13.2.7.

Development Packaging

For developments in the cloud on SAP BTP, you have several options for deploying your various developments. For applications with multiple technology components, which is often the case with SAP Fiori or UI5, application programming interfaces (APIs), or workflows, for instance, an application can be deployed as a multitarget application (MTA). Through the packaging of an application as an MTA, you can deploy the application in SAP BTP. This MTA file can also be used in CI/CD pipelines, which we'll discuss in Section 13.2.7.

An MTA is a package that can contain multiple application and resource modules, created with different technologies and that may require different runtimes. However, these modules share a common lifecycle. The MTA package contains the modules, descriptions, and interdependencies to other modules, services, and interfaces. MTAs can be deployed via SAP HANA extended application services, advanced model (SAP HANA XSA); SAP BTP, Cloud Foundry environment; and SAP BTP, Neo environment. The resulting file that is deployed is an *.mtar* file.

Transport Options

The Change and Transport System (CTS) is a well-established process for moving ABAP objects and configuration changes from one system to another. A few years ago, SAP came up with an enhanced version of CTS, called CTS+, to implement similar transportation concepts for non-ABAP objects (e.g., Java, SAP HANA artifacts, etc.). You can use CTS+ to move objects between different SAP BTP subaccounts or accounts. You must, however, set up your CTS+ system in an SAP NetWeaver 7.4 Support Package Stack (SPS) 10 or higher environment or in SAP Solution Manager 7.2.

A typical integration for CTS+ and SAP BTP would involve the following steps:

1. Set up SAP BTP accounts/subaccounts.
2. Establish an HTTPS connection between the CTS+ server and your SAP BTP subaccounts.
3. Configure the Transport Organizer Web UI.
4. Configure application types.
5. Create virtual systems in CTS+ to map all SAP BTP accounts.
6. Create transport routes between virtual systems.

A typical transport process using CTS+ for SAP BTP development objects involves the following steps:

1. Create an MTA.
2. Create a transport request.
3. Attach the MTA to a transport request.
4. Release the transport request.
5. Import the transport request into the target system(s).

During transport or deployment from development to other environments, the traditional tool of CTS or CTS+ has been augmented with Git-enabled CTS (gCTS). Then comes the option of Cloud Transport Management, which is a service available in SAP BTP. We'll cover these in the following sections.

Git-Enabled Change and Transport System

gCTS is used for deployment of ABAP code in SAP BTP and on-premise SAP S/4HANA 1909 and later. While the base of this tool is SAP's standard transport management, it integrates with GitHub and can leverage some of the interesting functions of GitHub, beyond code versioning, like branching and merging. Some configurations are needed to connect the ABAP packages to the repositories in GitHub which then becomes the central repository. Multiple repositories are supported and then projects can be organized in different repositories of the same Git server, for a particular customer. Also, gCTS supports both ABAP development as well as customizing. gCTS can be integrated with CI tools, which is covered in some detail in Section 13.2.7 when we discuss project "Piper". Finally, there is an SAP Fiori app for various configurations like accessing common Git functions. The Git functions are cloning the remote Git repository to the ABAP system, committing ABAP objects, pushing the commits to the remote repository, and pulling changed objects to target systems. Figure 13.7 shows the build information using gCTS on SAP BTP.

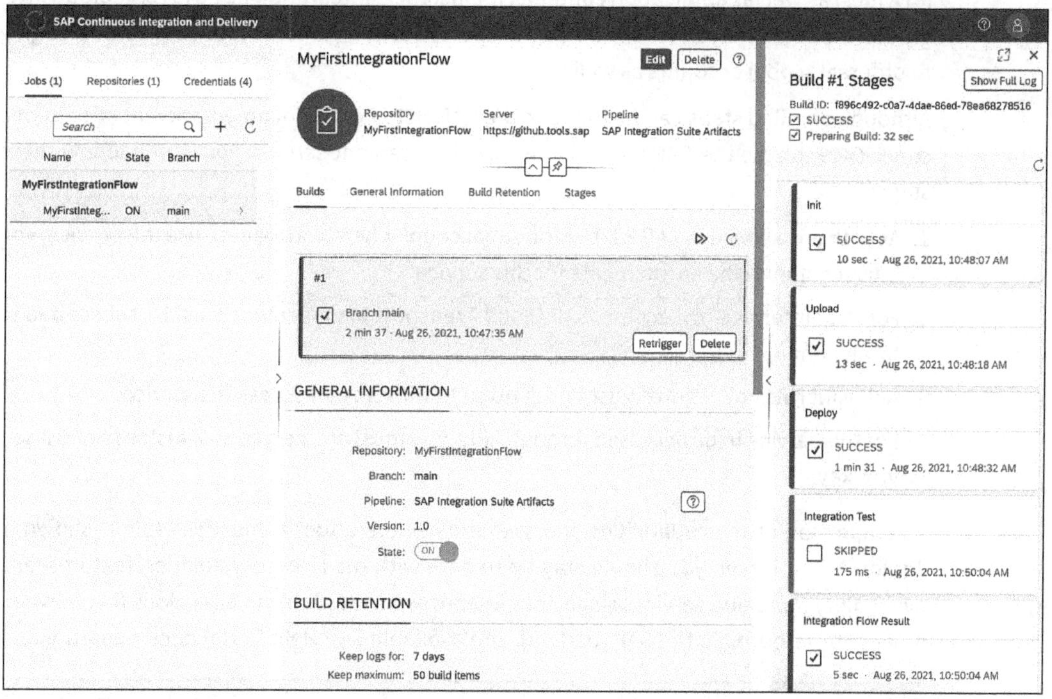

Figure 13.7: Showing the Build Process in SAP BTP Using gCTS

The major steps for gCTS setup are:

1. Set up the gCTS SAP Fiori app for use in the ABAP system where you want to configure gCTS.
2. Enable the gCTS in the ABAP system through this app.
3. Decide and set up authentication between the ABAP system and the Git repository.
4. Clone the remote Git repository to the ABAP system, thus ensuring that the content of the remote repository is accessible for your ABAP system by importing it to the ABAP system.

5. Configure the repositories. For example, disable the default parameter of automatic push-pull (depending on the situation).
6. ABAP objects should use packages which have configuration “managed by gCTS.”
7. Perform development in ABAP using common SAP tools like ADT.
8. Push the changed objects in the repository.
9. The target system pulls the changed objects.

SAP Cloud Transport Management

This service from SAP BTP focuses on transporting objects in SAP BTP, Neo environment, or SAP BTP, Cloud Foundry environment. This service can transport both developments (i.e., MTA files) as well as application contents, for instance, integration flows for SAP Cloud Integration for data services. It can handle transports between different subaccounts belonging to different global accounts as well.

Although detailed steps can be found in the SAP Help Portal documentation at *http://s-prs.co/v550409*, to use the SAP Cloud Transport Management service, you would follow these steps:

1. As with any service in SAP BTP, in the subaccount where you want to use this service, you must configure the entitlements for this service.
2. For user interface (UI) access, SAP Cloud Transport Management must be subscribed to by an admin.
3. Relevant role collections must be set up so that users can access the service.
4. For this service to be accessed through APIs, you must create a service instance and a service key.

A comparison of the capabilities and scenarios where a tool is most suitable is shown in Table 13.1. However, you should stay up to date with the latest capabilities, features, and availability (for cloud services since not all services are available in all regions for all hyperscalers) by referring to the SAP roadmap and available SAP Help Portal documentation.

Areas	CTS/CTS+	gCTS	SAP Cloud Transportation Management
Available as a cloud service	No	No	Yes
Supporting cloud/on-premise transport	Both	Both	Cloud only (SAP BTP, Cloud Foundry environment and SAP BTP, Neo environment)

Table 13.1: Capabilities of the Various SAP Transport Mechanisms Available

Areas	CTS/CTS+	gCTS	SAP Cloud Transportation Management
Supported technology stack	ABAP/non-ABAP development-specific content	ABAP development specific-content	Non-ABAP (development and application-specific contents)
Integration with CI tool (Jenkins)	Not supported	Supported (also possible to integrate with Microsoft Azure DevOps)	Supported
Integration with SAP Solution Manager	Supported	Supported	Not supported
Integration with SAP Cloud ALM	Not supported	Not supported	Supported
Supports distributed development	Not Supported	Supported	Not supported
Adoption	Widely used and very mature but cannot handle several SAP cloud solutions; not suitable from a DevOps perspective	Usage is just starting, and maturity is medium to high (because of the underlying SAP Cloud Transport Management service). This option is quite suitable for DevOps, but the scope of the technology is small.	Low usage so far; expected to pick up to handle the deployment of SAP cloud products.

Table 13.1: Capabilities of the Various SAP Transport Mechanisms Available (Cont.)

From a DevOps perspective, depending on the complexity of the customer landscape, whether primarily on-premise, primarily cloud based, or a hybrid, a combination of tools like gCTS or CTS+ with the SAP Cloud Transport Management service may be required. Once the development or customization is ready for deployment, several things must be done before moving to production. These actions, defined as part of the release process, can include quality assurance/static code quality checks through tools as well as dynamic checks using mostly automated and some manual testing. Finally, the transports should be integrated with change management tools like SAP Solution Manager or SAP Cloud ALM. These activities could be orchestrated through the CI tool, which is the glue tying together the process to make it automatic and efficient. Now, let's explore quality checks in more depth.

In summary, when it comes to choosing the right transport tool, it really depends on whether your landscape is on-premise, cloud-based, or hybrid. For traditional on-premise environments, CTS or CTS+ is typically the primary choice, with CTS+ handling non-ABAP objects alongside standard ABAP transports. If you're working with modern, pipeline-driven ABAP development, gCTS is a better fit because it integrates ABAP changes with Git and enables seamless CI/CD workflows. On the other hand, for cloud-native content on SAP BTP, SAP Cloud Transport Management is the right solution, offering a fully managed, cloud-based way to move artifacts across subaccounts without relying on an onpremise system. In most real-world scenarios, organizations use a combination of these tools and tie them together through SAP Solution Manager's Change Request Management (ChaRM) to ensure consistent governance across a hybrid landscape.

13.2.5 Quality Assurance

Quality engineering is a subject on its own with wide application. From a DevOps perspective, quality must be embedded in each phase of the application lifecycle and, like everything else, automated as much as possible. In fact, some measurements of DevOps KPIs are related to quality and indicate how successful the DevOps program is, for example, the KPIs shown in Table 13.2.

KPI	CTS/CTS+
Deployment frequency	For the primary application or service you work on, how often does your organization deploy code?
Lead time for changes	For the primary application or service you work on, what is your lead time for changes (i.e., how long does it take to go from a code commit to the code successfully running in production)?
Mean time to recover (MTTR)	For the primary application or service you work on, how long does it generally take to restore service when a service incident occurs (e.g., unplanned outage or service impairment)?
Change failure rate	For the primary application or service you work on, what percentage of changes results either in degraded service (e.g., leads to service impairment or service outage) or subsequently requires remediation (e.g., requires a hotfix, rollback, fix forward, or patch)?

Table 13.2: DevOps KPIs for Measuring Quality

In this section, we'll cover quality assurance from a development perspective. Testing is one pillar of quality assurance, while code quality check (in terms of best practices, security, scalability, etc.) is a second pillar.

Testing concepts have always existed in the SAP world, including automated testing. Now, two new concepts have gained popularity in the SAP DevOps community—TDD and shift-left testing. As described earlier in Section 13.1.1, these concepts were adopted in the non-SAP community much earlier; only in the last 3 to 4 years have these concepts gained momentum in the SAP arena.

However, when we talk about testing from a DevOps perspective, we mean testing at all stages of development including unit testing, ideally in an automated fashion. If we test more than just a sample scenario, instead testing all scenarios and all parts of the code that produce some output, we can prevent defects from seeping into the next environment, thereby saving time. This efficiency is achieved through writing code that automates the testing of code as soon as the code is released or committed to Git. Ensuring code quality is an important step for DevOps because the focus is not just on faster development of functionality, but also better-quality code.

Some example unit test tools include the following:

- ABAP Unit framework where ABAP Unit is the runtime tool for checking the functional correctness of a program.
- JUnit is another choice for Java programs.
- QUnit is used for UI5 unit testing.

Examples of a unit test tool script for each of these technologies are shown in Figure 13.8 and Figure 13.9, respectively.

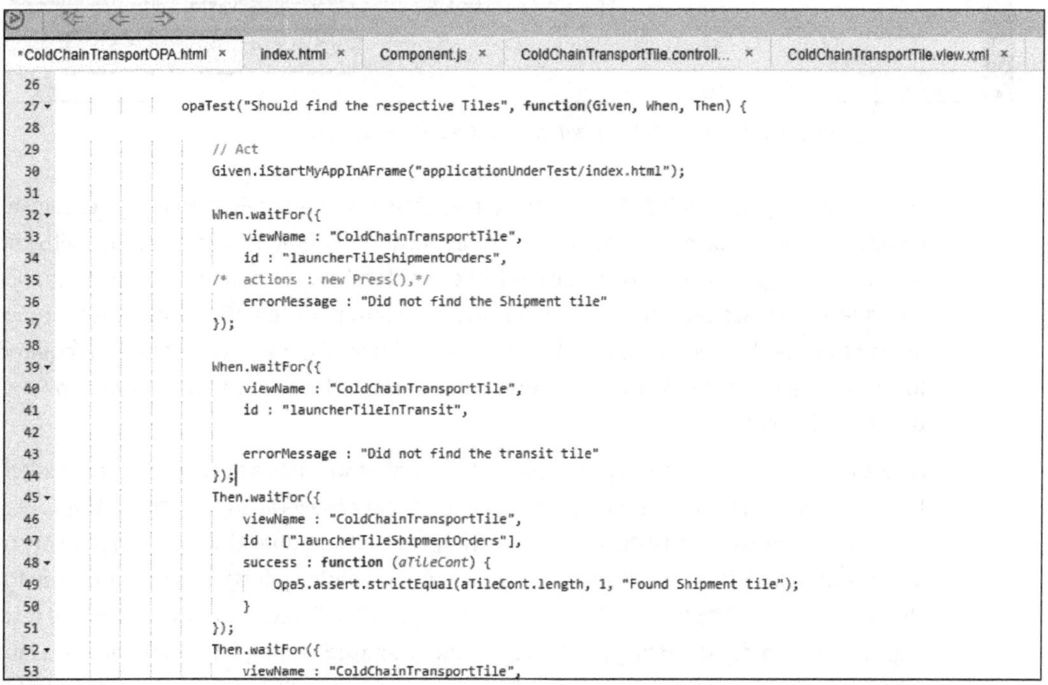

Figure 13.8: OPA5 Sample Code for Frontend Testing

From an ABAP code quality perspective, the ATC is a powerful framework with many tools providing the functionalities required from a quality perspective. For instance, the Code Inspector can perform static code analysis of ABAP code, while SAP Code Vulnerability Analyzer conducts vulnerability assessments of static code. The ATC features additional checks specific to SAP HANA databases as well as data model perspectives for SAP S/4HANA. The ABAP Unit framework is actually also part of the overall ATC umbrella.

A best practice is to include an ATC check once a developer performs a code check-in or requests a transport release once development is done. This step should be included in the activities triggered by the pipeline.

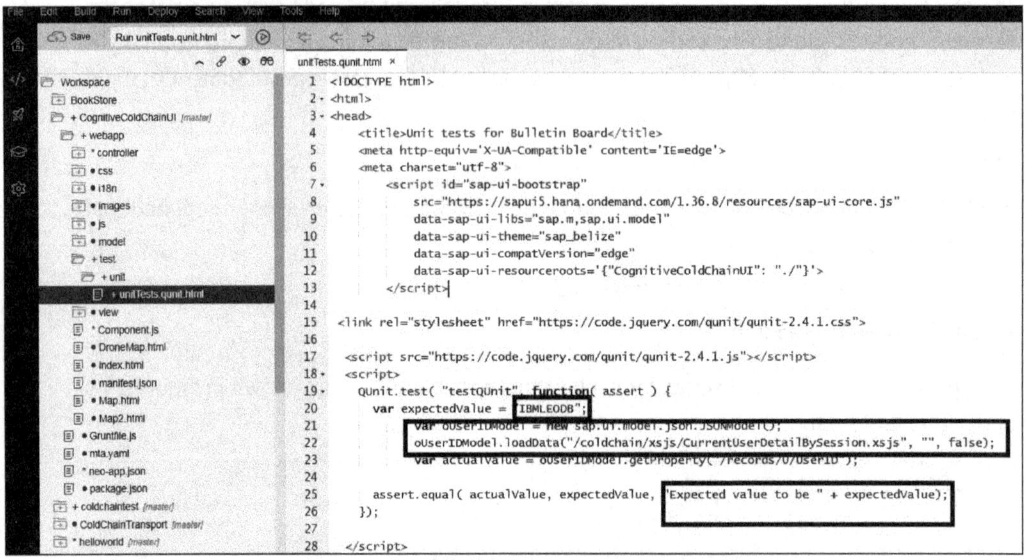

Figure 13.9: QUnit Code for Automated Unit Testing of JavaScript

Other tools may be required for testing at later stages of the application lifecycle, such as integration tests that ensure the complete application is functioning correctly, with the necessary integration between backend and frontend working as expected. Scenario-based testing is also relevant, as it verifies the application's integration with other components of the landscape for expected outcomes. Additionally, regression testing is required for maintenance scenarios to ensure that other parts of the application continue to function after changes.

All these tests require some type of test tool, several of which are available on the market. However, SAP has been partnering with Tricentis to cover several SAP products. This tool is powerful for integration testing and regression test automation. UIVeri5, an open-source JavaScript testing framework aimed at system and scenario UI testing, also works well for UI checks in an SAPUI5 frontend; however, this approach will require more end-to-end testing. Another tool for UI testing is OPA5, a JavaScript-based testing framework designed for component testing with SAPUI5. However, we don't recommend using this tool for end-to-end scenarios due to certain gaps, such as in handling authentication.

13.2.6 Monitoring and Feedback

For the "Ops" part of DevOps, you need to operate the application landscape and utilize any opportunity to automate its processes because often these activities are quite repetitive. To ensure that the applications are all running fine, with availability according to their service level agreements (SLAs) or to see if any issues or failures in the interfaces have occurred, proactive monitoring is required. Also, as soon as any failure occurs anywhere, you must be alerted. A logging feature should capture data to help you identify the issue. Finally, some reports and analytics should be available.

This monitoring and alert framework thus must exist across technologies so that you can access different views of the technology and monitor and measure metrics at a process level, at an application level, at a product level, and at the overall landscape level. Monitoring applications can be achieved at various levels. SAP BTP has three services relevant for operations:

- The SAP Monitoring service for SAP BTP provides a layer of low-level monitoring by measuring container-related metrics (e.g., CPU, memory, data usage). Then, this service can provide traces and logs to show you what happened to the application after deployment during a particular period. SAP BTP has an integration with the Kibana Dashboard, which is a tool for visualization, for analyzing logs, for conducting time-series analytics, and for application monitoring, all of which can be displayed.
- The SAP Alert Notification service for SAP BTP provides proxies for alerts coming from SAP BTP and from hyperscalers.
- SAP Automation Pilot automates repetitive system maintenance tasks.

From a monitoring point of view, SAP's monitoring service can be augmented by additional external monitoring and application logging solutions, all integrated with internal and external services and third-party application monitoring systems like Splunk, Dynatrace, Catchpoint, or Prometheus, for a comprehensive view of applications. Dynatrace, for example, can also be attached to SAP BTP and offers several visualizations of applications and services that show interdependencies among components or services.

Further, in SAP BTP, Cloud Foundry environment, an application logging service within SAP BTP works alongside the Loggregator system and is used for streaming log messages and container metrics. This component uses the Elastic Stack (ELK) suite, which constitutes Elasticsearch, Kibana, and integrations. (You can find more details at *https://www.elastic.co*.)

For alerts, this service works on SAP BTP, Neo environment, or SAP BTP, Cloud Foundry environment and any hyperscaler. These alerts are of the publish-subscribe type and can be sent to any channel of choice (i.e., Microsoft Teams, Jira, ServiceNow, Slack, etc.); what actions should be performed based on an alert will depend on the scenario. Some simple configuration is required in the subscription model for this service so that the receiving system can interpret alerts and understand the data passed through alert events.

SAP Automation Pilot has ready-to-use content for typical scenarios like database restarts or SAP HANA updates and currently supports the scripts in Perl, Shell, Node.js, or Python. This service can perform process orchestration as well. Thus, if any system activity does not support the scripting languages we just mentioned, but has APIs, then these APIs can be used for orchestration as well. This service has a set of commands (e.g., `UpdateHANAServiceInstance`), and scripting is required only if the command you need is not available. In the orchestration process, SAP Automation Pilot is integrated with the SAP Alert Notification service. You can also include manual decision steps, and thus, the services will stop and wait for manual inputs. You can also schedule the automation program or sequence runs.

Feedback is an essential component of the DevOps cycle to make your applications better by removing bugs and adding features to improve user satisfaction. Apart from incidents or tickets, users can now provide feedback through various communication channels like Microsoft Teams or Slack channels, for instance, for a particular product, application, or solution. Some automations are possible in terms of ticket/issue categorization. However, the next step is typically a manual process through which incidents categorized as bugs or feature requests are selected and prioritized in the backlog.

[+] **Tip**

More details about monitoring services and their relevant configuration settings are found in the following SAP Help Portal documentation links:

- For application logging: *http://s-prs.co/v550410*
- For SAP Alert Notification service: *http://s-prs.co/v550411*
- For SAP Automation Pilot: *http://s-prs.co/v550412*

13.2.7 Continuous Delivery Options in SAP BTP

When we talk about CD, we're generally referring to both CI and CD. Earlier in Section 13.2.4, when we covered the integration capabilities of various transport tools, you may have wondered why all that is required, whether with SAP Solution Manager/SAP Cloud ALM or with a CI tool.

In this section, you'll learn about *pipeline engineering*, which is a concept newly adopted in SAP. Then, we'll proceed with the tooling options available in SAP BTP for enabling CI to implement such pipelines. This approach helps in automating deployments. Further integration with change management tools helps you control your deployments as well.

Pipeline Engineering

Now an important aspect of the DevOps process, pipeline engineering entails translating the overall development process, including quality assurance, approvals, and notifications.

Depending on an agreed-to process for a project or in a support scenario for different applications, pipelines must be set up. For example, the flow diagram shown in Figure 13.10 involves such a deployment process flow, using gCTS and a CI tool like Jenkins.

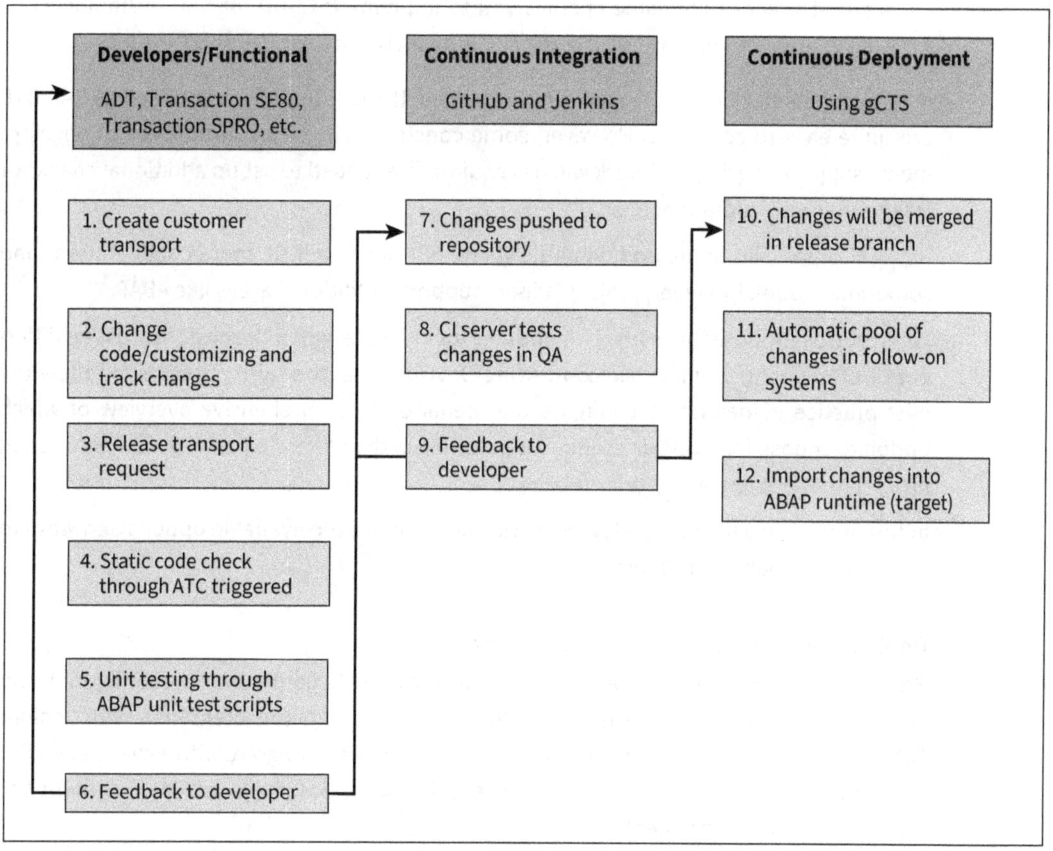

Figure 13.10: Example Deployment Process Using gCTS

SAP BTP provides different kinds of solutions, both services and contents, to enable DevOps CI/CD for SAP-centric technologies. Let's explore the options SAP BTP provides for enabling pipelines next:

- **SAP Continuous Integration and Delivery**
  Using this service, you can configure and run predefined CI and CD pipelines that build, test, and deploy your code changes. The following technologies and models are supported by this service:
  - SAP Cloud Application Programming Model
  - SAP Fiori
  - Cloud Integration
  - Container-based applications projects

- **Project "Piper"**
  We'll cover this option in more detail in the next section.
- **Continuous Integration and Delivery Best Practices Guide**
  This best practice guideline enables you to implement CI/CD for SAP technologies by using CI pipelines on any CI/CD stack and is available at *http://s-prs.co/v550413*.

In terms of the flexibility of the solutions provided, the CI/CD services provided by SAP BTP are quite easy to configure. However, some constraints in terms of the type of developments supported exist, and work will be required if you need to set up additional checks or integration with additional tools.

Project "Piper" has some additional flexibility but needs a little more configuration (and sometimes code). However, project "Piper" supports additional areas like ABAP.

When you cannot use these first two options, for example, when the customer uses a different CI/CD pipeline tool like Microsoft Azure DevOps, then the third option is to follow the best practice guidelines to configure the scenario. A comprehensive overview of which option to choose for a certain scenario is provided in the SAP Help Portal documentation at *http://s-prs.co/v550414*.

In this section, we'll discuss a few scenarios and explore the available options and ways to configure CI pipelines for these scenarios.

Deployment Automation: Project "Piper"

Project "Piper" is an open-source project that provides CI content for preconfigured Jenkins pipelines for certain scenarios around SAP technology for integration with Jenkins (which is a third-party CI tool). The contents can be used to integrate with a client's Jenkins infrastructure and also adapted for client-specific scenarios. Project "Piper" content has the following two components:

- A shared library at GitHub containing a description of the steps, scenarios, and utilities required to use Jenkins pipelines
- A set of Docker images useful for implementing best practice processes

Project "Piper" includes contents for an SAP BTP ABAP environment (colloquially referred to as "Steampunk") pipeline as well as a general-purpose pipeline. The ABAP pipeline supports SAP BTP ABAP environment, while the general-purpose pipeline supports MTA files and SAP Cloud Application Programming Model applications.

The pipeline for ABAP involves several predefined steps that can be executed in the Jenkins server and that leverage APIs of the ABAP environment. Project "Piper" features a shared Jenkins library containing reusable assets built for this purpose.

As shown in Figure 13.11, which shows the build pipeline steps, project "Piper" leverages several APIs, such as the following:

- The Cloud Foundry command-line interface (CLI) to spawn/delete ABAP environments on SAP BTP, Cloud Foundry environment automatically (through scripts)
- The ABAP Git repository management for SAP BTP ABAP environment service for connecting to Git and to pull code into the target system
- ATC to run checks and unit test scripts

Figure 13.11 shows part of the project "Piper" scenario ready for SAP BTP ABAP environment. Jenkins pipelines can be configured. Each pipeline has stages which in turn each have steps. Each stage of the pipeline can be extended so that you can cater to customer-specific scenarios.

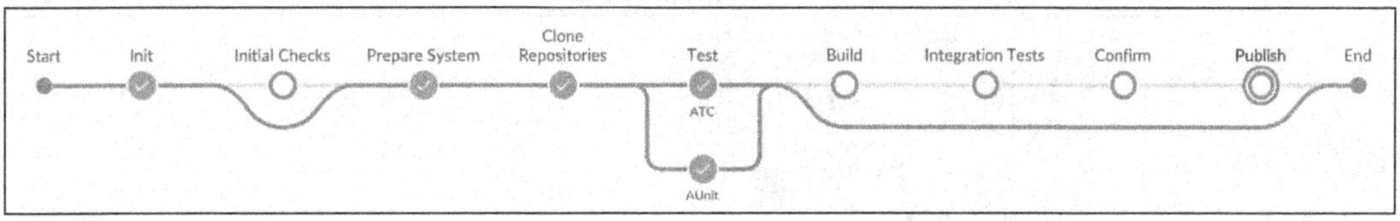

Figure 13.11: Jenkins Pipeline Flow for ABAP on Steampunk CI/CD Scenario

More details about configuration, extension, and example scenarios are provided in the documentation for project "Piper" at *https://www.project-piper.io/guidedtour/*.

Integrating with Change Management Solutions

Managing changes in software (both development and configuration changes) across multiple types of products (i.e., SaaS, PaaS, or on-premise) hosted across different hyperscalers or in multiple data centers, and even across multiple technologies, can be quite challenging. Although not an IT requirement, business processes may require such change management solutions because processes may run across multiple products and platforms. An example landscape from the perspective of a typical large client is shown in Figure 13.12.

SAP Solution Manager is the main change control solution that can handle this integration, while SAP Cloud ALM caters to more cloud-centric solutions. For transporting or deploying various technologies, we covered several solutions from SAP in Section 13.2.4, like CTS, CTS+, gCTS, SAP Cloud Transport Management, and even non-SAP CI tools like Jenkins. These transport tools must be controlled through the change management process and through change management tools so that you can have complete auditability and are compliant with legal requirements from, for example, the Sarbanes-Oxley Act. This auditability requires that your transport and CI tools have good integration with change management tools like SAP Solution Manager, and this integration does exist today through public APIs.

Thus, any change in any technology can be managed by any transport management tool and integrated into a change document within the implementation service for change request management with SAP Solution Manager and is centrally controlled.

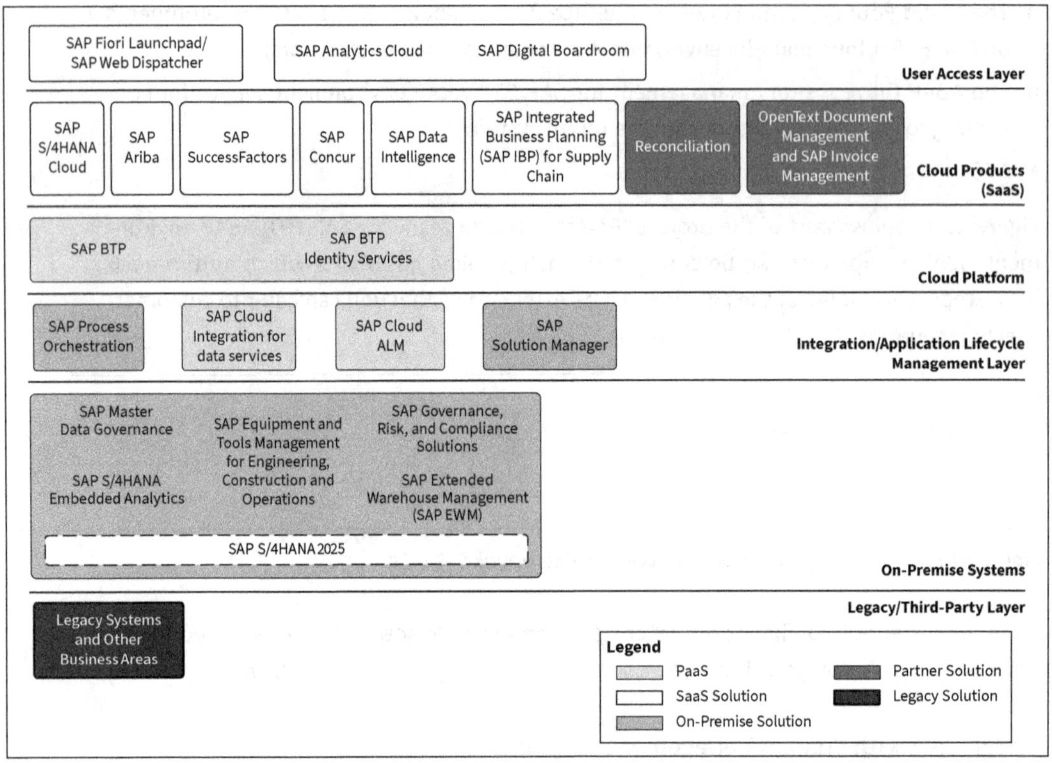

Figure 13.12: Hybrid Landscape with Application Lifecycle Management Using SAP Solution Manager and SAP Cloud ALM

The major steps for setting up this kind of integrated scenario for change management include the following:

1. Write code (in any IDE).
2. Push the code to a Git server (source code management).
3. The SAP DevOps services for change management or SAP BTP can call the CI server (Jenkins).
4. The SAP DevOps services client can connect with SAP Solution Manager through an OData service. Thus, the code (perhaps after some iterations of testing) will be deployed in the dev server as part of step 3 and in parallel can be attached to a change document (and to a transport request) in SAP Solution Manager in step 4.
5. The transport request can be in CTS+, SAP Cloud Transport Management service, or gCTS.
6. A handover occurs from the CI server to the change management process in SAP Solution Manager (and eventually in SAP Cloud ALM in the future).

13.3 Best Practices for DevOps in SAP

Let's now consider some best practices that you can apply for DevOps in a cloud-native development environment and, to some extent, in the ABAP world, such as the following:

- DevOps projects generally have several phases: the think, code, manage, and learn phases. In the continuous planning phase (the think phase), as the project continues, note that several MVPs for new and improved features might be under development at the same time in different phases of the method. For example, while some stakeholders formulate the next MVP (in the think phase), developers may be coding another MVP (in the code phase), while the operations team is monitoring several deployed MVPs (in the manage phase). Finally, the whole team can learn about deployed MVPs through customer feedback and analytics (in the learn phase).
- A common practice in CI is refactoring because, when a new function is added, other code affecting other functions will be impacted. We talked about tools for change impact analysis earlier in Section 13.2.1.
- While setting up DevOps tooling for a project, tools must be identified based on the technology of the applications being built. Often, a toolset may already be set up as part of the DevOps process. You'll initially set up tools with a view toward serving a specific project, but then the tool's usage may be extended to other projects. Thus, a *toolchain* is a set of tools that enable the rapid development, testing, and deployment of an application that is scalable, resilient, and made for the cloud. A toolchain needs to be chosen that suits the way your project team works and the type of applications they are developing.
- Fast-changing business needs now can be addressed with the help of the cloud, and your teams can continuously deliver new features of the highest quality. Teams will also need tools that support a repeatable process to deliver the same code to test and production environments. Teams can use infrastructure as code techniques to automatically provision identical test and production environments. This concept is less common in SAP environments where an environment is usually provisioned for each application. However, an environment may be provisioned for a particular testing period. The Deliver phase is about understanding and using these tools to ensure a repeatable, automated delivery process that results in a fully tested production-ready apps.
- To achieve CD in a consistent and reliable way, the team must create and configure a delivery pipeline to divide the software delivery process into stages. Ideally, the delivery pipeline moves the code forward through each stage automatically with minimal human intervention. In practice, some additional quality checks must be maintained as code passes to production. The goal is for the final stage of the pipeline to produce code that is ready for production. In Jenkins, the delivery pipeline can be created automatically through code, for example, using Apache Groovy. To manage workflows, tools like IBM UrbanCode Deploy can be useful.
- In the manage phase, a monitoring solution detects problems across the data centers where the application is deployed. If automated monitoring detects a failure in a data

center that hosts the application and no failover location has been set up, the users of the application will experience an outage. To reduce outage times, your team can implement a backup instance of the application in another data center with failover automation. These kinds of services are normally part of the cloud platform where the application is hosted.

- Further, services that the production application depends on must be monitored to ensure that the team knows when dependent services are experiencing outages. Some service outages result in what appear to be application outages to end users. To some extent, service monitoring can be set up depending on where these services are hosted. For example, if hosted on an SAP BTP service, the platform will perform the monitoring of these services.
- Finally, tools like SAP Solution Manager or SAP Focused Run feature general dashboards that can include other application-specific dashboards and reports that show detailed outage data and metrics to indicate whether an application meets the defined SLA.

[»]

Note

The benefits achieved from DevOps depend on which phases you implement. In our experience, in general, tooling and setting up a DevOps platform and the adoption of specific tools can take 6 to 8 months. Then, you can engage in a continuous improvement cycle.

Also, the adoption of DevOps principles may require time because mindsets must change. While you should not expect immediate returns right after a DevOps implementation, adopting DevOps will reduce development and test efforts drastically, and the return on investment in a DevOps implementation should be achieved within a year. One major and immediate benefit for development, starting with CI/CD pipelines, is that you can do development and deployment in an Agile way, through automated pipelines that include automated quality checks, unit testing, and deployment to the next environment. All these activities would otherwise take up manual effort and time and make the cycle dependent on various teams working in silos. This automation can thus result in a lot of efficiency.

More benefits are reaped over a longer term, such as the following:

- Agility in meeting business requirements helps improve customer (end user) satisfaction.
- Agility in meeting business requirements also improves business efficiency since changing business needs can be addressed using this method.
- Continuous feedback helps you consider issues flagged by key users so you can take prompt action to mitigate issues, again improving customer satisfaction.
- Reduction in new product development timeline using continuous business planning, which can be quite high.

13.4 DevOps Tooling Integration for SAP and Hybrid Landscapes

So far, we've explored the various tooling options that you can use to implement DevOps in an SAP-centric landscape. Figure 13.13 shows the DevOps services available from SAP BTP, which includes, as discussed throughout this chapter, several tools for application lifecycle management and monitoring, automation, and more.

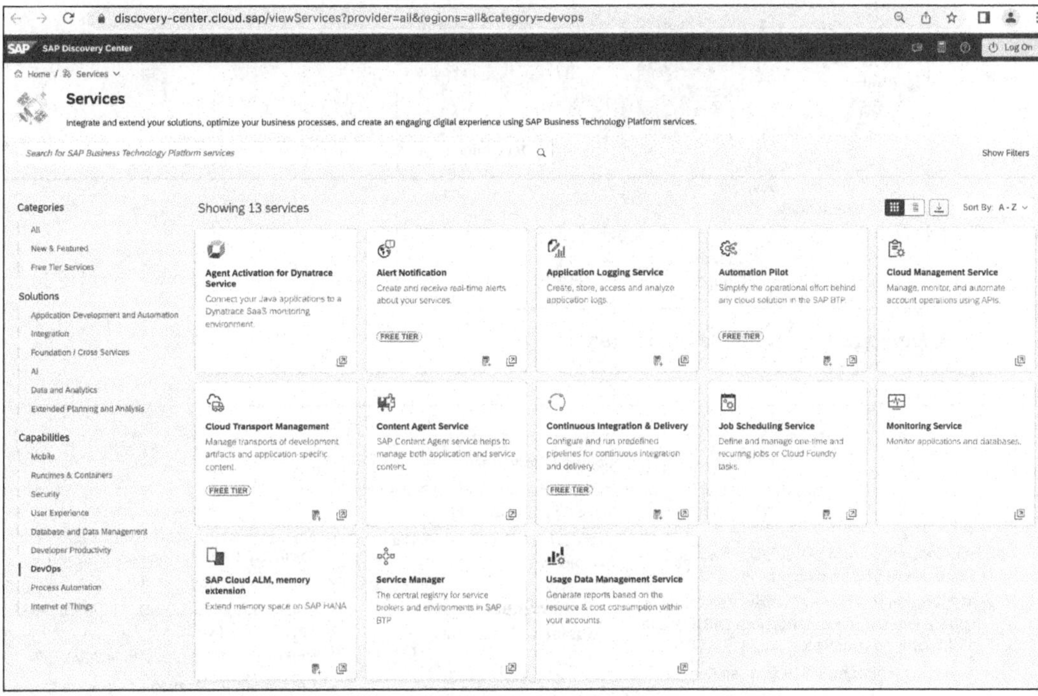

Figure 13.13: SAP Discovery Center showing SAP BTP DevOps Services

Let's now focus on the entire process for DevOps adoption. In simple terms, if you need to implement the basics of DevOps, you must plan which tools will be used and how these tools must be set up. Once this information is determined, you can implement the tooling and integration so that your development and support teams can start using them. A simple phase-wise adoption cycle is shown in Figure 13.14.

Some complications may arise, and some choices may need to be made in each step, especially from a tooling point of view. Figure 13.15 shows some tool options for different types of landscape, whether SAP BTP-based or a hybrid (on-premise and cloud combined). For some SAP products, DevOps is not yet mature, while others are more tuned toward DevOps adoption, such as cloud-native development in SAP BTP.

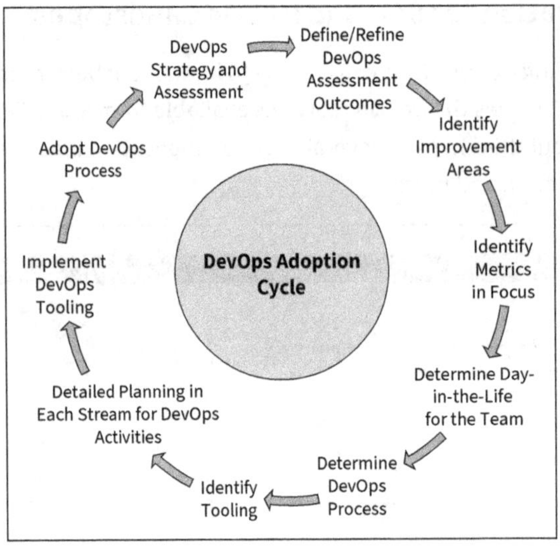

Figure 13.14: DevOps Adoption Steps

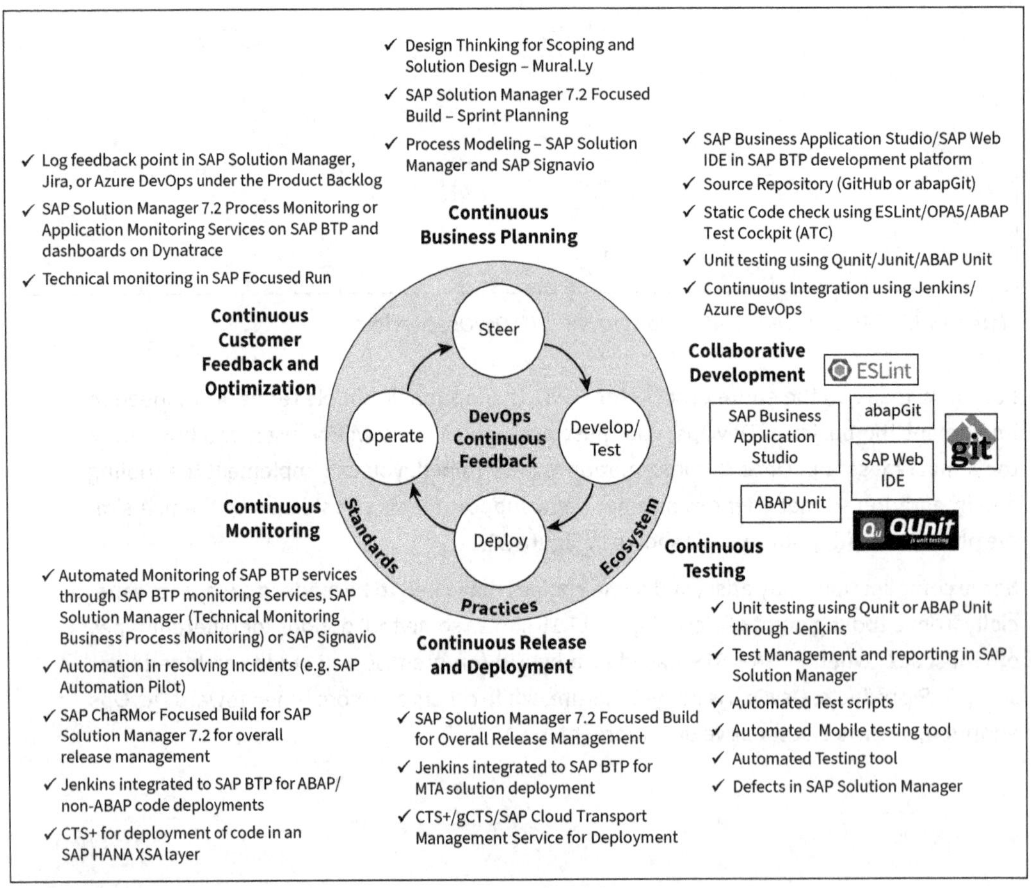

Figure 13.15: DevOps Tooling Options in an SAP-centric Hybrid Landscape

Now comes defining the steps of the process, which in turn is typically designed with more than one deployment pipeline, as in the simple example shown in Figure 13.16.

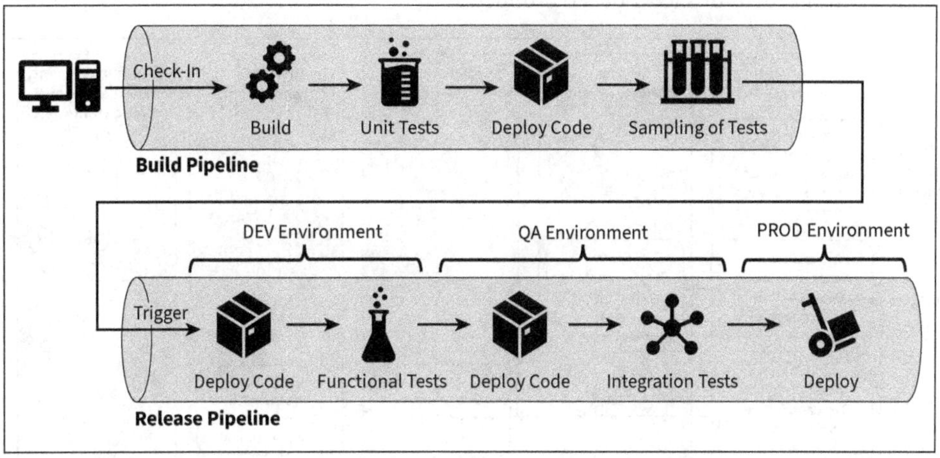

Figure 13.16: Multiple Pipelines

Note that we have two pipelines: one for build and another for release. These pipelines show the stages of the process, and you can drill down into more detailed steps to show the specific tasks for quality checks, testing, and more.

From the pipeline execution view, some metrics about the application's health also emerge, for example, how much time was required to deploy the application to an environment. The execution result of a Jenkins pipeline is shown in Figure 13.17.

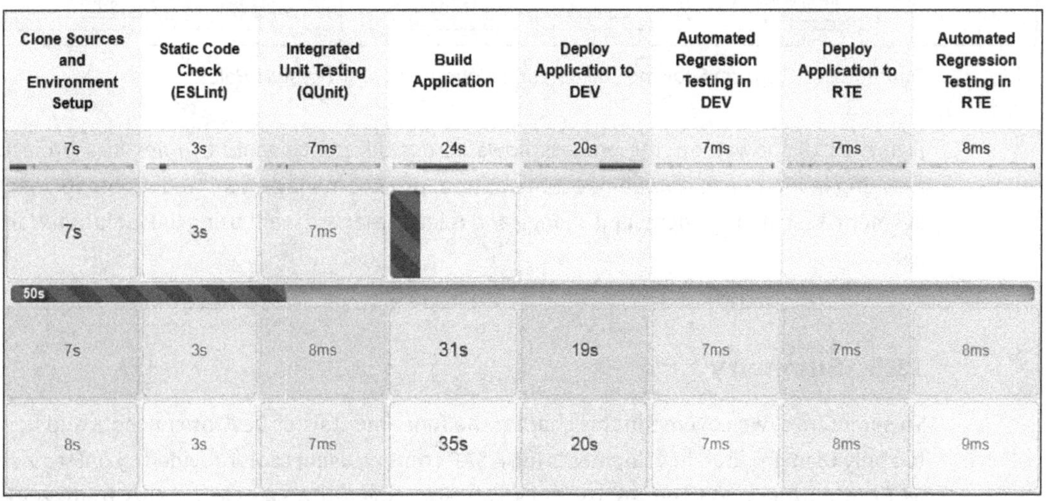

Figure 13.17: Jenkins Pipeline: Execution View

The execution view shows a log of every step of the process designed in the pipeline and includes how long executing each step required.

Finally, as shown in Figure 13.18, SAP Solution Manager can play a significant role across multiple phases of the application lifecycle. SAP Cloud ALM is coming to play a similar role, though with a focus on cloud-based products.

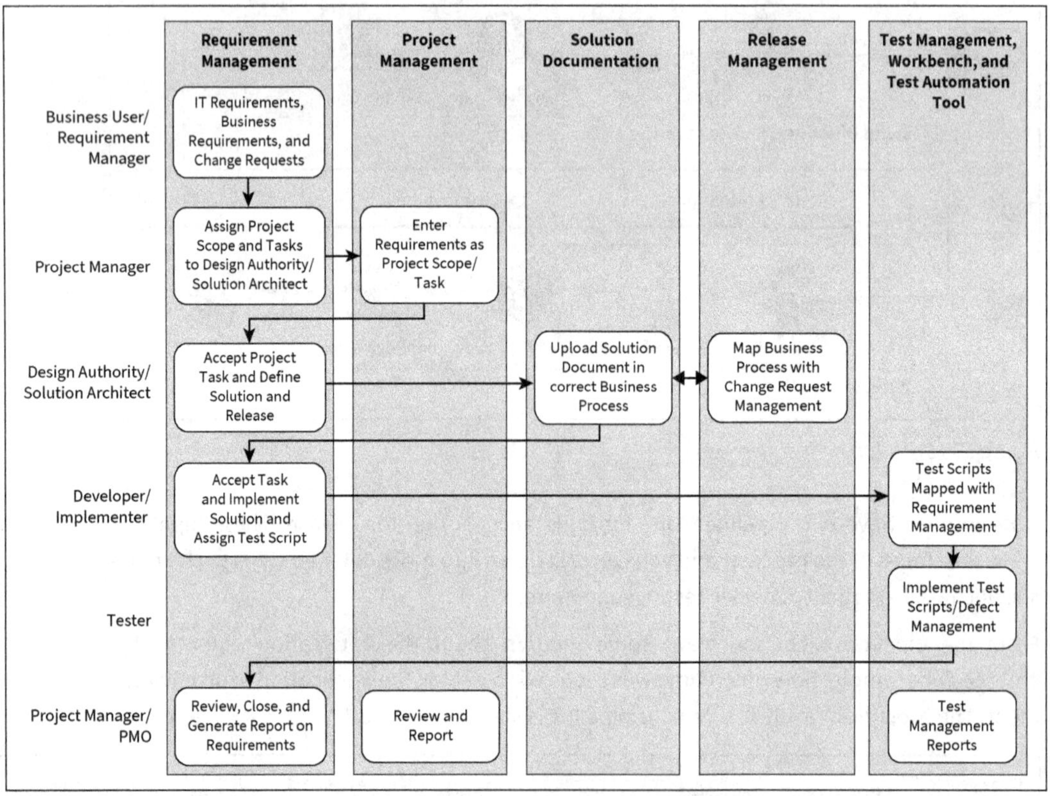

Figure 13.18: Diagram Showing SAP Solution Manager's Many Possible Roles

Figure 13.18 shows how the process flows across different agents or roles in a project, resulting in integrated solution documentation, project management, requirements management, test management, and change and release management using SAP Solution Manager.

13.5 Summary

To summarize, we covered in this chapter the fundamentals of DevOps concepts and how to apply them in cloud developments in an SAP context, using tools provided by SAP as well as from partners and third-party tools. We also covered a sample DevOps framework, touching on the concepts of CI and CD, various application lifecycle management tools, and how they form the core fabric beneath the implementation of DevOps. We also shared several best practices around requirements management, CD, overall DevOps adoption practices, tooling, pipeline engineering, tools and services from SAP BTP, and example DevOps tooling for diverse complex landscapes that mix cloud and on-premise, SAP and non-SAP systems. Finally, you learned about the benefits of DevOps for development.

DevOps is a new way of working where development and operations occur in iterative cycles, and as a result, business requirements are translated to IT solutions more quickly and more efficiently. Many concepts are new to the SAP world but have been in use in the non-SAP world, and you should try to leverage the lessons from this "other" world to avoid making the same mistakes. Thus, while a large chunk of work involves implementing the right toolchain, the process and the people aspects also play huge roles in the adoption and subsequent success of a DevOps project. Some additional roles have also evolved. Apart from the DevOps architect/engineer, a site reliability engineer must ensure that application failures are avoided.

Not all companies can adopt DevOps fully or all at once. Most often, DevOps is a journey, usually inspired by expectations about easy adoption and/or of fruitful returns on investment. The right expectations need to be managed along the DevOps journey when setting up the tooling platform. Ideally, the tooling platform should offer robust tools that cover most technologies—both SAP and non-SAP, on-premise and cloud based. You can invest in an entire stack of tooling on your own, or you can leverage cloud-based tooling available from various vendors. The benefits of following DevOps have been established enough that many mature IT organizations have adopted this paradigm, and the SAP portfolio features many tools and integrations with leading market tools and products to help your organization move in this direction.

As an architect, you should define an SAP BTP application lifecycle management strategy before starting your first project to avoid future disruptions. Remember that SAP BTP is still new, and a proper application lifecycle management strategy is a key success factor for any journey to the cloud.

Chapter 14
Security

The focus of this chapter is to provide an overview of authorization and authentication services provided by SAP Business Technology Platform (SAP BTP) for securing applications built in SAP BTP. Application programming interfaces (APIs) are available for developers to implement these services in service providers.

SAP BTP provides the Identity Provisioning service and Identity Authentication service for securing applications in the cloud. These applications can interact with services built on the cloud or with services built on on-premise systems. The SAP Authorization and Trust Management service, part of extended services for user account and authentication (XSUAA), provides for the authentication of user identities in SAP BTP through trust management between service provider and identity provider. Authorizations use technical roles at the application level, which can be aggregated into business-level groups and role collections.

SAP provides Identity Provisioning for identity lifecycle management, especially for managing users, roles, and groups based on System for Cross-Domain Identity Management (SCIM) industry standards. Identity Authentication provides services for authentication, single sign-on (SSO), integration for various user stores, and self-services such as user self-registration. In addition to SAP-provided services, SAP BTP can be configured to use delegated authentication with external user stores in combination with Identity Authentication or external identity providers.

In this chapter, we'll cover various options for authentication, SSO, access management, authorizations for applications developed in SAP BTP, web security in SAP BTP, and Joule security.

14.1 SAP BTP Authentication Options

Applications in SAP BTP do not store user information. Instead, applications running on SAP BTP use identity providers as the user base. These users can be managed through an external corporate identity provider, an SAP-provided Identity Authentication tenant, or SAP ID. As shown in Figure 14.1, an Identity Authentication tenant can be used along with an existing corporate user store, like Microsoft Entra ID for authentication. We'll discuss these options in this section.

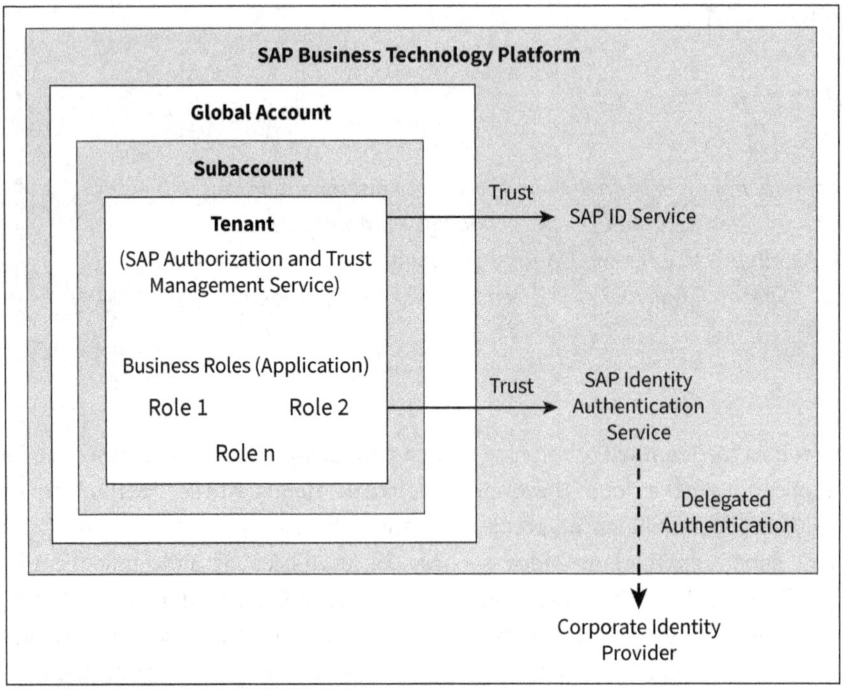

Figure 14.1: Overview of Identity Providers

14.1.1 Identity Authentication

Identity Authentication is the SAP BTP-provided solution for identity lifecycle management for applications built purely using cloud services or built using hybrid services (both cloud and on-premise). This service provides capabilities for authentication, SSO, on-premise integration, and self-services such as user self-registration. Various industry-standard SSO protocols like Security Assertion Markup Language (SAML) 2.0, OAuth 2.0, and X.509 certificates are supported by the service. In this section, we'll provide an overview of the activities required for setting up Identity Authentication in SAP BTP.

[»] **Note**

Kerberos authentication is also supported when users access SAP BTP applications from a corporate network.

[»] **Note**

Identity Authentication can also act as the identity provider or can delegate authentication to a corporate identity provider.

For using Identity Authentication, a tenant must be obtained. Each tenant represents an instance of the Identity Authentication service, with its own configuration and its own data.

Business applications in SAP BTP that have protected resources and that require authentication delegate the authentication process to Identity Authentication. In this scenario, SAP BTP acts as a service provider, and the Identity Authentication server acts as an identity provider. A trust setting must be established between a service provider and an identity provider. When users request an application, they are directed to the Identity Authentication logon page for credentials. The trust configuration can be set up between SAP BTP and an Identity Authentication tenant through the SAP BTP cockpit. Click on **Trust Configuration** from the subaccount menu, as shown in Figure 14.2.

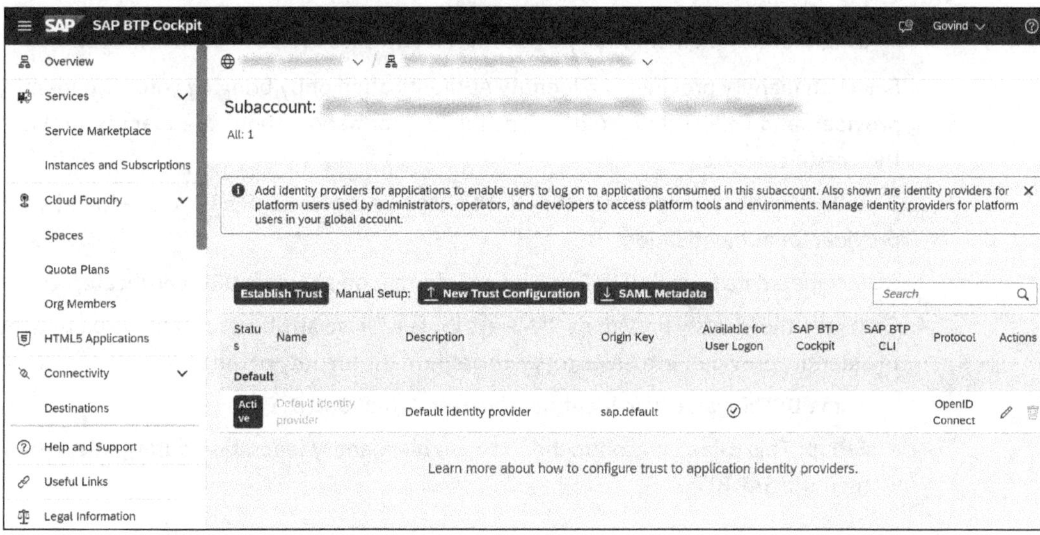

Figure 14.2: Trust Configuration Page in SAP BTP

Click on the **Establish Trust** button to manually establish trust between the Identity Authentication tenant and an identity provider, as shown in Figure 14.3.

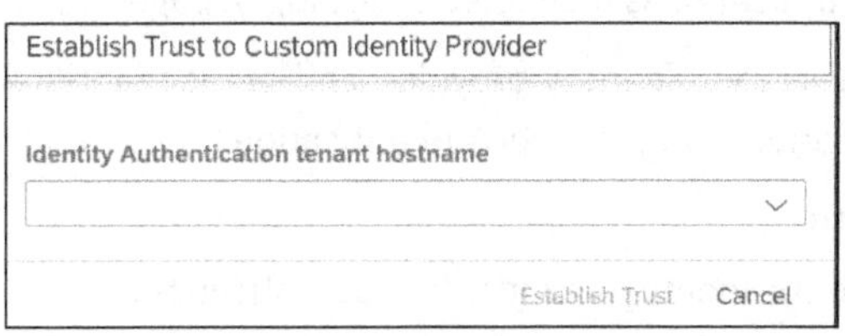

Figure 14.3: Providing Details about the Identity Authentication Tenant

Trust is established with the Identity Authentication tenant, as shown in Figure 14.4.

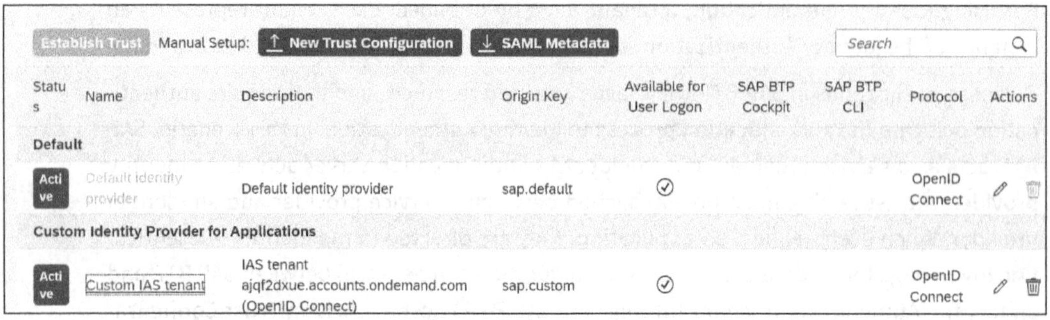

Figure 14.4: Trust Configuration with the Identity Authentication Tenant

From a high-level implementation perspective, follow these steps to establish trust:

1. Select an identity provider, via Identity Authentication or by bringing your own identity provider, and collect its metadata, including information about the identity provider URL, certificate, etc.
2. Bind the identity provider to SAP BTP. This step configures SAP BTP to use the identity provider for authentication.
3. Collect metadata from SAP BTP, including information about its URL, certificate, etc.
4. Configure the identity provider's SAML attributes: These attributes should be passed, by the identity provider, in a SAML token to confirm the identity of the user.
   - **NameID**: This value help identifies the user ID that SAP BTP may use.
   - **Group**: This value is recommended to enable identity federation during role assignments in SAP BTP.

14.1.2 Solution Overview for User Authentication via the SAP BTP User Store

In this authentication scenario, SAP BTP acts as a service provider, and Identity Authentication acts as the identity provider. SAP BTP and the identity provider require trust settings so that any authentication request from an application is redirected to the logon page.

Figure 14.5 shows the interactions between user, identity provider, and service (application) with an implementation of authentication using Identity Authentication.

❶ A user requests access to the application.

❷ The application redirects the request to the identity provider for authentication.

❸ The identity provider challenges the user for credentials.

❹ The user provides credentials.

❺ The identity provider authenticates user credentials and issues a SAML assertion token.

❻ The request returns to the service with the SAML assertion for fulfillment.

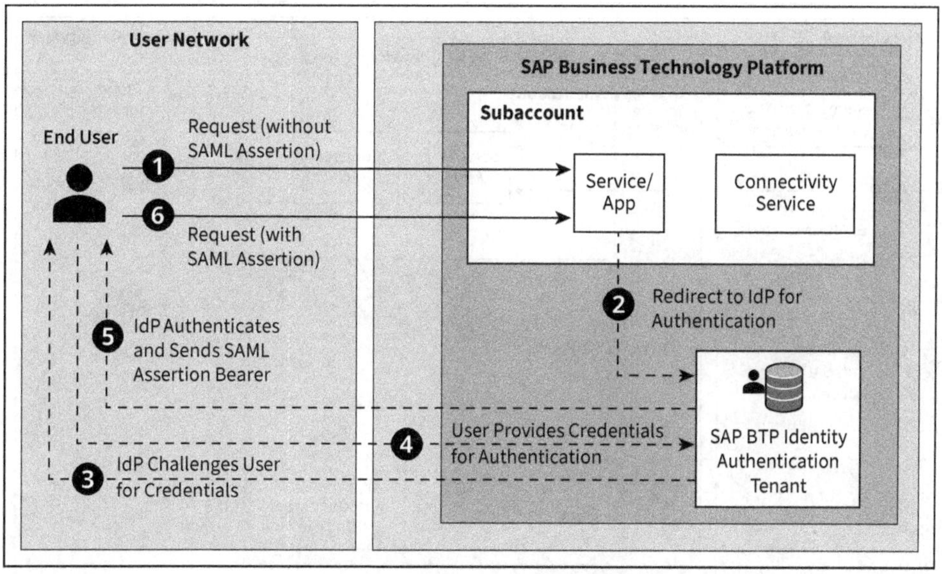

Figure 14.5: Solution Overview: User Authentication through User Store in Identity Authentication

14.1.3 Solution Overview for User Authentication Using a Corporate User Store

In this authentication scenario, SAP BTP uses an existing Lightweight Directory Access Protocol (LDAP) system within your organization's own network along with Identity Authentication. The corporate user store acts as a repository for user identities, and Identity Authentication acts as an identity provider. Authentication requests are proxied to the corporate user store via a cloud connector. SAP BTP and the identity provider should have a trust setting so that any authentication request from an application is redirected to a logon page.

Figure 14.6 shows the authentication of an SAP BTP application using an on-premise user store and SSO.

❶ A user requests access to the application.

❷ The request is redirected by the application to the identity provider for authentication.

❸ The identity provider challenges the user for credentials.

❹ The user provides credentials.

❺ The request is forwarded to the identity provider and is then proxied to the corporate user store through the cloud connector for authentication.

❻ The identity provider authenticates user credentials and issues a SAML assertion token.

❼ The request returns to the service with the SAML assertion for fulfillment.

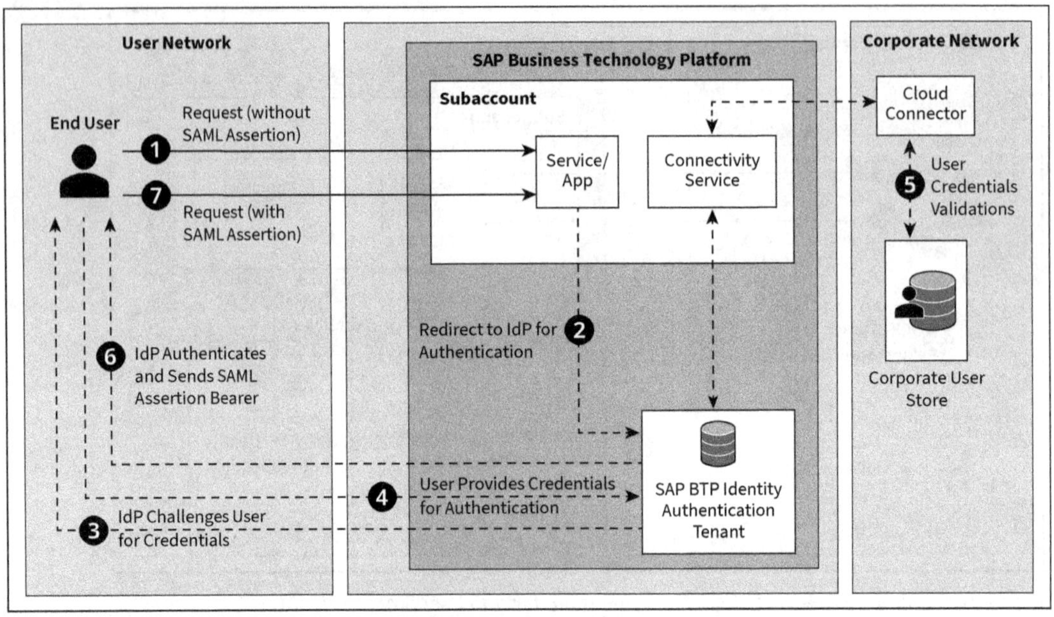

Figure 14.6: Solution Overview: User Authentication Using a Corporate User Store

14.1.4 Solution Overview for User Authentication Using Your Own Identity Provider

In this authentication scenario, SAP BTP acts as a service provider, but you are bringing your own identity provider. Trust settings must be established between the service provider and your identity provider. Once this setup is completed, all the services in SAP BTP will be authenticated through your identity provider. Figure 14.7 shows the authentication of an SAP application using this approach.

❶ A user requests access to the application.

❷ The request is redirected by the application to the identity provider for authentication.

❸ The identity provider challenges the user for credentials.

❹ The user provides credentials.

❺ The identity provider authenticates user credentials and issues a SAML assertion token.

❻ The request returns to the service with the SAML assertion for fulfillment.

[»] **Note**

Any identity provider that supports the SAML 2.0 protocol can be used for establishing the trust between the service provider subaccount and the identity provider.

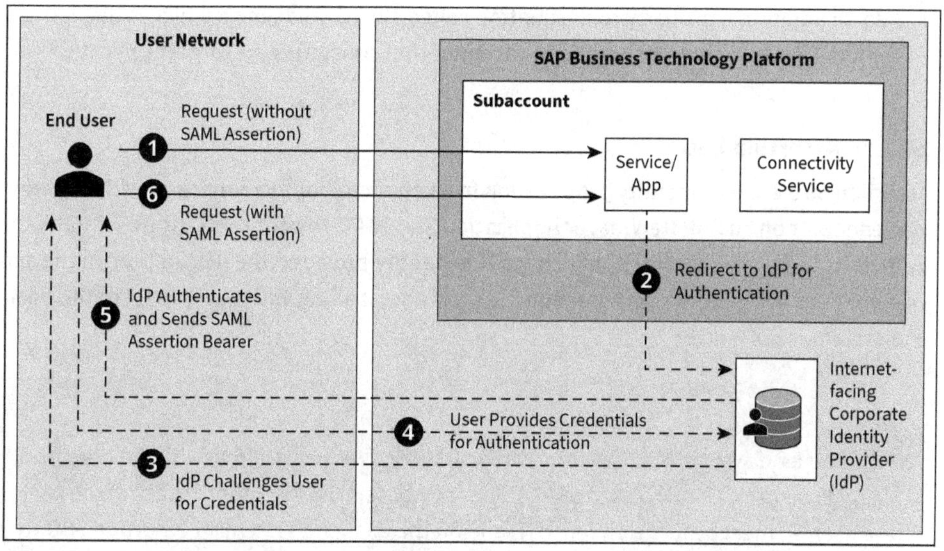

Figure 14.7: Solution Overview: User Authentication with Your Own Identity Provider

> **Note**
>
> SAP ID is the default identity provider service in SAP BTP. Trust relationships between SAP BTP and the SAP ID user store are created automatically.

14.2 SAP BTP Single Sign-On

This section discusses the various authentication types (identity propagation) and their suitability for service endpoints on the cloud and on-premise. These service endpoints could provide access to either SAP or non-SAP systems.

14.2.1 Application-to-Application Single Sign-On

Application-to-application single sign-on enables identity propagation from applications in a service provider to other applications consumed in the service provider account. This propagation could involve applications in the same service provider account or in multiple service provider accounts. Enabling application-to-application single sign-on in the SAP Destination service provider account allows users to consume resources in the SAP Destination service provider account without having to pass user credentials again.

To set up application-to-application single sign-on, follow these steps:

1. Establish trust between two subaccounts by adding the local provider name from the first account to the second account.
2. Create a new destination to the application and set the **Authentication** to **AppToAppSSO**.

3. Add an additional property to destination called `saml2_audience`, which is the name of the local SAML 2.0 provider of the subaccount that consumes the target application.

14.2.2 Assertion SSO

Assertion SSO enables identity propagation from applications in a service provider to a service endpoint on an SAP NetWeaver Application Server (AS) system. Once the user has been authenticated on the service provider, via the identity provider, the destination on the service provider will generate an assertion SSO and pass it along with the request to the backend system.

[»] **Note**

If the backend system is on-premise, then all service requests, along with the assertion SSO, will be proxied through the SAP Connectivity service. Also, the identity of the user (the user ID) must be same on SAP BTP and on the backend system for assertion SSO to work.

From a high-level implementation perspective, to set up assertion SSO, follow these steps:

1. Create the certificate to be configured with the destination configuration on SAP BTP.
2. Get the backend Secure Sockets Layer (SSL) system certificate.
3. Add the backend SSL system certificate to the SAP Connectivity service trust store. If the backend system resides in the cloud, then this certificate must also be added to the SAP BTP trust store.
4. Map the backend system in the SAP Connectivity service.
5. Enable the backend system parameters.
6. Ensure that the "SAPAssertionSSO" certificate is trusted on the backend system. The issuing server must be added to the corresponding access control list (ACL).

[+] **Tip**

In the SAP Connectivity service, you must select **None** for the **Principal Type** field.

14.2.3 Principal Propagation

Principal propagation enables SSO between SAP BTP and backend systems (SAP and non-SAP) through the SAP Connectivity service. This SSO propagation is based on an X.509 client certificate generated by the SAP Connectivity service. Communication between the SAP Connectivity service and a backend system can use HTTPS or remote function calls (RFCs) with secure network communication (SNC).

The SAP Connectivity service requires prior configuration for principal propagation to work. These settings are **CA Certificate** and **Principal Propagation**.

From a high-level implementation perspective, follow these steps:

1. Create the certificate authority (CA) certificate and the system certificate to be configured with the SAP Connectivity service.
2. Get the backend SSL system certificate.
3. Add certificates to the SAP Connectivity service.
4. Select the **Principal Propagation** attribute to use in the cloud connector.
5. Map the backend system in the cloud connector.
6. Enable backend system internet communication manager (ICM) parameters.
7. Enable the backend service for certificate-based authentication.
8. Establish a trust relationship with the SAP Connectivity service certificate on the backend system.
9. Map the certificate user to a user on the system.

14.3 Access Management on SAP BTP

Access management for applications deployed in SAP BTP is based on web access for business users and API access for business systems. Applications are deployed in their runtime containers and can use the OAuth 2.0 protocol for authentication implementation.

14.3.1 Key Concepts in User Authentication in SAP BTP

Here are some key concepts in user authentication implementation in SAP BTP:

- Authentication using SAML 2.0: SAML 2.0 is an XML-based open standard used to facilitate the exchange of authentication and authorization data of a principal (user) between an identity provider and a service provider. The exchange of data is performed using *bearer assertions*, and the integrity of the assertion is protected by means of XML encryption and an XML signature. Federated identities can be managed using the SAML 2.0 protocol.
- Authorization with OAuth 2.0: SAP BTP applications use the concept of microservices wherein an application can have multiple independent runtime containers. OAuth 2.0 orchestrates the management of delegation of authorizations to these services based on scopes.
- JSON Web Token (JWT): JWT is an open standard for securely transmitting information between two parties. Information exchanged can be trusted as the token is digitally signed with the private key of the authorization server through the user account and authentication service (UAA). The token contains information specifying the issuer, the subject, the expiration time, and any additional custom information required by the consumer.

14.3.2 User Account and Authentication Service

UAA is the service provided by SAP BTP for user authentication and authorization. There are two instances of the UAA service:

- For platform users like developers and administrators: This type of users can use various tools for administering services in the SAP BTP cockpit and use application development tools like SAP Business Application Studio. These users are authorized for some services provided by the platform, and their interaction with UAA is limited to authentication for the platform.
- For business users: Authentication and authorizations are managed by XSUAA. This component provides a simple programming model for authentication and authorizations in business applications.

UAA uses OAuth 2.0 for the delegation of authorizations for an application. In the context of an OAuth flow, UAA provides scopes, role templates, and attributes to applications deployed in the runtime of SAP BTP, Cloud Foundry environment. If a business user has a certain scope, both an access token and a refresh token with this scope are issued. These tokens enable the user to run the application, while the scopes are used to define the model that describes who has the authorization to start an application that runs in SAP BTP.

The UAA service provides a programming model for developers to define templates that can be used to build role and authorization models for business users accessing applications deployed to the runtime of SAP BTP, Cloud Foundry environment. In the OAuth 2.0 workflow, the UAA acts as an OAuth server.

14.3.3 Web Access

Applications are authenticated using the OAuth 2.0 protocol. With this protocol, the UAA service is an OAuth 2.0 authorization server, the application router (AppRouter) is an OAuth 2.0 client, and the application logic running in the Node.js and Java backend services is an OAuth 2.0 resource server. We'll cover these topics in more detail in the following sections.

Application Router

The AppRouter serves as the single entry point to business applications using microservices. The AppRouter is a Node.js app, publicly available in Node Package Manager (npm). The purpose of the AppRouter includes the following activities:

- Serving static content
- Authenticating users
- Dispatching requests to backend applications (microservices)

SAP Authorization and Trust Management Service

Extended services for user account and authorization (XSUAA), found within the SAP Authorization and Trust Management service, is a service for managing authentication and authorizations for business applications in SAP BTP. XSUAA can establish trust with SAP ID or Identity Authentication, thus acting like an OAuth server and creating JWTs. An example XSUAA authentication sequence is shown in Figure 14.8:

1. The browser sends its first request to the application, which is intercepted by the AppRouter. The AppRouter considers this request as an initial request because a session ID is not included.
2. The AppRouter redirects the request to the XSUAA component. The URL for the redirect is taken from the variable VCAP_SERVICES of the AppRouter environment.
3. The browser sends the request to the XSUAA component. XSUAA considers this request as an initial request because a JWT is not included.
4. XSUAA redirects the request to the SAML 2.0 identity provider. The URL for the redirect is taken from the configuration maintained in the trust relationship between XSUAA and the SAML 2.0 identity provider.
5. The browser sends the request to the SAML 2.0 identity provider. The SAML 2.0 identity provider considers this request as an initial request because authentication was not provided.
6. The SAML 2.0 identity provider challenges the user to enter credentials.
7. The user enters the credentials and sends a request with these credentials.
8. The SAML 2.0 identity provider authenticates the user, creates a SAML 2.0 bearer assertion, and includes it in the response. The response is redirected back to XSUAA.
9. The browser sends the request with the SAML 2.0 bearer assertion to the XSUAA component. XSUAA considers this request as authenticated and uses the information of the bearer assertion to create a JWT with an authorization code. XSUAA keeps the JWT and includes the authorization code in its response. The response is redirected to the AppRouter.
10. The browser sends the request with the authorization code to the AppRouter. The AppRouter considers this request as authenticated and uses the authorization code to request the JWT from XSUAA. The AppRouter uses `clientid` and `clientsecret` to authenticate itself against XSUAA.
11. The XSUAA component retrieves the JWT associated with the authorization code and includes it in its response.
12. The AppRouter now has the JWT and creates a session ID and assigns it to the JWT. The JWT is added to the browser request and forwarded to the microservice.
13. The AppRouter includes the session ID in the response and routes the response back to the browser.

The subsequent browser requests include the session ID. The AppRouter derives the JWT from the session ID, adds it to the subsequent request, and forwards the request to the microservice.

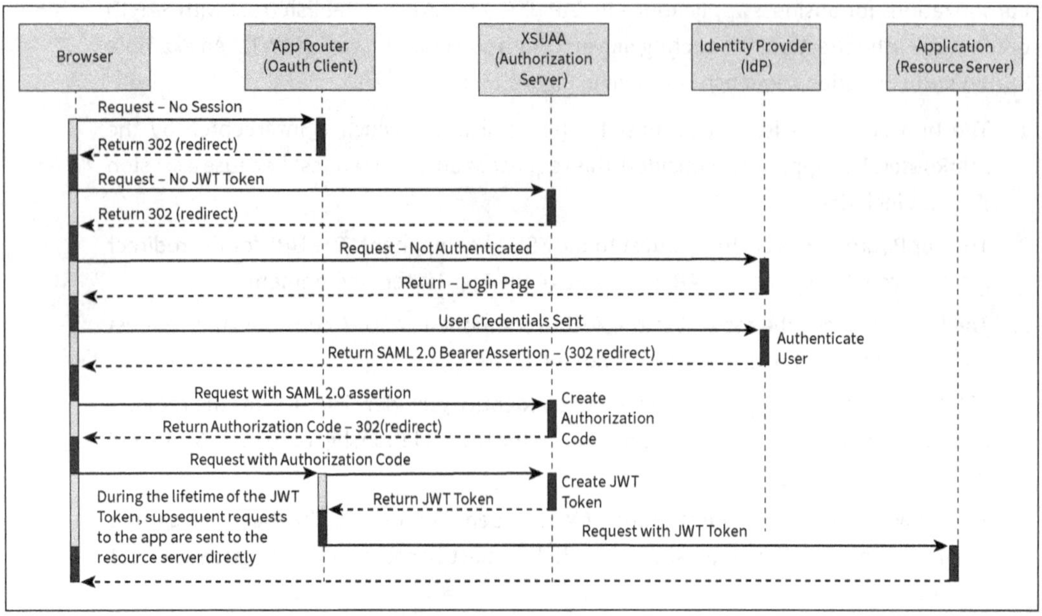

Figure 14.8: XSUAA Authentication Sequence

Note
The JWT is never sent to the browser. Instead, it is assigned to the request by the AppRouter through an authorization code, which ensures that user identity is secured implicitly.

The authentication process is triggered by the AppRouter component, which is configured in the design-time artifact *xs-app.json*.

Authorization management is based on the deployment security descriptor (*xs-security.json*), where access is authorized.

14.4 Authorizations for Applications on SAP BTP, Cloud Foundry Environment

Applications with content deployed to backing services like databases require not only authentication but also appropriate authorizations. In this section, we'll provide an overview of security artifacts used for authorizations and their implementation.

14.4.1 Access Control

An authorization policy can be divided into two phases:

- Authorization: Defined in the deployment security descriptor (*xs-security.json*), where access is authorized.
- Policy enforcement: The general rules by which requests for access to resources are either approved or rejected.

14.4.2 Security Artifacts for Authorizations

Developers store authorization information as design-time role templates in the security descriptor file *xs-security.json*. Using the XSUAA service broker, they deploy security information to the XSUAA service. The administrators view the authorization information in role templates, which they use as part of the runtime configuration. The administrators use the role templates to build roles, which are aggregated in role collections. The role collections are assigned, in turn, to business users.

The tasks performed by developers to set up authorizations for applications include the following:

- Specify the security descriptor file containing the functional authorization scopes for your application.
- Create role templates for the application using the security descriptor file.
- Create a service instance from the XSUAA service using the service broker.
- Bind the service instance to the application by including it in the *manifest.json* file.
- Deploy the application.

The tasks performed by administrators to set up authorizations for applications include the following:

- Use an existing role or create a new one using role templates.
- Create a role collection and assign roles to it.
- Assign the role collections to SAML 2.0 user groups.
- Assign the role collection to the business users provided by the SAML 2.0 identity provider.

14.4.3 Authorization Enforcement

Authorization enforcement is performed in the following application components:

- **AppRouter**
  After successful authentication, the AppRouter checks the authorizations based on scopes for the application and routes requests to the relevant runtime containers.
- **Application containers**
  JWTs from the AppRouter contain user and scope details. This token is validated using the XSUAA advanced container security API.

- **SAP HANA database**
  Applications using backend database services provided by SAP BTP have their own dedicated database schema and technical user to access the content of the schema. Appropriate privileges can be added to this technical user so that the application can access the database artifacts.

14.4.4 Scopes, Role Templates, and Attributes

Scopes are functional authorizations that are assigned to users, and they define the actions that can be performed within a service. Scopes are used for authorization checks by the AppRouter and by the application's HDI container.

Attributes define the application entities that a user may access. These attributes can be dynamic in nature, and their values can be instantiated according to defined access rules. The security descriptor file has attributes based on entities whose value is not known at design time.

A role template describes a role and any attributes that apply to the role. Role templates must be instantiated using role-building tools as values for the attributes associated with the role template to be instantiated.

As stated previously, authorizations for business applications on SAP BTP are driven by role templates, scopes, and attributes defined in the security descriptor file. The relationships between these objects are shown in Figure 14.9.

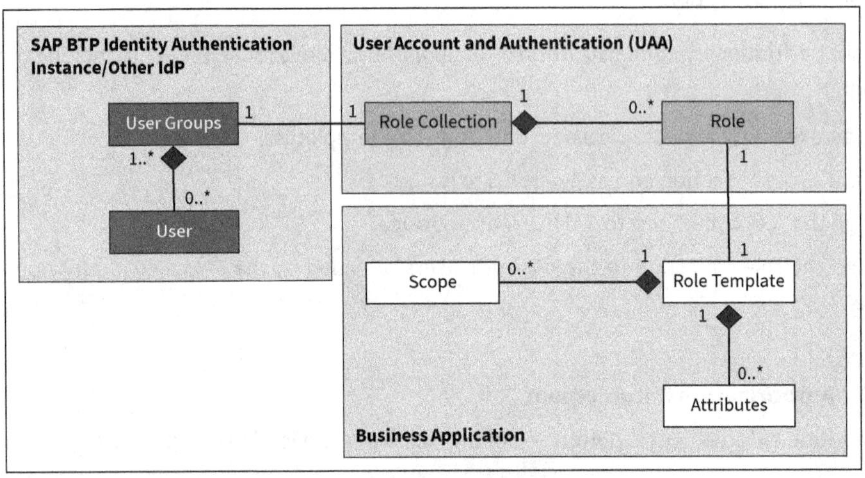

Figure 14.9: Relationship between Authorization Artifacts in SAP BTP

14.5 Web Security for Applications in SAP BTP

In addition to security mechanisms implemented through authentication and authorization services, another important security concern is that applications must also defend themselves against web-based attacks. In this section, you'll learn about the various tools available for guarding your applications against web-based attacks.

14.5.1 Protection Against Cross-Origin Resource Sharing

Cross-origin resource sharing (CORS) permits web pages from other domains to make HTTP requests to an application domain. In a normal scenario, such a request will be automatically refused by the web browser's security policy.

In an SAP BTP application, you can enable cross-origin support in the AppRouter by setting the `CORS` environment variable.

14.5.2 Protecting Against Cross-Site Request Forgery

Cross-site request forgery (CSRF) is a type of malicious attack that tricks a user into sending unintended requests. For example, an attacker can trick an authenticated user into clicking a link to update their personal information.

During AppRouter configuration, an option is available to specify if a route needs protection for CSRF tokens. Protection for a CSRF token can be specified in the *xs-app.json* file using the code shown in Listing 14.1.

```
{
...
  "routes": [
      {
        "source": "^/sap/ui5/1(.*)$",
        "target": "$1",
        "destination": "ui5",
        "csrfProtection": true
      },
...
}
```

Listing 14.1: CSRF Token in the xs-app.json File

14.5.3 Protecting Against Click-Jacking Attempts

A click-jacking attack tricks web users into performing an action they did not intend, typically by rendering an invisible page element on top of the action the user thinks they are performing.

During AppRouter configuration, `whitelistService` can be enabled to prevent click-jacking attacks, using the following code:

```
{
  "whitelistService": {
    "endpoint": "/whitelist/service"
  }
}
```

A list of allowed host names and domains can be maintained using the `CJ_PROTECT_WHITELIST` environment variable.

If the whitelisted service is enabled in the AppRouter, each time an HTML page must be rendered in a frame, the whitelist service is used to check if the parent frame is allowed to render the content in a frame.

14.5.4 Protecting Against Cross-Site Scripting

Cross-site scripting (XSS) is a security vulnerability in web applications. Attackers can use these vulnerabilities to inject arbitrary code into an application's client frontend. This code is rendered by a victim's web browser and is interpreted within the victim's current authentication.

To help protect applications against XSS attacks, SAP provides the Node.js package `@sap/xss-secure`, which includes APIs to encode CSS, HTML, JavaScript, URLs, and even XML code.

14.5.5 Setting Up Timeouts for Application User Sessions

Applications can be configured to automatically log off from the UAA service if the user is inactive for a defined time. This feature can be achieved by setting the environment variable `SESSION_TIMEOUT` to a predefined number of minutes.

14.5.6 Securing Node.js Applications

Node.js includes various security-related libraries, such as the following:

- *@sap/xsenv*: A utility that enables the easy setup of (and access to) environment variables and services (for example, CORS or CSRF)
- *@sap/xss-secure*: A utility that provides protection against XSS
- *@sap/xssec*: XS advanced security API for Node.js, which includes the XS advanced container security API for Node.js

14.5.7 Securing Your Java Applications

SAP HANA includes various security-related Java libraries, such as the following:

- `xs-env`: A utility for setting up and accessing XS advanced environment variables and services

- `audit-java-client-api`: A client that enables applications to use the XS advanced audit log service
- `java-container-security`: A convenient Java wrapper for the security context of an authentication

14.6 Joule Security

Identity and access management (IAM) is one of the most critical parts of adopting Joule as an AI solution. Joule is embedded across SAP's product stack and implements enterprise-grade security through a comprehensive, multilayered approach that ensures secure access, data protection, and compliance. As depicted in Figure 14.10, Joule's security foundation is built upon four key components: SAP BTP as the secure hosting environment, SAP Build Work Zone for semantic navigation control, SAP Cloud Identity Services for authentication and authorization, and integrated SAP applications as secure data sources.

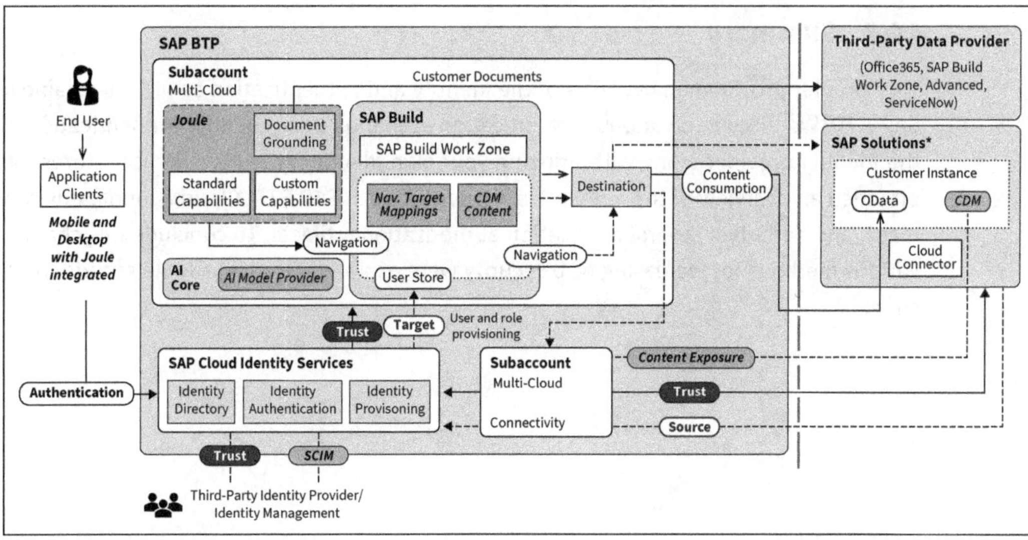

Figure 14.10: Secure Joule Integration Architecture

The primary component for enterprise-grade security that Joule relies on is SAP Cloud Identity Services, which acts as a secure gateway for accessing SAP applications and Joule within an ideal landscape. It is also one of the most critical prerequisites for Joule adoption. Joule leverages Identity Authentication for centralized user authentication, including integration with corporate identity providers through either SAML or OIDC (OpenID Connect) protocols. SAP products across the landscape must use the same SAP Cloud Identity Services tenant with a consistent global user ID across systems. Additionally, Joule does not maintain its own authentication system; rather, it relies on inheriting user permissions from the SAP system, ensuring consistent access control. On the other hand, SAP Build Work Zone provides the necessary semantic layer that maps user roles replicated from the SAP system to available content and navigation targets. In this context, role replication helps establish the authorization context that determines which content, applications, and data users can access through Joule's conversational interface.

In the case of third-party identity providers, SCIM acts as a critical bridge between third-party identity providers and SAP Cloud Identity Services for Joule integration. SCIM enables third-party identity providers (such as Microsoft Active Directory, Okta, or PingOne) to integrate with SAP Cloud Identity Services through standardized APIs. The authentication flow also requires that third-party identity providers establish trust with SAP Cloud Identity Services using OIDC protocols, as SAML-based trust is not supported for Joule integration.

Overall, the secure Joule integration architecture portrayed in Figure 14.10 ensures enterprise-grade protection while providing a seamless user experience across the SAP ecosystem. The security measures are designed with SAP's existing enterprise security policies in mind.

14.7 Summary

This chapter provided an overview of the identity and authentication services available in SAP BTP. We discussed various authentication scenarios using Identity Authentication as the identity provider along with bringing your own identity provider and user stores. We provided an overview of the authentication sequence in SAP BTP, Cloud Foundry environment, and described several application authorization artifacts. To conclude this chapter, various methods for increasing web security were explored alongside Joule security measures.

Chapter 15
Application Monitoring

Application monitoring plays a critical role in the success of a hybrid cloud environment. A successful monitoring environment provides continuous monitoring of applications by triggering alerts, actions, and notifications for all exceptions. A solid monitoring environment should log all critical actions taken and analyze this information in a meaningful way to enable better decision-making.

Over any period of time, an organization could build tons of cloud applications on SAP Business Technology Platform (BTP), which, without a proper strategy to monitor them, would be inefficient. SAP BTP provides a proper infrastructure and many out-of-the-box monitoring solutions to establish an accurate and fine-grained cloud monitoring system.

In this chapter, we'll describe the application monitoring services available in SAP BTP (using the Neo environment as an example), which will cover both Java and database monitoring. You'll also learn how to build your own monitoring applications based on SAP BTP's monitoring application programming interfaces (APIs). For SAP BTP, Cloud Foundry environment, we'll cover application logging, real-time analysis, and visualization options.

15.1 Monitoring Java Applications

You can monitor critical metrics about your Java applications directly in the SAP BTP cockpit. These metrics can be helpful for monitoring the performance of your applications by providing granular information. A list of available metrics is shown in Table 15.1.

Metric	Value
Used Disk Space	What percentage of the total disk space is currently used
Requests per Minute	The number of HTTP requests processed by the Java application in the last minute
CPU Load	What percentage of the CPU is used on average in the last minute
Disk I/O	How many bytes per second are currently read or written to the disk

Table 15.1: Java Application Monitoring Metrics

Metric	Value
OS Memory Usage	What percentage of operating system memory is currently used
Heap Memory Usage	What percentage of heap memory is currently used
Average Response Time	The average response time, in milliseconds, of all requests processed in the last minute
Busy Threads	The current number of threads that are processing HTTP requests

Table 15.1: Java Application Monitoring Metrics (Cont.)

Note

The Java application monitoring shown in this section is based on SAP BTP, Neo environment.

To display these metrics, navigate to **Applications • Java Applications**, as shown in Figure 15.1. A list of your Java applications will be shown on the right.

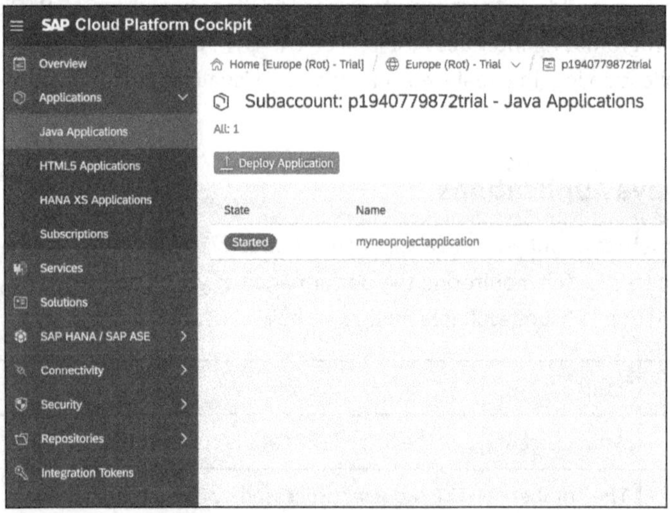

Figure 15.1: Java Application

Once you click on an application's name, you can view different metrics about that particular application under the **Monitoring** menu item, as shown in Figure 15.2.

Under **Monitoring**, you'll find four additional menu options:

- **Processes**: This option displays the current monitoring metrics data.

- **Logging**: This option is used for any type of application logging. You can set the log levels by clicking the **Configure Loggers** button.
- **Application Monitoring**: Historical data of all the monitoring metrics can be viewed graphically. You'll have the option of selecting various time ranges for this purpose.
- **JMX Console**: This option can help you to perform Java Management Extensions (JMX)-based monitoring for your Java application.

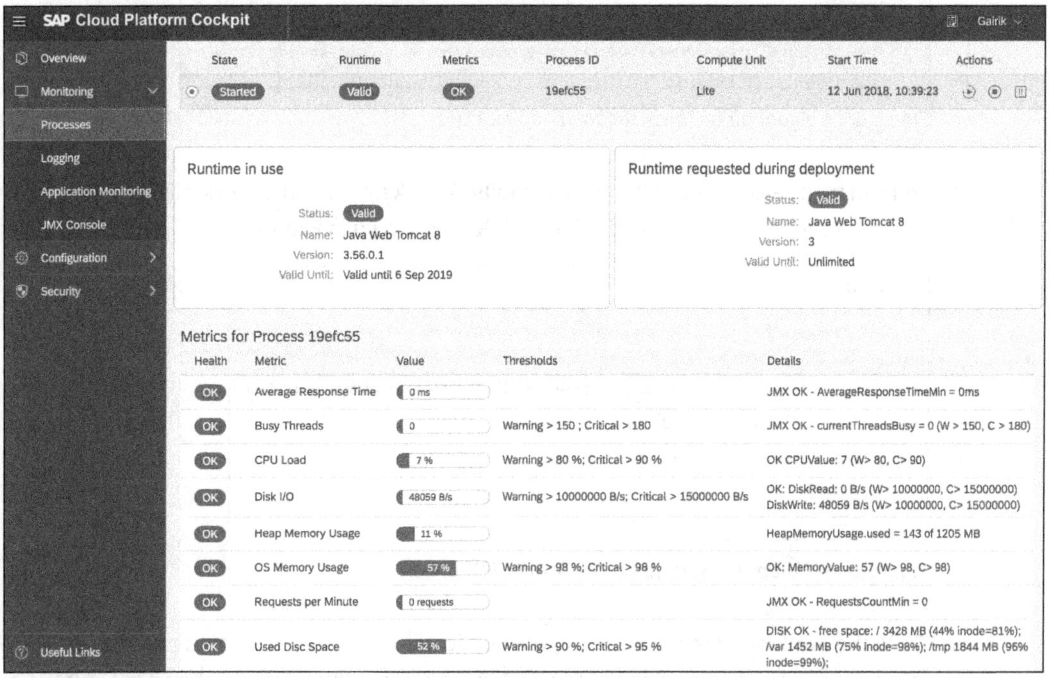

Figure 15.2: Java Application: Monitoring Details

> **Note**
>
> JMX is a standard API for monitoring anything related to Java. JMX can be used to deploy, start, stop, and dynamically configure Java applications, among other things. JMX has been available in Java SE since Java 5, and in Java EE, since Java EE 1.4. Various performance and monitoring metrics can be extracted for further analysis using JMX.

You can also set up an availability check for a Java application using the SAP BTP cockpit. Navigate to your Java application and select **Create Availability Check** in the availability check panel, as shown in Figure 15.3.

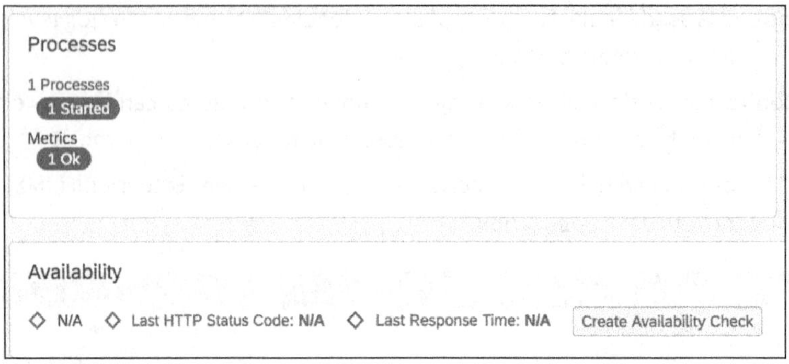

Figure 15.3: Availability Check for Java Applications

You can then select the URL for the availability check from the dropdown list and also maintain values for warning and critical thresholds, as shown in Figure 15.4.

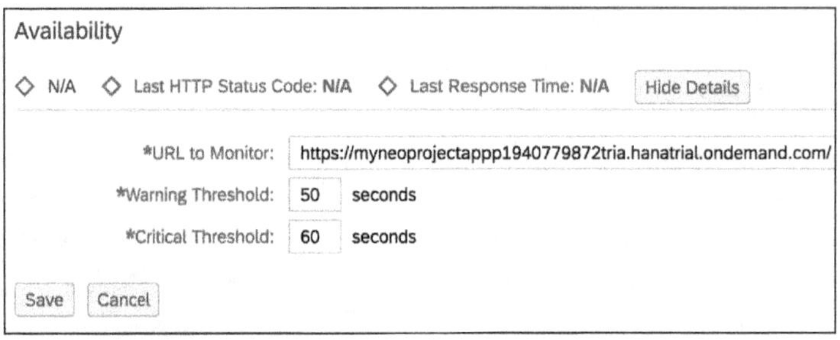

Figure 15.4: Values for Setting Up the Availability Check

Once the availability check is set up, the application's availability status and the last HTTP response code will be seen in the top part of the availability panel, as shown in Figure 15.5.

Figure 15.5: Java Application: Availability Status

15.2 Monitoring Database Systems

With respect to the databases available in SAP BTP, you can monitor the database itself (i.e., an SAP HANA database) as well as SAP HANA extended application services (SAP HANA XS) applications running on an SAP HANA database. In the SAP BTP cockpit, you can monitor the current state of the databases as well as see performance trends for your databases. The default metrics for database systems are shown in Table 15.2.

Metric	Value
CPU Load	The percentage of the CPU that is used on average in the last minute
Disk I/O	The number of bytes per second that are currently being read or written to the disk
Network Ping	The percentage of packets that are lost to the database host
OS Memory Usage	The percentage of the operating system memory that is currently being used
Used Disk Space	The percentage of the local disk space of the operating system that is currently being used

Table 15.2: Default Database Monitoring Metrics

[«]

Note

SAP HANA Cloud is monitored using the SAP HANA Cloud application subscription (added by default after an instance of SAP HANA Cloud is created in SAP BTP subaccount).

Assuming that you already have an SAP HANA multitenant database container (MDC) database created in your SAP BTP account, you can then create a database system for your database in the SAP BTP cockpit. A *database system* is a set of tools required for better monitoring and administering existing databases. An SAP BTP trial account does not provide the option to create a database system. You may need an enterprise account to perform this operation. Once you create a database system, you can access it by navigating to **Database Systems**, as shown in Figure 15.6.

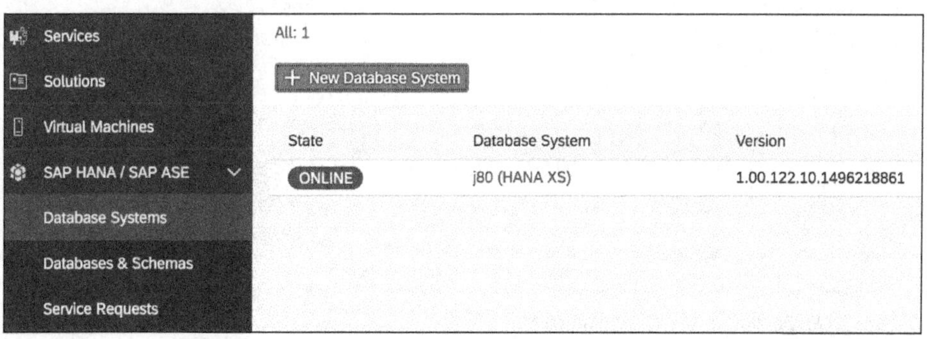

Figure 15.6: Database Systems in the SAP BTP Cockpit

Clicking on the database system will take you to the screen shown in Figure 15.7.

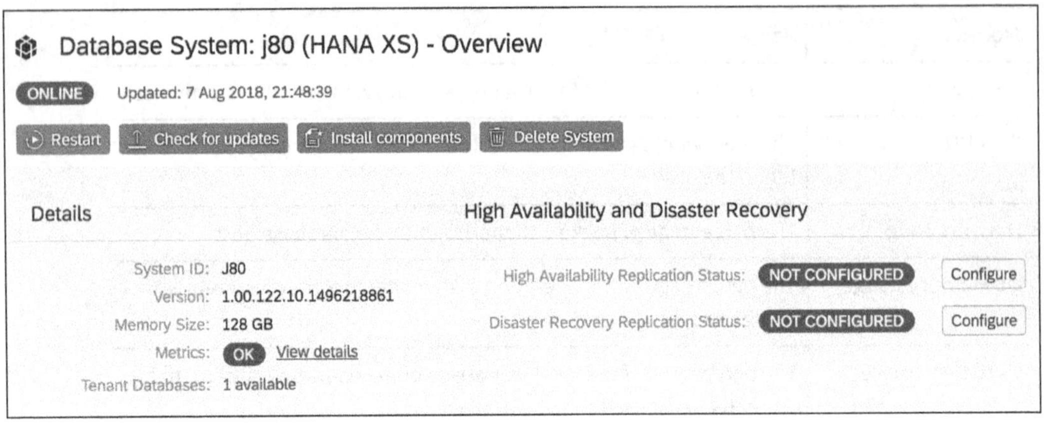

Figure 15.7: Details of the Database System

The **Current Metrics** section displays details about database metrics, as shown in Figure 15.8. Most of these metrics are self-explanatory and easy to understand.

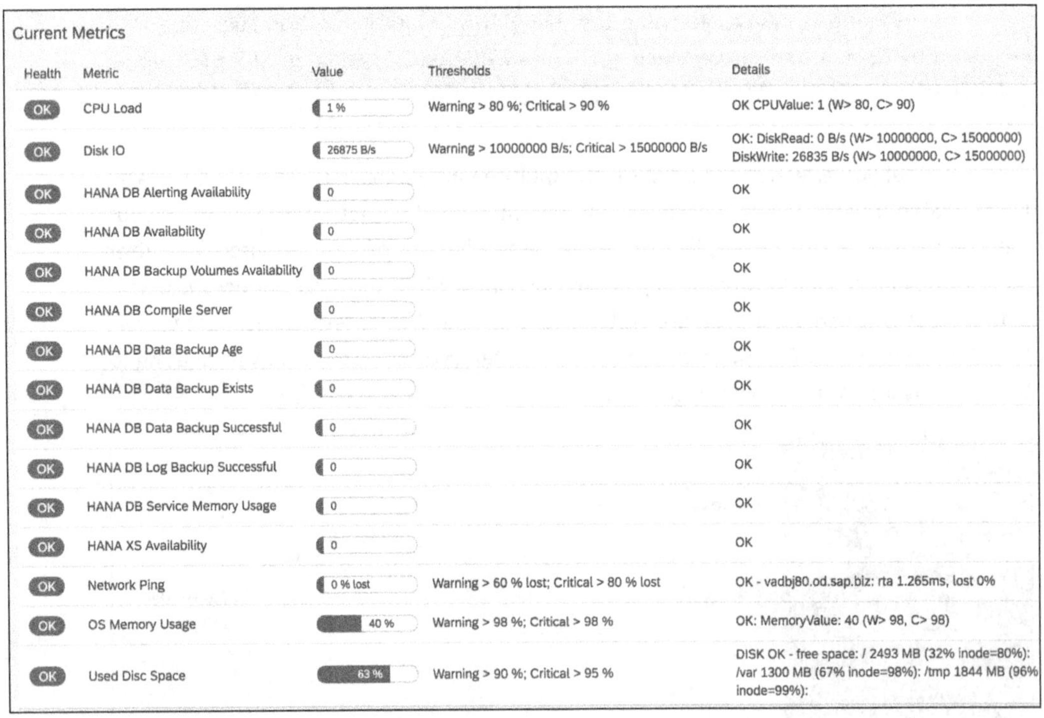
Current Metrics

Health	Metric	Value	Thresholds	Details
OK	CPU Load	1 %	Warning > 80 %; Critical > 90 %	OK CPUValue: 1 (W> 80, C> 90)
OK	Disk IO	26875 B/s	Warning > 10000000 B/s; Critical > 15000000 B/s	OK: DiskRead: 0 B/s (W> 10000000, C> 15000000) DiskWrite: 26835 B/s (W> 10000000, C> 15000000)
OK	HANA DB Alerting Availability	0		OK
OK	HANA DB Availability	0		OK
OK	HANA DB Backup Volumes Availability	0		OK
OK	HANA DB Compile Server	0		OK
OK	HANA DB Data Backup Age	0		OK
OK	HANA DB Data Backup Exists	0		OK
OK	HANA DB Data Backup Successful	0		OK
OK	HANA DB Log Backup Successful	0		OK
OK	HANA DB Service Memory Usage	0		OK
OK	HANA XS Availability	0		OK
OK	Network Ping	0 % lost	Warning > 60 % lost; Critical > 80 % lost	OK - vadbj80.od.sap.biz: rta 1.265ms, lost 0%
OK	OS Memory Usage	40 %	Warning > 98 %; Critical > 98 %	OK: MemoryValue: 40 (W> 98, C> 98)
OK	Used Disc Space	63 %	Warning > 90 %; Critical > 95 %	DISK OK - free space: / 2493 MB (32% inode=80%): /var 1300 MB (67% inode=98%): /tmp 1844 MB (96% inode=99%):

Figure 15.8: Database Monitoring Metrics

The **Metrics History** can also be viewed in the same screen for specific periods. Periods can range from the last 12 hours to any custom date range, as shown in Figure 15.9 and Figure 15.10.

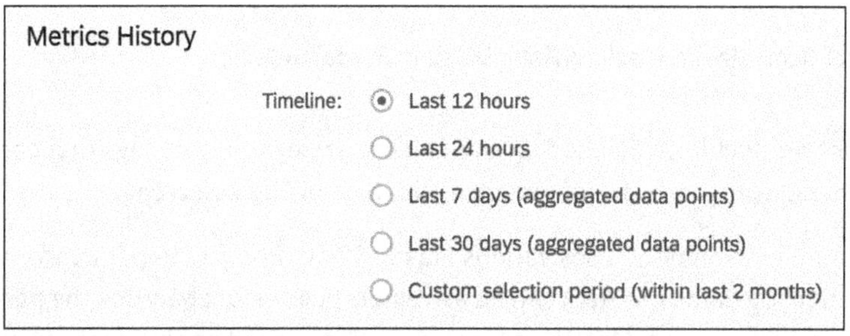

Figure 15.9: Date Range Selection to Display Metrics History

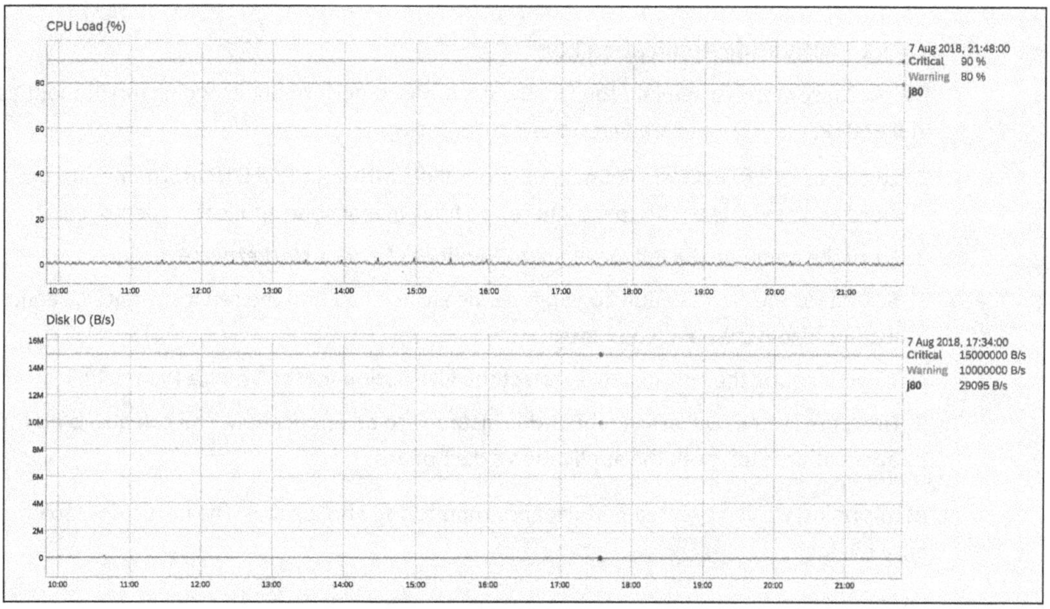

Figure 15.10: Sample Metrics History in Graphical Display

15.3 Monitoring Cloud Foundry Applications

Many tools are available for monitoring applications in SAP BTP, Cloud Foundry environment. However, in this book, we'll focus on Kibana, which comes with SAP BTP, Cloud Foundry environment by default and can be configured easily. Let's look at some key terms before going into SAP BTP, Cloud Foundry environment application logging in detail:

- **Elastic Stack (ELK) suite**
  This group of products is an open-source solution for searching, analyzing, and visualizing data from a live system.
- **Elasticsearch**
  Elasticsearch is a product from Elastic Stack to search, analyze, and store data.

- **Kibana**
  Kibana is a product from Elastic Stack to visualize data in real time.
- **Logstash**
  Logstash is another product from Elastic Stack to collect, parse, and store logs that can be used by the other Elastic products, for instance, Kibana and Elasticsearch.

SAP BTP, Cloud Foundry environment, uses various Elastic Stack products to parse, store, and visualize application log data. This data can be extremely helpful for analyzing the performance of an application and for taking the necessary actions based on this analysis. In the following sections, we'll describe how to activate and visualize this application log.

15.3.1 Activating Application Log

Follow these steps to activate the Elastic Stack-based performance logging functionalities in SAP BTP, Cloud Foundry environment:

1. Log on to SAP BTP, Cloud Foundry environment, with your SAP BTP username and password and navigate to the space where you have at least one application deployed.
2. From the menu on the left, navigate to **Services • Service Marketplace**.
3. Search for the "Application Logging" service. Once you find the service, create a service instance from the navigation menu.
4. While creating the new instance, select the **Lite** option for the **Service Plan** field.
5. Next, the system will ask you bind this instance to an application. You can also bind the application later, from the application's dashboard.

At this point, we've activated performance monitoring and logging. The next step is to visualize the log.

15.3.2 Visualizing Application Log

You can visualize the log in one of two ways: directly using the log-specific URL or navigating to the log from the **Logs** menu item of the application in SAP BTP, Cloud Foundry environment. Some considerations to keep in mind for visualizing application logs include the following:

- If you've activated application logging correctly, then you can use the URL pattern *https://logs<CFDomain>* (e.g., *https://logs.cf.eu.hana.ondemand.com*) to navigate to the Kibana dashboard.
- Alternatively, you can open the application from the SAP BTP cockpit in SAP BTP, Cloud Foundry environment. You'll find the **Logs** menu in the navigation panel. Click the **Open Kibana Dashboard** button, as shown in Figure 15.11.

Your default Kibana dashboard view will look similar to the screen shown in Figure 15.12.

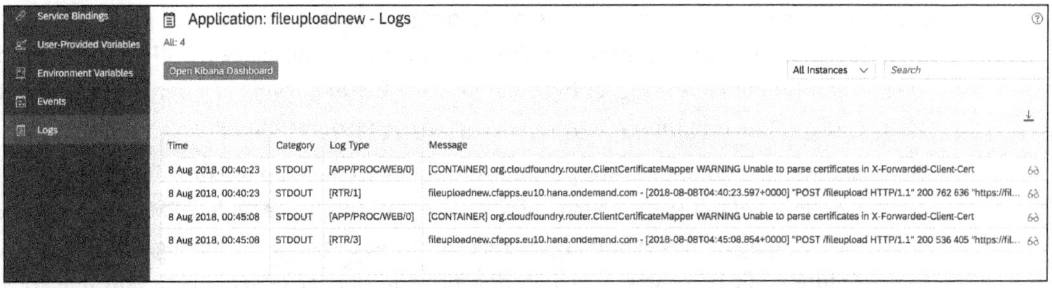

Figure 15.11: Opening the Kibana Dashboard from an Application

Figure 15.12: Kibana Dashboard

The Kibana dashboard offers various dashboard display options to meet your specific needs. You can display API usage, API calls and statistics, backend calls, API average response times, and much more. You can also configure your Kibana visualization as needed based on the available fields. More details about using Kibana can be found at *www.elastic.co/guide/en/kibana/current/index.html*.

Additionally, in SAP BTP, Neo environment, you can even connect your Java application's Dynatrace software as a service (SaaS) monitoring environment. Setting up the Dynatrace service enables the injection of the Dynatrace agent configuration into all of your applications so you can start collecting monitoring data and various metrics, events, and end-to-end traces. Finally, this approach can send all the data to your monitoring environment, which will run in data centers hosted by Dynatrace.

When we look at logging on Cloud Foundry, the standard SAP Application Logging service gives us an out-of-the-box Kibana dashboard, which works well for basic troubleshooting. However, it's important to be aware of its limitations, especially around retention and cost.

Logs are retained for only seven days, and storage is governed by service plan quotas—if those limits are exceeded, logs can be dropped, so monitoring usage is critical. If you need longer retention or expect higher log volumes, SAP Cloud Logging is a better option, as it offers configurable retention with a scalable, pay-per-resource model.

That said, Kibana alone may not be sufficient in all scenarios. For compliance and long-term retention requirements, advanced observability, enterprise-wide SIEM integration, or unified monitoring across hybrid landscapes, it makes sense to stream logs to third-party tools. Solutions like Dynatrace or Datadog provide deeper observability and analytics, while platforms like Splunk allow you to centralize SAP BTP logs with the rest of your enterprise infrastructure for security and incident management.

15.4 Monitoring APIs

Now that we've looked at various metrics for monitoring an application, you can consider building your own monitoring tools based on information available from SAP BTP. In this section, we'll explain how to access application monitoring and logging data through various APIs so that you can build your own visualizations. In this section, we'll only focus on how to access the data. With this data, you can use your own visualization tools (e.g., SAP Fiori) to build your own monitoring applications.

15.4.1 REST APIs for Java Applications

SAP Business Accelerator Hub can serve as your single place for searching, exploring, testing, and understanding any SAP API. SAP Business Accelerator Hub can be accessed through at *https://api.sap.com/*. Once opened in your web browser, select an API from the navigation panel on the left and then click on **View more** to search for "SAP BTP Monitoring", as shown in Figure 15.13.

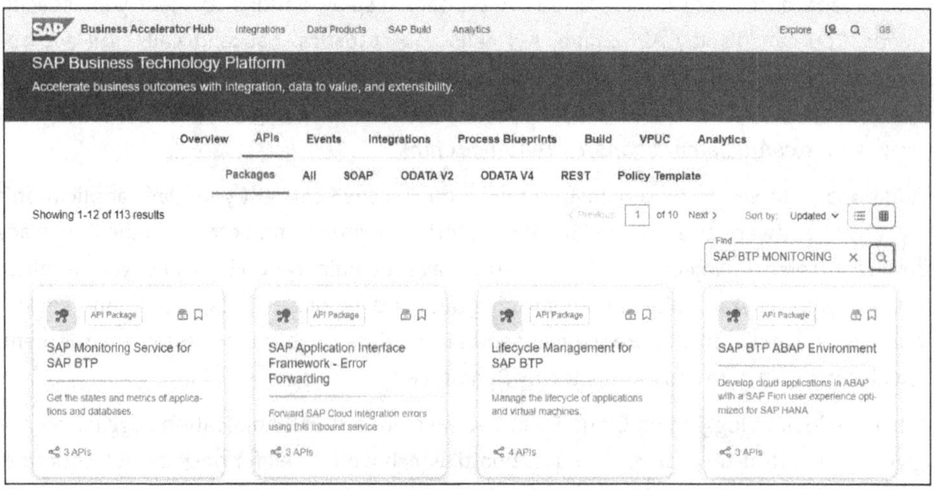

Figure 15.13: SAP BTP Monitoring in SAP Business Accelerator Hub

Figure 15.14 shows the three APIs available in SAP Business Accelerator Hub for the SAP Monitoring service for SAP BTP (Neo environment, in this case).

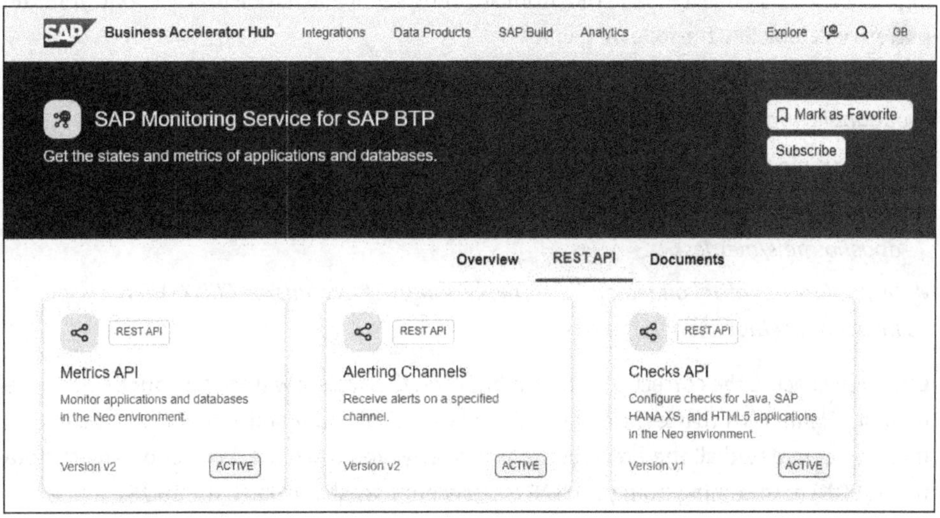

Figure 15.14: Two Versions of Monitoring APIs

If you click on **Metrics API**, you'll find various APIs to access process data, application metrics, process states, and application states, as shown in Figure 15.15. We'll only explain the application metrics-related APIs in this section.

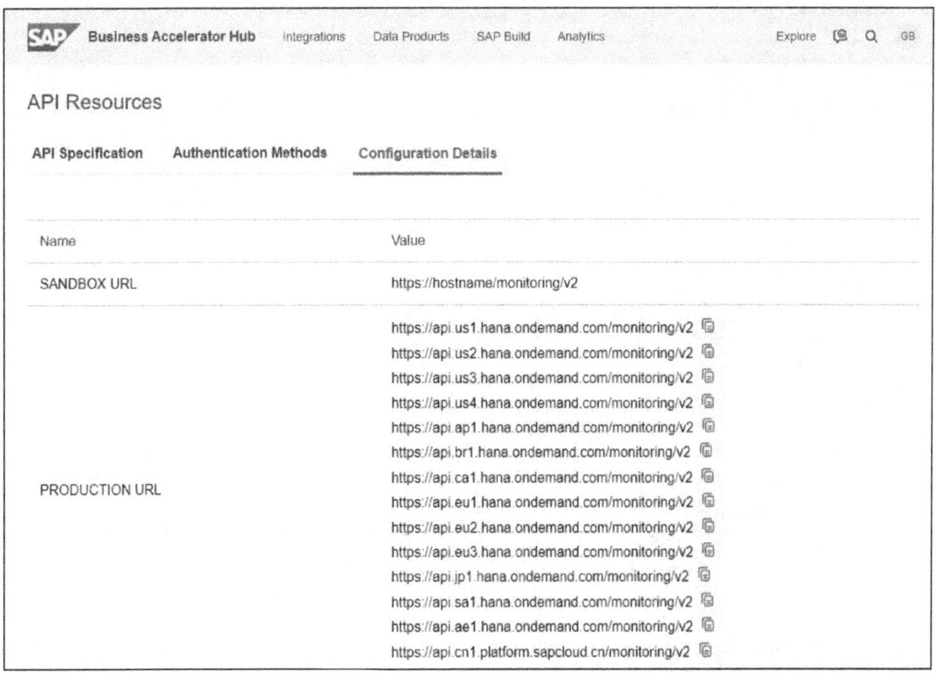

Figure 15.15: Application Metrics API

The URL pattern for is *https://api.<host>/monitoring/v2*. Version 2 is more secure and is protected by OAuth 2.0 client credentials. Now that you know the URL pattern for the Monitoring v2 API, let's try to access the URL from any browser. If you have a trial account, your URL pattern will look like the following link:

https://api.hanatrial.ondemand.com/monitoring/v2/accounts/<your SAP ID>trial/apps/<your application name>/metrics

Some example URLs include the following:

- *https://api.hanatrial.ondemand.com/monitoring/v2/accounts/sNNNNNNNNNNtrial/apps/iotmms/metrics*
- *https://api.hanatrial.ondemand.com/monitoring/v2/accounts/pXXXXXXXXXXtrial/apps/myneoprojectapplication/metrics*

Once you execute the correct URL in your browser, the system will ask for your SAP BTP credentials. Remember to use the email ID linked to your account and your SAP BTP password. If you've completed all the previous steps properly, you'll see the JavaScript Object Notation (JSON) data coming from your SAP BTP account, as shown in Figure 15.16.


```
[
  {
    "account": "p1940779872trial",
    "application": "myneoprojectapplication",
    "state": "Ok",
    "processes": [
      {
        "process": "5d8f8f332b97b04fee7ad308105288df0dcb18d2",
        "state": "Ok",
        "metrics": [
          {
            "name": "Used Disc Space",
            "state": "Ok",
            "value": 48.0,
            "unit": "%",
            "warningThreshold": 90.0,
            "errorThreshold": 95.0,
            "timestamp": 1533787954000,
            "output": "DISK OK - free space: / 3730 MB (48% inode=82%); /var 1524 MB (79% inode=98%); /tmp 1844 MB (96% inode=99%);",
            "metricType": "rate",
            "min": 0.0,
            "max": 8063.0
          },
          {
            "name": "Requests per Minute",
            "state": "Ok",
            "value": 1.0,
            "unit": "requests",
            "warningThreshold": 0.0,
            "errorThreshold": 0.0,
            "timestamp": 1533787955000,
            "output": "JMX OK - RequestsCountMin = 1 ",
            "metricType": "performance",
            "min": 0.0,
            "max": 0.0
          },
          {
            "name": "Disk I/O",
            "state": "Ok",
            "value": 19333.0,
            "unit": "B/s",
            "warningThreshold": 1.0E7,
            "errorThreshold": 1.5E7,
            "timestamp": 1533787955000,
            "output": "OK: DiskRead: 12005 B/s (W> 10000000, C> 15000000) DiskWrite: 46185 B/s (W> 10000000, C> 15000000)",
            "metricType": "performance",
            "min": 0.0,
            "max": 0.0
          },
          {
```


Figure 15.16: JSON Data Coming Through the API

Now, you can access the performance monitoring data for your Java application externally through the API. Go ahead and build some cool dashboards or graphs using your favorite visualization tools.

15.4.2 REST APIs for Database Systems and SAP HANA Extended Application Services

Just as you can access Java application monitoring metrics through APIs, you can also access the metrics of your database systems and SAP HANA XS applications. In this case, basic authentication is not supported—OAuth 2.0 client credentials only.

Example URL patterns for database systems and SAP HANA extended application services, advanced model (SAP HANA XSA) applications include the following:

- *https://api.hana.ondemand.com/monitoring/v2/accounts/<subaccount_name>/dbsystem/<database_system>/metrics*
- *https://api.hana<trial>.ondemand.com/monitoring/v2/accounts/<subaccount_name>/instances/<XS_instance_name>/apps/<XS_app_name>/metrics*

Note

Since trial accounts do not have the option of setting up database systems, the URL does not contain the string "trial."

15.5 Monitoring Applications with Notifications

Notifications at the right time are an essential aspect of application monitoring on any platform, especially for business-critical applications running in cloud environments. Notifications trigger subsequent events and help you avoid manual tracking of monitoring information on an on-demand basis. As shown in Figure 15.17, the SAP Alert Notification service in SAP BTP allows you to raise instant alerts whenever an issue arises in your cloud application or its dependencies regardless of the underlying runtime environment, whether Neo or Cloud Foundry.

The SAP Alert Notification service even offers the ability to subscribe to various events coming from the platform and can handle custom alerts generated by your applications including notifications from hyperscalers and from Dynatrace. The notifications generated by the SAP Alert Notification service can further be pushed to various channels, according to your preferences, like email or to a Slack channel/Microsoft Teams space for the DevOps team responsible for a certain app or even for further persistency.

The service also offers a wide spectrum of built-in events out of the box, which include events related to application lifecycle, cloud integration, mobile services, transport management, quota, pricing, SAPUI5 adaptation projects, and SAP HANA Cloud services along with a provision for event extension. While setting up the SAP Alert Notification service, you can basically define the conditions for the trigger, define the delivery channel to deliver an event, and even define the action for undelivered events, along with other standard parameters.

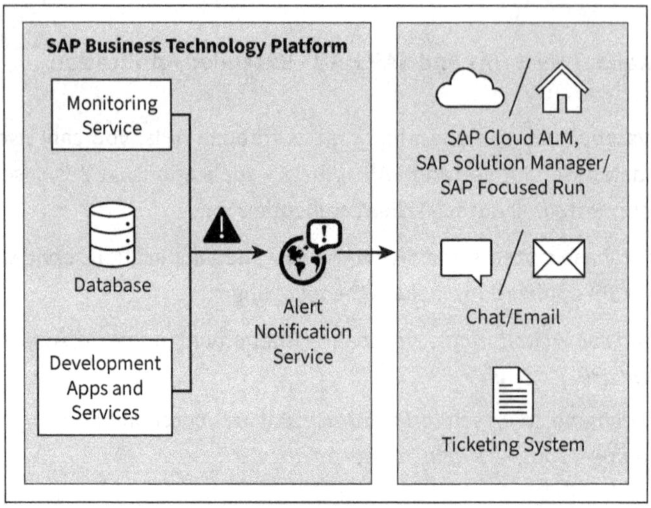

Figure 15.17: SAP BTP: SAP Alert Notification Service

15.6 SAP Automation Pilot

After establishing the monitoring and notification aspects of a DevOps cycle, the next important step is to establish automated actions to resolve the alerts and events notified by the SAP Alert Notification service. When an application failure occurs, which can be any kind of unexpected downtime of either microservices, databases, dependent services, or networks, or reconfigurations of solutions, it is critical to execute various commands, scripts, service calls, or even application restarts. However, the traditional approach driven by manual intervention makes the process of repetitive DevOps tasks slower and can delay pursuing the right action. The SAP Automation Pilot service can come into the picture to automate multiple DevOps-related tasks with ease. According to SAP's official definition, the primary goal of SAP Automation Pilot is to simplify and automate complex manual processes to minimize the operational effort behind any cloud solution in SAP BTP. On SAP BTP, SAP Automation Pilot, especially when combined with the Alert Notification service, can easily scale up your application management for your business-critical cloud functionalities.

In the following sections, we'll cover SAP Automation Pilot concepts and provide an overview of how it works.

15.6.1 Concepts

SAP Automation Pilot comes with predelivered content that covers the broad area of DevOps tasks for SAP BTP, Neo environment and SAP BTP, Cloud Foundry environment. Before we can configure SAP Automation Pilot on SAP BTP, let's briefly introduce some key components for SAP Automation Pilot, such as the following:

- **Data types**
  The service supports the typical data types (i.e., Boolean, number, string, JSON object, and JSON array) as required for defining the data type of value parameters inside key-value pair combinations.
- **Input**
  This component represents the required input data for commands in the form of a key-value pair.
- **Commands and catalogs**
  As shown in Figure 15.18, the system comes with predefined, ready-to-use automation content (known as *commands*), organized in catalogs, that can handle any scenario, including, for example, an SAP HANA database update, full application stack restart, or even a simpler operation like invoking a REST API or triggering a script. You can even build your own custom commands without any coding. Also, when you subscribe to the service, by default, you'll receive two catalogs out-of-the-box—**Welcome** (welcome-TXXXXXXRY) and **Recommended Actions** (RA-TXXXXXXRY). The Welcome catalog contains example inputs and commands that explain all of the basic SAP Automation Pilot mechanisms and the **Recommended Actions** catalog contains commands that showcase example application and database system recovery procedures. Interestingly, all these commands are derived from the most common customer use cases.

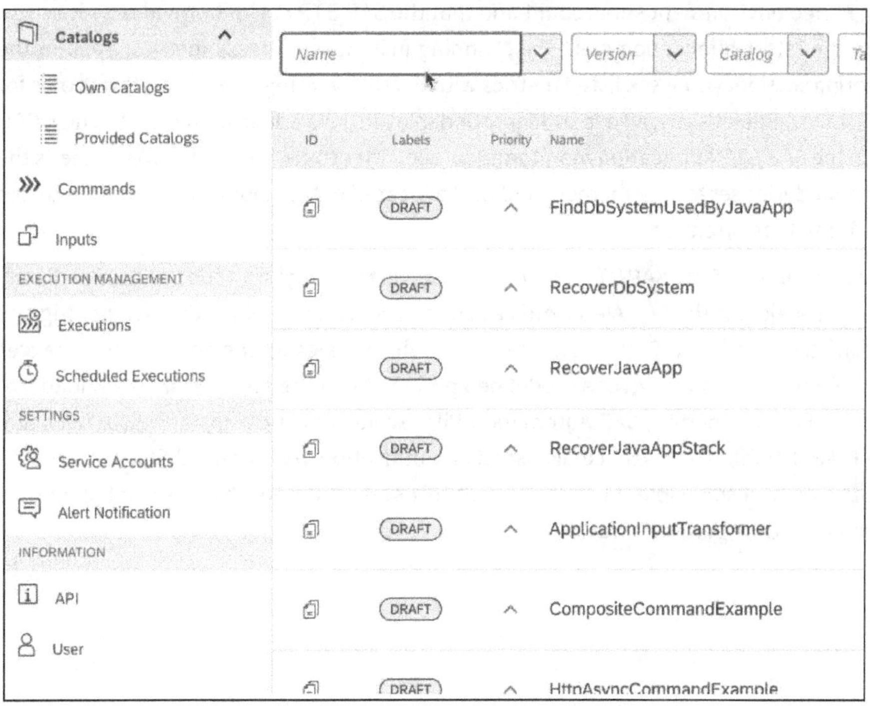

Figure 15.18: Default Catalog: Most Common Recommended Actions

- **Execution**
  This component represents the execution state of a command and can be in various states like **Running, Paused, Failed, Finished, Aborted, Suspended**, and so on. At a maximum, any execution can run no more than 21 days and use no more than 3 MB of memory each.
- **Scheduled execution**
  Autopilot executions can be scheduled to be executed at different times (i.e., once, yearly, monthly, weekly, daily, hourly, etc.).

Let's now illustrate how SAP Automation Pilot can be useful for monitoring and managing deployed applications in the context of the Java application that we deployed in earlier sections.

It's important to be aware that the Neo environment is currently in maintenance mode. Because of this, SAP strongly recommends that all new developments be built on the Cloud Foundry environment or the ABAP environment. This ensures long-term support, access to the latest SAP BTP services, and alignment with SAP's overall multicloud strategy.

15.6.2 How SAP Automation Pilot Works

To understand the applicability of SAP Automation Pilot in combination with the SAP Alert Notification service, let's discuss a possible scenario in the context of Java applications deployed in SAP BTP, Neo environment. Recall that our Java application is running in an SAP BTP, Neo environment subaccount and that the SAP BTP cockpit provides a nice overview of the latest application state (i.e., running instances, process metrics, etc.). In the monitoring section, you'll see listed metrics with default warning and critical thresholds for each instance. But let's say, all of a sudden, our deployed Java application reports a high CPU spike in the SAP BTP application monitoring screen. After some time, this spike exceeds the upper boundaries set for the memory and CPU allocated to the deployed instance and thus needs immediate attention.

As shown in Figure 15.19, SAP BTP's monitoring services spring into action and immediately send an error alert to the SAP Alert Notification service instance, which then further triggers a preconfigured action in SAP Automation Pilot, which scales up the application instances and monitors performance over a predefined period. When the application's CPU load stabilizes for a certain period, SAP Automation Pilot scales down the instances. In between, the SAP Alert Notification service also sends a notification via a defined Slack channel to notify the team responsible. In the course of this scenario, note that no manual input is required from the operator's side.

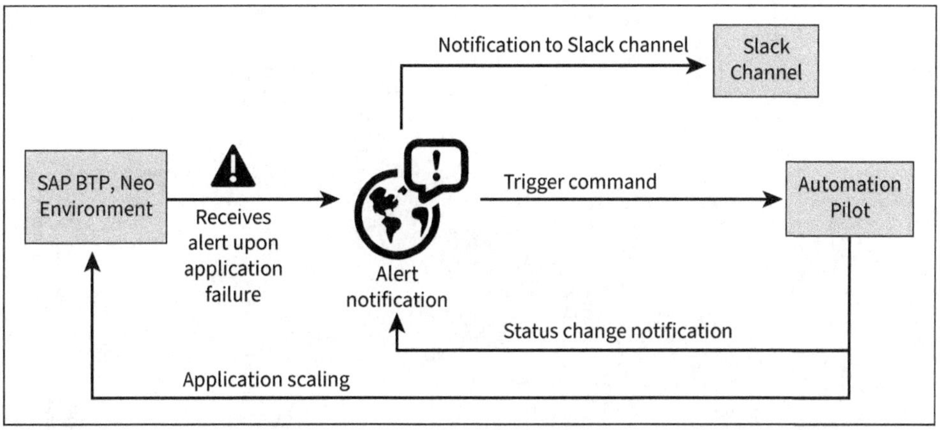

Figure 15.19: Integration Between SAP Automation Pilot and SAP Alert Notification Service

15.7 Summary

In this chapter, we looked at the various monitoring and logging capabilities of both SAP BTP, Neo environment and SAP BTP, Cloud Foundry environment. Additionally, we showed you how DevOps-related tasks, like alerting and triggering actions, can be easily automated without the need for manual intervention simply by integrating the SAP Alert Notification service and SAP Automation Pilot. Even if an administrator is responsible for setting up some of these tools, you should be familiar with these tools to monitor your applications. A common mistake is to consider performance monitoring and application logging functions only after an application has gone live. Understanding how to use these tools will help you prepare for better application monitoring well in advance.

PART IV
Supporting Tools and Services

Chapter 16
SAP Workflow Management

Since SAP Workflow Management was retired in 2023, existing workflow content must be enhanced and adapted before being deployed via SAP Build Process Automation. In this chapter, you'll learn all about developing workflows using SAP Business Application Studio along with its features and functions available as part of the current offering followed by a discussion about potential use cases that you can develop using the workflows. You'll also learn about migrating your existing content to SAP Build Process Automation.

Workflows have been in use in SAP landscapes, more specifically in SAP ERP, for many decades now. A useful tool for orchestrating complex interdependent actions, some workflows are automated while others require user intervention.

Over the last few years, SAP's offering in this space has evolved significantly and SAP Build Process Automation is now the strategic successor for building approvals and process execution instead of SAP's workflow engine. Organizations who have existing workflows built in SAP Business Application Studio can deploy them to SAP Build Process Automation with minor modifications, but for new workflow projects, SAP recommends using SAP Build Process Automation's no-code process builder instead. The workflow editor in SAP Business Application Studio is still available and can be leveraged to build workflows that require extensive custom coding, complex service integrations, or advanced ABAP/JavaScript development, which are not available through no-code tools.

In this chapter, we'll explore the features and capabilities of SAP Workflow Management that remain available, along with some use cases that can still be developed using the workflow editor in SAP Business Application Studio. Workflows created and modified with the SAP Business Application Studio workflow editor can continue to be edited within SAP Business Application Studio and deployed to an SAP Build Process Automation service instance. SAP Workflow Management is no longer available as a standalone service for renewal or new subscriptions. To use the workflow editor inside SAP Business Application Studio, simply create a new dev space with the **Workflow Management** extension enabled; it is listed under **Additional SAP Extensions** when creating a dev space.

This chapter covers features of a cloud-based workflow, including the process model, configuration, workflow triggering, and workflow monitoring. You'll also learn how to create complex workflows by orchestrating multiple subflows, user interfaces (UIs), and integrations, ultimately achieving process visibility.

16.1 Overview

Cloud-based workflows are modeled using Business Process Model and Notation (BPMN), currently version 2.0, so that the business processes are modeled following a standard graphical notation style to represent the flow. A standard modeling notation ensures that a model can be understood across the entire enterprise. Workflows are composable, which means you can create several subflows and compose them together to form the final workflow. The triggers of a workflow can include manual action or various automation events, including conversational artificial intelligence (AI) such as chatbots, other robotic process automation (RPA) features, and even microservices.

SAP Business Application Studio serves as the primary development environment for creating workflows within SAP Workflow Management. It provides a web-based BPMN editor and comprehensive development tools for modeling, testing, and deploying workflow solutions to the SAP Build Process Automation runtime.

This section begins by covering SAP Workflow Management's key capabilities, and subsequently addresses topics such as limitations, process models, user interface, workflow configuration, workflow triggering, rules, and process visibility.

16.1.1 Key Capabilities

Let's explore the key capabilities that SAP Workflow Management offers within its current scope, including the following:

- You can create workflows using industry-standard BPMN 2.0 notation using the web-based editor within SAP Business Application Studio.
- You can use the workflow editor inside SAP Business Application Studio by simply enabling **Workflow Extension** while creating your dev space.
- You can create workflows as part of multitarget applications (MTAs) with managed app routers for Cloud Foundry deployment
- You can model various gateways and events including start events, end events, intermediate message events, intermediate escalation events, and intermediate timer events.
- Various types of tasks including user tasks, service tasks, mail tasks, and script tasks to implement comprehensive workflow logic are available.
- Workflows created inside SAP Business Application Studio can be migrated to SAP Build Process Automation with minor adjustments.
- You can establish outbound connectivity to external systems and services through service tasks.
- You can consume workflow service APIs to search for custom workflow attributes and find respective workflow instances at runtime

- You can leverage business rules and "eventing" inside the workflow design, which can be from any system and exposed through Event Mesh.
- Deployed workflows as well as tasks can be monitored through the Monitor app.

16.1.2 Limitations and Considerations

Since SAP Workflow Management was replaced with its successor product called SAP Build Process Automation, there is a set of limitations and constraints including:

- Workflows developed in Business Application Studio cannot be imported or edited directly within SAP Build Process Automation's visual interface and can only be modified inside the SAP Business Application Studio workflow editor.
- Some workflow service features are not available in SAP Build Process Automation, including workflow scopes for technical user API access and all APIs related to workflow definitions.
- Existing workflows must be enhanced and adapted before being deployed to SAP Build Process Automation.

16.1.3 Process Model

One way to start building a workflow is by importing prebuilt workflow content. Another alternative in case no such scenarios are available in existing content packages is to build a process model from scratch.

Once the business process you want to implement in the workflow is defined in any modeling tool, this process must be detailed out to ensure that you know the inputs, outputs, decision points, where a particular process step should occur, its trigger, and what marks the end of the process. Corresponding technical details must be considered: For example, to execute a particular step in SAP S/4HANA, you must know what API is available. Armed with these details, you can model a workflow using SAP Business Application Studio.

In SAP Business Application Studio, graphical tools allow for easy configuration of a workflow and also make it easy to modify existing workflows by inserting or deleting steps or creating process variants. These features make your workflows completely flexible, without the complexity of code.

Figure 16.1 shows the SAP Workflow Management perspective in SAP Business Application Studio, highlighting the following:

❶ The components that exist under a workflow project.

❷ Workflow elements that can be dragged and dropped into the middle pane of the workflow editor.

❸ For each element, the actions that can be added (by right-clicking on that element).

❹ For each element, complete details about definition, data mapping, and more can be configured in the pane on the right.

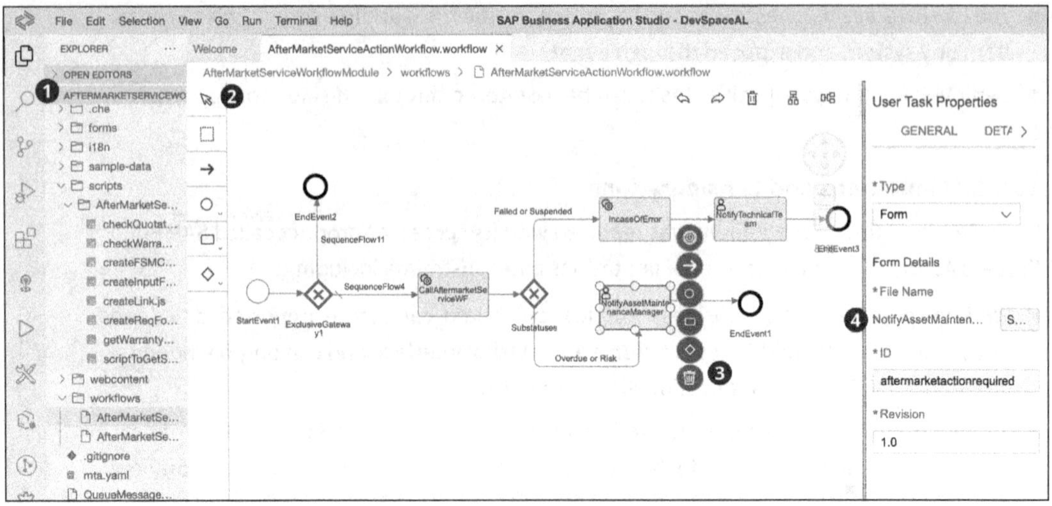

Figure 16.1: Showing the Workflow Editor in an SAP Business Application Studio Space

You can create your own workflow template to make your workflows reusable via the workflow monitor once the workflow's steps have been defined. The standard content still includes a default process variant, and other process variants can still be configured, but the Process Flexibility cockpit functionality has been integrated into SAP Build Process Automation as part of the transition from the deprecated SAP Workflow Management service.

16.1.4 User Interfaces

A particular workflow step can involve a user interaction, called a *user task*. Workflows provide options for creating simple forms as a frontend, for creating SAPUI5 components, or for creating an SAP Fiori elements UI. Alternately, if you have complex logic that needs to be created or emulated along with a UI, then you can create a microservice with a UI component (i.e., Node.js) and embed the business logic for required processing. Figure 16.2 shows a UI that can be created for a workflow.

Form editors are embedded into the workflow editor itself. These fields can be editable or read only. SAPUI5 applications, of course, provide more flexible design options but involve more effort. Other technologies can be used for the UI apart from SAP UI5, for example, JavaScript.

For user tasks, a boundary timer event can be attached so that, when this task is not completed within a stipulated time, an alternate flow is triggered (e.g., when a particular approval needs to take place by a certain time). These timer events are of two types: canceling and non-canceling boundary events. When a canceling boundary event occurs, the user task it's attached to is canceled if the given time limit is exceeded, which is useful, for example, in an escalation flow. For a non-canceling boundary event, the event is triggered but does not cancel the user task, which is useful, for example, for sending reminder emails to the user to take action, when the stipulated time has elapsed.

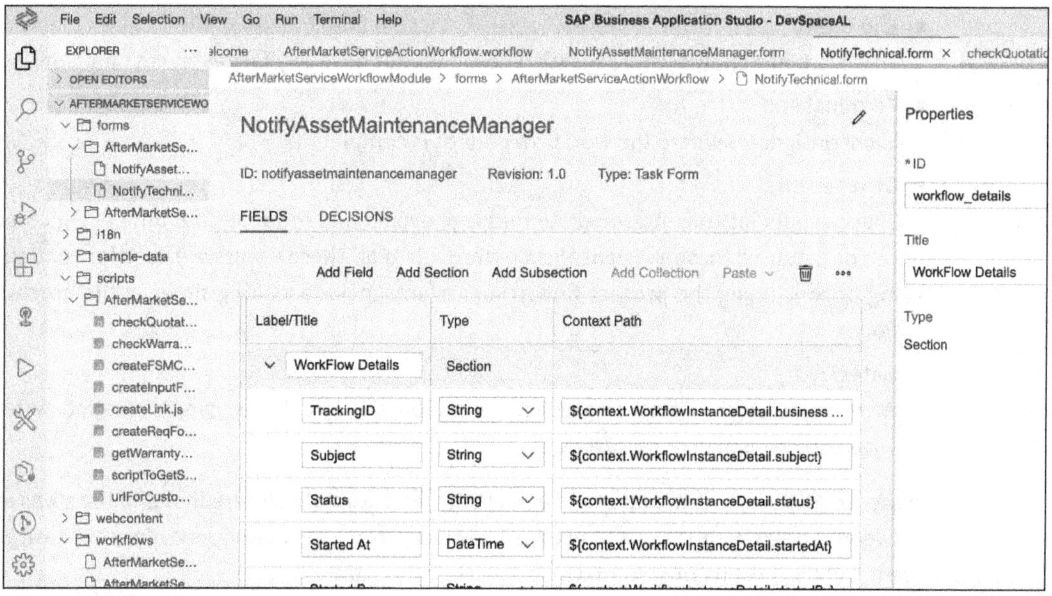

Figure 16.2: Sample UI in Workflow Project (Workflow Form)

16.1.5 Configuring Workflow Steps

Once a workflow project has been created through a process template or directly in SAP Business Application Studio, its process steps must be created and configured through the workflow editor. Each step can be one of the following types of tasks:

- **User tasks**
  These tasks are performed by human users and can be accessed from the inbox. For a specific user task, you must define who is responsible for this task, which could be one or more users or user groups or even based on some rule configured.
- **Service tasks**
  These tasks are performed by systems, for instance, an external service call. These tasks are started immediately when the flow reaches that object.
- **Script tasks**
  Script tasks are, as the name suggests, flow objects that have scripts that are executed when the process flow arrives at that task.
- **Mail tasks**
  These tasks are for sending emails to one or more recipients through SAP BTP destinations. The template for emails can be created and reused, and data from the workflow context can be fed into the email.

Some of the following elements are generally required in a workflow step:

- **Start event**
  Every workflow has a start event that designates the trigger of the workflow. No incoming flow exists for a start event.
- **End event**
  This event designates where the workflow does not have any further event or action.
- **Terminal end**
  Event designates where the workflow ends or is completed.
- **Other events**
  These events include intermediate message events and intermediate time events that can be captured through Event Mesh or through APIs. These events are used for continuing or concluding the process flow. Timer events include waiting times in the process flow.
- **Gateways**
  When branching has been created for a workflow step, a gateway can be inserted. Gateways can be parallel or exclusive.

In the workflow editor, the start and end events are automatically created and the steps in between can be chosen from the palette and dragged into the editor pane. For each task or event, details must be maintained in the pane on the right.

During the start of a workflow, you'll need to pass some parameters as the initial context, which can be passed on to subsequent steps. The context should be the minimum set of information that needs to be shared with a user or obtained from the user that is required from a flow perspective. In case a larger chunk of data is required, and if calculations are to be performed on this data, that data should be kept in a separate system, perhaps the source system itself (e.g., an SAP S/4HANA system) and retrieved through API calls. This approach will make the workflow lighter and improve its performance.

Principal propagation is used to retrieve information about the user who started the workflow or did a particular step (like approve).

16.1.6 Triggering the Workflow

A few mechanisms exist for triggering SAP workflows, for instance, the following options:

- Workflows can be triggered through events.
- Workflows can be triggered from another microservice, which can include a UI.
- Workflows can be triggered from chatbots, which can call the workflow API.
- Workflows can also be triggered from non-SAP systems by calling the public API available for the workflow runtime, after maintaining a little configuration in the third-party application.

Let's dig a bit more deeply into these scenarios and how workflows are triggered.

Considering the first scenario of events, the configuration of events themselves can be performed in Event Mesh so that, whenever an event is published in Event Mesh, the workflow is triggered. One way to set this trigger up is by putting the code for instantiating the workflow in a JavaScript file in an SAP Cloud Application Programming Model project, which is then deployed in SAP BTP, Cloud Foundry environment. Thus, a condition for the workflow trigger could be some specific events occurring. Understandably, Event Mesh and other workflow services must be available and configured as prerequisites for this use case.

From a UI perspective, which can be part of a microservice, similarly, you can instantiate a workflow by calling the workflow API with `POST /v1/workflow-instance`. This action can exist in the form of a button like a **Submit** button or similar.

The same concept of an API call is leveraged when triggering of a workflow from a chatbot, where again a certain step may include the triggering of a specific workflow to handle a specific user request processed through the chatbot.

Note that all these workflow APIs are representational state transfer (REST)-based. While calling an API, the complete URL of the workflow must be provided, including the API host URL and the application ID.

However, sometimes, a mismatch exists between the supported protocols among various non-SAP systems (and sometimes involving SAP systems as well), which will require an integration layer like Cloud Integration.

[«]

Note

You can find APIs for workflows for SAP BTP, Cloud Foundry environment in SAP Business Accelerator Hub at *http://s-prs.co/v550415.*

Details about the configuration required, including security aspects, are covered in the SAP Help Portal documentation, available at *http://s-prs.co/v550416.*

Some good blogs are available as well for detailed steps about configuring a workflow trigger from an SAP Fiori app in SAP BTP, Cloud Foundry environment; for instance, *http://s-prs.co/v550417.*

16.1.7 Configuring the Rules

Business rules are pieces of simple value-based logic that can be configured using the business rules engine. This separate service and the configured rules can be consumed by SAP Workflow Management through APIs. The rules engine can be accessed from the SAP Workflow Management cockpit or from your SAP BTP subaccount.

These rules could relate to validation or decision steps, could help a user choose the right process variant, or could even call an RPA bot, based on key attributes from the vocabulary.

Rules management has transitioned from the standalone Manage Rule Projects app to SAP Build Process Automation. The Manage Rule Projects app is now available within SAP Build

Process Automation and supports importing existing business rules projects from SAP Workflow Management. In the Manage Rule Projects app, you'll need to create a rules project, within which you'll identify the data elements used in the rules, the rule services, and rulesets and, finally, define the rules themselves.

In the rule service, you must provide the rule service name that can then be called from whichever application is consuming the rule. In this context, you'll need to define the vocabulary to identify which data element is the input and which is the output. Individual rules can be created that contain decision tables, as shown in Figure 16.3.

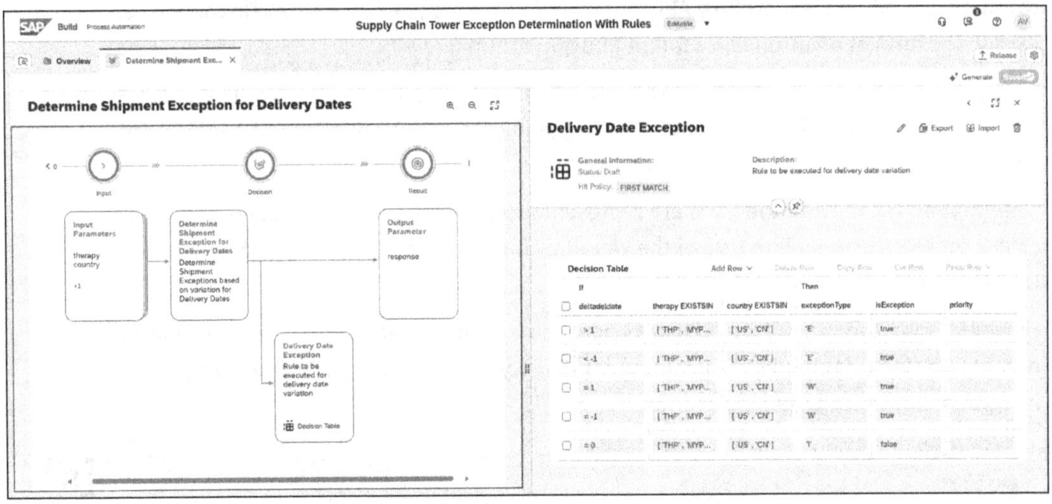

Figure 16.3: Decision Table Rule Sample

Another type of rule is called a text rule, which could include logical expressions, as shown in Figure 16.4. In this example, our rule has a condition involving two parameters (temperature and vibration), and a trigger occurs when these values exceed both thresholds you've set. If this condition is met, then the event described under the **Then** heading is triggered.

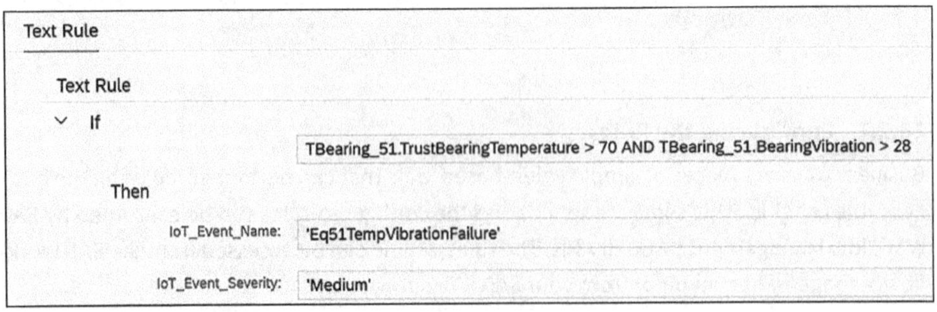

Figure 16.4: Text Rule Sample

Figure 16.5 shows an example of a script used in a workflow that consumes the rules configured earlier and passes the rules' contextual parameters to the next workflow steps.

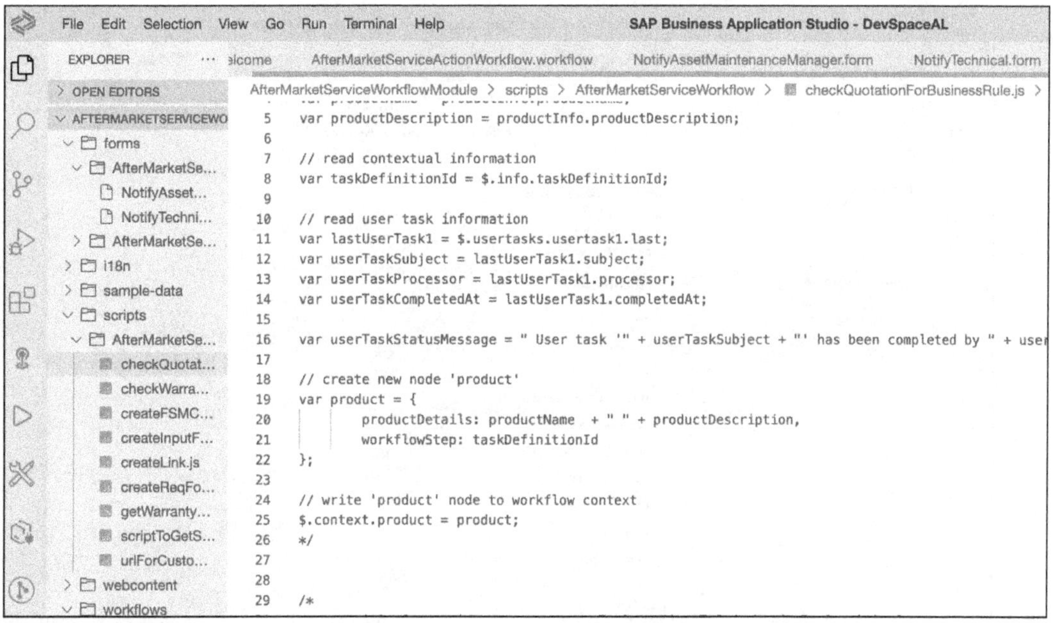

Figure 16.5: Code Showing Consumption of Business Rules in Workflow Script Step

16.1.8 Process Visibility

This unique feature of SAP Workflow Management is designed for creating dynamic dashboards. Consider some processes that run in a hybrid environment—on-premise and on the cloud, SAP and non-SAP. Information is gathered, in real time, from various process steps through events of the workflow, which can arise in all these disparate systems. Context links these events together, and according to your configuration, end-to-end process KPIs from different transactional systems can be calculated and displayed in a graphical format. Business events from transactional systems for SAP SaaS solutions (i.e., SAP Sales Cloud, SAP SuccessFactors, or SAP S/4HANA Cloud) or from any non-SAP applications (whether in the cloud or on-premise) can also be published to Event Mesh, as the event broker, and consumed by the process visibility feature. These transactional systems provide REST APIs published to Event Mesh. The events triggered in the process visibility feature can be viewed via the Event Acquisition app.

Process visibility also provides options for configuring actions. Thus, for a particular insight, your business users may have one or more recommended actions, for example, restarting an error status instance. For a workflow, several visibility scenarios can be configured, with each having a graphical representation (i.e., a bar chart, a donut chart, or a line chart), as shown in Figure 16.6.

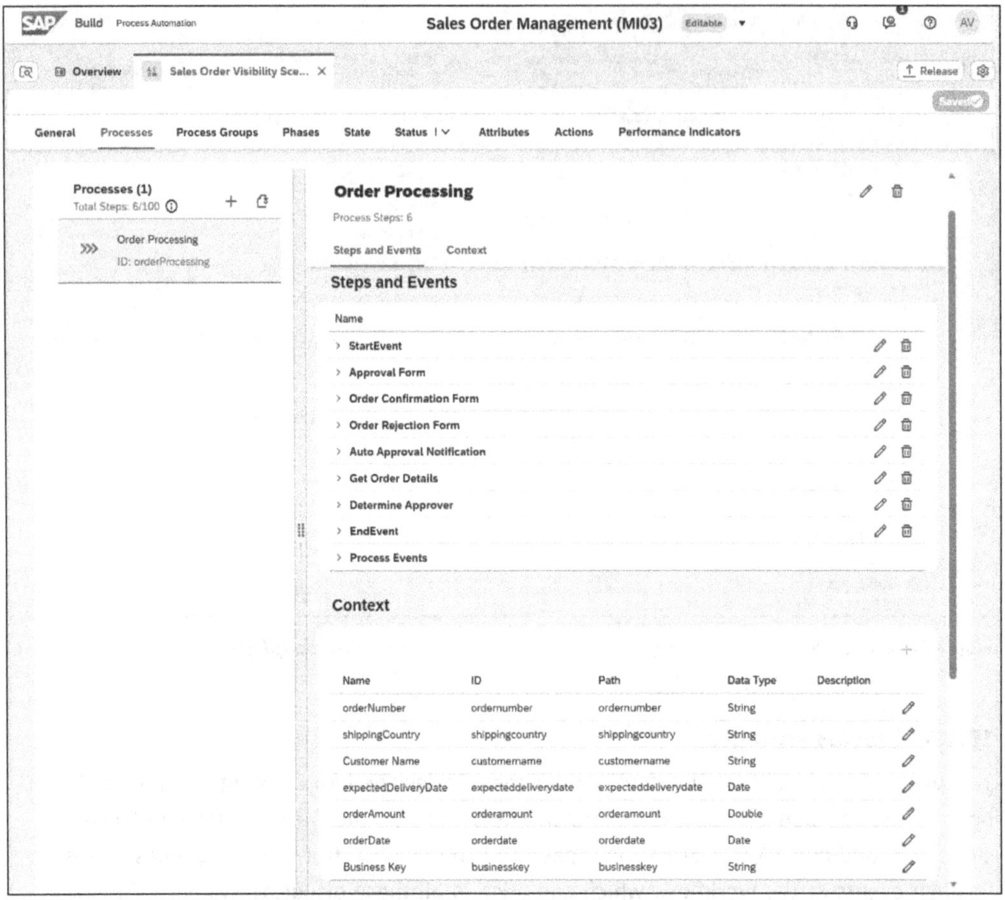

Figure 16.6: Process Visibility Scenario: Configurations

Process visibility scenarios from SAP Workflow Management need to be imported into SAP Build Process Automation business process projects and redeployed. The new implementation provides improved interfaces, enhanced monitoring features, and integrated experience within SAP Build Process Automation. Workflows developed in SAP Business Application Studio can be added as process participants in process visibility scenarios within SAP Build Process Automation. You can move existing visibility scenarios from SAP Workflow Management by exporting them as *.zip* files and importing them into business process projects in SAP Build Process Automation. Once imported and deployed, you can access and manage visibility scenarios through the **Control Tower • Environments • Visibility Scenarios** tab, where you can monitor processes, access dashboards, and configure notifications.

16.2 Integrations

In this section, we'll cover two aspects of integration for workflows—workflow-specific integrations and intelligent services.

16.2.1 Integration Specific to Workflow Steps

Workflow-specific integrations can be divided further into two divisions:

- The workflow can be triggered, started, or stopped from another application, as described in Section 16.1.6.
- The workflow can connect process steps to different tasks, including user tasks or actions performed in another system altogether.

With the second option, workflows can have two-way integration.

The first aspect is dealt with using the Workflow API, which can provide broad access to the various objects within the workflow. You can change the assigned user or change the status of a workflow using the Workflow API, which provides access to several endpoints like the user tasks instance, task definition, workflow definition, workflow instance, and more. Workflows can be modeled for callbacks.

An SAP cloud workflow instance can be started from another application and can be instantiated by calling the Workflow API. These REST-based APIs allow for the tight integration between tasks on SAP BTP and the workflow capability and uses OAuth 2.0 as the authorization mechanism.

Currently, the APIs available as part of the SAP Workflow Management service, in addition to the Workflow API, include the following:

- Process Flexibility API, for orchestrating workflows across different systems
- Process Visibility API, for pushing events, managing process visibility, and fetching scenario instances
- Rule Authoring API, for creating the business rules
- Rule Execution API, for deploying, searching, and using created business rules
- Inbox API, for retrieving workflow tasks assigned to a particular user (used to incorporate them in the SAP Fiori inbox)

16.2.2 Integrating Intelligent Services

Another aspect is to integrate with one or more of the following functions:

- Machine learning algorithms: Any workflow step can call a machine learning algorithm to perform predictive analysis and show the user suggestions for the best next step.
- Chatbots or conversational AI: Either a workflow process can be triggered from a chatbot or the workflow can trigger a chatbot to engage in some interactive user activity.
- Integration with SAP Build Process Automation: Any step of a workflow can trigger an automation, like updating a purchase order (PO), through SAP Build Process Automation-based automation.

[»] **Note**

For more details about workflow configurations, refer to *http://s-prs.co/v550418*.

16.3 Model Architecture for Workflow on SAP Business Technology Platform

In this section, we'll explore a sample use case and describe all the steps for modeling a workflow in SAP BTP, from model to implementation and architecture of the solution. Your own scenarios or use cases may be more complex, requiring additional functionality beyond using SAP workflows. Let's consider an example scenario.

Let's say you have been monitoring your machines/equipment through a monitoring dashboard, developed in SAP BTP on top of an Internet of Things (IoT) setup. This dashboard displays the real-time values of the operational parameters of machines. If parameter values exceed specified threshold limits, an event is triggered, based on business rules defined, which creates a service request automatically in SAP S/4HANA and sends a notification to the monitoring dashboard application (which is a microservice).

In a real-life scenario, a set of parameters is monitored, and if one or more parameters cross a threshold limit, an alert is raised and captured in the application. This application, in turn, checks when the machine's last maintenance was performed and when the next servicing is due. If servicing is not due within a stipulated time and the alert concerns a critical parameter (both configured in the application), a service order is automatically created (via an API) in SAP S/4HANA, triggering the workflow start event. This workflow sends an email notification to the service provider asset manager, who must then review or modify the service order created in SAP S/4HANA, which is passed in the workflow's context parameters. The service provider asset manager submits a change, and the workflow approval step is generated for the customer, especially if the machine is not in warranty. Some additional steps may be required in between; for example, you may need to generate a quote for servicing the machine. If the quote exceeds a certain amount, an approval step may be necessary; if not, the service order is simply released to SAP S/4HANA. An example process model is shown in Figure 16.7.

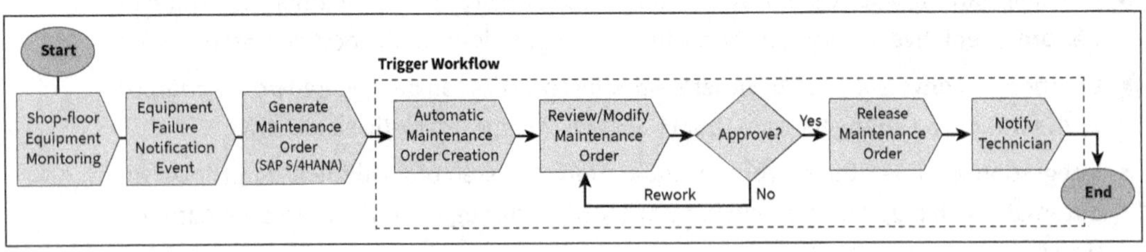

Figure 16.7: Process Flow of a Complex Client Use Case

The architecture of such a solution is shown in Figure 16.8. The interactions of the events, the APIs with the transactional systems, and the controlling systems like the workflow can be orchestrated through API Management and Event Mesh.

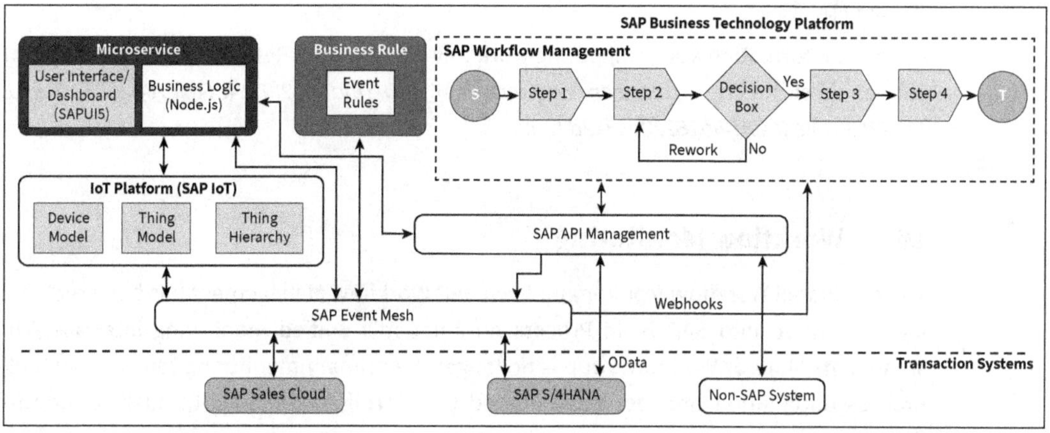

Figure 16.8: Architecture of a Complex Use Case

In our example, the monitoring of the machine is operationalized through the Cumulocity IoT platform via the SAP Asset Performance Management solution, which will also use master data from the SAP S/4HANA system. If the Cumulocity IoT platform isn't available, you can take advantage of any IoT platform offering REST-based endpoints for integration. The event generated when some parameter values cross the configured thresholds will be trapped by Event Mesh and will be captured by the microservice to take the next steps. The microservice in turn can invoke the business rules engine and finally trigger a workflow.

16.4 Workflow Package

As described at the beginning of this chapter, when you create a workflow from scratch, you're creating a multitarget application (MTA) since the project can include, apart from workflows, other components like the UI modules, scripts, business logic (Node.js modules), business rules, and any related components.

SAP Build Process Automation has transitioned from the traditional MTA deployment model used in SAP Workflow Management to a unified project-based approach. The deployment process in SAP Build Process Automation significantly simplifies the traditional MTA approach by providing a unified project-centric model that eliminates the complexity of managing multiple modules and build dependencies while preserving the ability to import and deploy legacy MTA packages.

One of the primary aspects of migration is to modify the *mta.yaml* file to reference the SAP Build Process Automation service instead of the legacy workflow service by changing the service name from `workflow` to `process-automation-service`, changing the service plan to `standard`, and including the `deployUsageScenario` as a workflow parameter in service-key configuration. In addition to the above-mentioned changes, we also need to make authorization and security-related changes along with destination and connectivity related updates.

For more information about migrating workflow artifacts to SAP Build Process Automation you can check out *https://help.sap.com/docs/PLM_EPD/1bfdb8b2f0264b66a818a2a889ee8c31/17b2243829fe4c49848a2625ff21812d.html.*

16.5 Workflow Monitoring

The traditional Workflow Monitor app from SAP Workflow Management has been replaced and integrated into SAP Build Process Automation's unified monitoring interface. The standalone Monitor Workflows app is no longer the primary monitoring tool, as SAP Build Process Automation provides consolidated monitoring capabilities through its Control Tower and integrated monitoring application.

During the transition from SAP Workflow Management to SAP Build Process Automation, existing workflow instances and monitoring data are automatically preserved. Users can continue to access historical workflow instances through SAP Build Process Automation's monitoring tools without requiring data migration, as long as they use SAP Build Process Automation in the same consumer account.

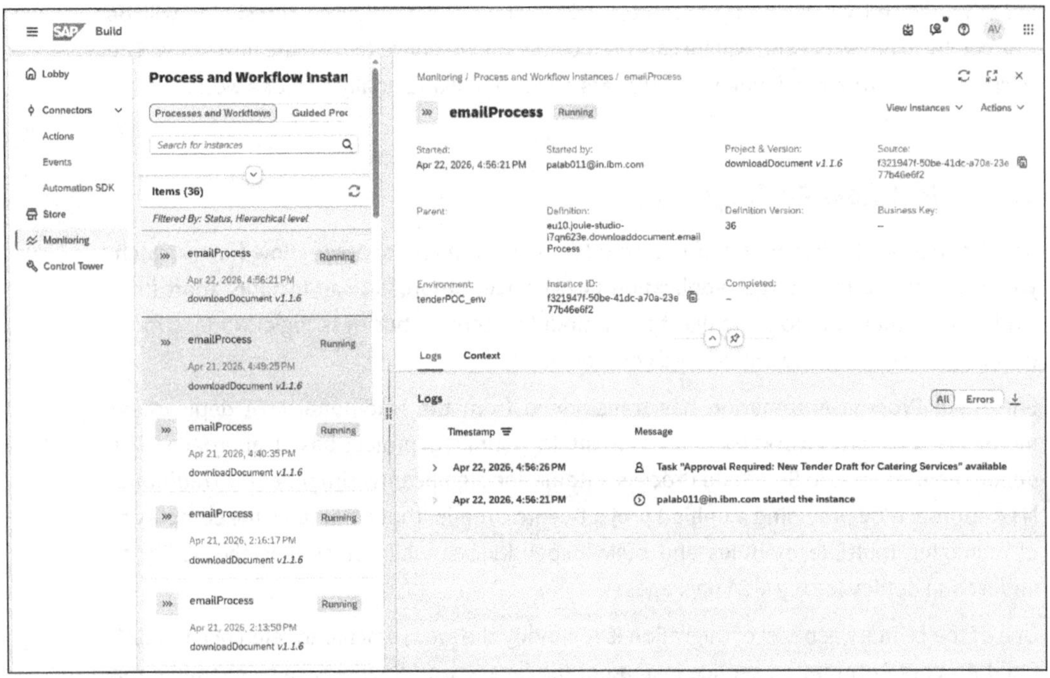

Figure 16.9: Monitoring App

The monitoring functionality has been enhanced and integrated into the broader SAP Build Process Automation ecosystem. The new monitoring includes visibility scenarios, process analytics, and cross-system workflow tracking that extends beyond traditional workflow monitoring

As depicted in Figure 16.9, monitoring now covers processes, automations, decisions, and visibility scenarios, in addition to traditional workflows.

16.6 Summary

In this chapter, we provided a complete overview of workflows in the cloud, along with some relevant use cases, features, and functions offered as part of the workflow functionality inside SAP Business Application Studio. With SAP Workflow Management now retired, we have discussed the latest changes in the overall offering followed by the migration process required to move content from SAP Workflow Management to SAP Build Process Automation. In the next chapter, we'll provide detailed information about SAP Build Work Zone.

Chapter 17
SAP Build Work Zone

As we transition from backend development concepts to delivering complete business solutions, SAP Build Work Zone plays a key role in uniting applications, data, and user experiences within a single environment. This chapter serves to bridge the gap between technical development and end-user interaction, demonstrating how your work is ultimately presented and consumed within the SAP ecosystem.

When examining modern organizations today, it is important to understand the complexity of their digital environments. In a typical enterprise, employees, managers, and even external partners use many different systems daily. Some systems come from SAP, others from non-SAP vendors, and each is designed to solve a specific business problem such as HR management, finance, procurement, projects, or collaboration. Although each tool may work well individually, the overall user experience often becomes confusing and inefficient. People are not always sure where to begin their workday, which system to log in to for approvals, or where to find important business information. As a result, users waste time switching between applications, searching for links, and recalling multiple usernames and passwords instead of focusing on their actual work.

This is exactly the problem SAP Build Work Zone is designed to solve. It provides a single, central entry point where employees can start their day. Instead of opening separate systems for HR, finance, project management, and communication, users log in once and see a personalized homepage created specifically for their role. From this single screen, they can access SAP applications like SAP SuccessFactors or SAP S/4HANA, as well as non-SAP tools such as Microsoft Teams or company-specific custom applications. The system automatically shows only what is relevant to the user's job, making it easier for them to understand priorities and complete tasks. On top of that, users can personalize their workspace further by adding widgets that show approvals, notifications, news, or frequently used links, creating a work environment that supports their daily activities.

This chapter is organized into three parts: an introduction to SAP Build Work Zone, an explanation of its key concepts, and an overview of the SAP Build Work Zone editor. Together, these sections provide both the background and the practical knowledge needed to understand and use the solution.

17.1 Introduction to SAP Build Work Zone

As organizations adopt SAP S/4HANA, SAP SuccessFactors, SAP Ariba, SAP Concur, and a growing landscape of cloud and third-party tools, user experience fragmentation has become one of the most common pain points in enterprise IT. Even technically sound implementations fail to deliver value when users cannot easily access applications, tasks, and data.

SAP Build Work Zone addresses this problem by providing a single, role-based digital entry point. Built on SAP Business Technology Platform (SAP BTP), it unifies the former SAP Launchpad service and SAP Work Zone. From a delivery perspective, SAP Build Work Zone is most successful when introduced alongside SAP S/4HANA transformations, cloud migrations, or employee experience initiatives. In such programs, it acts as the digital front door to enterprise processes without requiring any changes to backend systems.

Another important aspect to understand is how SAP Build Work Zone is developed and maintained. Because it's based on low-code and no-code principles, business users, not just IT professionals, can design, update, and manage content using drag-and-drop tools. Pages can be adjusted quickly as business needs change, which is especially important during digital transformation programs. SAP Build Work Zone also includes SAP's AI solution, Joule, which adds intelligence to the digital workplace. With Joule, users can ask questions in natural language, quickly find information, get summaries of documents, and complete routine tasks faster, supporting decision-making and increasing productivity.

Finally, it is useful to know that SAP Build Work Zone is available in different editions to suit different organizational needs. The SAP Build Work Zone, standard edition focuses mainly on providing a central launchpad for applications and tasks, helping users access their tools efficiently. The SAP Build Work Zone, advanced edition goes a step further by offering a complete digital workplace experience, including collaboration features like team workspaces, blogs, discussion forums, and company-wide news and updates. The platform works on both desktop and mobile devices through the SAP Mobile Start app, allowing users to stay connected.

[»]

Note

We recommend positioning SAP Build Work Zone as an experience layer rather than a portal replacement. This approach improves executive alignment and avoids unnecessary technical debates.

17.2 Key Concepts

As discussed, SAP Build Work Zone is delivered in two primary editions that address different enterprise needs, as follows:

- **SAP Build Work Zone, standard edition (earlier called SAP Launchpad service)**
  This is mainly about access. It gives users a central, role-based home page where they can see and launch all the applications and tasks they are authorized for. So, if a user logs in, they immediately see only what is relevant to their role.
- **SAP Build Work Zone, advanced edition (earlier called SAP Work Zone)**
  This goes a step further. In addition to application access, it focuses on collaboration and engagement. Here, users don't just open apps—they can work together, share documents, post updates, and maintain knowledge articles.

Now, inside the advanced edition, we have something called workspaces. You can think of a workspace as a team room. Teams use it to share information, documents, and updates. Workspaces can be public, private, or even external if you want to collaborate with partners or suppliers. Each user also gets a personal workspace for their own use. Inside a workspace, you'll typically find a feed for updates and announcements, forums for discussions, calendars for events, and a knowledge base where structured information is stored using forms.

Next are workpages. Workpages are dashboard-style pages created using a simple drag-and-drop editor. You don't need deep technical skills to build them. On a workpage, you can place application tiles, widgets, and UI integration cards that show live data. The idea is to give users a single, visually clean page where they can see everything they need at a glance.

In the background, the advanced edition uses something called the digital workplace service. You can think of this as the engine behind SAP Build Work Zone. It handles administration, manages user profiles, and provides APIs for collaborative content like feeds, blogs, and discussions.

From a technology perspective, everything runs on SAP BTP, Cloud Foundry environment and follows SAP Fiori design principles. There are different ways content gets into SAP Build Work Zone, as follows:

- Federation: Systems like SAP S/4HANA expose their apps automatically using a common data model.
- Manual integration: Admins manually configure apps such as SAPUI5, Web Dynpro, or simple URL-based apps.
- Content packages: Prebuilt bundles (cards, workflows, etc.) created in SAP Business Application Studio and imported into the system.

For on-premise systems, SAP uses the cloud connector. This creates a secure tunnel between your system and SAP BTP—there is no direct internet exposure. For user management and authentication it uses the following:

- Identity Authentication to manages login
- Identity Provisioning which syncs users from systems like Microsoft Entra ID or SAP SuccessFactors

Both of these services are mandatory for a proper setup.

UI integration cards are small, reusable cards that show live business data—for example, open sales orders or pending approvals. They fetch data through APIs using destinations in SAP BTP and display it directly on workpages.

Everything is also available on mobile through the SAP Mobile Start app. Users can log in using fingerprints or face recognition. They get a single inbox for all tasks and notifications across their SAP systems.

[zB]

Example

A professional services company implemented SAP Build Work Zone, advanced edition to standardize onboarding across regions. HR teams built onboarding journeys combining SAP SuccessFactors data, policies, learning links, and discussion areas, while IT maintained security and governance.

17.3 SAP Build Work Zone Editor

SAP Build Work Zone is based on a clear and scalable content hierarchy. A business site represents the overall digital workplace. Within a site, spaces organize content by business domain, pages focus on specific tasks or scenarios, and sections group applications, widgets, and integration cards. This structure mirrors how users think about their work rather than how systems are technically designed. When modeled correctly, it allows consultants to map business processes directly into intuitive navigation. Let me explain the SAP Build Work Zone editor, which is also commonly called the workpage editor or page designer.

In simple terms, this editor is a low-code, WYSIWYG tool, meaning that what you design on the screen is exactly what users will see. You don't need to write code to use it, which is why both IT administrators and business users can comfortably work with it.

This editor is mainly used to create workpages. You can think of workpages as landing pages—either for the main site or for specific workspaces—where users start their day and access apps, data, and information. The editor uses a grid-based layout. You build a page by adding rows, and each row can have up to six columns. This gives you a lot of flexibility to control how content is placed on the page. Everything works through drag and drop. You simply pick a component, place it on the page, move it around, resize it, and align it visually—no technical skills are required.

Inside the editor, you have access to a widget catalog with many ready made components. For example, you'll find standard widgets like banners, text blocks, and images; content and document widgets that connect to document repositories, forums, or knowledge base articles; and collaboration widgets such as polls, surveys, and news feeds to keep users engaged. You just select what you need and configure it directly on the page.

One useful feature is that changes are saved automatically while you work. You can store the page as a draft for review or testing, and once everything looks good, you publish it so it becomes visible to users.

The editor also supports UI integration cards. These cards show live data from backend systems like SAP S/4HANA using OData services. The good part is that you can configure these cards directly in the editor. You also have control over the layout details, such as making a row full width, adjusting spacing and padding, and setting background colors or images for individual sections.

In newer versions, the editor even includes AI assistance through Joule, which can help generate text, summarize documents, or quickly create content right within the editor. You can open the editor from different places, depending on what you're working on: from the main home page by clicking the pencil (edit) icon—this requires admin roles like Company Admin or Page Content Admin; from within a workspace, where workspace admins can customize their own landing pages; or with the site menu editor, which is used to link these workpages to the main navigation menu.

To complete this discussion, the next sections look at how content and access are structured in SAP Build Work Zone. We first explain core building blocks such as spaces, pages, cards, widgets, and content packages, then cover how users, roles, and permissions control access. Finally, we discuss content integration and how data and applications from different systems are brought together.

Note

Always design navigation around business domains such as finance, HR, or operations instead of system names.

Example

In a consumer goods organization, finance managers struggled during month-end close due to fragmented reporting and approvals. A dedicated finance space in SAP Build Work Zone combines KPI cards, approval inboxes, and links to SAP S/4HANA applications, significantly improving decision speed.

17.3.1 Spaces, Pages, Cards, Widgets, and Content Packages

Let's explain how SAP Build Work Zone is designed internally. At its core, SAP Build Work Zone uses a modular design. That means instead of building everything as one large, complex structure, SAP Build Work Zone breaks things down into small, logical building blocks. These blocks can then be reused and combined depending on business needs.

Broadly, you'll hear about spaces and pages for structure, cards and widgets for interaction, and content packages for deployment.

Let's start with spaces and pages, because they form the backbone of navigation. Think of a space as a top-level folder. Each space represents a business area or role, like finance, HR, or supply chain. These spaces show up in the main navigation menu, so users can easily switch between business areas. Inside each space, you have pages.

Pages are where users actually do their work. This is the screen where applications, tiles, and cards are displayed. Each page is divided into sections, which help organize content based on what users need for their daily activities.

Now, the important part is security and personalization. Everything here works on a role-based model. Administrators assign spaces to specific business roles. Users only see the spaces and pages that match their assigned roles. So, every user gets a clean, relevant view without any unnecessary content.

Next are UI integration cards. UI integration cards show live data at a glance. Cards are small, reusable UI components built using SAPUI5. Their main purpose is to show live business data in a compact and easy-to-read way. Instead of opening a full SAP application, a user can quickly check KPIs, see recent approvals, and view lists, charts, or timelines. Technically, each card is defined using a *manifest.json* file. This file tells the system how the card should look and which backend OData or REST APIs to call for data.

There are different card types, such as analytical cards (charts), list and table cards, and timeline cards. Usually, these cards are built in SAP Business Application Studio and delivered as part of a content package, so they can be reused across systems and landscapes.

Now let's talk about widgets. Widgets are the building blocks you use when designing pages in the workpage editor, especially in the advanced edition. They allow you to add different types of content and interaction to a page, such as:

- Collaboration widgets like feeds, forums, polls, and surveys.
- Content widgets like knowledge base articles, wiki pages, blogs, images, and videos.
- Business widgets like task lists from SAP Task Center, search widgets, or business record views.

One key difference here is self-service. While spaces and pages are mostly configured by administrators, business users can use widgets inside workspaces to create and personalize their own collaboration areas without IT involvement.

Finally, all of this content is usually delivered using content packages. A content package is basically a ZIP file that bundles related objects together, making deployment and transport much easier. A package can include UI integration cards, workspace templates, workflows, and sometimes even preconfigured roles. There are two main types of content packages, as follows:

- Standard or global packages, provided by SAP or partners and installed directly through the channel manager
- Custom packages, developed internally using SAP Business Application Studio and uploaded manually to the SAP BTP subaccount

Administrators manage these packages through the administration console, where they can install, update, or remove them as needed.

17.3.2 Users, Roles, and Permissions

Let's now discuss security in SAP Build Work Zone, because once you see how it's structured, it actually makes a lot of sense. The most important thing to understand is that SAP clearly separates identity from permissions. These two are handled separately.

First, there is the question of who the user is (identity management). Everything begins with user identity, which is managed through SAP Cloud Identity Services. When a user attempts to log in, Identity Authentication manages the process. This is where authentication occurs, including passwords, MFA, and even single sign on if the company uses systems such as Microsoft Entra ID or Okta. Identity Provisioning operates in the background. Its job is to keep users and groups in sync by pulling them from systems such as SAP SuccessFactors or LDAP into the SAP Build Work Zone environment. At this stage, SAP knows who the user is but does not yet know what they can access.

Second, you have what the user can access (role-based access control). Authorization is handled in multiple layers, starting from the platform and moving toward content and collaboration:

- **Platform level: SAP BTP subaccount**
  Before a user can even open SAP Build Work Zone, they must be assigned the right role collections in the SAP BTP cockpit. Think of these as entry passes:
  - `Workzone_User` lets the user access the site.
  - `Workzone_Admin` gives full administrative access.
  - `Workzone_Area_Admin` allows managing specific content areas.

  If a user doesn't have these roles, they simply cannot enter SAP Build Work Zone at all.
- **Content level: Spaces and pages**
  Once the user is inside SAP Build Work Zone, the next question is: What content should they see? This is controlled through spaces and pages, which are mapped to business roles. Here's how it works: Administrators map SAP BTP role collections to specific spaces, and each space usually represents a business area like finance, HR, or sales. If a user doesn't have the Finance role, they won't see the finance space or any of its pages.

  The same logic applies to applications: Apps appear in the app launcher only if the user's role allows it. If you don't have the right role, you won't see the app, so every user sees only what's relevant to their role.
- **Collaboration level: Workspaces (SAP Build Work Zone, advanced edition)**
  If you're using the SAP Build Work Zone, advanced edition, there's one more layer called workspaces. Workspaces have their own local permissions, separate from spaces and pages:
  - Public workspaces can be viewed by anyone with basic SAP Build Work Zone access.

  - Private workspaces require an invitation.
  - External workspaces are meant for partners or vendors and give them very limited access.

  This allows teams to collaborate safely without exposing the entire portal.

Access alone isn't enough—SAP also controls what actions users can perform and what users are allowed to do (functional permissions). Inside the SAP Build Work Zone administration console, permissions are broken down into specific roles, for example:

- Company administrator: Full control over the entire portal, settings, themes, and all workspaces
- Area administrator: Manages a specific section, like an HR area
- Support administrator: Read-only access for monitoring and troubleshooting
- Page content administrator: Can design pages using the workpage editor, but cannot change security
- Workspace coordinator: Manages members and content within a specific workspace

As such, two users might see the same page, but only one of them may be able to edit it.

Let's tie it all together with a simple flow:

1. The user logs in via Identity Authentication.
2. Identity Authentication passes user attributes and groups (for example, Manager_Group) to SAP BTP.
3. SAP BTP maps that group to a role collection like Workzone_Manager.
4. SAP Build Work Zone reads that role and does the following:
   - Displays the correct spaces.
   - Shows the relevant pages.
   - Loads the right UI integration cards, such as a manager dashboard.

Everything the user sees is driven by this chain. On a technical level, SAP Build Work Zone checks OAuth scopes (like `Workzone.Read` or `Workzone.Admin`) whenever an API is called. In the SAP Build Work Zone, advanced edition, the digital workplace service maintains its own permission data for workspaces and keeps it synchronized with SAP BTP roles. This ensures security is enforced consistently everywhere.

17.3.3 Content Integration

SAP Build Work Zone supports secure access for external users through identity federation and role-based authorization. This makes it suitable for partner and supplier portals without custom development. Let's look at content integration in SAP Build Work Zone, because this is what actually makes the platform useful to end users.

At a high level, content integration is the way SAP Build Work Zone brings together applications, business data, and collaboration content from different systems—both SAP and non-SAP—into one single place. The goal is simple: users should not have to jump between multiple systems to get their work done. To achieve this, SAP uses three main integration approaches. Each one is meant for a different kind of requirement:

- **Content federation in which backend systems expose their content**
  The first and most scalable approach is called content federation. Here, systems like SAP S/4HANA take responsibility for exposing their own apps and roles. SAP Build Work Zone doesn't recreate anything—it simply consumes what the backend system provides. Technically, this works using something called the Common Data Model (CDM). You can think of CDM as a standard language that describes applications, roles, and UI elements like spaces and pages. Here's how it works, step by step:
  - In the backend system, an admin selects which roles and apps should be shared
  - The backend system converts this into a CDM document
  - SAP Build Work Zone connects to this using a design-time destination to read metadata and a runtime destination to launch the actual app

  The channel manager in SAP Build Work Zone fetches this content, and admins assign those roles to users. The biggest benefit here is that the backend system stays in control. Any updates, fixes, or lifecycle changes are handled in SAP S/4HANA itself, which makes this approach clean and scalable.
- **Manual integration for custom or non-SAP applications**
  The second approach is manual integration. This is typically used when you want to integrate individual SAPUI5 or Web Dynpro applications, third-party web tools, or custom applications built on SAP BTP that don't support federation. In this case, the administrator manually configures the application in the content manager and defines the following:
  - The application type (SAP Fiori, SAP GUI for HTML, or URL-based)
  - The routing, meaning where the app actually runs (a URL or an SAP BTP destination)
  - The visual details, like tile title, icon, and semantic objects for navigation

  This approach gives you a lot of flexibility, but it requires more manual setup and maintenance, so it's best for smaller or specialized use cases.
- **Content packages for ready-made business scenarios**
  The third approach uses content packages. Think of a content package as a ready-to-use bundle that delivers a complete business scenario in one go. A typical content package may include UI integration cards, workflows, and workspace templates. From a technical point of view, these packages are built in SAP Business Application Studio, exported as a ZIP file, and installed into SAP Build Work Zone using the administration console.

  A good real-world example is HR content packages for SAP SuccessFactors. They provide prebuilt cards for org charts, leave requests, time tracking, approvals, and more.

Connectivity and security are common for all approaches. No matter which integration method you use, some core components are always involved. The cloud connector creates a secure tunnel to on-premise systems like SAP S/4HANA or SAP ERP, without exposing them to the internet. Principal propagation ensures the user's identity is securely passed to the backend system, so users only see data they are authorized to access.

SAP Task Center is very important from a user perspective—it collects tasks and approvals from all connected systems and shows them in one unified inbox.

[zB]

Example

A manufacturing company replaced email-based supplier communication with an SAP Build Work Zone-powered portal that provided order visibility and shared documentation, significantly reducing supplier inquiries.

17.4 Summary

SAP Build Work Zone, in simple terms, is a single starting point for users where they can access all their applications, tasks, and collaboration tools from one place. It runs on SAP BTP and shows content based on a user's role, so everyone only sees what's relevant to them. There are two versions: the standard edition, which mainly works like a central launchpad for applications, and the advanced edition, which adds collaboration features like workspaces, forums, feeds, and document sharing. So, it's not just about opening apps, but also about working together.

SAP provides a low-code/no-code workpage editor, which works on a simple drag-and-drop concept. Both IT admins and business users can use it to build grid-based pages that act like dashboards. On these pages, you place UI integration cards and widgets, which show live business data such as KPIs, lists, or approvals. These cards are built using SAPUI5 and pull real-time data from backend systems like SAP S/4HANA through APIs, so users can see important information without opening full applications.

Finally, everything is kept secure and well controlled. Applications are connected using content federation, where backend systems expose their apps and roles using the CDM, and SAP Build Work Zone simply consumes them. User login and identity are handled by SAP Cloud Identity Services, while access to spaces and pages is controlled using role-based permissions on SAP BTP. For on-premise systems, the cloud connector creates a secure connection and ensures the user's identity is passed all the way to the backend. This way, users only see the data and content they are authorized to access, end to end.

Chapter 18
SAP Build Process Automation

In this chapter, we'll cover SAP Build Process Automation, its components that orchestrate various business processes, and how it can be integrated with different types of applications.

By the time you read this chapter, you should have a good understanding of the SAP Build portfolio, where SAP offers a unified, low-code platform on SAP Business Technology Platform (SAP BTP) that enables users of all skill levels to build enterprise apps, automate business processes, and design business sites using easy, visual drag-and-drop tools, prebuilt content, and AI assistance. This unified solution bundle combines essential tools such as SAP Build Code, SAP Build Process Automation, and SAP Build Work Zone.

At the end of 2022, due to limitations such as multiple tools, overly IT-centric automation, and a fragmented user experience, SAP transitioned SAP Intelligent Robotic Process Automation's functionality to SAP Build Process Automation. This shift enabled process-centric and event-driven automation on SAP BTP, business-user-friendly citizen development, and clean-core alignment.

In this chapter, we will cover the basic capabilities of SAP Build Process Automation, including business process orchestration, business rules framework, forms, AI services, and attended/unattended robotic process automation—all within a unified low-code platform designed to orchestrate and automate end-to-end business processes. We will explore how, through deep integration with SAP Integration Suite and enterprise applications, SAP Build Process Automation coordinates human and system activities while enabling intelligent bots that leverage AI to understand context, process documents, make decisions, and execute actions across both SAP and non-SAP landscapes.

18.1 Basics of SAP Build Process Automation Design

Before we explore all the components, we should note that SAP Build Process Automation combines the functionality of several solutions, including SAP Workflow Management, SAP Intelligent RPA, SAP Business Rules Management, and AI based document processing. When integrating SAP or non-SAP systems or orchestrating processes, it provides visual tools for designing business processes, automations, and decisions. SAP Build Process Automation includes various SDKs such as Excel, Outlook, BAPI, automation, workflow, rules, and document information extraction, along with built in monitoring and governance, empowering business users and IT teams to digitize processes faster and more securely.

In the following sections, we'll discuss SAP Build Process Automation setup and its lobby. We'll also explore how to set up and use desktop agents and their different modes. Once we have met all the prerequisites, we will see how to create a project in the SAP Build lobby and explore the available artifacts. Additionally, we'll walk through the packaging and deployment of a project in SAP Build Process Automation.

18.1.1 Tenant Setup for SAP Build Lobby

Now, let's try to understand the required setup for SAP Build Process Automation on SAP BTP. To use any service on SAP BTP, you first need to obtain the necessary subscription. First, you can search for the service inside the **Service Marketplace**. Once you subscribe to the service plan, it appears under **Instances and Subscriptions**, and you can easily navigate to the SAP Build lobby by clicking the **Go to Application** icon next to the **Application** title, as shown in Figure 18.1.

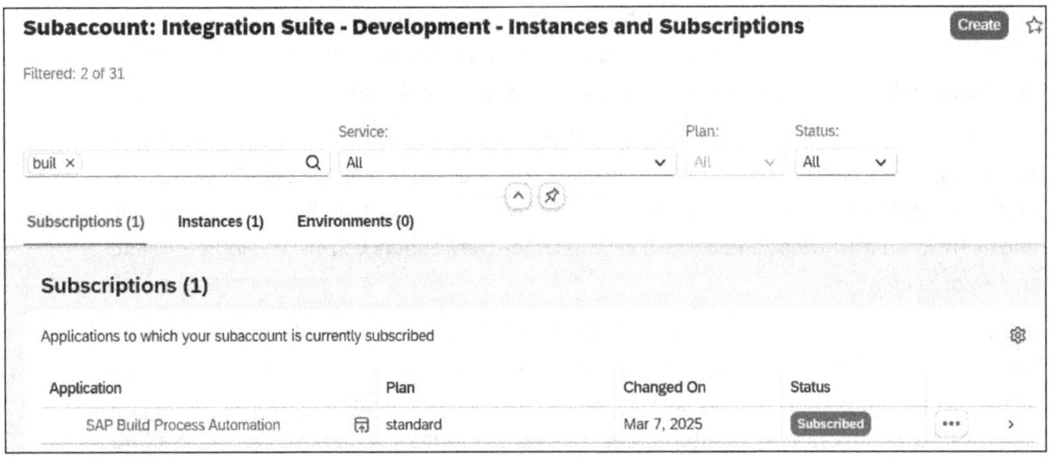

Figure 18.1: SAP Build Process Automation Instance

Go to Application takes you to the SAP Build lobby where you will find options to create a new project, enhance or customize an SAP template, and configure and deploy a project, as shown in Figure 18.2.

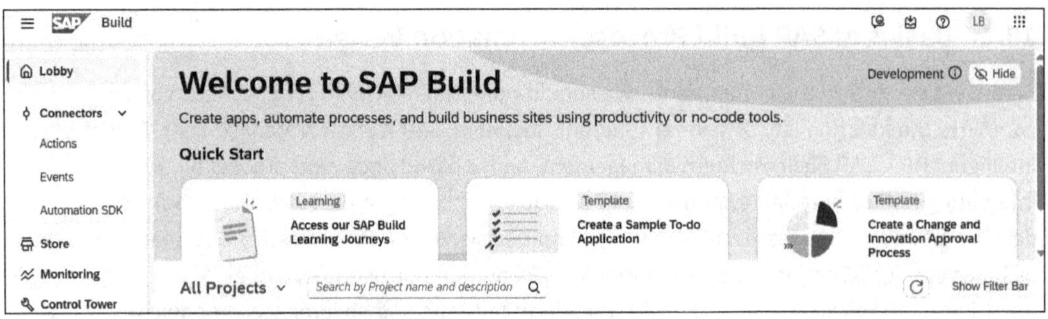

Figure 18.2: SAP Build Lobby

Before we start working on an SAP Build Process Automation project, it's important to understand the available tools and infrastructure, key design components, and best practices of working with SAP Build Process Automation.

Let's start with the tools and infrastructure first. As soon as you log in to the SAP Build lobby you'll see the following:

- **Control Tower**
  This is a central hub for managing tenant configuration, backend configuration, agent configuration, and more, as shown in Figure 18.3.

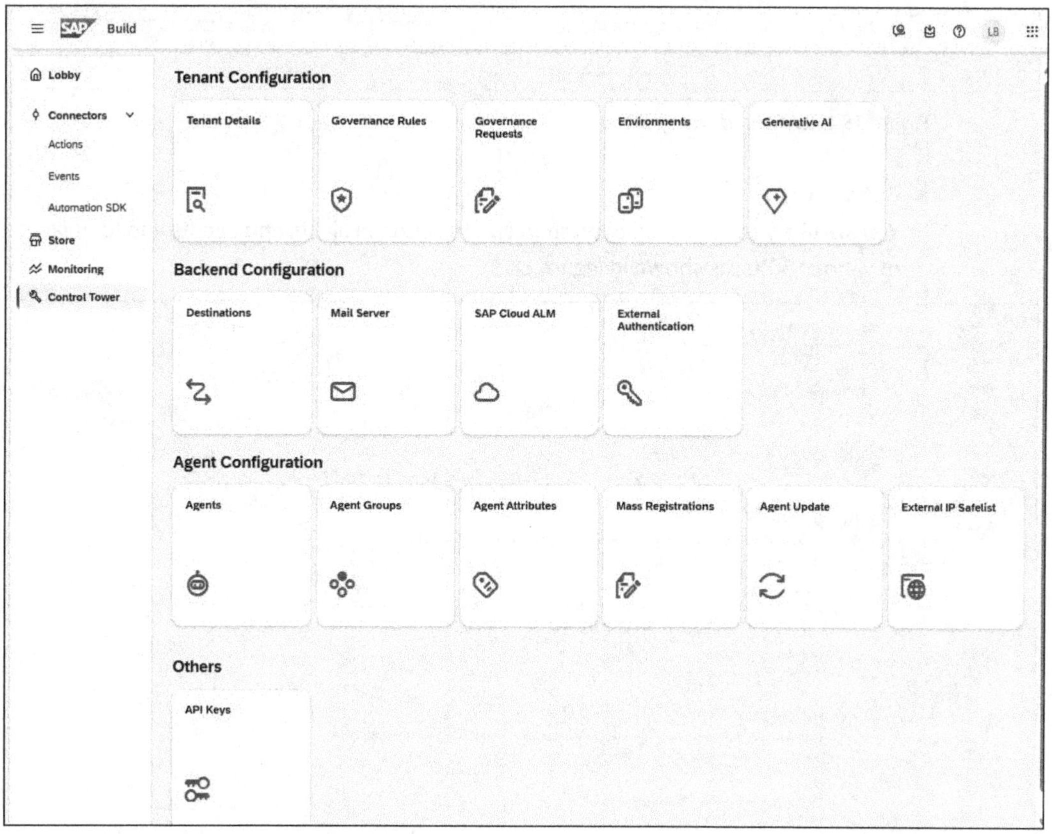

Figure 18.3: SAP Build Lobby Control Tower

- **Store**
  This is an SAP-provided repository that contains prebuilt content, including industry-specific templates, which can be imported and copied to accelerate design and development of a solution, as shown in Figure 18.4.

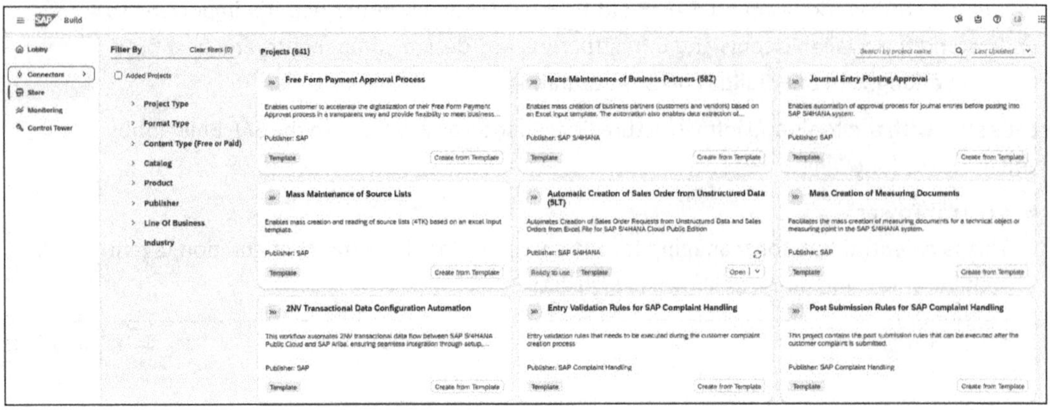

Figure 18.4: SAP Build Lobby Store

- **Connectors**
  You are likely to use this tool often to create action projects and events and to make use of various SDKs, as shown in Figure 18.5.

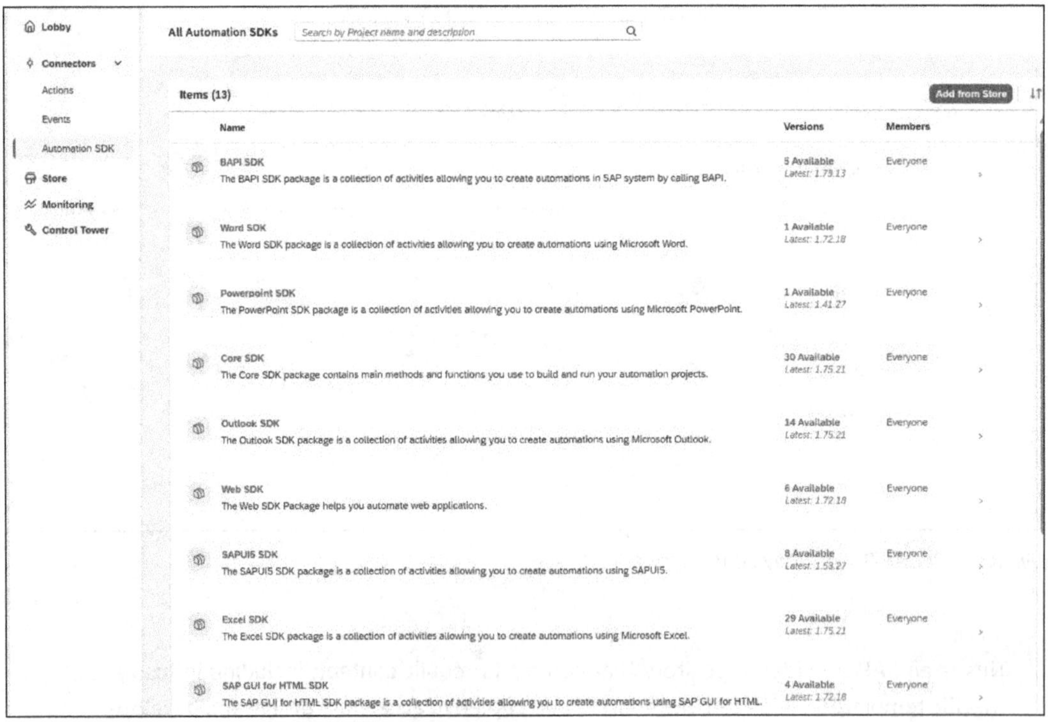

Figure 18.5: SAP Build Lobby Connectors Automation Jobs

- **Monitoring**
  This component offers a built-in dashboard where we can monitor the status and health of deployed process instances, jobs, and events, as shown in Figure 18.6.

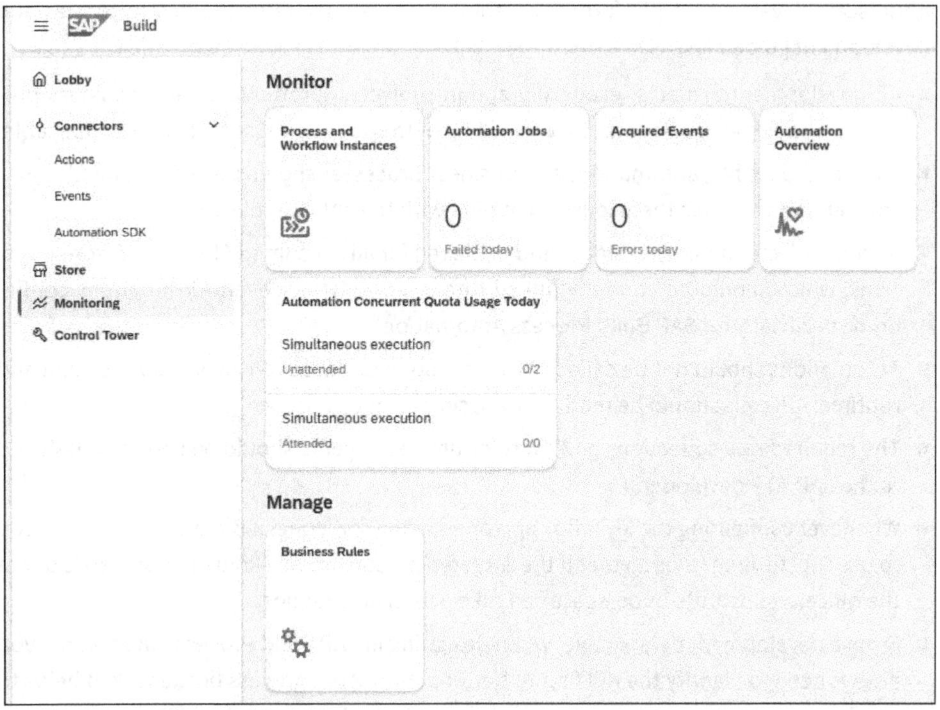

Figure 18.6: SAP Build Lobby Monitoring

Now that you understand the required tools and infrastructure, another important part of designing an automation process is understanding the development environments which can be used to build your solutions. Following are some of the available development environments:

- **Process builder (core)**
  This is the central, visual tool within SAP Build Process Automation for building workflows and forms (including SAPUI5) and orchestrating all processes.
- **SAP Business Application Studio**
  This development studio isn't part of the SAP Build lobby. This is a separate instance which can be used for developing custom SAPUI5 user interfaces such as approval forms and apps that are then embedded into SAP Build Process Automation processes.
- **Joule Studio**
  This is used for building custom AI agents that can plan and execute complex business scenarios by integrating with SAP Build Process Automation for advanced automation.
- **Cloud Studio (UiPath)**
  This is a web-based studio from UiPath, accessible within SAP Build. It is used to design and manage RPA bots for tasks like document extraction.

The next things you need to know about designing automation processes are the best practices. Before starting any development work—whether on a low-code/no-code platform or a full-scale development platform—you should be well aware of the best practices and development guidelines, as follows:

- All development artifacts, especially action projects, automations, and business processes, should be developed in such a way that they can be reused wherever applicable.
- Whenever possible, automations and business processes should be logically divided into smaller parts to make them loosely coupled with the entire application.
- Action projects should be created and deployed for all consumed APIs from different systems, which should be correctly linked through SAP BTP-relevant destinations configured specifically for SAP Build Process Automation.
- Automations should not be called within a loop to optimize API calls in both design and runtime; API calls should be modularized, simple, and concise.
- The required role collections and roles for different agents should be correctly assigned to the SAP BTP developer.
- Whenever configuring the My Inbox app or emailing someone, it is always recommended to use substitution rules so that if the targeted person (for any kind of approval) is out of the office, a substitute or delegate can take action on a request.
- From a development perspective, when designing in SAP Build Process Automation, you always need to identify the KPI first to better pinpoint the process bottleneck. It helps to configure the context parameter and attributes from the process visibility perspective in order to avoid rework.
- Business rule management should be used when dealing with financial business data types like currency and quantity.
- When developing a workflow, it's always recommended to keep the flow clear, aligned, and readable. The description should be consistent and meaningful.
- Ensure the reusability of blocks/tasks in decision-making flows to keep the workflow as simple as possible.
- From a process visibility point of view, data types need to be defined and derived correctly so that the analytics generated for users are error-free.

18.1.2 Projects

Before we continue with how to create a project and other related details, we must understand some prerequisites, such as destination creation specifically for SAP Build Process Automation and the required role collections.

[!]

Destination Configuration

While defining the destination for an SAP S/4HANA system, the following are the properties which you must maintain in the SAP Destination service:

- sap.build.usage=odata_gen
- HTML5.DynamicDestination=true
- sap.cloud.service.alias=spa
- AppgyverEnabled=true
- WebIDESystem=<SID>
- sap.applicationdevelopment.actions.enabled=true
- sap.processautomation.enabled=true
- sap.cloud.service=com.sap.spa.processautomation
- sap-client=<clientid>

Roles and Authorizations

The developer's user ID needs to be assigned to a set of role collections. If you go to the role collection assignment section, you will see many **ProcessAutomation** and **Form**-related role collections which need to be assigned. It's always recommended that you refer to SAP's help documentation for the latest information about role collections. Based on personas, custom role collections should be created and assigned to users to control authorizations across the solution.

Now that all the prerequisites are in place, in this section we'll focus on creating the different artifacts for an automation project and on deploying the automation project in SAP Build lobby. SAP Build Process Automation comes with various types of automations and associated artifacts, as shown in Figure 18.7.

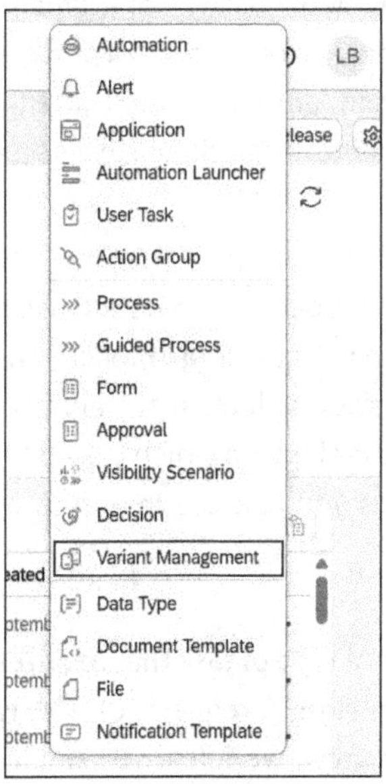

Figure 18.7: Automation Project Artifacts

Each component is unique and serves a distinct purpose in building robust automations. It is critical to have a good understanding of each component when designing and building an enterprise grade automation application.

Let's look at each of these options in turn:

- **Automation**
  An automation is one of the most useful and important artifacts that helps to orchestrate different tasks and activities. It is primarily powered by robotic process automation to handle all kinds of repetitive and manual tasks that a person would otherwise perform manually in the system. This artifact is designed to perform activities such as reading Excel files, checking mail, and extracting data from attachments. In SAP Build Process Automation, automation can be created using the built-in cloud studio in the SAP Build lobby, where a visual, low-code/no-code editor allows you to drag and drop the components you need.
- **Business Process**
  In SAP Build Process Automation, a business process is a graphical representation of a sequence of tasks, activities, and decisions designed to achieve a specific business goal—such as sales order creation or inbound delivery creation—by orchestrating different workflows across SAP and non-SAP systems and incorporating various human tasks and forms efficiently and quickly. This repeatable data flow is designed to automate manual steps, enhance consistency, and connect enterprise applications without extensive coding.
- **Alert**
  Alert automation is a very common and essential technique that helps you raise automatic alerts when specific events occur in any business process, keeping you informed of potential problems or issues and triggering actions to resolve them quickly. For example, you could create an alert that is raised when a certain number of errors occur in a process to further investigate and prevent them from happening again, or raise an alert when a process takes longer than a specified amount of time to complete in certain business scenarios.
- **Application**
  The application launch feature is a valuable tool for SAP Build Process Automation users who want to automate processes involving direct interaction with external applications from within a process during execution. Essentially, it allows you to launch any application from within a process, regardless of the platform or technology on which the application is built, providing great flexibility in the applications you can use in your processes.
- **User Task**
  In the context of SAP Build Process Automation, a user task is a type of task that requires human intervention to complete, such as approving or rejecting a request, providing input or feedback, completing a transaction, initiating a new workflow, or similar

actions. When a workflow reaches a user task, the corresponding recipients will see the task appear in their automation Inbox.

- **Action Group**
  Action groups within SAP Build Process Automation are used to call BAPIs. To call and execute BAPIs from SAP Build Process Automation, you must add the BAPI SDK as a dependency. Action groups can execute BAPIs by establishing connections to the SAP system using connection types such as SSO, basic, or custom to automate business processes.
- **Form**
  Forms are essential for any automation, especially when manual intervention is required to make decisions based on various data points through an interface. Forms in the workflow can be used to collect user input, display information, and enable users to take actions. Some common examples related to forms include collecting user input, such as name, address, or email address, to populate workflow variables or trigger other actions.
- **Approval**
  An approval form within SAP Build Process Automation is used to collect information from users who are approving or rejecting tasks. The approval form can be used to present contextual information related to a business process. For example, a purchase requisition approval form may include fields such as order, material, and amount, along with an input field for comments and options for the approver to either approve or reject. Information gathered from an approval step can further be leveraged to determine whether to proceed to the next step or terminate the process, and can also help identify potential problems in the process, allowing for appropriate actions to be taken.
- **Decision**
  At the beginning, we mentioned that SAP has unified SAP Workflow Management, SAP Business Rules Management, and many other offerings through SAP Build Process Automation. Previously, SAP Business Rules Management was a separate capability within SAP Workflow Management. To leverage it, we had to define and deploy a rule independently to invoke it within a workflow. Now that this functionality is offered as a decision inside SAP Build Process Automation, citizen developers and business users can define rules such as credit limits and expense amounts directly within SAP Build Process Automation.
- **Data Type**
  Data types represent the definition of a data structure that can be used to store data in a process. In SAP Build Process Automation, data types are used to define the format and content of data that can be exchanged between different parts of a process. Available data types include **Number**, **String**, **Date**, **DateTime**, **Password**, etc. You can either define data types manually or import them directly using Excel.
- **Document Template**
  In SAP Build Process Automation, a document template defines the structure of a document that can be used to extract data from it. Document templates are an important part

of SAP Build Process Automation because they define the format and content of documents extracted from a process. This makes it easier to understand and manage the data extracted from documents while ensuring that the data is consistent and accurate. Document templates and Document Information Extraction (DOX) are two powerful tools that can be used together to extract data from documents in SAP Build Process Automation.

- **File**
  Files are an important part of SAP Build Process Automation that allow you to store and manage data—such as plain text, YAML, JSON, or XML—in a way that is easy to access. You can use files to store data used in your automation processes, including input data, output data, and configuration data.

The **Overview** section shows basic information, such as project name and description, about automation and business process projects, as shown in Figure 18.8.

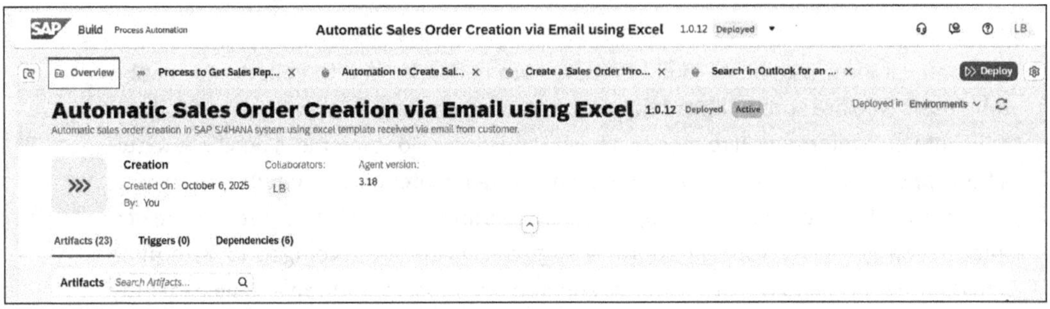

Figure 18.8: Overview of an Automation Project

Environment variables are properties used by the automation solution during execution. You can create variables with the same name in multiple environments, so when a project is packaged and deployed to an environment, the variables defined in that environment are used. A package imported into multiple environments requires the same environment variables to be created in all of them.

Like any other project, an SAP Build Process Automation project has certain dependencies that need to be added in the dependencies section. We can add as many business process project dependencies or action project dependencies as required. As part of the business process project dependencies, we include a wide range of dependencies such as the BAPI SDK, Google Authorization SDK, Google Vision AI SDK, Outlook SDK, Excel SDK, and many more. When we add a package as a dependency, we can use automation artifacts, environment variables, alerts, applications, and data types from the added package within the automation project. Additionally, third-party SDKs and custom or standard action projects designed for service integration can also be added as dependencies in automation projects.

18.1.3 Agents

When we talk about agents in SAP Build Process Automation, you may get confused about the difference between them and Joule Agent. From the perspective of SAP Build Process Automation, agents refer to the desktop agent, which is a software component used to run automation projects, launch the application, read screen information, enter data, and execute automation locally in the system. On the other hand, Joule Agent is a process orchestrator which can be built in the Joule Studio.

The desktop agent is a mandatory component in SAP Build Process Automation. You must configure the agent and connect it to the SAP tenant where SAP Build Process Automation is enabled. It is not possible to create or execute automations without the desktop agent being installed. It serves as the essential bridge between the cloud-based SAP Build Process Automation and the local desktop environment where automations get executed.

To configure the desktop agent, follow these steps:

1. Go to the **Control Tower** of the SAP Build lobby and then click on the **Agent** tile. Once you are on the **Agent** page, click on **Go to Download Page** in the right top corner, as shown in Figure 18.9.

Figure 18.9: SAP Build Lobby Control Tower: Agent

2. Follow the instructions given on the agent download page. First you need to download the installer based on your machine type (Windows or MacOS).
3. Enable the browser extension for SAP Build Process Automation and then connect your tenant to the desktop agent.
4. Once these steps are completed, go to the desktop agent itself and activate the tenant for which you want to use the desktop agent and set the mode. The desktop agent operates in two modes. It works in attended mode where human intervention is required to trigger the process, mostly in the build and testing phases. The desktop agent also works in unattended mode in the production environment when no human intervention is expected.
5. Once the tenant is activated and the mode is set in the desktop agent, register the agents in the SAP Build Process Automation **Control Tower** by clicking on **Register Agent**, as shown in Figure 18.10.

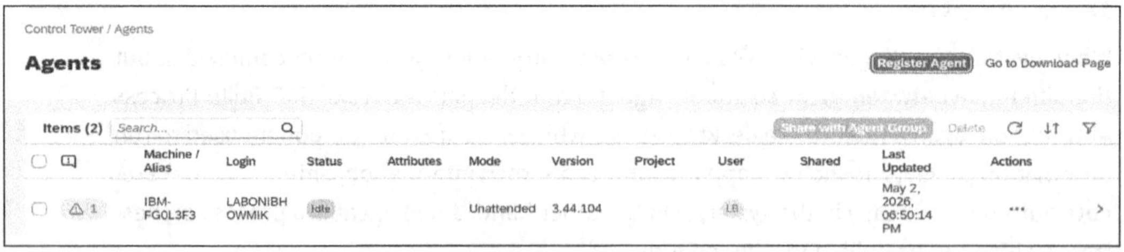

Figure 18.10: SAP Build Lobby Agent Registration

As an official SAP component, the desktop agent integrates natively with SAP Build Process Automation tenants and requires registration with your SAP BTP subaccount. SAP provides official documentation, support, and regular updates for it through their standard product maintenance lifecycle.

Before moving on, let's take a quick sneak peek at the licensing model, cost components, and ways to optimize the cost of the desktop agent:

- **Licensing model**
  The desktop agent is included as part of SAP Build Process Automation licensing and does not require separate licensing fees. The cost is embedded within the overall SAP Build Process Automation subscription model, which follows a user-based pricing structure with standard user and advanced user tiers. The cost can be calculated via SAP Discovery Center at *https://discovery-center.cloud.sap/protected/index.html*.
- **Cost optimization and best practices**
  - Use unattended mode for production deployments to avoid per-user charges and leverage connection-based pricing instead.
  - Consolidate automation execution on fewer agents, when possible, to reduce connection counts. You should deploy agents across multiple machines for high availability and load distribution.
  - Leverage the included API calls and storage limits before making additional capacity purchases.
  - Consider running agents in virtual environments for background activities to optimize infrastructure costs.
  - Implement proper error management within automation and monitor execution logs.

18.1.4 Recorder

The recorder provides the capture capability that allows users to capture applications and record automation workflows. The terms are used interchangeably throughout SAP Build Process Automation documentation, where capture describes the action performed by the recorder tool.

Capturing is the primary method used to bring application screens into the design environment so that bots can interact with them. Multiple capture modes, including manual capture mode, automatic capture mode, and capture on hover, are available. Core capture capabilities include the enhanced screen auto capturing capability in the recorder that automatically initiates capture when screen changes are detected, improving the user experience during recording.

This functionality helps create automated workflows and enables the system to understand and replicate user actions on various applications through screen recording and element detection.

The capture functionality supports multiple technologies including Windows applications, SAP GUI for Windows, and various web applications (SAP Web GUI for HTML, SAPUI5, SAP SuccessFactors, SAP Business ByDesign, and SAP Concur).

Captures previously existed in SAP Intelligent RPA. During the migration process, existing automations and capture capabilities from SAP Intelligent RPA continue to work seamlessly in SAP Build Process Automation, maintaining continuity for organizations that had already invested in SAP Intelligent RPA.

18.1.5 Packages

Packages in SAP Build Process Automation are deployable and executable units that bundle together all your developed artifacts—such as automations, processes, forms, and business rules—into a single, non-editable single entity.

These packages serve as containers for managing version control, and deployment across different environments. There are various ways that we can create a package in SAP Build Process automation. Let's quickly check what the options are:

- Migration, where package containing all SAP Workflow Management project artifacts get migrated to SAP Build Process Automation
- New project development in SAP Build Process Automation, where a package can be created in various ways, as follows:
  - You can import prebuilt packages from the SAP Build catalog. There are 500 SAP applications and industry-specific, prebuilt, ready-made solutions accessible through SAP Business Accelerator Hub and the SAP Build lobby store. These packages significantly accelerate development by providing templates and can be deployed and used for clients with minimal configuration and effort.
  - Custom automation packages are built from scratch using the process builder tool, containing custom business process, business rules, automations, and visibility scenarios tailored to specific organizational needs.
  - Hybrid mode allows us to create a package that combines prepackaged content with custom workflows for optimal results.

18.1.6 Deployment

Now that we have our artifacts and packaging ready, let's see how we can deploy this project. To deploy a project, there are some prerequisites which we must check, as follows:

- The project must be error free and have the **Released** status
- Destinations for other integrated systems must be successfully configured and connected.
- API trigger must be added, and the API key needs to be configured.
- Users need to have the roles required deploy the project in the shared environment assigned to them.

Once all these are done, click the **Release** button located in the top right corner. Once the package is released, you can deploy the project by clicking the **Deploy** button in the top right corner in the correct deployment environment. This environment can be common to multiple packages or specific to just this one, which will be defined in the **Environment** section under the **Control Tower** of the SAP Build lobby.

18.2 Orchestrating Automation Processes

Process orchestration in SAP Build Process Automation coordinates with different systems, complex business processes across those systems, and people, and automates end-to-end, intelligent, and scalable business processes. This entire process of orchestration enables organizations to:

- Break down large processes into smaller, manageable subprocesses and phases.
- Automate each component individually but connect through a correlation between the processes.
- Reassemble them into a seamless, unified workflow ensuring efficient execution of cross-functional, enterprise-wide processes.

Let's go into some more detail to understand what are the core capabilities which help SAP Build Process Automation orchestrate process, as follows:

- Workflow management, which controls task sequencing, dependencies, and execution flow and provides real-time monitoring and actionable insights by monitoring the task in real time.
- Task automation (RPA integration) which embeds robotic process automation (RPA) within workflows, supports both attended and unattended bots, and automates repetitive system tasks alongside human activities.
- Human task integration triggers human interactions like approvals and form submissions. Enables smooth handoffs between users and automation to ensure decision-making at key process stages.

- Event-driven processing supports multiple triggers like API-based, event-driven, scheduled, and manual, and enables dynamic and real-time process execution.

Some of the orchestration components which play a pivotal role in the process orchestration of SAP Build Process Automation are as follows:

- **Process builder**
  Provides a central design environment to model workflows with forms, approvals, automations, and rules.
- **Decision logic (business rules)**
  Decision tables drive conditional routing, control approvals, branching, and next steps
- **Forms and approvals**
  Capture business data and enable structured, multilevel approvals.
- **API integrations**
  Connects SAP and non-SAP systems for seamless system interaction within workflows through action projects.

Now let's see how SAP Build Process Automation coordinates the process flow:

1. First, a process is triggered via APIs, events, schedules, or manual actions.
2. This starts the orchestration lifecycle using task routing and tasks are automatically assigned based on business rules, roles, and hierarchies, which are managed through a centralized inbox.
3. Another important aspect of process flow is the parallel processing of other tasks to support simultaneous execution of multiple activities to ensure synchronization across branches.
4. SAP Build Process Automation also handles exceptions and risks in the process and triggers escalation workflows when needed.

Following are the benefits of this kind of process orchestration:

- **Improved business efficacy**
  Process orchestration rejects manual handoffs and reduces process cycle times by automating end-to-end workflows that extend into multiple systems and departments.
- **Corrected process stability**
  Orchestrated processes ensure standardized execution paths, reducing variability and errors while maintaining compliance with business rules and regulatory requirements.
- **Greater operational responsiveness**
  The low-code orchestration environment enables rapid adaptation of business processes to changing requirements, allowing organizations to respond quickly to market demands and operational needs.

18.3 Attended and Scheduled Bots

When the desktop agent runs in *attended mode* where users manually trigger automations and can interact with them during execution, it functions like an attended bot. On the other hand, the desktop agent operates in *unattended mode* when automations execute automatically according to predefined schedules without human intervention or are triggered externally via an API.

Now, let's see what the best use cases for these kinds of bots are. Human-centric processes are the most suitable cases, where humans are needed in the loop during process execution to make decisions to proceed further, to trigger the process manually, or where the user needs assistance during the process. Some common examples of attended bots are as follows:

- Triggering the employee termination or onboarding process
- Assisting with sales quote generation by fetching pricing, discounts, and stock availability
- Approving the generated sales quote draft.
- Providing desk assistance to customer service attendants

Similarly, let us now examine the use cases for scheduled bots. Scheduled bots are mostly suitable when an automation or business process won't be triggered randomly or on demand, and no human involvement is required. These use cases are mostly fully automated, need to be triggered at particular times, and are ideal for batch and bulk processing.

Scheduled bots run as background processes on any virtual machine or server. Some of the most common use cases for scheduled bots are as follows:

- Used for mass and bulk upload like mass sales contract creation
- Upload master data and create business partners in the system
- Executing month end reports for invoicing or report generation

From the process orchestration point of view, both kinds of bots add significant value to the process. Attended bots are used in human-centric workflows, while scheduled bots are used in fully automated flows; both can be combined into a single orchestrated process by using the following:

- Human decisions to provide required human judgment.
- Bot execution for repetitive and batch-oriented tasks where no human input needed.
- System integrations to assist users or to execute processes in the background.

18.4 Imbuing Intelligence in Bots

In this section, we will see how we can incorporate intelligence and agentic AI in SAP Build Process Automation. When intelligence and agentic AI come together with SAP Build

Process Automation, it works as an autonomous AI system that can perceive its environment, make decisions, and act independently to achieve specific goals within business processes.

Unlike traditional automation that follows predefined rules, agentic AI leverages machine learning, natural language processing, and advanced analytics to learn from provided data, predict outcomes, and take proactive actions.

SAP Build lobby offers Joule Studio, which helps to build custom Joule skills. Joule and AI services available on SAP BTP enable the following in SAP Build Process Automation:

- Document understanding
- Predictions and recommendations
- Decision support
- Take actions autonomously
- Decide next steps based on context
- Interact with tools/APIs like a digital worker

SAP Build Process Automation integrates AI capabilities through multiple layers, including embedded generative AI features, integration with SAP AI services, and support for third-party AI solutions. Following are the available AI components in SAP Build Process Automation:

- **Machine learning models**
  You can integrate with SAP AI Core and SAP Business AI to access both pretrained and custom LLMs. These models can be used for use cases such as document classification, sentiment analysis, and business-specific predictions tailored to organizational needs.
- **Conversational AI**
  Conversational AI integrated with Joule enables natural language interaction for process automation. Here, users can describe workflows and business processes in plain language using the interactive interface, and Joule can assist in generating automation steps, simplifying process design and management for developers.
- **Document information extraction services**
  AI-powered document processing enables automatic extraction of key data from unstructured documents. This capability supports scenarios such as extracting data from various unstructured email bodies, attachments, or other documents, as well as identifying anomalies and missing information to improve data quality.
- **Decision intelligence**
  AI-driven decision-making enhances the orchestration process by analyzing runtime data, criticality, and historical data. The agentic AI capabilities of SAP Build Process Automation enable intelligent task routing, dynamic approvals, and optimized decision outcomes within business processes.
- **Autonomous decision making**
  Different Joule Capagents can analyze multiple scenarios in real-time, evaluate, and assign tasks to the appropriate personas without human intervention.

Please note the following points as SAP Build Process Automation is still evolving and getting matured:

- Not all SAP Build Process Automation artifacts can be generated using generative AI. For detailed information please refer to the *Known Limitations for Generative AI* documentation on SAP Help.
- Quotas and restrictions apply to generative AI functions and require careful capacity unit planning based on your SAP Discovery Center calculation.
- AI model availability varies by region and data center, with specific rate limits per tenant.
- Organizations must ensure compliance with applicable laws, regulations, and corporate policies when using AI models.

18.5 Integration

Regarding integration, SAP Build Process Automation truly becomes a powerful offering. It doesn't just automate tasks; it connects and orchestrates processes across the entire landscape. It provides comprehensive integration capabilities that enable seamless connectivity across hybrid environments, linking SAP and non-SAP systems through various protocols and methods. The core integration framework of SAP Build Process Automation is as follows:

- Built on SAP Business Technology Platform (SAP BTP) with native integration to SAP services
- Utilizes SAP Connectivity service for external system connections
- Supports multiple integration modes including synchronous and asynchronous messaging

It uses the following connection types and protocols:

- REST APIs and OData services for modern application integration for various SaaS offering of various cloud platforms
- SOAP web services for legacy system connectivity
- File-based integrations (XML, CSV, JSON formats)
- Database connections via JDBC adapters
- HTTP/HTTPS protocols for secure communication

Let's now look at the different kinds of integrations which SAP Build Process Automation supports. In general, for all kinds of standard and custom APIs, OData services are consumed via SAP Build action projects. Action projects can be reused in different automation and business process projects.

Event Mesh should be leveraged for all kinds of event-based integration. Standard and custom business events can be triggered for real-time process automation through Event

Mesh. Both SAP systems and non-SAP systems can also post events to Event Mesh. Some examples of SAP-native system integration are as follows:

- **SAP S/4HANA integration**
  - Direct API connectivity through Graph and SAP Business Accelerator Hub.
  - Using OData destinations and action projects, the ABAP RESTful application programming model is also accessible.
  - Event-based integration is done using Event Mesh.
- **SAP SuccessFactors integration**
  There are many prebuilt automations which expedite integration for various HR processes.
- **Other common integration for SAP solutions**
  - SAP Document Management service connectivity
  - Integration with SAP Cloud ALM for operations automation

Some examples of third-party system integrations are as follows:

- **Cloud services**
  - Microsoft 365 integration (Outlook, Excel, SharePoint, OneDrive)
  - Google Workspace connectivity for document and email automation
  - Third-party cloud applications via Open Connectors
  - API specifications upload and custom API action creation
- **On-premise systems**
  It's common to have to integrate with on-premise SAP and non-SAP systems, as follows:
  - Legacy system connectivity through the desktop agent
  - Database integrations via JDBC connectors
  - File system access for document processing
  - SMTP integration for email communications

As we are talking about integration, some other important topics like security, error handling, data exchange protocols and formats also need to be discussed:

- **Authentication methods**
  - SAP Cloud Identity Services is used as the primary identity provider for identity authentication
  - Supports OAuth 2.0 and SAML federation
  - Service key-based authentication is used for system-to-system communication
  - Windows Generic Credentials is used for on-premise system access
- **Security protocols**
  - TLS encryption is used for all communications, especially for integration with banks

- Role-based access control and authorization can be implemented in the different layers of applications
- Secure credential storage and management
- Data protection compliance with GDPR requirements
- **Data exchange formats which are supported**
  - JSON for modern API communications
  - XML for structured data exchange
  - CSV for tabular data processing
  - OData for SAP system integration, especially for cloud integration
  - EDI formats for B2B communications
- **Transformation capabilities**
  - Built-in data mapping and transformation tools are available in the SAP Build lobby.
  - Document information extraction using AI/ML for unstructured data
  - Format conversion between different data types is possible
  - Custom transformation logic is built through scripting
- **Monitoring and management**
  - Process and workflow instances can be monitored in the SAP Build lobby's monitor
  - Control tower for agent and system status tracking
  - Real-time execution logs and error tracking
  - Integration with SAP Cloud ALM for comprehensive monitoring
- **Error handling**
  - Automated exception handling and retry mechanisms
  - Comprehensive logging and trace collection
  - Alert notifications for failed integrations
  - Diagnostic file collection for troubleshooting

18.6 Summary

SAP Build Process Automation is a low-code/no-code (citizen developer) platform that enables organizations to design, automate, and optimize business processes. It combines workflow management and robotic process automation into a single unified solution, allowing both business and IT users to collaborate on automation initiatives. Let's wrap up our chapter on SAP Build Process Automation by recalling the following features:

- Offers seamless integration with SAP and non-SAP systems (cloud and on-premise)
- Supports APIs, connectors, and event-driven triggers
- Enables end-to-end, cross-application workflows

- Centralizes monitoring and management via the SAP Build lobby
- Is designed for both citizen developers and professional developers
- Enables business users to automate without coding, while IT ensures governance
- Automates repetitive manual tasks, digitizes approval workflows, and handles high-volume processes

Chapter 19
Joule

In this chapter, we'll explore how businesses can aspire to become intelligent enterprises by leveraging SAP's latest AI solution, Joule, across their enterprise systems. Starting with a product overview, we will dive deep to understand its key capabilities, components, features, background processes, and underlying architecture.

It has been some time since generative AI swept through the enterprise world, and the early phase—when the primary challenge was identifying how to leverage generative AI in the context of enterprise solutions—has largely passed. Today, this space is evolving at an unprecedented rate, with new AI models and capabilities being introduced constantly, making AI adoption across enterprise solutions imperative for survival. Driven by this paradigm shift, enterprise applications now focus more on context-driven or context-aware intelligence that is comparable to human decision-making.

In this chapter we are going to introduce Joule, its key capabilities, its building blocks, and how it works in the background.

19.1 Introduction to Joule

As depicted in Figure 19.1, today the significance of a digital companion within enterprise applications—which not only answers queries but also offers real-time feedback and assistance—is greater than ever before. Integration of AI assistants into everyday enterprise applications is now helping enterprise users to not only make faster decisions but also to focus more on productive work instead of navigating through multiple screens and transactions. In addition to achieving operational efficiency, embedded AI assistants or agents are also helping enterprises to survive rapidly changing regulatory requirements and market conditions.

Stringent go-to-market timelines to stay competitive are another key factor driving companies to adopt embedded AI offerings within their SaaS solutions or platforms instead of building from scratch. Integrated through a simple conversational user interface, these AI solutions harness the power of large language models (LLMs) to process user queries and generate relevant responses based on a comprehensive analysis of extensive enterprise datasets.

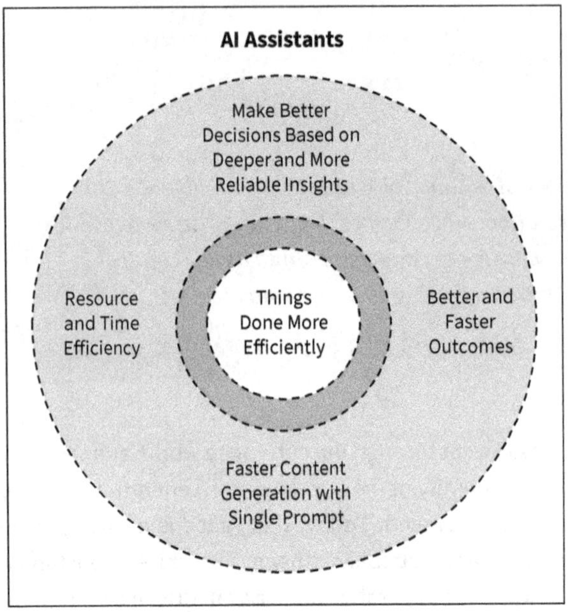

Figure 19.1: AI Assistants

SAP has steadily advanced its embedded offerings over the past few years, and based on this foundation, SAP has introduced its first ever generative AI solution, Joule, to make the user experience more intuitive and productive. According to SAP's official definition, Joule is a generative AI solution embedded throughout SAP's cloud enterprise portfolio, delivering proactive and contextualized insights from SAP applications and third-party sources.

One goal of Joule is to fundamentally transform the way we interact with SAP solutions. What sets Joule apart is its ability to understand user context and generate or gather relevant information through simple, natural language prompts, enabling business users to focus more on productive work instead of repetitive tasks. With Joule, business users can make faster, more reliable decisions based on useful insights generated from enterprise data. With a better understanding of underlying business data and context, Joule supports improved outcomes by tailoring responses and suggestions to meet enterprise user goals. Joule is not limited to the periphery of business applications or users and offers a wide spectrum of capabilities, as shown in Figure 19.2.

Joule not only brings value to enterprise users but also accelerates the development of enterprise solutions by providing a wide range of tools and capabilities. It has already been rolled out across multiple PaaS and SaaS solutions. For business users, it enables smarter decision-making by providing embedded assistance through products such as SAP S/4HANA Cloud Private Edition, SAP S/4HANA Cloud Public Edition, SAP SuccessFactors, SAP Ariba, SAP Analytics Cloud, and others. It also accelerates the delivery of enterprise solutions with productivity accelerators like SAP Joule for Developers and Joule Studio. Additionally, multilanguage capabilities make it easier for global teams to work with SAP systems in their preferred languages, enhancing accessibility and adoption. Joule also offers AI-assisted document grounding that helps avoid ambiguities and hallucinations when recommending critical business decisions based on business data.

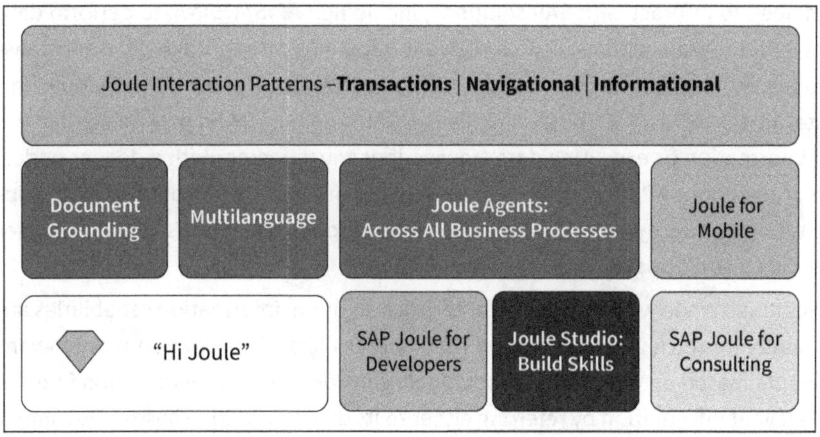

Figure 19.2: Joule Capabilities

Now that you understand the significance of Joule in the context of enterprise applications, in the next section, we are going to dive deeper into its capabilities.

19.2 Capabilities

Joule is transforming at a significant pace and offering a wide spectrum of capabilities, as shown in Figure 19.3, that can interpret user inputs into actionable intelligence and insights, which not only can drive human decisions, but also improve operational efficiency for both business consultants and developers.

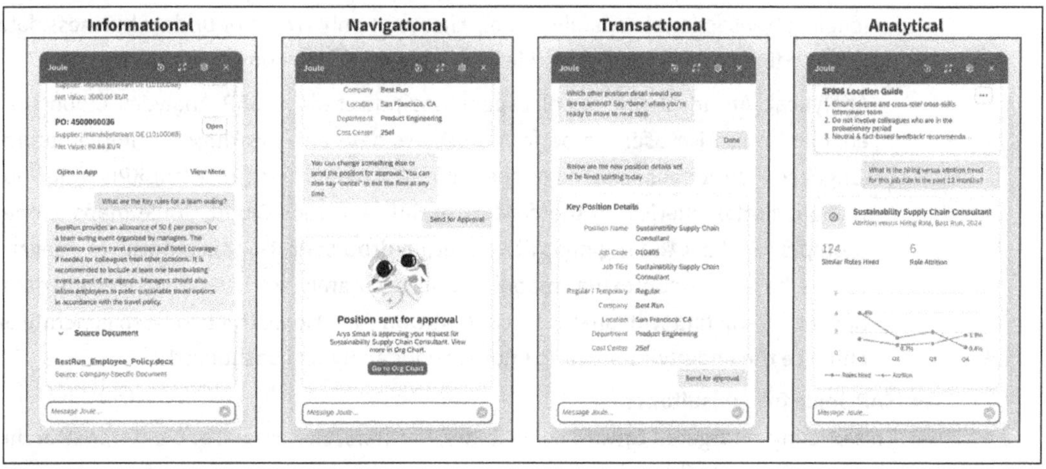

Figure 19.3: Joule Capabilities Deep Dive

Let's dive deep into each of these capabilities and understand what they are all about:

- **Joule for business users**
  Joule is specially tailored for enterprise users who manage day-to-day business operations and need to interact with SAP solutions, including SAP S/4HANA, to perform data reporting, execute transactions, search relevant data, navigate through different transaction codes, or find simple information such as material or plant details. Joule can accelerate all these tasks through simple natural language prompts, delivering the required information or executing data transactions much faster. With a deeper understanding of the core SAP foundation and underlying business data context, Joule supports better outcomes by tailoring responses to enterprise user needs. Joule is further segmented into additional categories, which include:
  - Informational: While working with an enterprise solution, information capabilities are quite handy for retrieving information from sources like SAP Help Portal or relevant documents. Based on the given prompt, Joule can contextually relate to and find the most relevant information by referring either to its own knowledge base or document sources.
  - Navigational: As the name implies, this set of capabilities from Joule helps business users, especially new ones, find information on how to navigate or execute various transactions within SAP systems like SAP S/4HANA through simple prompts, instead of memorizing transaction codes and navigating multiple screens.
  - Transactional: Relevant for enterprise users who perform data transactions frequently, Joule can execute these transactions on your behalf using simple prompts. For example, sales order creation requires navigating to Transaction VA01 and entering various header and item level data points, including sales organization, distribution channel, division, material number, quantity, and other relevant details for the products or services being sold. For business users engaged in various activities, sourcing all relevant information by navigating to different transactions can be challenging; transactional capabilities help them not only create or update business data but also find contextually related information needed to complete tasks or steps.
  - Analytical: An advanced set of capabilities powered by SAP Analytics Cloud that enables enterprise users to perform business data analysis through simple natural language instructions, instead of spending effort identifying relevant KPIs or formulating complex queries. In the context of SAP Analytics Cloud, Joule maps simple prompts like **"List the regions with a margin drop over 8% this month"** to relevant stories based on contextual mapping and accurately retrieves the relevant data. Beyond interacting with analytical content, you can also instruct Joule to generate or enhance the underlying script to improve your analytical dashboard.
- **SAP Joule for Consultants**
  This offering is targeted toward consultants, partners, customers, and enterprise architects working on building intelligent enterprise solutions or cloud transformation projects, providing them with all the required information and the latest offerings in the SAP

technology space. SAP Joule for Consultants, as shown in Figure 19.4, provides proprietary access to SAP knowledge sources such as SAP Help, 100 SAP Certifications, and over two terabytes of expertly curated SAP Community content, ensuring trustworthy, domain-specific information—this is where Joule differs from other LLM bots like ChatGPT, which rely more on outdated internet sources. For precise and relevant responses, Joule utilizes a RAG (retrieval-augmented generation) mechanism that retrieves the most relevant information from a large database or collection of documents and generates an answer based on it. Additionally, SAP Joule for Consultants relies on SAP's Global AI Ethics Policy, which is powered by the highest security and ethical standards. SAP Joule for Consultants can be especially instrumental during the discovery phase of a project to drive solution positioning, align best practices, enable faster decision making, and translate business language into SAP process language.

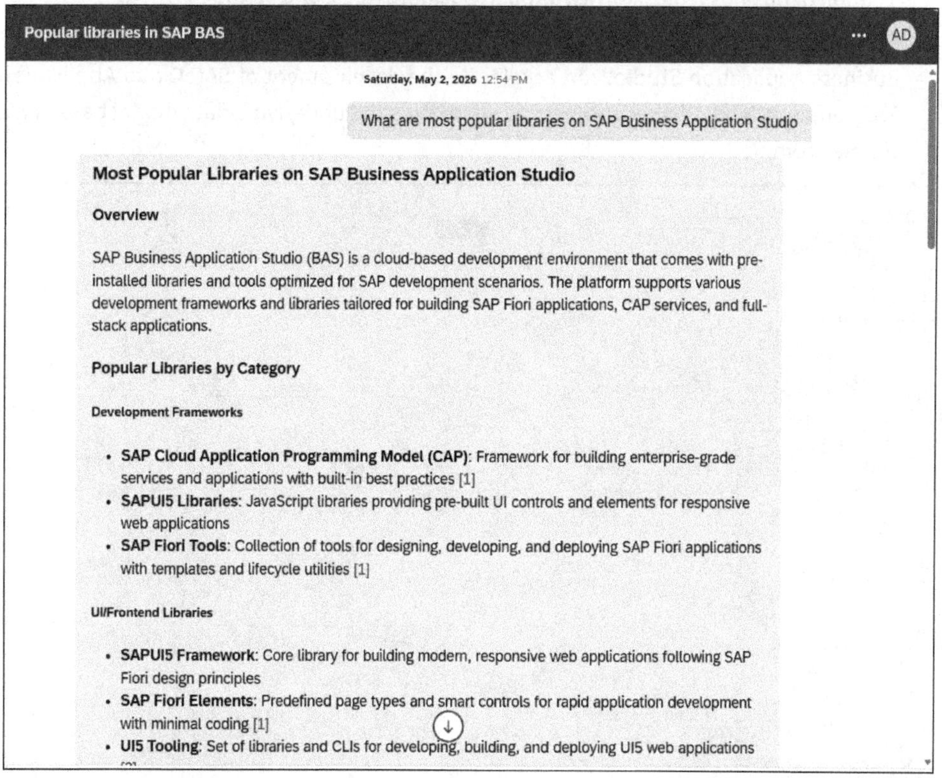

Figure 19.4: SAP Joule for Consultants

- **SAP Joule for Developers**
  This offering includes a set of design time capabilities that enhance developer productivity and proficiency by generating code snippets and artifacts within SAP Build Code, SAP Build Process Automation, SAP BTP ABAP environment, and SAP S/4HANA Cloud. In this context, Joule greatly assists the developer community in building new enterprise apps or extensions by accelerating the generation of relevant and accurate code during design

time in a faster and smarter way through natural language–based instructions. Instead of starting from scratch, developers can use Joule to get a head start, significantly reducing go-to-market time by minimizing the considerable effort required at each stage of the application development lifecycle, including development, testing, and documentation. Key capabilities of SAP Joule for Developers include:

- Code generation, autocompletion of existing code, comment generation, code review and refactoring of existing code for improved quality.
- Full stack cloud application generation based on SAP Cloud Application Programming Model and ABAP core data services (ABAP CDS), along with unit tests, and application logic development.
- Enhance automation based on summarization of overall processes along with generation of business rules, forms, applied business rules, and scripts.

Figure 19.5 shows a glimpse of how SAP Build Code and Joule, embedded inside SAP Business Application Studio, can accelerate the development of SAP Cloud Application Programming Model-based projects by generating the underlying data model based on a simple prompt.

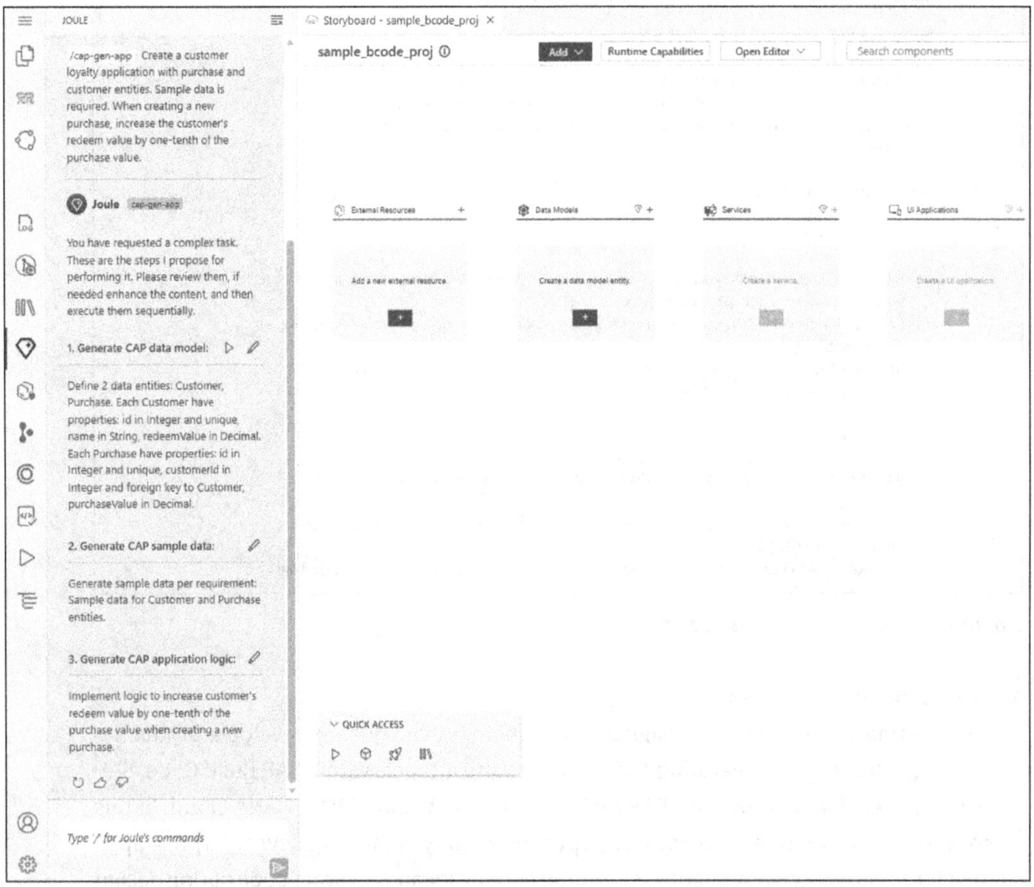

Figure 19.5: Joule Embedded Inside SAP Business Application Studio

- **Joule Studio**
  Joule Studio is SAP's newest addition to the SAP Build platform for creating, extending, and managing the Joule skill builder and custom AI agents. As an integral part of the SAP Build platform, Joule Studio provides a unified low-code experience to both create custom enterprise agents and extend Joule's out-of-the-box capabilities through custom skills. In this context, a skill represents a basic action or "verb" that enables Joule to perform rule-based tasks. For example, when an end user enters prompts like "Get the current stock", "Create a Sales Order", or "Fetch current exchange rates", Joule maps the request to an equivalent skill to retrieve relevant data from backend systems. A skill is a reusable and reliable functionality for Joule or Joule-based AI agents. SAP has already published a vast library of over 1,600 prebuilt skills, and you can use Joule Studio to create your own.

  Because Joule Studio is low-code driven, citizen developers can easily use Joule Studio to create any type of enterprise-grade AI agents or skills, requiring only integration with backend systems like SAP S/4HANA or other third-party systems through principal propagation, supported by SAP BTP. In addition to development capabilities, Joule Studio offers robust governance features, including agent lifecycle management, monitoring, and enterprise-grade security. Figure 19.6 shows a glimpse of the newly introduced Joule Agent and the skill builder capability as part of the Joule Studio project creation window inside the SAP Build lobby.

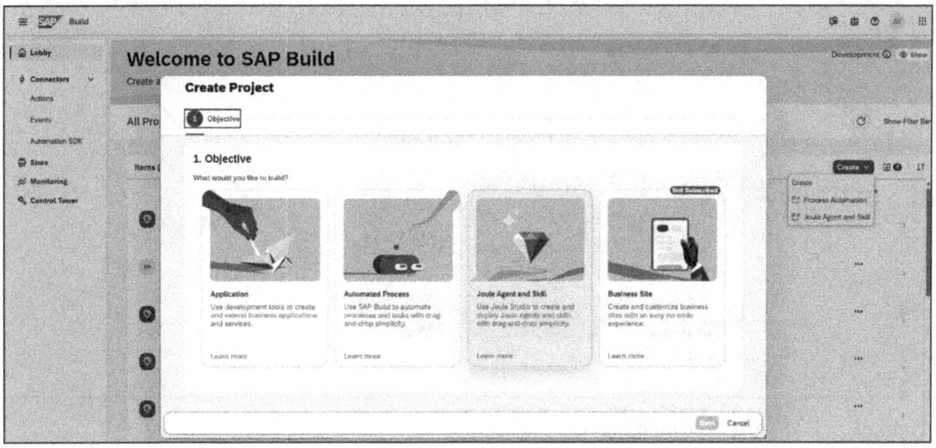

Figure 19.6: Joule Studio

Now that you understand what Joule is all about, in the next section we are going to dive a bit deeper to understand how Joule works.

19.3 Key Process and Components

Joule is evolving rapidly, with new capabilities introduced every cycle, demonstrating the significance of the AI assistant and its deep integration into the entire SAP ecosystem. Its understanding of the current user context, combined with a strong emphasis on data privacy and accuracy, makes it an ideal candidate for building enterprise-grade applications. But how does Joule work? Let's dive a little deeper to understand the underlying process.

As depicted in Figure 19.7, the core process behind Joule is triggered by the prompt entered by the end user through Joule's SAP Fiori-compliant conversational user interface. Upon receiving the user prompt, the underlying Joule engine first enriches the input with relevant context by referring to a catalog of business scenarios and grounding the response in verified enterprise knowledge. In this context, the scenario catalog is a collection of metadata for the available scenarios, functions, and skills of SAP cloud applications. The knowledge catalog combines both customer-owned and SAP-owned knowledge that Joule accesses through RAG. While processing the input prompt, in addition to contextual enrichment, Joule also considers the user's role(s) and permissions to ensure they do not access information they are not authorized to view. Finally, the generated response is filtered and presented to the user with enterprise-grade security and data privacy, while Joule also keeps track of chat history and context throughout the conversation.

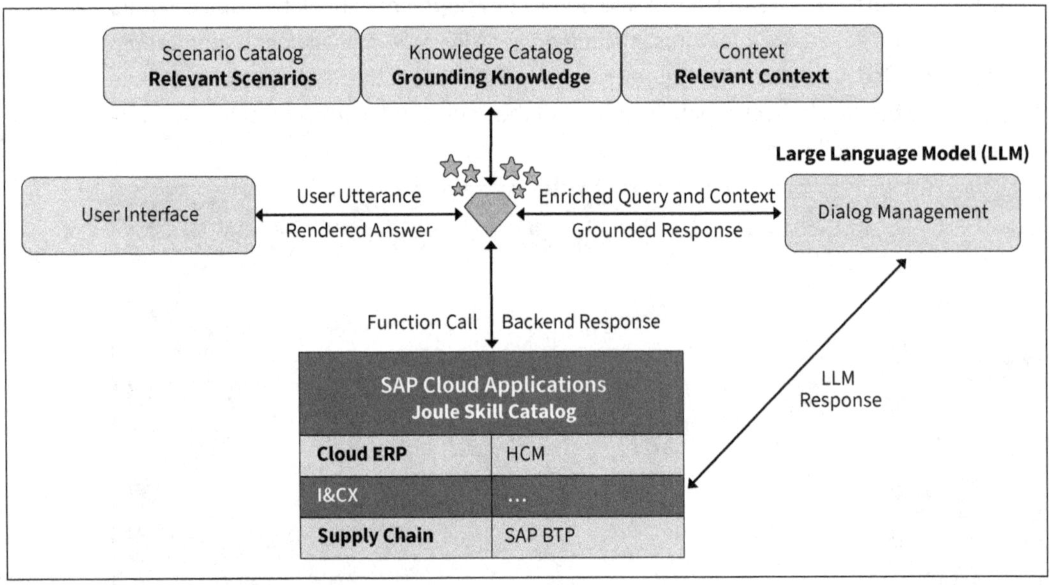

Figure 19.7: Joule Background Process

For more details related to Joule integration you can refer to *https://help.sap.com/docs/joule/integrating-joule-with-sap/introduction?locale=en-US.*

19.4 Joule Extensions and Agents

Joule Agents will play a key role in making enterprises driven by SAP systems more autonomous. To accelerate this journey, SAP has already published over 30 agents and more than 2,500 skills across various product stacks, spanning a wide spectrum of lines of business

(LOBs). Many of these out-of-the-box AI agents and skills serve basic needs, but Joule Studio plays a critical role in tailoring them to specialized behavior, governance, and integrations.

Joule Studio is embedded inside the SAP Build lobby, allowing you to initiate your AI agent creation journey by setting up a **Joule Agent and Skill** project, as shown in Figure 19.8.

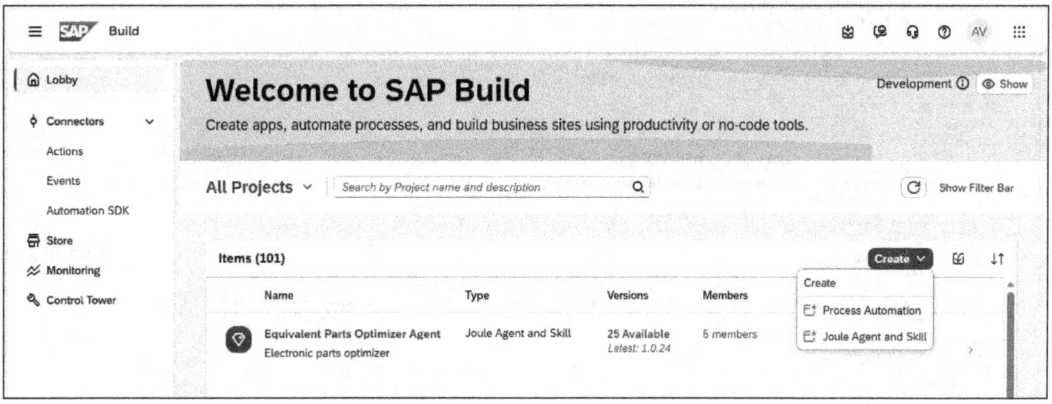

Figure 19.8: Joule Agent and Skill Project

As the next step, you can start creating the required **Joule Skill** and **Joule Agent** inside the project, as shown in Figure 19.9. In this context, skills play a critical role in integrating Joule Agents with any kind of backend APIs or services for secure execution. According to SAP's official definition, a Joule skill is a discrete, task-focused capability that Joule can execute through its conversational interface to interact with SAP applications.

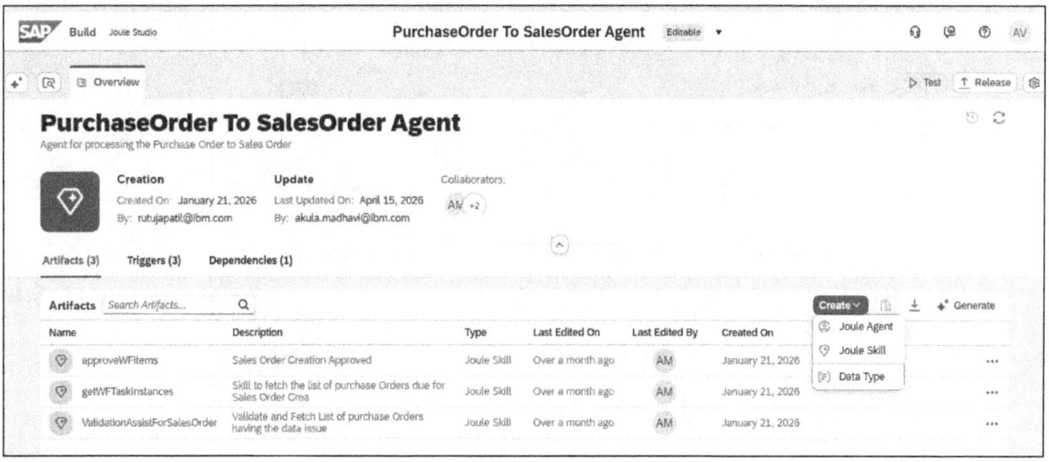

Figure 19.9: Joule Agents and Skills

Inside a Joule skill, as shown in Figure 19.10, you can start by defining the required input and output parameters, along with a description that better defines the skill for the AI agent.

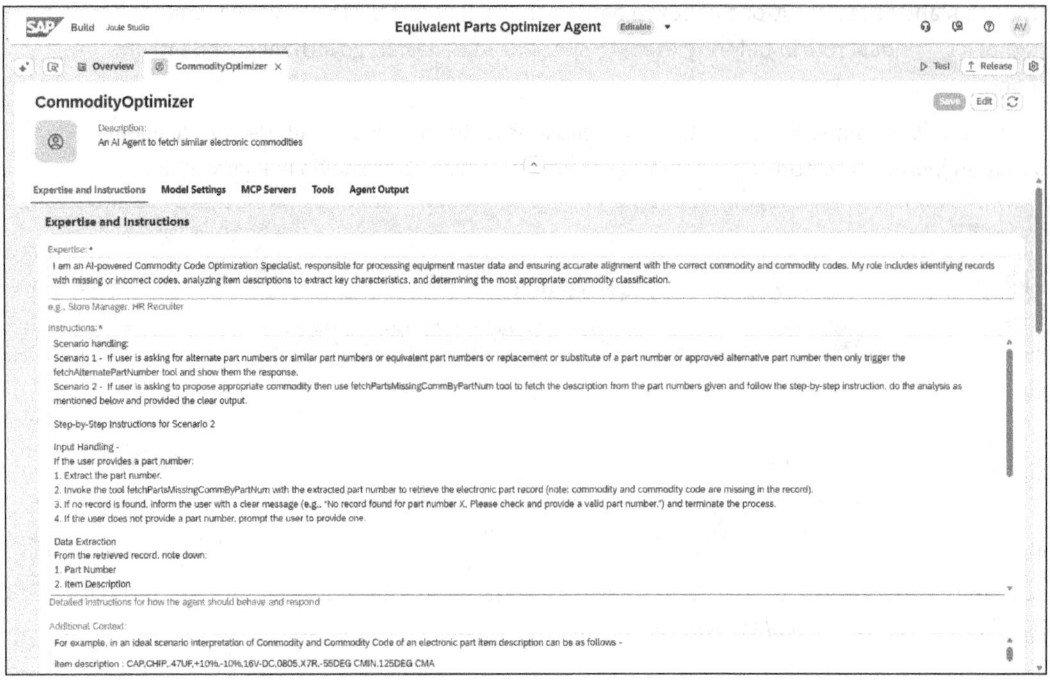

Figure 19.10: Joule Skills

While creating the skill, as part of the configuration you can integrate it with a wide spectrum of capabilities including action invocation, automations, processes, and even other Joule skills, as shown in Figure 19.11.

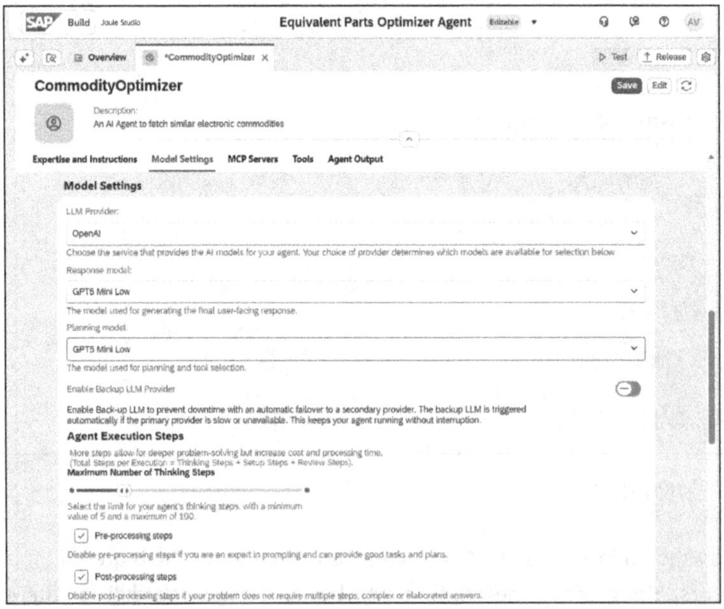

Figure 19.11: Skill Configuration

Once all relevant skills are ready, you can initiate the agent creation which only requires the **Expertise and Instructions** statement shown in Figure 19.12.

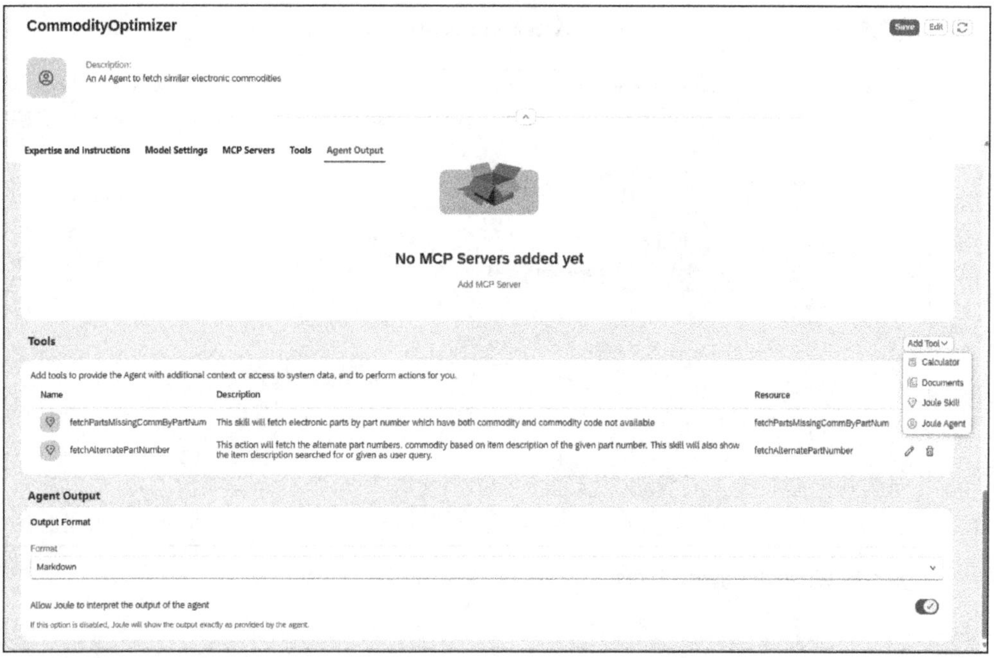

Figure 19.12: Joule Agents

In addition to instructions, expertise, and context, you also need to configure the model settings, as shown in Figure 19.13. Model settings are all about picking a brain for the agent; in this case, SAP provides a wide spectrum of foundation models which you can configure as your response and planning models.

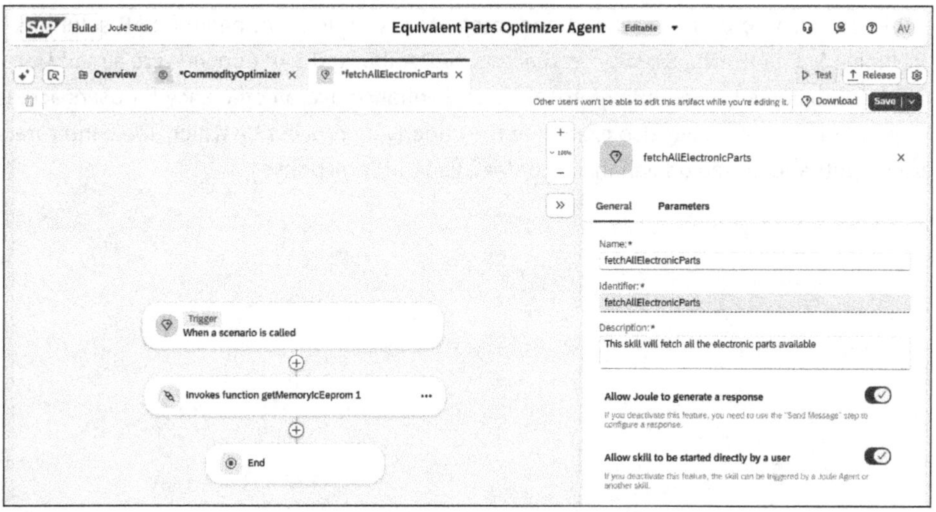

Figure 19.13: Joule Agents: Model Configuration

Finally, inside the tools segment shown in Figure 19.14, you can start adding the required tools such as Joule skills, documents, and other Joule Agents.

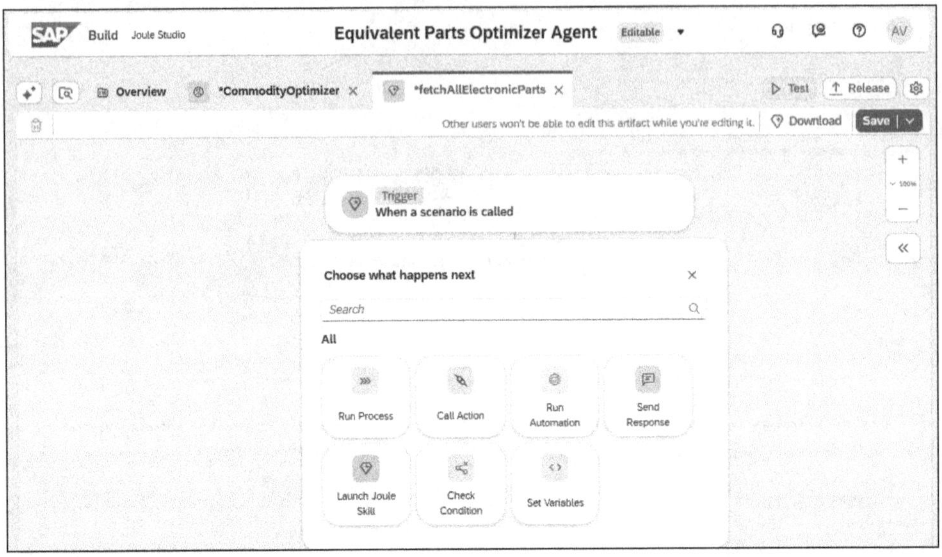

Figure 19.14: Joule Agents: Tools

19.5 Summary

As we wrap up the chapter, let's quickly recap what we have learned so far. Joule is one of SAP's groundbreaking innovations that will transform the future of enterprise solutions through advanced natural language processing techniques. Joule will not only bring value to business consultants but also enhance productivity with offerings like SAP Joule for Consultants, SAP Joule for Developers, and Joule Studio. Being a low-code/no-code offering, Joule can also be easily extended and integrated with Joule AI agents. SAP solutions—including SAP S/4HANA, SAP SuccessFactors, SAP Ariba, and SAP Concur—are already integrated with Joule to streamline processes and enhance overall efficiency for users across these platforms. We have also examined the underlying process by which Joule interprets user inputs and, based on enhanced context, generates responses.

Chapter 20
SAP Business Data Cloud

SAP Business Data Cloud is a data solution that unifies and governs SAP and third-party data with a business data fabric, enabling a trusted data foundation for advanced analytics and artificial intelligence (AI). The data foundation provides high-quality data with unified semantics and business context, making it AI ready. It leverages a delta sharing mechanism to connect with other open ecosystems through zero-copy data sharing.

SAP Business Data Cloud is a software as a service (SaaS) data platform which provides access to harmonized and governed data sets from SAP and third-party systems. These data sets are aligned with business *domains* and provide *data as products* for self-service data infrastructure, governance and ownership. SAP Business Data Cloud accelerates the AI journey by providing AI-ready data and faster time to insights. It also connects with open data ecosystems such as SAP Databricks, SAP Snowflake, and Google Cloud to enable true data democratization.

This chapter will cover the architecture of SAP Business Data Cloud, its capabilities, data products, intelligent applications, and AI/ML scenarios using SAP Databricks.

20.1 Architecture

SAP Business Data Cloud was developed to provide organizations with a modern data fabric of trusted data for analytics, machine learning, and current and future AI-powered applications. The design principles employed in SAP Business Data Cloud development are as follows:

- One domain model: SAP Business Data Cloud harmonizes domain specific data from various source systems.
- Data product: Data is a product with its own lifecycle. This ensures that all versions of the same entity are merged, providing high-quality domain data. Activation results in the creation of pipelines for data ingestion and transformation, semantically enriched data models, and dashboards.
- Open data ecosystem: Organizations with data platforms such as SAP Databricks, SAP Snowflake, BigQuery, and Microsoft Fabric can integrate them with SAP Business Data Cloud using SAP Business Data Cloud Connect. This integration uses delta sharing protocol with zero-copy data sharing to avoid persistence of SAP Business Data Cloud data in external data platforms and persistence of external data in SAP Business Data Cloud.

- Scalable data storage: SAP Business Data Cloud uses SAP HANA Cloud, data lake for data storage. SAP HANA Cloud, data lake separates computing power from storage, making the solution scalable and able to handle huge amounts of data.

Figure 20.1 shows the architecture of SAP Business Data Cloud according to the design principles discussed earlier. There are three layers in the architecture: data provisioning, data transformation, and harmonization and consumption of curated data by applications and tools.

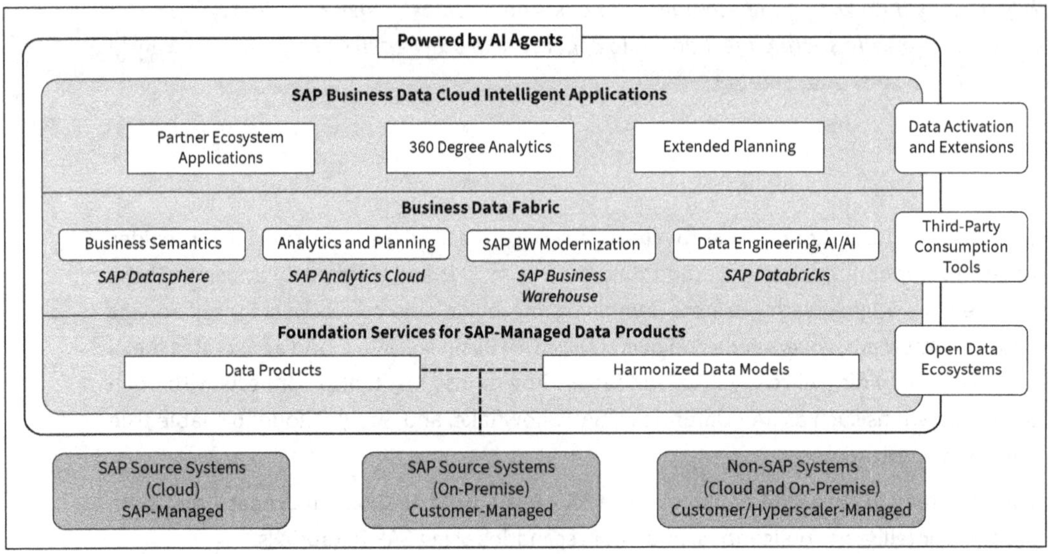

Figure 20.1: Architecture of SAP Business Data Cloud

> **Note**
>
> SAP Snowflake is not shown in the business data fabric in Figure 20.1, as it is not generally available as of June 2026.

The core of the SAP Business Data Cloud data platform is the business data fabric. According to SAP, business data fabric is an architecture that turns all your data and context into one governed foundation, serving as the trusted knowledge core for every enterprise application and AI agent.

The business data fabric in SAP Business Data Cloud offers the following capabilities:

- Intelligent computing power: Provides *elastic scale* computing power for analytics, advanced analytics and AI workloads across multiple runtimes provided by SAP HANA Cloud, SAP Databricks, and SAP Snowflake.

- Semantic layer: Combine data from various sources to create a harmonized and governed data model with consistent semantics across business domains. Knowledge graphs provide data lineage to support cross-domain contextual understanding.
- Data products: Contains self-contained curated data sets which are governed, reusable, discoverable, and easily consumable.
- Data management: Includes multiple data integration patterns like batch, real time, and data federation alongside a catalog of governed data assets, business domain-based data ownership, and AI agent-based data lifecycle management.
- Machine learning and AI: Contains custom ML and AI workloads, complex pipelines, and zero-copy data sharing within open data ecosystems.
- Data visualization and planning: Consume curated data sets for analytics, reporting, and business planning.

Foundation services are the core operational backbone for SAP-managed data products. These services are created when a data product delivered by SAP is activated. They help with data ingestion from source systems, data transformation and harmonization, sharing with open data ecosystems using Delta Sharing and discovering metadata using the Open Resource Discovery (ORD) protocol. SAP HANA Cloud, data lake is the storage used by foundation services

Note
Custom-managed data products have custom data acquisition layers, semantic layers, and visualization layers. Data storage happens within object stores managed by SAP Datasphere

SAP Business Data Cloud offers out-of-the-box, ready-to-run intelligent applications for many lines of business (LoBs), all completely managed by SAP. These applications are delivered as prebuilt data ingestion pipelines, analytical models, dashboards, and workflows. Infused with AI and ML, these applications learn, reason, and act within the context of business data.

20.2 Capabilities

SAP Business Data Cloud contains many services like SAP HANA Cloud, SAP Datasphere, SAP Analytics Cloud, SAP Databricks, SAP BW/SAP BW/4HANA, private cloud edition, SAP Business Data Cloud Connect, and SAP Master Data Governance. These services (or components) are conceptually organized into four layers, as shown in Figure 20.2.

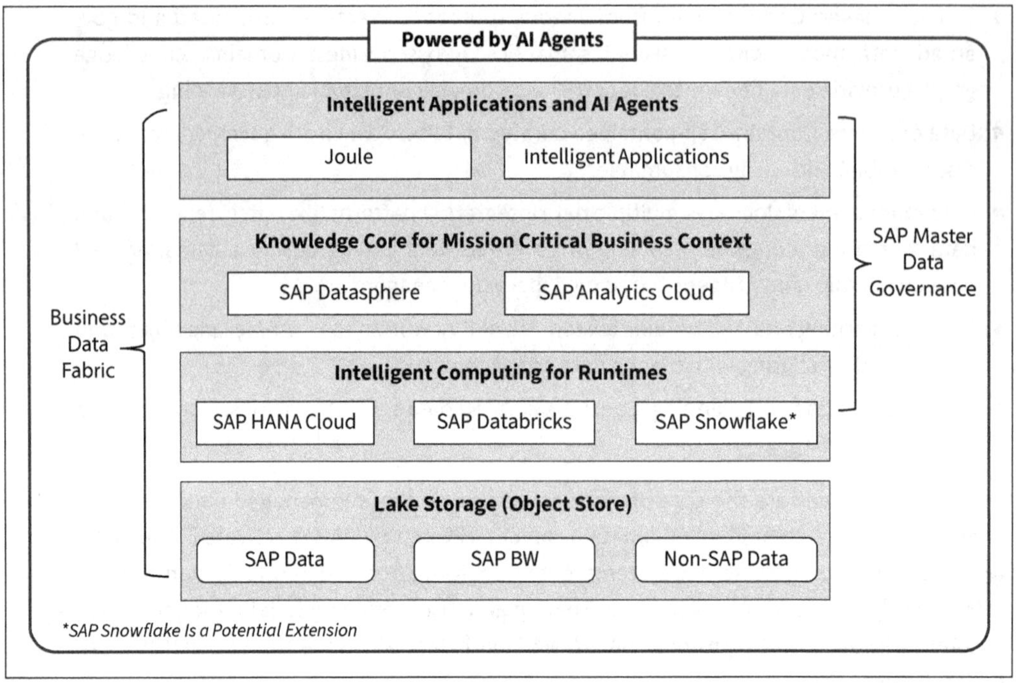

Figure 20.2: Conceptual Layers of SAP Business Data Cloud

[»]

Note

SAP Analytics Cloud and SAP Datasphere services were moved from the SAP BTP services catalog to the SAP Business Data Cloud service catalog at the beginning of 2026.

Let's look at each layer in a little more detail:

- **Lake storage**
  The lake storage layer is the physical data storage layer where data from both SAP and non-SAP sources is stored and managed using the architecture principles of SAP HANA Cloud, data lake. Lake storage is high-performance, unified storage for structured, semi-structured, and unstructured data.
- **Intelligent compute**
  Intelligent compute provides the computing power for the workloads associated with data lakes, data warehouses, ML, and data engineering. Workloads are dynamically and intelligently managed based on cost and performance tradeoffs. It consists of the following:
  - SAP HANA Cloud: Supports multimodel processing—relational, spatial, graph, vector, and time series—across data lake, data warehouse, and ML workloads. SAP HANA Cloud's built-in libraries—Predictive Analysis Library (PAL) and Automated Predictive Library (APL)—support building ML and advanced analytics models.

  - SAP Databricks: Supports complex data engineering and ML/AI workloads through the Spark engine.
  - SAP Snowflake: As an alternative to SAP Databricks, it also supports ML/AI workloads.

  SAP HANA Cloud has already been covered in Chapter 5. We will cover SAP Databricks in Section 20.6.

- **Knowledge core**

  The knowledge core layer enhances data in SAP Business Data Cloud with semantically rich business context for consumption by analytical applications and AI. It consists of the following:

  - SAP Datasphere: Provides a modeling layer on top of data products, both SAP-managed and custom-managed, for creating semantically rich analytical models. It also serves as a central tool for analytical roles and data access control for role-based data access. It provides tools for data integration with multiple data sources.
  - SAP Analytics Cloud: Provides data visualization and planning capabilities in SAP Business Data Cloud. It serves as a central tool for business users to gain business insights and execute planning functions. It has built-in AI-driven natural language processing for data discovery and ad-hoc analysis.

- **Intelligent applications and AI agents**

  Intelligent applications are low-code/no-code or pro-code dashboards powered by AI that are delivered and managed by SAP. SAP Analytics Cloud is the core component used in building these applications. These applications can be extended. SAP-managed live data connections link analytical models in SAP Datasphere to SAP Analytics Cloud.

  Joule is SAP's AI-powered solution to help with task automation and provide intelligent business insights through a conversational interface and predictive analytics. Joule can leverage the business data fabric of SAP Business Data Cloud in any SAP or custom application. SAP Analytics Cloud has embedded Joule capabilities.

20.3 SAP Business Data Cloud Cockpit

The SAP Business Data Cloud cockpit is the UI-based interface for deployment, governance, and data product exploration. SAP Business Data Cloud administrators use it to compose the SAP Business Data Cloud landscape, configure and manage various components, and administer roles and users. Data scientists and data analysts use it to explore and activate SAP intelligent applications, data packages, and custom and partner system data products.

The SAP Business Data Cloud cockpit can be used to create one or more formations in SAP Business Data Cloud. The administrator can select SAP Business Data Cloud components to include in each cloud formation. Some components, such as SAP Datasphere and SAP Analytics Cloud, are mandatory for each cloud formation.

Once the components are added to a cloud formation, secure and reliable connections can be made with data sources. Connections can be made only from the allowable SAP and non-SAP sources.

[+]

SAP for Me

The SAP Business Data Cloud cockpit redirects to SAP for Me for composing landscape and connections.

Discovery and installation of intelligent applications and data products are covered in Section 20.4 and Section 20.5.

20.4 Data Products

Data products are the building blocks of the SAP Business Data Cloud business data fabric and serve as a standardized way to share and consume curated business data and associated metadata across applications and domains. Metadata provides business semantics and context to the data set enabling advanced analytics and AI. The characteristics of data products are as follows:

- Securely and efficiently consumable via APIs or events with open ecosystem and business intelligence (BI) tools. Supported APIs are SQL, Delta Sharing, Event, REST, and OData.
- Discoverable through SAP's data product catalog in the SAP Business Data Cloud cockpit using the ORD protocol.
- Described with rich metadata and semantics.
- Reusable across various use cases like advanced analytics, ML, AI, planning and analytics, and building AI agents.
- Decentralized ownership based on business domain. Data products are managed by domain experts and not by a centralized IT team.
- Efficient data storage in modern object-store based data lake architecture, providing scalable storage.

The key attributes of data products are listed in Table 20.1.

Attribute	Description
Type	Primary (source oriented) or derived (consumer oriented)
Category	Categorizes the data set as master or transactional (business object or analytical)

Table 20.1: Attributes of a Data Product

Attribute	Description
Visibility	Public, private, and internal
Input ports	Integration dependencies with data input for lineage
Output ports	APIs or events through which data can be consumed
Responsible	Organization/team responsible for lifecycle management

Table 20.1: Attributes of a Data Product (Cont.)

Data products delivered by SAP are called SAP-managed data products, data products created by your own organization are called customer-managed data products, and data products created by SAP partners are called partner-managed data products.

Activation of SAP-managed data products is fully automated and is carried out in the background. Activation results in the artifacts being created within the foundation services, allowing for the following:

- Creation and management of data pipelines for data extraction, cleansing, and transformation
- Storage of data sets from data pipelines in the object store
- Creation of delta lake tables, an open data format on top of the object store

[+]

Custom-Managed Data Products

Data pipelines for custom-managed data products are completely managed in SAP Datasphere by organizations and data is stored in the SAP Datasphere object store as SAP HANA data lake files (HDLF) and consumed through the delta lake table. This delta lake table can be published in the SAP Business Data Cloud catalog for discovery and consumption across open ecosystems.

Data products in the SAP Business Data Cloud cockpit are discovered using the ORD protocol. This protocol, developed by SAP, enables services to publish standardized metadata for data products. ORD helps discover metadata, provides real-time metadata updates, and enables interoperability between the open ecosystem and SAP Business Data Cloud as both a consumer and publisher of data product metadata. Data product discovery through SAP Business Data Cloud is shown in Figure 20.3.

The steps for activation of data products and installation of intelligent content are as follows:

1. Ensure the connection between data sources and SAP Business Data Cloud components are in place.

2. Select the connection to the source system from which the data product needs to be activated.
3. Use search functionality in SAP Business Data Cloud to explore various data packages and intelligent applications, organized by LoB, in SAP source systems and select those you require.
4. Activate data packages and install intelligent content.

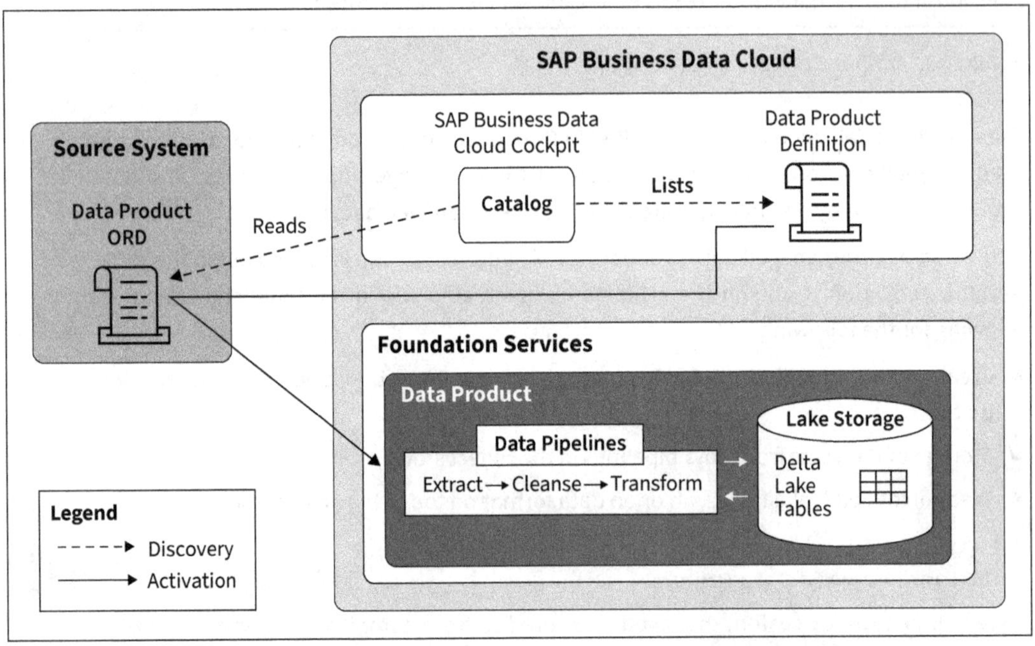

Figure 20.3: SAP-Managed Data Product Discovery Using the ORD Protocol

[»]

What Are Data Packages?

A data package is a technical grouping of all related data products. Organizations need to subscribe to SAP Business Data Cloud Intelligent Packages (a commercial wrapper for data products and intelligent applications) to install a data package or a data product within a package.

20.5 Intelligent Applications

Intelligent applications are AI-powered dashboards that provide insights and make recommendations to business users in real time. They are delivered and fully managed by SAP, ready to use through a simple and automated installation. These applications are built in SAP Analytics Cloud and consume analytical models in SAP Datasphere. These models are consuming SAP-managed data products

Groups of domain-specific intelligent applications are delivered via SAP Business Data Cloud Intelligent Packages. Some of the Intelligent Packages are as follows:

- Cloud ERP Intelligence
- Supply Chain Intelligence
- People Intelligence
- Spend Intelligence
- Travel and Expense Intelligence

Key components of intelligent applications are listed in Table 20.2.

Component	Description
Visualization object	SAP Analytics Cloud stories for dashboards and reports
Analytic models	Analytic models in SAP Datasphere that prepare and expose data from data products
Data products	Curated and business-semantic rich data sets fully managed by SAP
Foundation services	Backend services for data ingestion and transformation
Roles	Scoped roles generated for access to relevant spaces

Table 20.2: Components of Intelligent Application

The components of an intelligent application are shown in Figure 20.4.

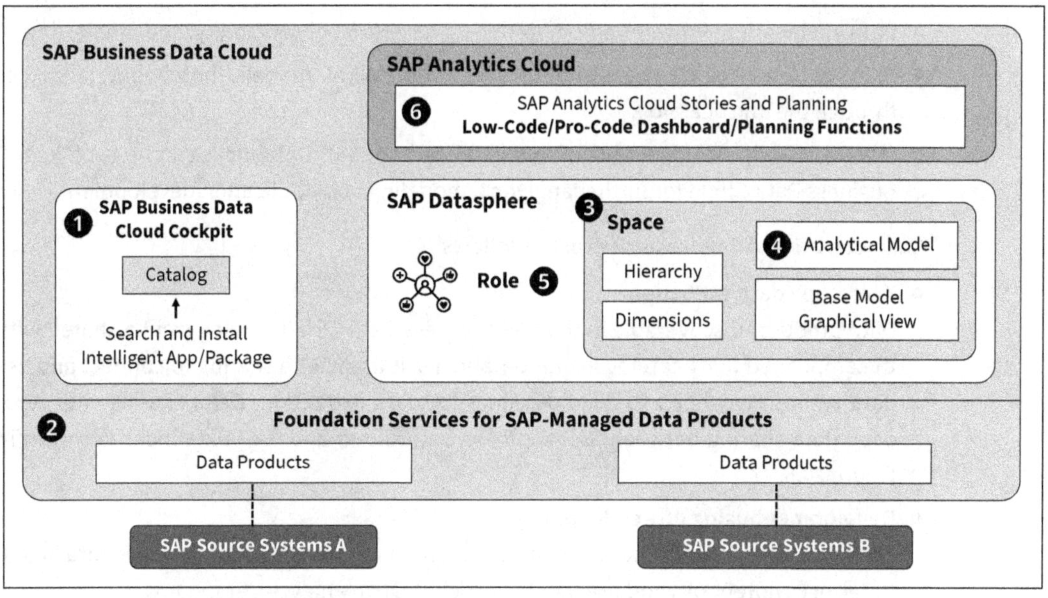

Figure 20.4: Installation of Intelligent Application

Here you will find the following:

❶ Discovery and installation of an intelligent application/package in SAP Business Data Cloud

❷ Generation and activation of associated data products in foundation services

❸ Generation of a space in SAP Datasphere

❹ Generation of base views and analytical models

❺ Generation of scope roles that provide access to a space

❻ Generation of intelligent applications

20.6 SAP Databricks

SAP Databricks is a data intelligence platform for data engineering and AI/ML workloads. It is a fully embedded original equipment manufacturer (OEM) component from Databricks in SAP Business Data Cloud. It allows you to seamlessly execute AI/ML algorithms on SAP Business Data Cloud data.

Using the delta sharing protocol, SAP Business Data Cloud data can be shared with SAP Databricks for processing and the results stored back in SAP Business Data Cloud. Zero-copy data sharing avoids redundancy of data in multiple components of SAP Business Data Cloud. Key features of SAP Databricks are as follows:

- AI/ML lifecycle support: Enables experimentation and production grade deployment of ML models including generative AI and large language models based on SAP Business Data Cloud data.
- Data engineering: Provides automated ETL processing, observability, and monitoring.
- Data governance: Manages and secures data, ML/AI models, notebooks, and files through the unity catalog.
- Integration with the SAP ecosystems: Results from SAP Databricks can be used in SAP Business Data Cloud for further analytical modeling and application development.

Characteristics of SAP Databricks are as follows:

- **Zero-copy data exchange**
  Data products discoverable in the SAP Business Data Cloud catalog can be shared with the embedded unity catalog in SAP Databricks. It uses delta sharing to connect and use data without any need to build pipelines for data extraction. Delta sharing with zero-copy data sharing between SAP Business Data Cloud and SAP Databricks is shown in Figure 20.5.
- **Development using pro-code tooling**
  You can create Apache Spark pipelines to blend data from SAP and non-SAP data in SAP Databricks notebooks and use SAP Databricks SQL to analyze data.

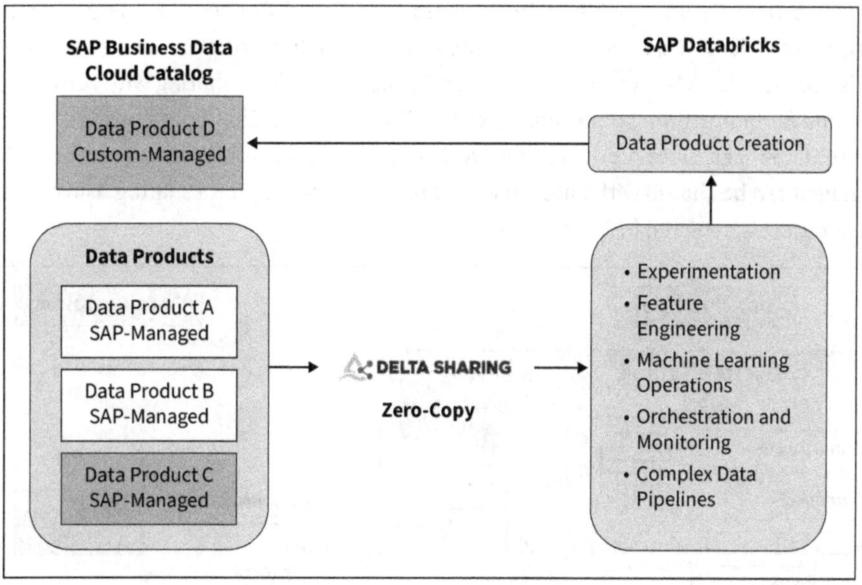

Figure 20.5: Delta Sharing with Zero-Copy Data Sharing

The components of SAP Databricks are as follows:

- Delta lake: This is an open-source storage framework for modern data lakehouse architecture. It layers on top of cloud storage like AWS S3, Azure Data Lake, and HDLF.
- Unity catalog: The unity catalog enables governance for structured and unstructured data, machine learning models, notebooks, dashboards, and files.
- Apache Spark: This ss an open-source analytics SQL engine for big data workloads. It includes many libraries for ML, streaming, graphs, and SQL.
- ML flow: This is a built-in open-source platform in SAP Databricks for the entire ML and generative AI lifecycle. It seamlessly tracks AI performance, version of ML/AI model versions, and experiments, and packages code for smooth enterprise-grade deployments.
- Notebook: The notebook is an interactive web-based tool that allows writing and execution of code, visualization of results, and documentation of work in one place. It is one of the primary tools for creating data science, ML, and data engineering workflows within SAP Databricks.

The predominant use cases for SAP Databricks are as follows:

- Data engineering: Build complex automated data pipelines for processing structured and unstructured data and blend it with trusted data from SAP Business Data Cloud.
- AI/ML: Build robust AI/ML models on SAP Business Data Cloud data in SAP Databricks notebooks and share them as data products with SAP Business Data Cloud for AI-driven decision making in your SAP ecosystem.
- Analytics: Explore and analyze large analytical loads for real-time analytics and visualization.

SAP Business Data Cloud provides SAP Business Data Cloud Connect to integrate with enterprise Databricks. This service helps organizations safeguard their investments in enterprise Databricks. SAP Business Data Cloud Connect uses delta sharing with zero-copy data sharing for bidirectional data sharing of SAP Business Data Cloud data products with enterprise Databricks. Data products that are SAP-managed, customer-managed, or partner-managed can be shared with enterprise Databricks. Zero-copy data sharing using delta sharing protocol is depicted in Figure 20.6.

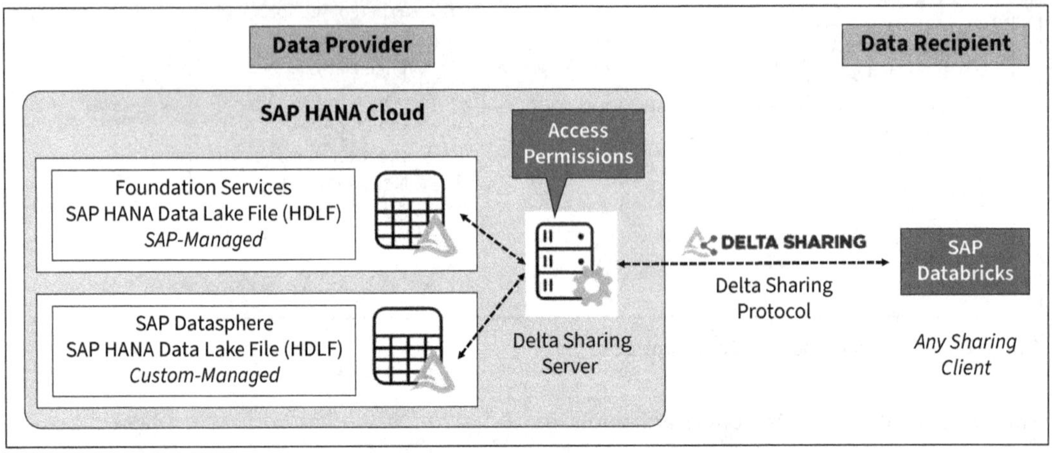

Figure 20.6: Delta Sharing with Enterprise Databricks using SAP Business Data Cloud Connect

Note

Data products must be explicitly shared with enterprise Databricks via the SAP Business Data Cloud catalog. This is a minor difference between SAP Databricks and enterprise Databricks.

20.7 Use Cases

We have discussed in detail delivering operational insights to business users through intelligent applications as a use case of SAP Business Data Cloud. In this section, we will explore additional use cases for SAP Business Data Cloud, which help organizations optimize their existing SAP BW investments and support Business AI with trusted data.

20.7.1 Modernizing SAP Business Warehouse with SAP Business Data Cloud

Customers with existing SAP BW systems have invested in developing a robust and trusted data warehouse that supports mission-critical applications across business processes. SAP Business Data Cloud enables the integration of SAP BW to leverage curated data sets for advanced analytics and AI/ML models.

The path to modernization of SAP BW is to *lift* organizations currently using on-premise SAP BW or SAP BW/4HANA to the SAP Business Warehouse, private cloud edition of SAP Business Data Cloud.

On-Premise Versions Supported for Modernization Path

Following are the versions currently supported for this modernization path:

SAP BW NetwWaver < 7.5: Upgrade to SAP BW on SAP HANA 7.5 SPS31

SAP BW NetWeaver 7.5 on any database/SAP HANA: Upgrade to SAP BW on SAP HANA 7.5 > SPS24

SAP BW/4HANA 1.0 and 2.0: Upgrade to SAP BW/4HANA 2023 and then lift your system to SAP Business Warehouse, private cloud edition

SAP BW/4HANA 2021: Update to the latest version and then lift your system to SAP Business Warehouse, private cloud edition

SAP BW/4HANA 2023: Ready to be lifted to SAP Business Warehouse, private cloud edition

Once SAP BW has been lifted to SAP Business Warehouse, private cloud edition in SAP Business Data Cloud, *shifting* can generate data products from SAP BW InfoProviders, as shown in Figure 20.7.

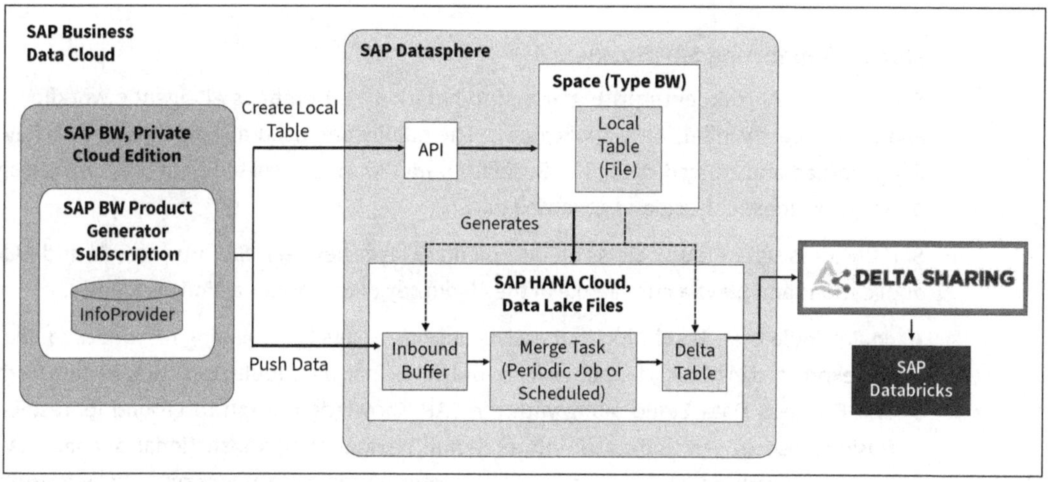

Figure 20.7: SAP BW Product Generator

Note

SAP Business Warehouse, private cloud edition is an SAP-managed cloud deployment of SAP BW or SAP BW/4HANA.

Data products generated based on SAP BW InfoProviders can be shared within an open ecoystem or SAP Databricks through the delta sharing protocol for AI/ML model development. This sharing of SAP BW data products is called *innovating*. The lift-shift-innovate approach in SAP BW modernization is shown in Figure 20.8.

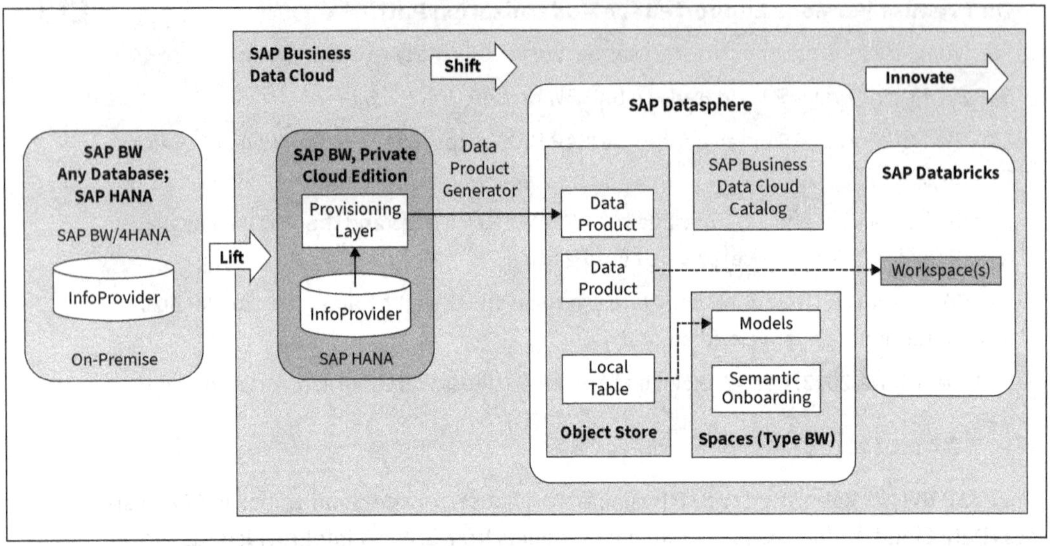

Figure 20.8: Lift-Shift-Innovate approaches in SAP BW modernization

20.7.2 Supporting SAP Business AI

SAP Business AI is an enterprise-grade AI foundation that embeds AI, agentic workflows, and Joule directly in business applications. The quality and accuracy of AI depend on having governed and trusted data. SAP Business Data Cloud is a trusted data platform which provides business context and governed data.

SAP Business Data Cloud acts as an intermediary layer between SAP Business AI and SAP applications and plays a critical role in the AI journey of any organization, as follows:

- Joule: Joule is an AI solution that assists business users by providing responses to queries expressed in natural language. To avoid hallucinations, Joule uses curated data from SAP Business Data Cloud along with the SAP Knowledge Graph to ground its results. Business user interaction can be informational, navigational, transactional, or analytical. Joule may coordinate AI agents to gather information for a query or execute a set of tasks.
- Supports AI agents: AI agents perform specific tasks either triggered by human queries or autonomously by other AI agents to achieve a goal. Decision making during task execution requires highly accurate, curated data. This data is provided by SAP Business Data Cloud

- Embedded AI: Intelligent applications infused with AI provide business users with insights that can make them more efficient. SAP Business Data Cloud provides embedded AI with the business context of the data as well as its metadata.
- Knowledge graphs: A knowledge graph connects business entities allowing you to discover relationships and generate facts. A knowledge graph learns from the data it's provided with and the accuracy of a knowledge graph depends on the quality of data in SAP Business Data Cloud.
- Supports AI engineering: AI engineering is the development of AI/ML models by data scientists. SAP Business Data Cloud provides tooling and runtimes for the development of AI/ML models. Tools are provided by SAP Databricks or AI Foundation and can be pro-code or low-code. This chapter briefly discussed one such tool from SAP Databricks: the notebook. AI/ML models can be shared as data products with SAP Business Data Cloud.

20.8 Summary

SAP Business Data Cloud is a data platform that supports analytics, advanced analytics, AI/ML, and complex data engineering, serving as a key pivot in supporting SAP Business AI. It is a true business data fabric based on modern data lakehouse and medallion architectures. It provides curated, governed, and contextually rich data sets for use in reporting, planning, and other open ecosystems. Delta sharing with zero-copy data sharing ensures that data products are virtually consumed within the recipient system, eliminating the need for complex data pipelines for data ingestion.

SAP Business Data Cloud offers out-of-the-box, ready-to-run intelligent applications for many LoBs, fully managed by SAP—from extraction to reporting. It also offers SAP BW, private cloud edition for modernizing on-premise SAP BW systems through the lift-shift-innovate approach. SAP Business Data Cloud unlocks the tremendous value of both SAP BW investments and the curated data sets within SAP BW.

Chapter 21
Conclusion and Roadmap

Congratulations on making it this far! This chapter looks into the future of SAP Business Technology Platform (SAP BTP).

As is now evident, enterprises' expectations of technology have changed significantly over the years. Technology is no longer merely a tool to record activities or automate tasks; it's how organizations build resilience, transform businesses, enhance customer experiences, improve employee engagement, strengthen partnerships, and explore new frontiers for the benefit of both the enterprise and the world. Modern enterprises should aspire to become autonomous enterprises through the use of technology. Thus, the IT landscape for enterprises no longer just contains ERP products but now includes products specialized for customers, employees, and vendors/partners, powered by automation and intelligence, made available in different consumption models (such as perpetual consumption or pay-as-you-go) through various delivery modes. This diversity has led to the emergence of platforms that provide the right tools and capabilities, enabling enterprises to focus on business specialization and enhance their differentiation from the competition. SAP has helped define this shift by introducing SAP BTP and now SAP Business AI Platform, providing the essential capabilities for intelligent enterprises to evolve.

In this book, we highlighted the transformative capabilities of SAP BTP, focusing on *why* to act rather than *how* (which is available in many published sources) and on how to choose the best way forward when faced with multiple options. In this final chapter, we look ahead to the future of SAP BTP and summarize some of the key concepts covered in this book.

21.1 The Autonomous Enterprise

At SAP Sapphire 2026 in Orlando, SAP introduced the term *Autonomous Enterprise* to describe enterprises that use humans and AI together to achieve their objectives accurately, efficiently, compliantly, and securely, while identifying new revenue opportunities in an uncertain business environment. This concept extends the capabilities of the previously introduced *intelligent enterprise* categorization. All enterprises aspire to be resilient, intelligent, and sustainable to achieve success. Intelligent enterprises are defined as enterprises who have agile and integrated business processes; use advanced technologies such as artificial intelligence (AI), machine learning, and the Internet of Things (IoT); and apply industry best practices. Intelligent enterprises work together with customers, partners, and employees to build flexible value chains while managing their environmental impact.

Autonomous Enterprises are data-driven and generate value from data residing both within and outside enterprise boundaries, encompassing all forms—from rigid and structured to fluid and unstructured—as well as streaming data generated by humans and machines. By applying AI and machine learning capabilities, enterprises can gain insight into what is happening, understand why, and predict what will happen in the future. Intelligence supports decision-making by suggesting the next best actions at the point of transaction, rather than after it has occurred, enabling enterprises to prevent negative outcomes.

The Autonomous Enterprise advances this further with extreme automation, driven by deep contextual intelligence, while maintaining adequate guardrails as required by regulators and stakeholders. It not only supports decision making but also enables automatic execution of decisions, freeing up humans for higher-value tasks while reducing errors. However, because wrong decisions based on low-quality data or inherent hallucinations in some AI models can cause significant disruption, it is paramount to have adequate governance and human control throughout the entire lifecycle of data management and execution performance—not only to meet regulatory compliance but also to ensure value for stakeholders and society at large.

While there have been significant advancements in AI technologies, foundational models, artificial general intelligence, and agentic AI capabilities, and most enterprises expect these technologies to drastically improve their business postures, the return on investment and success rate so far have not met expectations. A closer analysis of the root cause reveals the following key gaps:

- Generative AI models lack enterprise specific business and process context. As such, their reasoning and actions don't align with enterprise needs.
- Quality data is key to quality AI outcomes, so applying AI on top of disconnected, outdated systems reduce the efficacy of outcomes.
- The ability to monitor and control AI makes AI outcomes risky for the enterprise.

SAP came up with targeted offerings to address these gaps via the following:

- *SAP Business AI Platform* brings together SAP Business Technology Platform (SAP BTP), SAP Business Data Cloud, Business Transformation Management, and SAP Business AI. SAP's knowledge graph and data products help provide the context AI agents need and Joule Studio helps develop extensions and new agents. Domain-specific foundational models trained on SAP code, data, and processes help provide the right business context.
- *SAP Autonomous Suite* includes business applications infused with AI agents that support end-to-end business processes. These assistants and agents facilitate intelligent automation of business processes across finance, supply chain, procurement, human capital management, and customer experience.
- *Autonomous user experience* through Joule Work is a single interface to SAP software as well as an autonomous orchestrator that executes tasks based on contextualized insights.

However, Autonomous Enterprises and intelligent enterprises excel not just because of technologies. They excel because they have the following advantages:

- They are agile, flexibly powered by not only technology but also through organizational culture and processes. Thus, change management is key for their business workflows and keeping their employees, partners, and customers engaged and productive in the extended network.
- They are interconnected with their partners' enterprises, customers' enterprises, and wider networks to provide the flexibility to adjust business practices and processes quickly.
- They are automated, with as many activities automated as possible to improve speed, reduce errors, and free up human resources for tasks that cannot be automated.
- They are sustainable since, as responsible corporations, they want to ensure that scarce resources are protected for future generations.
- They are secure in their ability to fight not only external threats but also internal threats in a proactive manner before any untoward incidents occur. Intelligence protects the personal data of customers, employees, and partners as well as the business data of the enterprise itself and partner enterprises.

Top of the curve enterprises require specialized skills; therefore, special care must be taken to source, retain, and develop them. Employee expectations of a company have also changed significantly during these uncertain times. How companies support both physical and mental well-being, while managing skill gaps in the workforce, can have a significant impact.

21.2 Cloud Platforms

As discussed throughout this book, the platform business model on an open, secure hybrid multicloud is a crucial enabler of the autonomous enterprise. The platform business model facilitates the exchange of value between two different groups. The platform business model has emerged as a significant approach in which enterprises can provide and monetize value without being a producer or a consumer.

The PaaS model, on the other hand, has emerged as an option where developers can rent tools to develop from a provider via the cloud. Slowly, this model expanded beyond development into integration, automation, data, user experiences (UXs), and more. The cloud is a key enabler in this journey by providing the following advantages:

- Speed
- Flexibility
- A platform to innovate
- Lower total cost of ownership (TCO)

Enterprises are moving aggressively to the cloud as they modernize their application portfolios. They want to make their businesses context-aware, self-learning, and intelligent. Data has become the most important resource for business, which must secure their data and handle massive amounts of both structured and unstructured data. Thus, enterprises expect a cloud platform to provide not only better efficiency, scale, resilience, and cost savings but also to have AI embedded in the fabric and security at its core.

Cloud offerings also come in different variants based on the services offered, generally following these models:

- **Infrastructure as a service (IaaS)**
  Infrastructure like compute, storage, and network resources, among others, are offered as a service that you can use to install your choice of operating system, database, applications, and more.
- **Platform as a service (PaaS)**
  Several software tools, including their underlying infrastructure, are provided as a service, thus allowing you to develop intelligent applications and integrate them with other applications.
- **Application software as a service (SaaS)**
  Applications are offered as a service, including their underlying "everything," so that you can get their functionalities quickly without having to develop them.
- **Every(any)thing as a service (XaaS)**
  Besides the previous model, several combinations of software and services (e.g., testing as a service) are offered by innovative providers catering to different needs of enterprises looking for agility, flexibility, and speed.

The IaaS model comes in multiple variants as well:

- On-premise, where the hosting is in the client's data center
- Private, where the hosting is by a service provider in their data center, but the infrastructure is fully dedicated to them
- Public cloud, where the hosting is by service provider and infrastructure is shared

In this journey from your current state to a business model powered by latest digital technologies, your enterprise may have a hybrid environment (i.e., a mix of on-premise, private dedicated cloud, and public shared cloud resources). While most aspire for 100% public cloud, factors like regulations, compliance, and security force companies to keep some components on-premise or in private dedicated hosting.

When businesses adopt the cloud, simply moving their existing workload to the cloud is insufficient. You must modernize your application portfolio by taking advantage of the capabilities provided by the cloud. Applications built by leveraging the capabilities of the cloud and designed to run on cloud are called *cloud-native applications*. While some existing applications can be refactored to run on the cloud, the maximum value is realized from cloud-native applications, which provide you the ability to define new business models hitherto unthinkable.

Two key features of cloud-native applications are microservice architecture and containerization. Microservices are an architectural style where large applications are composed of smaller components each focusing on certain unique functions interacting with each other through application programming interfaces (APIs) hiding the underlying implementation. Microservices offer flexibility, scalability, and maintainability over monolithic architectural styles, all of which are important characteristics required to support digital transformations.

Another key requirement of a digital transformation is the ease and speed of deployment of applications so that business functionalities can be released faster. Containerization provides that ability in addition to improving the utilization of servers, when compared to virtual machines (VMs), by providing virtualization on top of an operating system. Containers can support bring your own language (BYOL)-type capabilities where you can deploy and run applications written in any language by encapsulating the environment required to run the applications along with the application itself. Containers have transformed the way applications can be deployed, run, and scaled up/down on variety of platforms. To manage and automate the deployment and running of containers, container management and orchestration products like Kubernetes play an important role.

Enterprises typically leverage PaaS offerings to introduce these capabilities quickly. They look for the following characteristics in their choice of PaaS:

- A strong secured foundation meeting regulatory requirement.
- Multiple deployment options (i.e., public cloud and private cloud) to provide flexibility.
- PaaS should provide control to the infrastructure layer even though, in most cases, you'll consume at a platform level to provide the flexibility to host multiple types of workloads and to optimize the usage of the infrastructure.
- PaaS should support a wide variety of runtimes to provide enterprises the right environment for their developmental needs.
- PaaS also should provide a variety of development environments, along with lifecycle tools, supporting approaches like DevOps, which allows for fast and efficient development.
- PaaS should have the ability to handle vast amount of data, analyze the data, gain insights, learn from the data, and infuse AI so that enterprises can develop intelligent applications. PaaSs should also have a rich portfolio of services both in terms of business functionalities and IT functionalities.

The cloud platform provides the agility to develop and integrate intelligent apps fast and efficiently.

To stay ahead of the curve in terms of innovation, enterprises often consume services from multiple cloud providers in different deployment options like on-premise, private dedicated cloud, and public shared cloud. A hybrid multicloud landscape is the norm.

While many of these concepts were thought of as relevant only for non-SAP technologies and not for SAP world, SAP is adopting and leading many of these initiatives through different foundations like the Cloud Native Computing Foundation (CNCF). SAP BTP provides a much-needed PaaS that allows an enterprise to develop cloud-native applications in an SAP-centric landscape and beyond. Together, all of this means that cloud platforms and AI together are a key enabler of the vision of an Autonomous Enterprise.

21.3 Digital Innovation

Innovation is the key driving force behind every successful enterprise. While innovation can and should occur across all spheres, digital technologies are driving transformative innovations within companies. The advent of AI, generative AI, and agentic AI technologies has opened up potential for new applications, business processes, and business models. Innovation is a culture at successful enterprises that involves not only technology but also the right environment.

While the advent of generative and agentic AI technologies makes adoption look easy, the lack of an adequate success rate highlights problems that enterprises need to address. It begins with identifying the right use cases to apply the appropriate technology. There has been a tendency to apply generative AI to all types of problems. When the right applications are chosen, the results have been phenomenal. Also, since significant change management will be required for employees, customers, and suppliers to adapt to the new way of working, enterprises should carefully weigh the potential returns and prioritize use cases accordingly. The platform model has made these technologies easily accessible and available to employees across various levels of an enterprise. This often leads to a mushrooming of prototype activities across the enterprise, resulting in the dissipation of resources. Hence, enterprises need a proper governance model to assess use cases against their value pools, prioritize them, and invest accordingly. Some of the early domains have included human capital management, customer contact centers, and others.

With the platform model and the early development of many of these technologies, their adoption often leads to security exposures. Hence, it's of absolute importance that proper architectural guidelines are established, vulnerabilities are assessed, and controls are put in place to prevent potential risks.

Given the rapid pace of technological advancement, enterprises need to maintain sufficient flexibility to easily switch, adopt, or adapt new technologies without discarding their existing investments. Hence, following a loosely coupled, interoperable architecture is advisable. In the agentic AI world, protocols like MCP (Model Context Protocol) and A2A (Agent2Agent) Protocol are increasingly enabling this interoperability. A key tenet of loosely coupled, interoperable models is the microservices/API-based architecture. Hence, a microservice-based architecture has become fundamental to digital innovation. An effective microservice design requires careful consideration and the use of best practices, such as the following:

- **Domain-driven design**
  You should define microservices according to the business domain. In a particular domain, you should not have a data model that supports all its subdomains in that domain, especially if the domain is quite big. Thus, we recommend independently modeling your domains and subdomains.
- **Granularity**
  A monolithic application should be broken down into a limited number of microservices and not into hundreds of microservices.
- **Scalability**
  Design your microservices for both horizontal and vertical scalability.
- **API-first microservice design**
  Depending on the business requirement, an API can be designed first and then the other components of the microservice, for instance, the user experience (UX), can be built later.
- **Type of service calls**
  Avoid synchronous inter-microservice calls. Use messages instead.

After proper design, microservices can be developed in an SAP BTP development environment as part of SAP Extension Suite and SAP Integration Suite. Developers can start developing quite quickly in SAP BTP without the need for elaborate installations of software as long as they have an account in SAP BTP. This approach allows you to keep your core ERP clean. Unlike in the past, when all developments to extend SAP standard functionalities were created in the core, which made the system inflexible, costly to maintain, and difficult to upgrade, the clean core approach recommends the side-by-side development of extensions outside the core. With this approach, the core will remain standardized to SAP standards, thus making your systems less costly to maintain, easy to upgrade, and flexible. SAP Build and Joule Studio enable you to develop integrated, SAP-compatible extensions outside the SAP core.

Security is a key requirement of any cloud platform, and SAP BTP is no exception. SAP BTP provides the Identity Provisioning and Identity Authentication services for securing applications in the cloud. Identity Provisioning facilitates the identity lifecycle management of applications including user management, role management, and more. Identity Authentication includes authentication, SSO, on-premise integration, and self-services (such as user registration). Various industry-standard SSO protocols like SAML 2.0, OAuth 2.0, and X.509 certificates are supported by the service.

Application monitoring plays a critical role for the success of a hybrid cloud environment. SAP BTP provides a number of monitoring capabilities and allows you to extend and develop your own monitoring capabilities using monitoring APIs. The built-in monitors in the SAP BTP cockpit cover both application and database monitoring including logging and alerting features. SAP BTP, Cloud Foundry environment includes tools from Elastic Stack (ELK) like Kibana, Elasticsearch, Logstash, and SAP Cloud ALM's capabilities. Besides monitoring, these tools allow you to analyze and visualize information for better preventive and corrective action. You can observe and monitor agents to see what decisions they're making and why; this information is becoming increasingly important to know and preserve for control and audit purposes.

One key expectation of a cloud platform is for integrated tooling to support accelerated development and release of applications. Support for DevOps is a key ask, and SAP BTP provides a number of tools to support DevOps. Joule Studio, along with its integration with development environments like Eclipse, VS Code, and GitHub, is bringing together these capabilities in a unified experience.

Enterprises generally expect their cloud platform to help infuse AI in everything that they do; analyze and make sense of huge volume of data; and bridge the physical–digital divide in the form of IoT solutions, blockchain, and more. SAP BTP also provides machine learning and AI capabilities. As part of SAP's commitment to help companies become Autonomous Enterprises, SAP wants to make AI capabilities pervasive and embedded in critical applications and business scenarios. Thus, SAP has added SAP AI Core services to SAP BTP to support full lifecycle management of AI scenarios, thus allowing AI services to be offered as standardized AI APIs for easy consumption. This approach allows for the development of custom AI scenarios using open-source frameworks. SAP BTP also provides SAP AI Launchpad for single-point access for all AI APIs.

As SAP BTP enriches its portfolio of offerings, its adoption is growing. Many organizations across the world have started on their digital innovation journeys using SAP BTP. A variety of use cases exist where SAP BTP has helped companies extend, connect, build, and integrate. These experiences have also highlighted certain gaps and issues with the platform that SAP is aggressively fixing.

21.4 SAP BTP Roadmap

SAP's PaaS portfolio has evolved significantly in its capabilities and its overall vision since its first launch. Starting as a cloud-based development platform, SAP BTP has transformed itself as the backbone for SAP-centric client IT landscapes. SAP BTP aims to provide all capabilities required by any enterprise to become intelligent. Thus, the platform has evolved from a technical platform to a business technology platform. We are certain that further evolution of the platform will occur in the future. SAP's official product roadmap is available at *https://roadmaps.sap.com/*. Regularly checking this link to determine the current status of SAP's products and of future products, including SAP BTP, is crucial. A quick glance at the SAP BTP roadmap is shown in Figure 21.1.

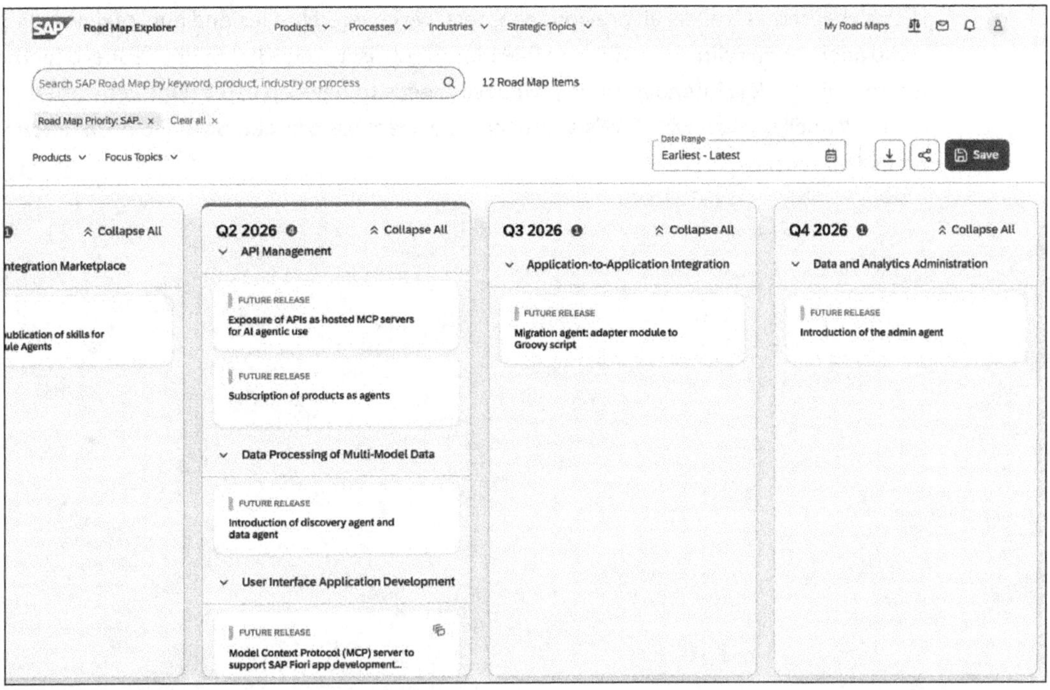

Figure 21.1: SAP Business AI in the SAP BTP Roadmap

There are separate roadmap items for SAP Business AI solutions across different domains, but these milestones are subject to change.

As made clear by the roadmap, SAP is bringing in a number of capabilities required by Autonomous Enterprises and allowing its partners to expand the platform's capabilities through partner solutions. SAP is trying to become more open, more interoperable, and more content-rich to help more enterprises gain business value from the technology and to enable business process innovation all around the globe.

21.5 Summary

Since its inception, SAP has supported enterprises in their digital journey—from recording transactions and automating actions to providing system insights. Now, SAP is focused on helping businesses further transform into intelligent enterprises that are agile, flexible, secure, intelligent, and prepared for the future. This requires a new architectural paradigm, new products, and new skills. Thus, SAP introduced SAP BTP, now integrated into SAP Business AI Platform, which embodies the new architectural paradigm and product capabilities required for this transformation. SAP is rearchitecting its product portfolio to align with the SAP Business AI Platform–centric focus of its vision. It's imperative that customers, system integrators, and partners also design and develop their extensions following these same principles

In this chapter, as well as all previous chapters, we covered the ins and outs of developing cloud applications with SAP BTP, and SAP Build, and discussed what might lie ahead for the future. This book is intended to help whoever needs to develop extensions get started on the right track to align with SAP's approach for a seamless solution portfolio of an Autonomous Enterprise.

The Authors

Govind Bajaj is an SAP business intelligence solution architect with more than 20 years of IT experience in managing and delivering complex data and analytics solutions. He has worked extensively with the design and development of advanced analytics and data migration solutions within SAP's business intelligence (BI) portfolio of products, both on-premise and in the cloud. His areas of interest include machine learning, solutioning and developing AI-ready data platforms, and agentic AI. Govind can be found on Twitter *@probajaj* and on LinkedIn at *www.linkedin.com/in/govind-bajaj-a5a7589*.

Arup Bhattacharya is an SAP S/4HANA transformation leader at IBM with more than 24 years of global delivery experience, specializing in large-scale, complex, and first-of-their-kind programs. He's an expert in migration strategies and generative AI-led optimization, consistently driving business continuity and delivery excellence across global clients.

Laboni Bhowmik is an SAP integration and automation architect working at IBM since 2007. She has more than 15 years of experience with SAP solutions and technologies, including SAP Manufacturing Integration and Intelligence, SAP Manufacturing Execution, SAP Overall Equipment Effectiveness Management, SAPUI5, and SAP BTP. Laboni has led several large-scale SAP manufacturing implementation projects for multiple global clients. She is also extensively involved in progressing IBM's SAP Digital Manufacturing and SAP Cloud Platform competency, including designing cutting-edge industry solutions, development, publications, and mentoring.

Avijit Dhar is a senior application architect at IBM India. He has more than 15 years of experience with leading the design and implementation of large-scale implementations, especially in the SAP manufacturing domain, and with growing IBM's SAP Internet of Things (IoT) and machine learning practice. He is a subject matter expert in user experience, IoT, machine learning, and other emerging technologies, and he has been involved in designing cutting-edge industry assets and solutions. Avijit has written two books with SAP PRESS: *SAP Cloud Platform: Cloud-Native Development* and *Machine Learning with SAP*. He has also written two E-Bites with SAP PRESS: *Introducing Predictive Maintenance with SAP Predictive Asset Insights* and *Hands On with SAP AppGyver*.

Dr. Anup K Ghosh is a distinguished engineer at IBM Consulting leading the development of SAP Business AI solutions and helping global Fortune 500 clients with AI-powered transformations. He's an active researcher with a number of patents and is a regular speaker at SAP TechEd and SAP Sapphire. Anup can be found on LinkedIn at *www.linkedin.com/in/anupkghosh*.

Gairik Acharya is a senior technical architect and associate partner at IBM with more than 22 years of IT experience. He is a recognized expert in ABAP, SAP HANA, OData, SAP Gateway, SAP S/4HANA, SAP Fiori, SAPUI5, and SAP Business Technology Platform. In his role as senior architect at the IBM SAP S/4HANA Center of Excellence group in North America, he has architected solutions for several global implementations for IBM's top-tier clients. He's a certified and recognized expert in SAP application consulting and is responsible for building new capabilities within the group. He's a frequent speaker at SAP TechEd, covering a variety of new technical areas.

Asidhara Lahiri is a managing director at Accenture's ATCI SAP practice, where she leads innovation and the enterprise architecture practice in SAP. She has worked as an enterprise architect at SAP India, creating roadmaps for customers' transformation journeys to achieve intelligent and modern SAP landscapes. At her previous organization, she helped build cognitive enterprises for SAP-centric landscapes using intelligent workflows and architected complex SAP implementations for global clients in industries such as industrial products, consumer products, oil and gas, airlines, and utilities. She has more than 26 years of experience, out of which 23 are with SAP solutions, including expertise in areas such as SAP Enterprise Architecture Framework, generative AI, SAP S/4HANA, SAP BTP, data and analytics, data governance, integration, and DevOps. In her roles as design authority, lead architect, practice lead, and CTO, she has been responsible for solution review/design, strategy creation, practice building, and innovation. Asidhara has coauthored two patents and also coauthored *SAP S/4HANA: An Introduction* (SAP PRESS 2025).

Index

C

D

E

F

G

T

U

V

W

X

Z